Sox2

Sox2

Biology and Role in Development and Disease

Edited by

HISATO KONDOH
Faculty of Life Sciences, Kyoto Sangyo University, Kyoto, Japan

ROBIN LOVELL-BADGE
The Crick Institute, London, UK

ELSEVIER

Amsterdam • Boston • Heidelberg • London
New York • Oxford • Paris • San Diego
San Francisco • Singapore • Sydney • Tokyo
Academic Press is an imprint of Elsevier

Academic Press is an imprint of Elsevier
125 London Wall, London EC2Y 5AS, UK
525 B Street, Suite 1800, San Diego, CA 92101-4495, USA
225 Wyman Street, Waltham, MA 02451, USA
The Boulevard, Langford Lane, Kidlington, Oxford OX5 1GB, UK

British Library Cataloguing-in-Publication Data
A catalogue record for this book is available from the British Library

Library of Congress Cataloging-in-Publication Data
A catalog record for this book is available from the Library of Congress

ISBN: 978-0-12-800352-7

Publisher: Shirley Decker-Lucke
Acquisition Editor: Shirley Decker-Lucke
Editorial Project Manager: Halima Williams
Production Project Manager: Julia Haynes
Designer: Matt Limbert

Typeset by TNQ Books and Journals
www.tnq.co.in

CONTENTS

LIST OF CONTRIBUTORS

Essam M. Abdelalim
Qatar Biomedical Research Institute, Qatar Foundation, Education City, Doha, Qatar; Department of Cytology and Histology, Faculty of Veterinary Medicine, Suez Canal University, Ismailia, Egypt

Natacha A. Agabalyan
Department of Comparative Biology and Experimental Medicine, Faculty of Veterinary Medicine and Alberta Children's Hospital Research Institute, Cumming School of Medicine, University of Calgary, Calgary, Alberta, Canada

Parth Armin
Department of Biology, University of Rochester, Rochester, NY, USA

Jessica Bertolini
Department of Biotechnology and Biosciences, University of Milano-Bicocca, Milano, Italy

Jeff Biernaskie
Department of Comparative Biology and Experimental Medicine, Faculty of Veterinary Medicine and Alberta Children's Hospital Research Institute, Cumming School of Medicine, University of Calgary, Calgary, Alberta, Canada

Ian Chambers
MRC Centre for Regenerative Medicine, Institute for Stem Cell Research, School of Biological Sciences, University of Edinburgh, Edinburgh, Scotland, UK

Kathryn S.E. Cheah
Department of Biochemistry, Li Ka Shing Faculty of Medicine, The University of Hong Kong, Hong Kong, China

G. Marius Clore
Laboratory of Chemical Physics, National Institute of Diabetes and Digestive and Kidney Diseases, National Institutes of Health, Bethesda, MD, USA

Mohamed M. Emara
Qatar Biomedical Research Institute, Qatar Foundation, Education City, Doha, Qatar; Department of Virology, School of Veterinary Medicine, Cairo University, Giza, Egypt

Rebecca Favaro
Department of Biotechnology and Biosciences, University of Milano-Bicocca, Milano, Italy

Andrew Hagner
Department of Comparative Biology and Experimental Medicine, Faculty of Veterinary Medicine and Alberta Children's Hospital Research Institute, Cumming School of Medicine, University of Calgary, Calgary, Alberta, Canada

Yasuo Ishii
Faculty of Life Sciences, Kyoto Sangyo University, Kyoto, Japan

Ian Jacobs
Department of Biomedical Genetics, University of Rochester Medical Center, Rochester, NY, USA

Ming Jiang
Division of Digestive and Liver Diseases and Columbia Center for Human Development, Department of Medicine, Columbia University, New York, NY, USA

Yusuke Kamachi
Graduate School of Frontier Biosciences, Osaka University, Suita, Osaka, Japan

Prasanna R. Kolatkar
Qatar Biomedical Research Institute, Qatar Foundation, Education City, Doha, Qatar

Hisato Kondoh
Faculty of Life Sciences, Kyoto Sangyo University, Kyoto, Japan

Wei-Yao Ku
Department of Biomedical Genetics, University of Rochester Medical Center, Rochester, NY, USA

Robin Lovell-Badge
The Crick Institute, London, UK

Jessica Mariani
Department of Biotechnology and Biosciences, University of Milano-Bicocca, Milano, Italy

Sara Mercurio
Department of Biotechnology and Biosciences, University of Milano-Bicocca, Milano, Italy

Balasubramanian Moovarkumudalvan
Qatar Biomedical Research Institute, Qatar Foundation, Education City, Doha, Qatar

Jonas Muhr
Ludwig Institute for Cancer Research, Department of Cell and Molecular Biology, Karolinska Institutet, Stockholm, Sweden

Nicholas P. Mullin
MRC Centre for Regenerative Medicine, Institute for Stem Cell Research, School of Biological Sciences, University of Edinburgh, Edinburgh, Scotland, UK

Silvia K. Nicolis
Department of Biotechnology and Biosciences, University of Milano-Bicocca, Milano, Italy

Sergio Ottolenghi
Department of Biotechnology and Biosciences, University of Milano-Bicocca, Milano, Italy

Raymond A. Poot
Department of Cell Biology, Erasmus MC, Rotterdam, Netherlands

Nilima Prakash
Helmholtz Zentrum München, Deutsches Forschungszentrum für Gesundheit und Umwelt (GmbH), Institute of Developmental Genetics, Germany; Technische Universität München, Lehrstuhl für Entwicklungsgenetik c/o Helmholtz Zentrum München, Germany; Hamm-Lippstadt University of Applied Sciences, Germany

Jianwen Que
Department of Biomedical Genetics, University of Rochester Medical Center, Rochester, NY, USA; Division of Digestive and Liver Diseases and Columbia Center for Human Development, Department of Medicine, Columbia University, New York, NY, USA

Waleed Rahmani
Department of Comparative Biology and Experimental Medicine, Faculty of Veterinary Medicine and Alberta Children's Hospital Research Institute, Cumming School of Medicine, University of Calgary, Calgary, Alberta, Canada

Karine Rizzoti
The Crick Institute, London, UK

Masanori Uchikawa
Graduate School of Frontier Biosciences, Osaka University, Suita, Osaka, Japan

Veronica van Heyningen
MRC Human Genetics Unit, Institute of Genetics and Molecular Medicine, University of Edinburgh, Edinburgh, GBR

Frederick C.K. Wong
MRC Centre for Regenerative Medicine, Institute for Stem Cell Research, School of Biological Sciences, University of Edinburgh, Edinburgh, Scotland, UK

Neng Chun Wong
Department of Biology, University of Rochester, Rochester, NY, USA; Division of Digestive and Liver Diseases and Columbia Center for Human Development, Department of Medicine, Columbia University, New York, NY, USA

Pin-Xian Xu
Genetics and Genomic Sciences, Icahn School of Medicine at Mount Sinai, New York, NY, USA; Developmental and Regenerative Biology, Icahn School of Medicine at Mount Sinai, New York, NY, USA

PREFACE

Sox2, which collectively refers to the *Sox2* gene and its encoded transcription factor SOX2, has a remarkable research history over a quarter of a century that marks the progress in our understanding of transcriptional regulation in higher organisms. The central importance of Sox2 in various biological processes such as embryogenesis, organogenesis, stem cell regulation, and diseases has also gained increasing attention. We thought it was timely to compile and organize our current knowledge on Sox2 in the form of a book, with comprehensive coverage from its molecular nature to organismal regulation. Thanks to the many specialists from various branches of Sox2 research who approved our idea and contributed chapters, we believe that our undertaking was successful. We hope that this book will become a useful resource for biomedical scientists of various disciplines, from students to professionals.

We missed one potential author who should have contributed to this book, the late Larysa Pevny, who passed away in 2012 at the age of just 47 years. She made important contributions to the study of Sox2, as you will see in many citations in various chapters. She also shared valuable mouse models produced by her with many laboratories around the world, which promoted Sox2 research. On this occasion, we would like to mention these contributions in tribute to her.

We once again thank the authors for their professional contributions, and Dr Jianwen Que and Dr Masanori Uchikawa for providing the beautiful figure panels for the front cover: immunostained embryonic trachea and lung (bottom left; see Chapter 17 Figure 3 for details) and enhanced green fluorescent protein fluorescence of a *Sox2-IRES-EGFP* knock-in E9 mouse embryo (bottom right). We also appreciate the patience and expert management of the editorial team of Academic Press/Elsevier, particularly Halima N. Williams, Elizabeth Gibson, and Julia Haynes, who made this undertaking possible.

Hisato Kondoh and Robin Lovell-Badge

PART 1

Basic Features of Sox2 Protein and Gene

CHAPTER 1

Historical Perspectives

Hisato Kondoh[1], Robin Lovell-Badge[2]
[1]Faculty of Life Sciences, Kyoto Sangyo University, Kyoto, Japan; [2]The Crick Institute, London, UK

A quarter of century has passed since the discovery of the first *Sox* gene, *SRY/Sry*. Shortly afterward, many related *Sox* genes encoding SOX family transcription factors were found to be distributed in the genome. The importance of their role in development and diseases has attracted growing attention. Among the *Sox* transcription factor genes, the role of *Sox2* has been highlighted mostly for its involvement in early developmental processes and organogenesis, and in particular for its central role in regulating a wide spectrum of stem cells.

In the investigation of various transcription factors involved in the developmental process, SOX2 research has always been on the leading edge and has provided a paradigm of their action from molecular to organismal dimensions. Through scientific processes in which basic problems have been answered concomitantly with the rise of new questions, we are in the position to grasp an overall view of *Sox2* and SOX2 functions across the dimensions. In this book, our current understanding is dismantled into individual dimensions for readers to synthesize them for their own study.

This chapter aims to familiarize readers with the history of SOX2 research over the past quarter century and highlights landmark findings and topics. We hope that readers will appreciate how the multifaceted functions *Sox2* are derived from the unique basic features of the SOX2 molecule and from multilayered *Sox2* regulation (Table 1).

DISCOVERY OF SOX2 AND OTHER SOX GENES PIONEERED BY SRY

The identification of *SRY/Sry* as a male-specifying gene marked a breakthrough not only in sex determination research but also in the area of genetic regulation of embryonic development (Gubbay et al., 1990; Sinclair et al., 1990). Shortly after this discovery, many genes sharing the High Mobility Group (HMG) box sequences similar to *Sry* were identified in the genome and were found to be expressed in embryos (Gubbay et al., 1990; Denny et al., 1992). These genes were named *Sox* (*Sry*-related HMG box) genes. Their HMG box sequences were similar to those of Lef/Tcf family transcription factors discovered around the same time, but *Sox* genes formed a clearly distinct gene group, as detailed in Chapter 6. The SOX proteins were characterized as deoxyribonucleic acid (DNA)-binding transcription factors because of their binding to (A)ACAA[A/T](G) sequences and their possession of activation or repression domains (Kamachi and Kondoh, 2013).

Sox2
http://dx.doi.org/10.1016/B978-0-12-800352-7.00001-3

Table 1 Chronological table of *Sox2* research

Topics	Year	Representative references
Discovery of *Sry/SRY*	1990	Gubbay et al. (1990) and
	1991	Sinclair et al. (1990)
Many *Sox* genes	1992	Denny et al. (1992)
	1993	
Tissue-specific expression of *SoxB1* genes	1994	
Identification of SOX2 regulatory targets: Requirement of partner factors; HMG–DNA complex structure	1995	Kamachi et al. (1995), Yuan et al. (1995), Werner et al. (1995), Uwanogho et al. (1995)
	1996	Collignon et al. (1996)
First summary of Sox research	1997	Pevny and Lovell-Badge (1997)
Sox1 knockout mice	1998	Nishiguchi et al. (1998)
	1999	
Final classification of *Sox* genes; *Sox2* in neural development	2000	Bowles et al. (2000) and Zappone et al. (2000)
SOX2–PAX6 interactions	2001	Kamachi et al. (2001)
	2002	
Sox2 knockout mice; *Sox2* in neural stem cells; Identification of *Sox2* enhancers; 3D structure of SOX2–partner–DNA ternary complex; *Sox2*-dependent congenital ocular diseases	2003	Avilion et al. (2003), Bylund et al. (2003), Graham et al. (2003), Uchikawa et al. (2003), Remenyi et al. (2003), Fantes et al. (2003)
Sox3 knockout mice	2004	Rizzoti et al. (2004)
Core regulatory circuits in human ES cells; SOX2 in inner ear development	2005	Boyer et al. (2005) and Kiernan et al. (2005)
iPS cells; *Sox2* in retinal development	2006	Takahashi and Yamanaka (2006) and Taranova et al. (2006)
Sox2 in endoderm development	2007	Que et al. (2007)
Core regulatory circuits in mouse ES cells, and miRNAs	2008	Chen et al. (2008) and Tay et al. (2008)
	2009	Xu et al. (2009)
Maternal *Sox2* activity	2010	Keramari et al. (2010)
SOX2–CHD7 interaction; SOX2 as a pioneer factor; *Sox2* in neuro-mesodermal bipotential precursors	2011	Engelen et al. (2011), Bergsland et al. (2011), Takemoto et al. (2011)
Sox2 in skin development	2012	Clavel et al. (2012)
	2013	
Sox2-positive cancer stem cells	2014	Vanner et al. (2014)

Remarkably, some *Sox* genes, in particular those with HMG box sequences closest to *SRY*, initially called a1 to a3 and now called *Sox1*, *Sox2*, and *Sox3*, respectively, and classified as *SoxB1* genes (Bowles et al., 2000), were found to be expressed in a highly tissue-specific manner in mouse embryos. This strongly suggests their involvement in the regulation of cell and tissue differentiation processes (Collignon et al., 1996;

Kamachi et al., 1998). A description of how these genes came to be named was provided by Lovell-Badge (2010). Expression data from the chicken version of *Sox1* to *Sox3* also emphasized the association of these genes with developmental processes (Uwanogho et al., 1995; Uchikawa et al., 1999).

In 1996, the *Drosophila Dichaete* gene (also called *fish-hook*), identified by mutants defective in embryonic processes, was found to code for a *Sox* gene (Nambu and Nambu, 1996; Russell et al., 1996) that is now classified as *SoxB1* (Phochanukul and Russell, 2010). These observations clearly indicated that *Sox2* and other *Sox* genes participate in developmental regulations not only in vertebrates but also in a wide range of animal species (Pevny and Lovell-Badge, 1997). Phylogenetic aspects of *SoxB1* gene evolution are given in Chapter 6.

SOX2 WITH DEFINED REGULATORY TARGETS, IN COOPERATION WITH PARTNER FACTORS

SOX2 was one of the transcription factors involved in the developmental processes whose regulatory target genes were identified earliest. Significant discoveries were made in 1995. Lisa Dailey and colleagues investigated *fibroblast growth factor 4* (*Fgf4*) activation in teratocarcinoma (and later embryonic stem (ES)) cell lines and found that SOX2 and OCT3 (a synonym of OCT4 and renamed as POU5F1 by the Mouse Genome Informatics Consortium) cooperate in the activation of the *Fgf4* enhancer bearing their juxtaposed binding sites (Yuan et al., 1995). We identified SOX2 as the major regulator of δ- and γ-*crystallin* genes specifically expressed in the lens (Kamachi et al., 1995), which indicates the involvement of SOX2 in lens development. Our study also indicated the requirement of cooperation of a second factor that differed according to the crystallin genes, which were later identified as PAX6 for the δ-*crystallin* gene (Kamachi et al., 1998, 2001) and MAF1 for the γ-*crystallin* gene (Rajaram and Kerppola, 2004). Thus, these pioneering studies not only indicated a wide range of SOX2 regulatory target genes but also that the transcriptional activation function of SOX2 is exerted only in concert with a partnering transcription factor, the combination of which also determines the regulatory target gene. This model was extended to cases of other SOX factors, described as the SOX-partner code (Kamachi et al., 2000), and validated in more recent studies, as discussed in Chapter 8.

MOLECULAR STRUCTURE OF SOX2 HMG AND ASSOCIATED DOMAINS INTERACTING WITH DNA AND PARTNER FACTORS

The three-dimensional molecular structures of the SOX HMG domain have been investigated from the beginning of *Sox* research. The findings indicated that the HMG domain of SOX2 and other SOX proteins consists of three α-helices in solutions with or without DNA, which bind DNA with two α-helices that interact with the minor groove of target DNA, bending it by widening its minor groove (Werner et al., 1995; Remenyi et al., 2003).

The three-dimensional structure of the SOX2 HMG domain protein bound to DNA, in particular in association with partner factors, was investigated by Remenyi et al. (2003) in their highly informative study. In representative cases of SOX2–partner interactions, the region of SOX2 around the C-terminal end of the HMG domain serves as the flexible interface with a variety of partner factors. This aspect of SOX–partner interaction is analyzed in Chapter 2. These structural analyses did not indicate how and in what order SOX2 and the partner factor interact with DNA. The dynamics of these interactions were investigated by G. Marius Clore's group (Takayama and Clore, 2012), as discussed in Chapter 3.

SOX2 FUNCTIONS IN THE EARLY DEVELOPMENTAL PROCESS, INVOLVING FUNCTIONAL REDUNDANCY WITH SOXB1 GENES AND MATERNAL FACTORS

SOX1 and SOX3, which belong to the same SOXB protein group, were found to be similar to SOX2 not only in the overall amino acid sequences but also in the expression patterns in embryos (Uwanogho et al., 1995; Collignon et al., 1996; Wood and Episkopou, 1999). This suggests that SOX1 to SOX3 share basic characteristics as transcriptional regulators and hence overlap in their functions in tissue where they are coexpressed. That is, knockout mice defective in one of three *SoxB1* genes would develop severe phenotypes only in tissues in which one of them is singly expressed. The first *SoxB1* gene inactivated in mice using the straightforward knockout technology was *Sox1* (Nishiguchi et al., 1998), in which the development of lens fibers was severely affected, where *Sox1* was singly expressed in the mouse. *Sox3* knockout mice were viable and mildly affected in the hypothalamopituitary axis (Rizzoti et al., 2004), presumably because these tissues require a high level of *SoxB1* activity (Zhao et al., 2012).

Zygotic *Sox2*-null homozygous mouse embryos derived from crossing heterozygous *Sox2*-defective parents were lethal and died around the time of implantation (about embryonic day 5.5) (Avilion et al., 2003). This is consistent with the observation that *Sox2* is the only *SoxB1* gene expressed before implantation and emphasizes the essential functions of SOX2 during early stages of embryogenesis. However, *Sox2* is expressed zygotically from early cleavage stages and is strongly expressed in both inner cell mass and trophectoderm in the preimplantation blastocysts; it raises the possibility that persistence of embryonic development to the peri-implantation stage in the absence of zygotic *Sox2* expression results from the contribution of maternal SOX2 or *Sox2* messages that were detected abundantly (Avilion et al., 2003). Later studies that inactivated both maternal and zygotic *Sox2* messenger ribonucleic acid (RNA) using siRNAs confirmed the essential functions of SOX2 during the cleavage stages and in the development of inner cell mass and trophectoderm (Keramari et al., 2010).

ROLES FOR SOX2 IN NEURAL AND ASSOCIATED TISSUES

Together with other SOXB1 factors, SOX2 is expressed in embryonic neural stem cells located in the ventricular zone. Counteraction of their activities by expressing a dominant-negative (transcriptionally repressing) form of SOX2 (Graham et al., 2003) or SOXB2 factors that act as transcriptional repressors (Uchikawa et al., 1999; Bylund et al., 2003) resulted in the failure of maintaining stem cells and premature neuronal differentiation, demonstrating that SOX2 together with SOX1 and SOX3 maintains the neural stem cell state.

Because *Sox2*-deficient embryos die during early stages of embryogenesis, the investigation of regulatory functions of *Sox2* at later stages requires more sophisticated approaches than simple inactivation of the gene. Conditional (cell and developmental stage-restricted) inactivation of the *Sox2* gene circumvents this problem and is used in various studies. In addition, the use of hypomorphic alleles of *Sox2* has been shown to be productive in identifying tissues sensitive to the expression level of SOX2 (or overall SOXB1 factors), as exemplified by the analysis of neural retina development that depended on SOX2 activity (Taranova et al., 2006). Another approach was to inactivate one of the *Sox2* gene-associated enhancers that regulate *Sox2* in a restricted domain of a tissue, such as the central nervous system (CNS). Two successful examples were inactivation of the 5′ enhancer (equivalent to N2 enhancer) of *Sox2*, combined with a null allele, which demonstrates an important regulatory role for SOX2 in the forebrain neurogenesis (Zappone et al., 2000; Ferri et al., 2004) and inactivation of the N1 enhancer that demonstrated the *Sox2*-dependent regulation of neural/mesodermal bipotential precursors in the trunk region (Takemoto et al., 2011).

SOX2 IN THE DEVELOPMENT OF NONNEURAL TISSUES

Among *Sox2*-expressing somatic tissue primordia, sensory primordia derived from the cephalic placodes develop subsequent to the neural tissues. In addition to the lens development discussed above, SOX2 is involved in regulation at multiple stages of inner ear development (Kiernan et al., 2005) that finally lead to the development of sensory hair cells (Ahmed et al., 2012a,b) and sensory neurons (Evsen et al., 2013). Chapters 12, 13, and 15 detail how SOX2 regulates the development of eye tissues and the inner ear.

Sox2 was also found to have important roles in a variety of additional tissues. Brigid Hogan's group focused on the regulatory function of *Sox2* in the anterior endoderm-derived organs such as the esophagus and lung, where *Sox2* was involved in multiple phases of organogenesis (Que et al., 2007, 2009). It was also discovered that the skin depended on the *Sox2* function (Clavel et al., 2012). Chapters 16 and 17 give an overview of these *Sox2*-dependent processes in the development of skin and lung tissues.

SOX2 IN THE STEM CELLS AND A POTENTIAL NEW ROLE FOR SOX2 IN CHROMATIN REGULATION

Various stem cells in vivo and in vitro express and depend on the activity of SOX2. The first description of core regulatory circuits involving SOX2 in human and mouse ES cells (Boyer et al., 2005; Chen et al., 2008) provided a paradigm in stem cell research. Besides ES cell lines derived from the blastocyst inner cell mass (Evans and Kaufman, 1981), epiblast stem cell lines from the epiblast of postimplantation embryos (Brons et al., 2007; Tesar et al., 2007) and neural stem cell lines from the neural stem cells in the embryonic or adult CNS (Conti et al., 2005) express SOX2 and depend on it, largely reflecting the expression of SOX2 in the derived in vivo stem cells. Recent studies indicate that even more varieties of in vivo stem cells, including cancer stem cells mentioned below, express SOX2.

The amazing discovery by Takahashi and Yamanaka, (2006) that the four-transcription factor gene cocktail consisting of *Sox2*, *Pou5f1*, *Klf4*, and *Myc* can lead to the formation of induced pluripotent stem (iPS) cells with characteristics similar to ES cells has had a strong impact on SOX2 research as well. However, the inclusion of *Sox2* and *Pou5f1* in the transcription factor gene cocktail was not surprising because SOX2 uses POU5f as a partner factor in ES cells, and they function by forming a heterodimer that activates *Sox2* and *Pou5f1* genes, resulting in the formation of co-activation loops for these genes, as discussed in Chapter 8.

However, the role of SOX2 must be more than pairing off with POU5F1. During iPS cell production, the exogenous transcription factor genes activate a wider variety of endogenous genes including *Nanog*, alter epigenetic signatures including cytosine methylation and histone modification patterns, and then must be turned off to be replaced by autoregulatory circuits consisting of endogenous genes, a gradual process that may take up to a month (Brambrink et al., 2008). Therefore, endogenous gene loci that are epigenetically silenced in fully differentiated cells must be forced open by the action of exogenous transcription factors.

The discovery that SOX2 binds strongly to chromatin remodeling factor CHD7 (Engelen et al., 2011), as discussed in Chapter 5, suggests a direct action of SOX2 in the chromatin remodeling process required for iPS cell production. The *Sox2*-dependent production of induced neural stem cells (Karow et al., 2012) may represent an analogous scenario. The observation that SOX2 likely functions as one of the pioneer factors, transcription factors that bind and mark genomic loci (usually enhancers) that are later activated via binding of transcription factor complexes (Bergsland et al., 2011), may also reflect manifestation of the analogous action of SOX2. These new features of

SOX2-dependent processes may be the basis of the fact that *Sox2* is frequently employed in regulating a wide range of stem cells, even cancer stem cells. Further studies along this line may provide new horizons in SOX2-dependent genome-wide regulation research.

REGULATION OF SOX2 ACTIVITY AT DIFFERENT LEVELS

Because *Sox2* is involved in a variety of processes in the developmental stages and tissues, it is unlikely that the gene is regulated by a simple set of enhancers. Indeed, using a systematic functional assay, Uchikawa et al. (2003) demonstrated that as many as 11 different enhancers exist that regulate *Sox2* in the neural and sensory tissues up to stage 15 (2.5 days of incubation) within the 50-kb chicken genomic span (equivalent of approximately 60 kb in mammalian genomes) encompassing the *Sox2* gene (Uchikawa et al., 2003). Specificities of these enhancers are distinct although there are some overlaps in the tissue domains. Extension of the genomic region of enhancer survey to 200 kb (equivalent of about 500 kb in mammalian genomes) identified a total of 27 neurosensory enhancers (Okamoto et al., 2015). More enhancer sequences candidates were predicted based on cross-species conservation. An important feature of these enhancers is that transcription factors and signaling systems involved in enhancer regulation are highly coordinated with the process in which the specific tissue is produced (Takemoto et al., 2011).

A combination of these enhancers may determine the level of *Sox2* expression of the primary transcript (*Sox2* is an intron-less gene), but investigations indicated the essential involvement of micro RNAs (miRNAs) in the posttranscriptional regulation of *Sox2* (Tay et al., 2008; Xu et al., 2009), as discussed in Chapter 4. The miRNA-dependent fine-tuning of *Sox2* expression level is essential for normal biological processes.

It is also known that SOX2 is subject to posttranslational modifications, as briefly discussed in Chapters 8 and 10. However, the impact of these modifications is assessed using cell lines; their evaluation in embryos and animals is much anticipated. In the case of SOX9, SUMOylation promotes participation in the inner ear development and inhibits it in neural crest development (Taylor and Labonne, 2005). It is possible that modifications of SOX2 protein also has an impact.

SOX2 AND DISEASE

Heterozygously, *SOX2*-deficient patients and those with hypomorphic *SOX2* mutations develop congenital diseases of varying spectra and penetrance, as discussed in Chapter 13. Eye malformations ranging from anophthalmia to microphthalmia, as originally investigated by FitzPatrick, van Heyningen, and associates (Fantes et al., 2003), and extended to anophthalmia–esophageal–genital syndrome (Williamson et al., 2006), are examples. The phenotypes of these patients resemble those defective in interacting partner factors. During the eye development, SOX2 interacts with its partners, such as PAX6 and OTX2, and the eye defects in patients with *SOX2* mutations and those with mutations in one of the partner transcription factor genes resemble each other. Another example is coloboma, heart defect, atresia choanae, retarded growth and development, genital abnormality, and ear abnormality (CHARGE) syndrome, which is caused by mutations in either *Sox2* or *Chd7* genes. These observations confirmed the tissue-dependent roles of SOX2–partner factor complexes (See Chapter 5).

An important finding for *Sox2* function is its involvement in oncogenesis. This research field is still developing and there is much more to be learned in the future. However, there are two major points. Dysregulation of *Sox2* is oncogenic in a context-dependent manner via gene amplification (Rudin et al., 2012), dysregulation of regulatory miRNAs (see Chapter 4), or other mechanisms (Boumahdi et al., 2014). It is also possible that ectopic activation of the enhancers is involved in some cases.

Another important aspect of involvement of *Sox2* is its expression in quiescent cancer stem cells, which is refractory to chemotherapy (Vanner et al., 2014). The process of quiescence of cancer stem cells may be analogous to quiescent retinal Mueller cells expressing *Sox2*, which is activated during the repair of injured retinal tissues but is lost when *Sox2* is inactivated (Surzenko et al., 2013). The enhancer responsible for the activation of *Sox2* in quiescent cancer stem cells is another subject to be investigated.

REFERENCES

Ahmed, M., Wong, E.Y., Sun, J., Xu, J., Wang, F., Xu, P.X., 2012a. Eya1-Six1 interaction is sufficient to induce hair cell fate in the cochlea by activating Atoh1 expression in cooperation with Sox2. Dev. Cell 22, 377–390.

Ahmed, M., Xu, J., Xu, P.X., 2012b. EYA1 and SIX1 drive the neuronal developmental program in cooperation with the SWI/SNF chromatin-remodeling complex and SOX2 in the mammalian inner ear. Development 139, 1965–1977.

Avilion, A.A., Nicolis, S.K., Pevny, L.H., Perez, L., Vivian, N., Lovell-Badge, R., 2003. Multipotent cell lineages in early mouse development depend on SOX2 function. Genes Dev. 17, 126–140.

Bergsland, M., Ramskold, D., Zaouter, C., Klum, S., Sandberg, R., Muhr, J., 2011. Sequentially acting Sox transcription factors in neural lineage development. Genes Dev. 25, 2453–2464.

Boumahdi, S., Driessens, G., Lapouge, G., et al., 2014. SOX2 controls tumour initiation and cancer stem-cell functions in squamous-cell carcinoma. Nature 511, 246–250.

Bowles, J., Schepers, G., Koopman, P., 2000. Phylogeny of the SOX family of developmental transcription factors based on sequence and structural indicators. Dev. Biol. 227, 239–255.

Boyer, L.A., Lee, T.I., Cole, M.F., Johnstone, S.E., Levine, S.S., Zucker, J.P., Guenther, M.G., Kumar, R.M., Murray, H.L., Jenner, R.G., Gifford, D.K., Melton, D.A., Jaenisch, R., Young, R.A., 2005. Core transcriptional regulatory circuitry in human embryonic stem cells. Cell 122, 947–956.

Brambrink, T., Foreman, R., Welstead, G.G., Lengner, C.J., Wernig, M., Suh, H., Jaenisch, R., 2008. Sequential expression of pluripotency markers during direct reprogramming of mouse somatic cells. Cell Stem Cell 2, 151–159.

Brons, I.G., Smithers, L.E., Trotter, M.W., Rugg-Gunn, P., Sun, B., Chuva de Sousa Lopes, S.M., Howlett, S.K., Clarkson, A., Ahrlund-Richter, L., Pedersen, R.A., Vallier, L., 2007. Derivation of pluripotent epiblast stem cells from mammalian embryos. Nature 448, 191–195.

Bylund, M., Andersson, E., Novitch, B.G., Muhr, J., 2003. Vertebrate neurogenesis is counteracted by Sox1-3 activity. Nat. Neurosci. 6, 1162–1168.

Chen, X., Xu, H., Yuan, P., et al., 2008. Integration of external signaling pathways with the core transcriptional network in embryonic stem cells. Cell 133, 1106–1117.

Clavel, C., Grisanti, L., Zemla, R., Rezza, A., Barros, R., Sennett, R., Mazloom, A.R., Chung, C.Y., Cai, X., Cai, C.L., Pevny, L., Nicolis, S., Ma'ayan, A., Rendl, M., 2012. Sox2 in the dermal papilla niche controls hair growth by fine-tuning BMP signaling in differentiating hair shaft progenitors. Dev. Cell 23, 981–994.

Collignon, J., Sockanathan, S., Hacker, A., Cohen-Tannoudji, M., Norris, D., Rastan, S., Stevanovic, M., Goodfellow, P.N., Lovell-Badge, R., 1996. A comparison of the properties of Sox-3 with Sry and two related genes, Sox-1 and Sox-2. Development 122, 509–520.

Conti, L., Pollard, S.M., Gorba, T., Reitano, E., Toselli, M., Biella, G., Sun, Y., Sanzone, S., Ying, Q.L., Cattaneo, E., Smith, A., 2005. Niche-independent symmetrical self-renewal of a mammalian tissue stem cell. PLoS Biol. 3, e283.

Denny, P., Swift, S., Brand, N., Dabhade, N., Barton, P., Ashworth, A., 1992. A conserved family of genes related to the testis determining gene, SRY. Nucleic Acids Res. 20, 2887.

Engelen, E., Akinci, U., Bryne, J.C., et al., 2011. Sox2 cooperates with Chd7 to regulate genes that are mutated in human syndromes. Nat. Genet. 43, 607–611.

Evans, M.J., Kaufman, M.H., 1981. Establishment in culture of pluripotential cells from mouse embryos. Nature 292, 154–156.

Evsen, L., Sugahara, S., Uchikawa, M., Kondoh, H., Wu, D.K., 2013. Progression of neurogenesis in the inner ear requires inhibition of Sox2 transcription by neurogenin1 and neurod1. J. Neurosci. 33, 3879–3890.

Fantes, J., Ragge, N.K., Lynch, S.A., McGill, N.I., Collin, J.R., Howard-Peebles, P.N., Hayward, C., Vivian, A.J., Williamson, K., van Heyningen, V., FitzPatrick, D.R., 2003. Mutations in SOX2 cause anophthalmia. Nat. Genet. 33, 461–463.

Ferri, A.L., Cavallaro, M., Braida, D., Di Cristofano, A., Canta, A., Vezzani, A., Ottolenghi, S., Pandolfi, P.P., Sala, M., DeBiasi, S., Nicolis, S.K., 2004. Sox2 deficiency causes neurodegeneration and impaired neurogenesis in the adult mouse brain. Development 131, 3805–3819.

Graham, V., Khudyakov, J., Ellis, P., Pevny, L., 2003. SOX2 functions to maintain neural progenitor identity. Neuron 39, 749–765.

Gubbay, J., Collignon, J., Koopman, P., Capel, B., Economou, A., Munsterberg, A., Vivian, N., Goodfellow, P., Lovell-Badge, R., 1990. A gene mapping to the sex-determining region of the mouse Y chromosome is a member of a novel family of embryonically expressed genes. Nature 346, 245–250.

Kamachi, Y., Kondoh, H., 2013. Sox proteins: regulators of cell fate specification and differentiation. Development 140, 4129–4144.

Kamachi, Y., Sockanathan, S., Liu, Q., Breitman, M., Lovell-Badge, R., Kondoh, H., 1995. Involvement of SOX proteins in lens-specific activation of crystallin genes. EMBO J. 14, 3510–3519.

Kamachi, Y., Uchikawa, M., Collignon, J., Lovell-Badge, R., Kondoh, H., 1998. Involvement of Sox1, 2 and 3 in the early and subsequent molecular events of lens induction. Development 125, 2521–2532.

Kamachi, Y., Uchikawa, M., Kondoh, H., 2000. Pairing SOX off: with partners in the regulation of embryonic development. Trends Genet. 16, 182–187.

Kamachi, Y., Uchikawa, M., Tanouchi, A., Sekido, R., Kondoh, H., 2001. Pax6 and SOX2 form a co-DNA-binding partner complex that regulates initiation of lens development. Genes Dev. 15, 1272–1286.

Karow, M., Sanchez, R., Schichor, C., Masserdotti, G., Ortega, F., Heinrich, C., Gascon, S., Khan, M.A., Lie, D.C., Dellavalle, A., Cossu, G., Goldbrunner, R., Gotz, M., Berninger, B., 2012. Reprogramming of pericyte-derived cells of the adult human brain into induced neuronal cells. Cell Stem Cell 11, 471–476.

Keramari, M., Razavi, J., Ingman, K.A., Patsch, C., Edenhofer, F., Ward, C.M., Kimber, S.J., 2010. Sox2 is essential for formation of trophectoderm in the preimplantation embryo. PLoS One 5, e13952.

Kiernan, A.E., Pelling, A.L., Leung, K.K., Tang, A.S., Bell, D.M., Tease, C., Lovell-Badge, R., Steel, K.P., Cheah, K.S., 2005. Sox2 is required for sensory organ development in the mammalian inner ear. Nature 434, 1031–1035.

Lovell-Badge, R., 2010. The early history of the Sox genes. Int. J. Biochem. Cell Biol. 42, 378–380.

Nambu, P.A., Nambu, J.R., 1996. The *Drosophila* fish-hook gene encodes a HMG domain protein essential for segmentation and CNS development. Development 122, 3467–3475.
Nishiguchi, S., Wood, H., Kondoh, H., Lovell-Badge, R., Episkopou, V., 1998. Sox1 directly regulates the gamma-crystallin genes and is essential for lens development in mice. Genes Dev. 12, 776–781.
Okamoto, R., Uchikawa, M., Kondoh, H., 2015. Sixteen additional enhancers associated with the chicken Sox2 locus outside the central 50-kb region. Dev. Growth Differ. 57, 24–39.
Pevny, L.H., Lovell-Badge, R., 1997. Sox genes find their feet. Curr. Opin. Genet. Dev. 7, 338–344.
Phochanukul, N., Russell, S., 2010. No backbone but lots of Sox: Invertebrate Sox genes. Int. J. Biochem. Cell Biol. 42, 453–464.
Que, J., Luo, X., Schwartz, R.J., Hogan, B.L., 2009. Multiple roles for Sox2 in the developing and adult mouse trachea. Development 136, 1899–1907.
Que, J., Okubo, T., Goldenring, J.R., Nam, K.T., Kurotani, R., Morrisey, E.E., Taranova, O., Pevny, L.H., Hogan, B.L., 2007. Multiple dose-dependent roles for Sox2 in the patterning and differentiation of anterior foregut endoderm. Development 134, 2521–2531.
Rajaram, N., Kerppola, T.K., 2004. Synergistic transcription activation by Maf and Sox and their subnuclear localization are disrupted by a mutation in Maf that causes cataract. Mol. Cell. Biol. 24, 5694–5709.
Remenyi, A., Lins, K., Nissen, L.J., Reinbold, R., Scholer, H.R., Wilmanns, M., 2003. Crystal structure of a POU/HMG/DNA ternary complex suggests differential assembly of Oct4 and Sox2 on two enhancers. Genes Dev. 17, 2048–2059.
Rizzoti, K., Brunelli, S., Carmignac, D., Thomas, P.Q., Robinson, I.C., Lovell-Badge, R., 2004. SOX3 is required during the formation of the hypothalamo-pituitary axis. Nat. Genet. 36, 247–255.
Rudin, C.M., Durinck, S., Stawiski, E.W., et al., 2012. Comprehensive genomic analysis identifies SOX2 as a frequently amplified gene in small-cell lung cancer. Nat. Genet. 44, 1111–1116.
Russell, S.R., Sanchez-Soriano, N., Wright, C.R., Ashburner, M., 1996. The Dichaete gene of *Drosophila melanogaster* encodes a SOX-domain protein required for embryonic segmentation. Development 122, 3669–3676.
Sinclair, A.H., Berta, P., Palmer, M.S., Hawkins, J.R., Griffiths, B.L., Smith, M.J., Foster, J.W., Frischauf, A.M., Lovell-Badge, R., Goodfellow, P.N., 1990. A gene from the human sex-determining region encodes a protein with homology to a conserved DNA-binding motif. Nature 346, 240–244.
Surzenko, N., Crowl, T., Bachleda, A., Langer, L., Pevny, L., 2013. SOX2 maintains the quiescent progenitor cell state of postnatal retinal Muller glia. Development 140, 1445–1456.
Takahashi, K., Yamanaka, S., 2006. Induction of pluripotent stem cells from mouse embryonic and adult fibroblast cultures by defined factors. Cell 126, 663–676.
Takayama, Y., Clore, G.M., 2012. Impact of protein/protein interactions on global intermolecular translocation rates of the transcription factors Sox2 and Oct1 between DNA cognate sites analyzed by z-exchange NMR spectroscopy. J. Biol. Chem. 287, 26962–26970.
Takemoto, T., Uchikawa, M., Yoshida, M., Bell, D.M., Lovell-Badge, R., Papaioannou, V.E., Kondoh, H., 2011. Tbx6-dependent Sox2 regulation determines neural or mesodermal fate in axial stem cells. Nature 470, 394–398.
Taranova, O.V., Magness, S.T., Fagan, B.M., Wu, Y., Surzenko, N., Hutton, S.R., Pevny, L.H., 2006. SOX2 is a dose-dependent regulator of retinal neural progenitor competence. Genes Dev. 20, 1187–1202.
Tay, Y., Zhang, J., Thomson, A.M., Lim, B., Rigoutsos, I., 2008. MicroRNAs to Nanog, Oct4 and Sox2 coding regions modulate embryonic stem cell differentiation. Nature 455, 1124–1128.
Taylor, K.M., Labonne, C., 2005. SoxE factors function equivalently during neural crest and inner ear development and their activity is regulated by SUMOylation. Dev. Cell 9, 593–603.
Tesar, P.J., Chenoweth, J.G., Brook, F.A., Davies, T.J., Evans, E.P., Mack, D.L., Gardner, R.L., McKay, R.D., 2007. New cell lines from mouse epiblast share defining features with human embryonic stem cells. Nature 448, 196–199.
Uchikawa, M., Ishida, Y., Takemoto, T., Kamachi, Y., Kondoh, H., 2003. Functional analysis of chicken Sox2 enhancers highlights an array of diverse regulatory elements that are conserved in mammals. Dev. Cell 4, 509–519.
Uchikawa, M., Kamachi, Y., Kondoh, H., 1999. Two distinct subgroups of Group B Sox genes for transcriptional activators and repressors: their expression during embryonic organogenesis of the chicken. Mech. Dev. 84, 103–120.

Uwanogho, D., Rex, M., Cartwright, E.J., Pearl, G., Healy, C., Scotting, P.J., Sharpe, P.T., 1995. Embryonic expression of the chicken Sox2, Sox3 and Sox11 genes suggests an interactive role in neuronal development. Mech. Dev. 49, 23–36.

Vanner, R.J., Remke, M., Gallo, M., et al., 2014. Quiescent sox2(+) cells drive hierarchical growth and relapse in sonic hedgehog subgroup medulloblastoma. Cancer Cell 26, 33–47.

Werner, M.H., Huth, J.R., Gronenborn, A.M., Clore, G.M., 1995. Molecular basis of human 46X,Y sex reversal revealed from the three-dimensional solution structure of the human SRY-DNA complex. Cell 81, 705–714.

Williamson, K.A., Hever, A.M., Rainger, J., et al., 2006. Mutations in SOX2 cause anophthalmia-esophageal-genital (AEG) syndrome. Hum. Mol. Genet. 15, 1413–1422.

Wood, H.B., Episkopou, V., 1999. Comparative expression of the mouse Sox1, Sox2 and Sox3 genes from pre-gastrulation to early somite stages. Mech. Dev. 86, 197–201.

Xu, N., Papagiannakopoulos, T., Pan, G., Thomson, J.A., Kosik, K.S., 2009. MicroRNA-145 regulates OCT4, SOX2, and KLF4 and represses pluripotency in human embryonic stem cells. Cell 137, 647–658.

Yuan, H., Corbi, N., Basilico, C., Dailey, L., 1995. Developmental-specific activity of the FGF-4 enhancer requires the synergistic action of Sox2 and Oct-3. Genes Dev. 9, 2635–2645.

Zappone, M.V., Galli, R., Catena, R., Meani, N., De Biasi, S., Mattei, E., Tiveron, C., Vescovi, A.L., Lovell-Badge, R., Ottolenghi, S., Nicolis, S.K., 2000. Sox2 regulatory sequences direct expression of a (beta)-geo transgene to telencephalic neural stem cells and precursors of the mouse embryo, revealing regionalization of gene expression in CNS stem cells. Development 127, 2367–2382.

Zhao, L., Zevallos, S.E., Rizzoti, K., Jeong, Y., Lovell-Badge, R., Epstein, D.J., 2012. Disruption of SoxB1-dependent Sonic hedgehog expression in the hypothalamus causes septo-optic dysplasia. Dev. Cell 22, 585–596.

CHAPTER 2

Three-dimensional Structure of SOX Protein–DNA Complexes

Prasanna R. Kolatkar[1], Balasubramanian Moovarkumudalvan[1], Essam M. Abdelalim[1,2], Mohamed M. Emara[1,3]

[1]Qatar Biomedical Research Institute, Qatar Foundation, Education City, Doha, Qatar; [2]Department of Cytology and Histology, Faculty of Veterinary Medicine, Suez Canal University, Ismailia, Egypt; [3]Department of Virology, School of Veterinary Medicine, Cairo University, Giza, Egypt

INTRODUCTION

SOX transcription factors (TFs) have been known to have important roles in all aspects of cell development. SOX proteins are known to participate in all aspects including the early embryonic states (SOX1, SOX2, and SOX3), ectoderm (SOX2), endoderm (SOX7, SOX17, and SOX18), and chondrogenic development (SOX5, SOX6, and SOX9). In addition, SOX2 has been a key player in induced pluripotent stem cells (iPSCs) (Abdelalim et al., 2014).

Of the 20 SOX factors known to date, only a few have structural data from X-ray crystallographic and nuclear magnetic resonance (NMR) methods. These include SOX2 (Williams et al., 2004; Reményi et al., 2003; Sahu et al., 2011), SOX4 (Jauch et al., 2012), SOX5 (Cary et al., 2001), SOX9 (Genomics, 2012), and SOX17 (Palasingam et al., 2009; Gao et al., 2013; Abe et al., 2007) (Table 1). In all cases, only the well-ordered high mobility shift (HMG) domain from the SOX family has been structurally studied. The N and C terminal (activation domain) portions have not yet been tractable using structural methods. SOX2 (Figure 1(A)) and SOX17 (Figure 1(B)) are the only members of the SOX family to have structures determined in both the deoxyribonucleic acid (DNA)-bound (Palasingam et al., 2009; Reményi et al., 2003; Williams et al., 2004) and free forms (Abe et al., 2007; Sahu et al., 2011; Gao et al., 2013). SOX2-bound forms are all in ternary complexes (Figure 2(A) and (B)) with Pit-1, Oct-1, and Oct-2 (POU) domains and DNA (Reményi et al., 2003; Williams et al., 2004). The rest of the SOX structures, such as SOX4 (Jauch et al., 2012) and SOX9 (Genomics, 2012), that have been studied have been bound to DNA (Figure 3(A)), except for SOX5 (Cary et al., 2001), which has only been studied in the unbound form (Figure 3(B)). The structures of the SOX proteins have answered many structure/function questions that previously existed.

One key question that previously existed was the relationship between the DNA bending angle (upon SOX binding) and cell functions such as transcriptional activity (Scaffidi and Bianchi, 2001). A key finding from all SOX-bound DNA structures (Figure 3(A)) is that they all share highly similar bending angles (approximately 90°) relative to the straight unbound

Sox2
http://dx.doi.org/10.1016/B978-0-12-800352-7.00002-5

Table 1 SOX three-dimensional structures to date

SOX HMG	Method	PDB code	DNA complex	References
mSOX2	X-ray	1GT0	Yes	Remenyi et al. (2003)
hSOX2	NMR	1O4X	Yes	Williams et al. (2004)
hSOX2	NMR	2LE4	No	Sahu et al. (2011)
mSOX4	X-ray	3U2B	Yes	Jauch et al. (2012)
mSOX5	NMR	1I11	No	Cary et al. (2001)
mSOX9	X-ray	4EUW	Yes	Genomics (2012)
mSOX17	X-ray	3F27	Yes	Palasingam et al. (2009)
hSOX17	X-ray	4A3N	No	Gao et al. (2013)
hSOX17	NMR	2YUL	No	Abe et al. (2007)
HMG-D	X-ray	1QRV	Yes	Murphy et al. (1999)
hSRY	NMR	1HRY 1HRZ	Yes	Werner et al. (1995)
mLEF-1	NMR	2LEF	Yes	Love et al. (1995)

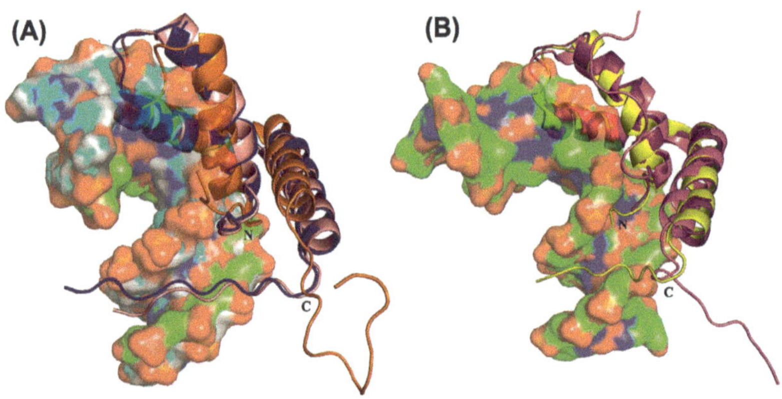

Figure 1 Superposition of (A) SOX2 proteins: DNA-bound mSOX2 (PDB code 1GT0; blue) and hSOX2 (PDB code 1O4X; pink) with unbound hSOX2 (PDB code 2LE4; orange) and (B) SOX17 proteins: DNA bound mSOX17 (PDB code 3F27; yellow) and hSOX17 (PDB code 2YUL; violet) with unbound hSOX17 (PDB code 4A3N; purple).

DNA (Palasingam et al., 2009). Thus, there is no observed difference between all SOX bound DNA bending angles to date, which suggests that there is no correlation between the bending angle and transcriptional activity. However, bending of the DNA could facilitate an interface for other cofactors that bound SOX proteins or adjacent sites.

SOX HMG DOMAIN STRUCTURE

SOX HMG (70–80 residue) structures have been studied at high resolution using both X-ray crystallography and NMR (Figure 3). Both mouse and human SOX proteins have been used for the structural studies and the high similarity of both human and mouse

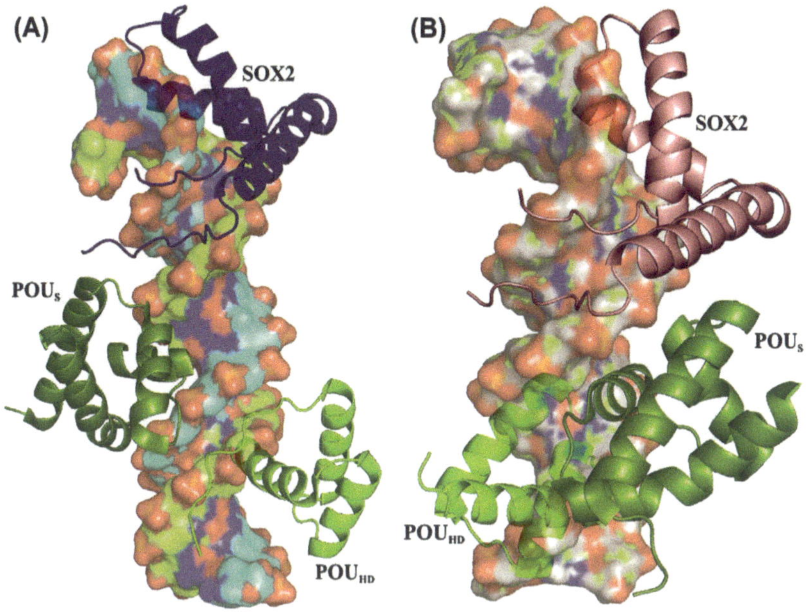

Figure 2 Diagram shows (A) crystal structure of an OCT1/SOX2/*Fgf4* ternary complex (PDB code 1GT0) and (B) NMR structure of an OCT1/SOX2/*HoxB1* ternary complex (PDB code 1O4X).

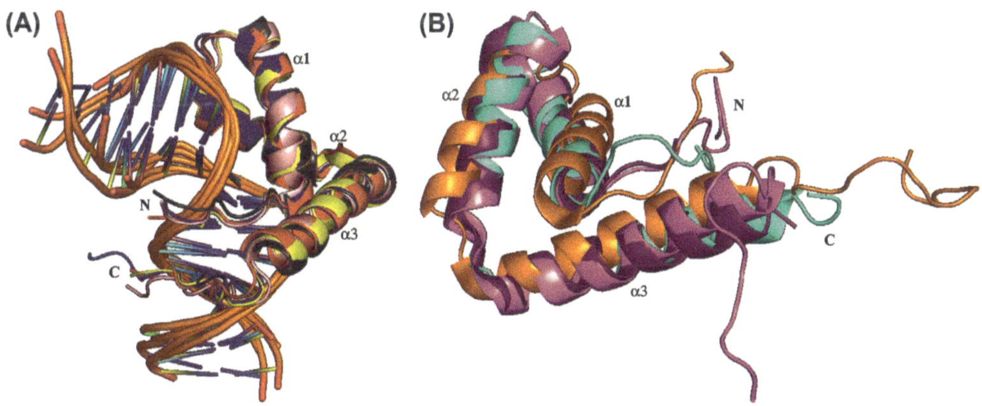

Figure 3 Superposition of (A) all DNA-bound SOX proteins: mSOX2 (PDB code 1GT0; blue), hSOX2 (PDB code 1O4X; pink), mSOX4 (PDB code 3U2B; red), mSOX17 (PDB code 3F27; yellow), and mSOX9 (PDB code 4EUW; gray) and (B) all unbound SOX proteins: hSOX2 (PDB code 2LE4; orange), mSOX5 (PDB code 1I11; cyan), hSOX17 (PDB code 2YUL; violet), and hSOX17 (PDB code 4A3N; purple).

forms make either a good model. SOX proteins have usually been described as L-shaped proteins composed of three alpha helices in which the first two helices are part of the large arm of the "L", whereas helix 3 along with a few N terminal residues forms the small arm. The key conserved residues are the hydrophobic residues forming the core of SOX proteins and the DNA binding residues (Figure 4).

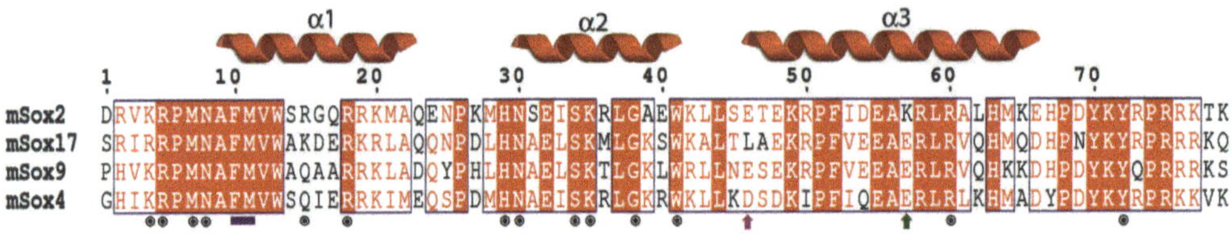

Figure 4 ***Structure-based sequence alignment of all mouse SOX proteins.*** Alpha helices are marked with a red helix above the alignment. The Phe–Met wedge is indicated with a blue bar below the alignment. The DNA interacting residues are marked by closed black dotted circles. The black arrow indicates the position (E57) at which the mutation can be made to convert SOX17 to SOX2 and both pink and black arrows show the mutation positions (E46 and K57) to convert SOX2 to SOX17. ESPript server was used for alignment (Robert and Gouet, 2014).

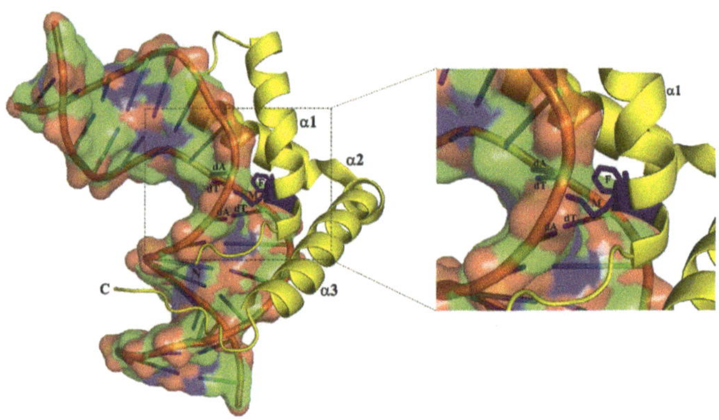

Figure 5 Hydrophobic side chains of SOX17, Phe10, and Met11 (FM wedge, shown in blue) of helix1 are inserted between the labeled nucleotides causing the bend in the DNA axis.

SOX proteins show little change in their overall structure upon binding of DNA. However, the bound DNA undergoes major bending facilitated by the so-called "finger." The finger is also known as the FM wedge owing to the presence of phenylalanine and methionine residues, which comprise this feature. These two key residues insert into the minor groove and spread a pair of bases apart, giving the approximately 90° bend (Figure 5). The C-terminal arms of all bound forms also adopt a highly similar position. Although they are highly flexible and relatively unstructured in the unbound state (Figure 3(B)), they form contact with DNA upon binding in many places and essentially wrap themselves around the DNA.

The structures of non-SOX HMG proteins such as SRY (Werner et al., 1995), HMG-D (Murphy et al., 1999), and LEF-1 (Love et al., 1995), all in their DNA-bound states, are available. Comparison of these DNA-bound structures with the HMG SOX17–DNA structure shows the extremely high similarity of their forms (Figure 6). In fact, these bound structures are more similar to each other than SOX2 or SOX17 when compared to their bound and unbound forms, respectively. Thus, the

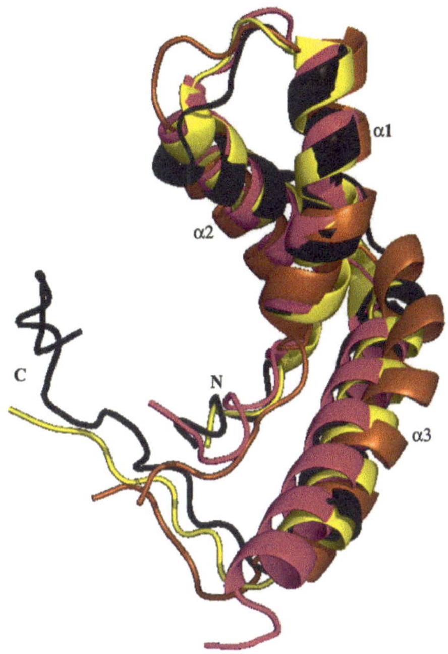

Figure 6 Superposition of SOX17 with other SOX-related proteins: mSOX17 (PDB code 3F27; yellow), hSRY (PDB code 1HRY; brown), HMG-D (PDB code 1QRV; magenta), and mLEF-1 (PDB code 2LEF; black).

mechanisms of binding and recognition of the SOX motif are virtually identical for this family of proteins.

MECHANISM OF MOTIF RECOGNITION FACILITATED BY STRUCTURES

Although individual SOX proteins bind to similar motifs, there are subtle propensities of different SOX proteins to bind slightly different motifs more favorably. The structures of DNA-bound SOX2 and SOX4 were shown to bind the primary (Figure 7(A)) motif (CTTTGTT), whereas DNA-bound SOX17 binds to the secondary motif (AATT-GTT) (Jauch et al., 2012). Primary motifs would have higher binding affinity, and thus secondary motifs could potentially function at higher concentrations of SOX. The secondary motif has two or more base differences relative to the primary motif in which the core TTGT is conserved but the flanking bases at positions 1 and 2 differ. Thus, small differences between the SOX proteins have a minor effect on element recognition.

The most profound effect for specific recognition, however, is clearly imparted by the relation of the co-motif (Figure 7(B)) binding of SOX (HMG) and OCT (POU) partners. The only known ternary structures, to date, have been OCT1–SOX2–DNA structures resulting from both X-ray crystallography (Figure 2(A)) (Reményi et al., 2003) and NMR (Figure 2(B)) (Williams et al., 2004) methods. These ternary complexes show how the

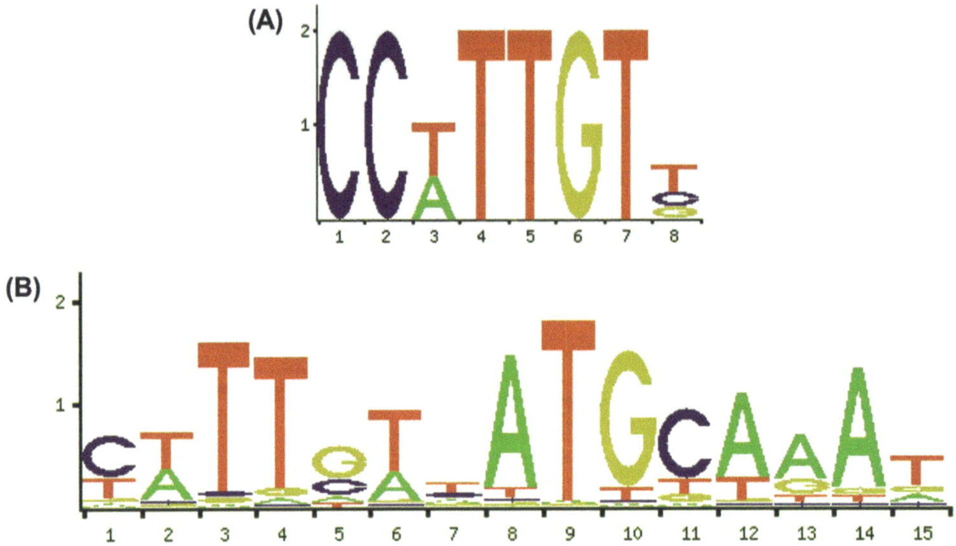

Figure 7 Sequence logo from the JASPAR database (Mathelier et al., 2014) showing the (A) SOX2 motif (B) canonical SOX_0bp_OCT co-motif.

OCT1-SOX2 protein–interaction interface is important for recognition of the specific co-motif. The extrapolated interaction surface for OCT4–SOX2 using the OCT1-comprising structures is also validated with mutagenesis studies that abrogate OCT4-SOX2 co-binding (Reményi et al., 2003). In addition, the OCT1–SOX2–DNA structures comprise both canonical motifs (no base pair between the two half-sites such as *HoxB1*) (Williams et al., 2004) and motifs with three additional base pairs separating the half-sites (*Fgf4*) (Reményi et al., 2003). Comparison of these structures shows that slight rearrangements of the OCT4–SOX2 interface are needed to accommodate the different co-motifs and thereby enact their specificity for binding and ensure the function for the respective genes. Individually, OCT4 and SOX2 would bind their own respective binding sites, but binding at the co-motifs is cooperative and thus functionally relevant sites are preferentially locked in through this cooperative mechanism, which the structures highlight (see also Chapter 3).

STRUCTURE-ASSISTED RE-ENGINEERING OF SOX MEMBERS

This binding of the cooperative co-motif is critical to understanding the mechanism of SOX17 binding and its role in endodermal differentiation as opposed to pluripotency for OCT4. An important question existed as to how such similar TFs (SOX family) with minimal sequence differences could be involved in completely different functional roles in development. A critical clue came from elucidation of the SOX17–DNA structure (Palasingam et al., 2009), which showed a dramatic difference in the charge surface area of SOX17 relative to SOX2: namely, a single residue at the OCT4–SOX2 interface (lys)

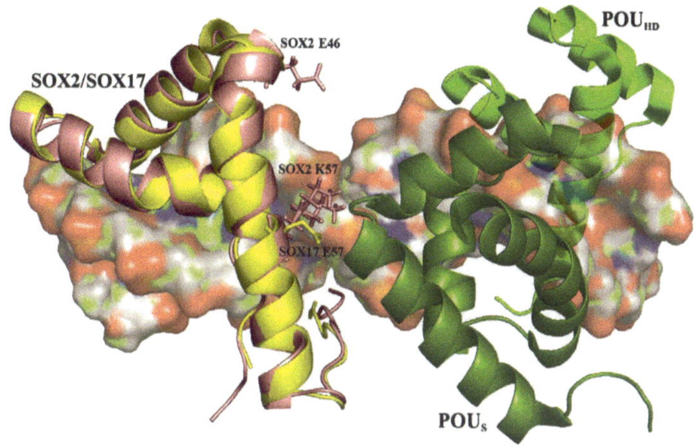

Figure 8 Structural model prepared with PyMOL (DeLano, 2002) using the structural coordinates for mSOX17 (PDB code 3F27; yellow) superimposed onto hSOX2 (pink) from the hSOX2/OCT1 on *HoxB1* DNA (PDB code 1O4X). OCT1-POU$_S$ and OCT1-POU$_{HD}$ are shown in dark and light green. SOX17 can be converted to SOX2 by introducing a point mutation at position E57 (SOX17^{E57K}) and SOX2 can be converted to SOX17 by introducing a double mutation at positions E46 and K57 (SOX2$^{E46L/K57E}$).

is negatively charged (glu) in SOX17 (when superposed onto the SOX2), whereas it is positively charged for SOX2.

Subsequent studies showed that SOX17 would in fact bind a different co-motif previously unknown (Aksoy et al., 2013a; Jauch et al., 2011). This motif is referred to as the compressed motif, because it has one less base pair between the two half-sites at position 7 (Figure 7(B)) of the canonical motif. OCT4 and SOX2 are unable to co-bind to this co-motif, whereas SOX17 and OCT4 cooperatively bind this motif. However, SOX17 can co-bind the canonical motif, albeit weakly compared with SOX2 binding. The structure and sequence comparison facilitated mutagenesis of the key residue, which forms the OCT4–SOX interaction. Mutating the key glutamic acid residue to a lysine residue transformed SOX17 into SOX2 functionally. SOX17^{E57K} (Figure 8) was able to bind the canonical motif in vitro (Jauch et al., 2011), and subsequent genomic studies (Aksoy et al., 2013a) showed that the mutated SOX17 was in fact binding predominantly the same locations as SOX2. Moreover, SOX17^{E57K} was able to substitute SOX2 in the original Yamanaka iPSC cocktail of TFs (Jauch et al., 2011). Surprisingly, there was even a significant increase in iPSC colony formation using the mutated SOX17.

The significant increase in iPSC colony formation was unexpected, and subsequent studies showed that the key driver for this accelerated process was the C-terminal activation domain (Aksoy et al., 2013b). Chimeric constructs using several SOX proteins and various C-terminal ends showed that certain activation domains such as those from SOX17 are more powerful in driving the iPSC-forming process. The fact that SOX2 is not the most efficient in producing iPSC could mean that evolution has chosen a fine balance between stemness and differentiation. Having a factor that can make iPSC but that can easily be regulated and channeled toward different cell fates is likely a key method for controlling a delicate pathway.

In contrast to the $SOX17^{E57K}$, the $SOX2^{K57E}$ mutations did not convert into an efficient endoderm forming TF (Aksoy et al., 2013a). Because the ternary structure of SOX17–OCT4–DNA is currently unknown, it had been proposed that other potential mutations are needed for converting SOX2 into SOX17. In fact, a subsequent study using molecular simulation showed that an additional change facilitated the full conversion of SOX2 into SOX17 (Merino et al., 2014). A tandem mutation comprising $SOX2^{E46L/K57E}$ (Figure 8) was necessary to transform SOX2 into binding the compressed motif cooperatively with OCT4. Interestingly, the second mutation site, $SOX2^{E46L}$, facilitates cooperative binding because Leu46 is buried within a hydrophobic pocket. Because this second mutation site when acting alone does not change the properties of SOX2, Lys57 is the prime driver for the primary OCT4 interaction on the compressed motif whereas the secondary mutation, Leu46, is responsible for conferring cooperativity in DNA binding. Leu46 forms favorable interactions with Tyr25 from OCT4, whereas Glu46 from either SOX2 or $SOX2^{K57E}$ is incapable of forming a similar favorable interface (Merino et al., 2014).

CONCLUSION

Although TFs such as SOX2 enable activities through protein interactions involving their HMG domain with other partners such as OCT4 (POU domain), other SOX family members such as SOX5, SOX6, and SOX9 have coiled-coil and dimerization domains that allow homodimerization and heterodimerization within this trio of SOX members (Han and Lefebvre, 2008; Lefebvre et al., 1998). Unfortunately, structural data for these domains are not available to date. However, the ability of different domains to form dimers shows the versatility of SOX factors in forming functional partnerships to carry out a highly diverse set of developmental programs.

In addition, SOX2 is known to homodimerize as well as make partnerships with HDAC1, HDAC2, and Sal4 using the HMG domain, but also regions N-terminal and C-terminal to the HMG domain (Cox et al., 2010). PAX6 (Kamachi et al., 2001), NPM1 (Niwa et al., 2009), and NANOG (Gagliardi et al., 2013) are also known to directly interact with SOX2. Thus, structures that include these regions will be of great benefit in understanding SOX2 interactions at the molecular level. A structure of the entire SOX2 molecule and not just domains will, of course, be the key to a holistic understanding of SOX partnerships with key developmental factors.

Structural information about SOX members is relatively sparse but critical insight into important mechanisms has been obtained through a detailed analysis of protein–protein and protein–DNA interactions involved in cell development. Specifically, understanding of the switch between pluripotency and cell development stages such as endoderm has been obtained through a combination of structural biology and biochemical analysis. In addition, genomic data (ChIP-seq), cell biology, and chimera generation have been used to validate this information (Jauch et al., 2011; Jauch and Kolatkar, 2013).

Large gaps in knowledge remain about SOX structures because there are only a few SOX structures (DNA-bound and unbound), and only two ternary structures from X-ray crystallographic and NMR data have been reported. In addition, only a handful of *cis*-elements (oligonucleotides) have been solved in complexes. A number of models have been constructed using simulations, such as for the $SOX2^{E46L/K57E}$ double mutant. Thus, a greater number of experimentally determined structures are needed to better understand the large amount of available genomic data. The structures will help to understand new mechanisms of SOX2 partnerships with the POU domain containing TFs as well as other partners. Additional structural data will help to analyze other SOX family member interactions and validate models that have been created. The SOX17 structure and $SOX17^{E57K}$ mutant information show how subtle changes result in drastic functional effects. Therefore, not only structures of native HMG domains from several SOX factors but also modified forms along with their cognate DNA will be needed to achieve complete understanding of cell developmental processes.

REFERENCES

Abdelalim, E.M., Emara, M.M., Kolatkar, P.R., 2014. The SOX transcription factors as key players in pluripotent stem cells. Stem Cells Dev. 23, 2687–2699.

Abe, H., Tochio, N., Miyamoto, K., Koshiba, S., Inoue, M., Kigawa, T., Yokoyama, S., 2007. Solution structure of the HMG box of human transcription factor SOX-17. RCSB. Protein Data Bank, unpublished data.

Aksoy, I., Jauch, R., Chen, J., Dyla, M., Divakar, U., Bogu, G.K., Teo, R., Leng Ng, C.K., Herath, W., Lili, S., Hutchins, A.P., Robson, P., Kolatkar, P.R., Stanton, L.W., 2013a. Oct4 switches partnering from Sox2 to Sox17 to reinterpret the enhancer code and specify endoderm. EMBO J. 32, 938–953.

Aksoy, I., Jauch, R., Eras, V., Bin, A.C., Chen, J., Divakar, U., Ng, C.K., Kolatkar, P.R., Stanton, L.W., 2013b. Sox transcription factors require selective interactions with Oct4 and specific transactivation functions to mediate reprogramming. Stem Cells. 31, 2632–2646.

Cary, P.D., Read, C.M., Davis, B., Driscoll, P.C., Crane-Robinson, C., 2001. Solution structure and backbone dynamics of the DNA-binding domain of mouse Sox-5. Protein Sci. 10, 83–98.

Cox, J.L., Mallanna, S.K., Luo, X., Rizzino, A., 2010. Sox2 uses multiple domains to associate with proteins present in Sox2-protein complexes. PLoS One 5, e15486.

DeLano, W.L., 2002. The PyMOL Molecular Graphics System. DeLano Scientific, Palo Alto, CA, USA. http://www.pymol.org.

Gagliardi, A., Mullin, N.P., Ying Tan, Z., Colby, D., Kousa, A.I., Halbritter, F., Weiss, J.T., Felker, A., Bezstarosti, K., Favaro, R., Demmers, J., Nicolis, S.K., Tomlinson, S.R., Poot, R.A., Chambers, I., 2013. A direct physical interaction between Nanog and Sox2 regulates embryonic stem cell self-renewal. EMBO J. 32, 2231–2247.

Gao, N., Jiang, W., Gao, H., Cheng, Z., Qian, H., Si, S., Xie, Y., 2013. Structural basis of human transcription factor Sry-related box 17 binding to DNA. Protein Pept. Lett. 20, 481–488.

Genomics, J.C.F.S., 2012. Crystal structure of a HMG domain transcription factor SOX-9 bound to DNA (Sox-9/DNA) from Homo sapiens at 2.77 Å resolution. RCSB. Protein Data Bank, unpublished data.

Han, Y., Lefebvre, V., 2008. L-Sox5 and Sox6 drive expression of the aggrecan gene in cartilage by securing binding of Sox9 to a far-upstream enhancer. Mol. Cell. Biol. 28, 4999–5013.

Jauch, R., Aksoy, I., Hutchins, A.P., Ng, C.K., Tian, X.F., Chen, J., Palasingam, P., Robson, P., Stanton, L.W., Kolatkar, P.R., 2011. Conversion of Sox17 into a pluripotency reprogramming factor by re-engineering its association with Oct4 on DNA. Stem Cells 29, 940–951.

Jauch, R., Kolatkar, P.R., 2013. What makes a pluripotency reprogramming factor? Curr. Mol. Med. 13, 806–814.

Jauch, R., Ng, C.K., Narasimhan, K., Kolatkar, P.R., 2012. The crystal structure of the Sox4 HMG domain-DNA complex suggests a mechanism for positional interdependence in DNA recognition. Biochem. J. 443, 39–47.

Kamachi, Y., Uchikawa, M., Tanouchi, A., Sekido, R., Kondoh, H., 2001. Pax6 and SOX2 form a co-DNA-binding partner complex that regulates initiation of lens development. Genes Dev. 15, 1272–1286.

Lefebvre, V., Li, P., De Crombrugghe, B., 1998. A new long form of Sox5 (L-Sox5), Sox6 and Sox9 are coexpressed in chondrogenesis and cooperatively activate the type II collagen gene. EMBO J. 17, 5718–5733.

Love, J.J., Li, X., Case, D.A., Giese, K., Grosschedl, R., Wright, P.E., 1995. Structural basis for DNA bending by the architectural transcription factor LEF-1. Nature 376, 791–795.

Mathelier, A., Zhao, X., Zhang, A.W., Parcy, F., Worsley-Hunt, R., Arenillas, D.J., Buchman, S., Chen, C.Y., Chou, A., Ienasescu, H., Lim, J., Shyr, C., Tan, G., Zhou, M., Lenhard, B., Sandelin, A., Wasserman, W.W., 2014. JASPAR 2014: an extensively expanded and updated open-access database of transcription factor binding profiles. Nucleic Acids Res. 42, D142–D147.

Merino, F., Ng, C.K., Veerapandian, V., Scholer, H.R., Jauch, R., Cojocaru, V., 2014. Structural basis for the sox-dependent genomic redistribution of OCT4 in stem cell differentiation. Structure 22, 1274–1286.

Murphy, F.V.T., Sweet, R.M., Churchill, M.E., 1999. The structure of a chromosomal high mobility group protein-DNA complex reveals sequence-neutral mechanisms important for non-sequence-specific DNA recognition. EMBO J. 18, 6610–6618.

Niwa, H., Ogawa, K., Shimosato, D., Adachi, K., 2009. A parallel circuit of LIF signalling pathways maintains pluripotency of mouse ES cells. Nature 460, 118–122.

Palasingam, P., Jauch, R., Ng, C.K., Kolatkar, P.R., 2009. The structure of Sox17 bound to DNA reveals a conserved bending topology but selective protein interaction platforms. J. Mol. Biol. 388, 619–630.

Reményi, A., Lins, K., Nissen, L.J., Reinbold, R., Schöler, H.R., Wilmanns, M., 2003. Crystal structure of a POU/HMG/DNA ternary complex suggests differential assembly of Oct4 and Sox2 on two enhancers. Genes Dev. 17, 2048–2059.

Robert, X., Gouet, P., 2014. Deciphering key features in protein structures with the new ENDscript server. Nucleic Acids Res. 42, W320–W324.

Sahu, S., Markley, J., Tonelli, M., Bahrami, A., Eghbalnia, H., 2011. Solution structure of the HMG box DNA-binding domain of human stem cell transcription factor Sox2. RCSB. Protein Data Bank, unpublished data.

Scaffidi, P., Bianchi, M.E., 2001. Spatially precise DNA bending is an essential activity of the Sox2 transcription factor. J. Biol. Chem. 276, 47296–47302.

Werner, M.H., Huth, J.R., Gronenborn, A.M., Clore, G.M., 1995. Molecular basis of human 46X,Y sex reversal revealed from the three-dimensional solution structure of the human SRY-DNA complex. Cell 81, 705–714.

Williams Jr., D.C., Cai, M., Clore, G.M., 2004. Molecular basis for synergistic transcriptional activation by Oct1 and Sox2 revealed from the solution structure of the 42-kDa Oct1.Sox2.Hoxb1-DNA ternary transcription factor complex. J. Biol. Chem. 279, 1449–1457.

CHAPTER 3

Dynamics of SOX2 Interactions with DNA

G. Marius Clore
Laboratory of Chemical Physics, National Institute of Diabetes and Digestive and Kidney Diseases, National Institutes of Health, Bethesda, MD, USA

INTRODUCTION

Sox proteins are members of the HMG-box family of architectural factors that bind to the minor groove of deoxyribonucleic acid (DNA), bending the DNA by 50–90° (Kamachi et al., 2000; Dailey and Basilico, 2001; Murphy et al., 2001) (see also Chapter 2). The Sox proteins generally act as part of various multitranscription factor complexes, thereby integrating a wide array of signaling pathways by way of combinatorial transcriptional control to provide temporal and cell-specific transcription regulation (Wolberger, 1999). An example of this phenomenon is provided by members of the Sox and Oct transcription factor families that interact with a variety of DNA promoter/enhancer elements to regulate transcription during embryogenesis and neural development (Kamachi et al., 2000; Dailey and Basilico, 2001) (see also Chapters 5 and 8).

Ternary complexes of SOX2 and OCT1 bound to the regulatory elements within the *HOXB1* promoter and fibroblast growth factor-4 (*FGF4*) enhancer have been solved by nuclear magnetic resonance (NMR) spectroscopy (Williams et al., 2004) and X-ray crystallography (Remenyi et al., 2003), respectively. The two promoters differ in the spacing between the SOX2 and OCT1 binding sites, thereby altering the relative orientations of the two proteins and the protein–protein interface in the two complexes (Figure 1). In the case of the *HOXB1* promoter, the SOX2 and OCT1 sites are immediately adjacent to one another, whereas they are separated by three base pairs on the *FGF4*-enhancer. As a result, the protein–protein interface is approximately double in size for the former ternary complex relative to the latter (Williams et al., 2004).

In this brief review, I will summarize our work on the dynamics of SOX2–DNA interactions in the context of specific and nonspecific SOX2·DNA complexes and specific SOX2·OCT1·DNA ternary complexes involving both *HOXB1* and *FGF4* DNA (Takayama and Clore, 2012a,b). These studies involve the application of NMR spectroscopy—in particular, three major techniques: line-shape analysis, *z*-exchange spectroscopy, and paramagnetic relaxation enhancement (PRE) spectroscopy—which will be briefly described in the next section. Using these methods, we are able to directly probe the mechanisms whereby these two transcription factors locate their

Sox2
http://dx.doi.org/10.1016/B978-0-12-800352-7.00003-7

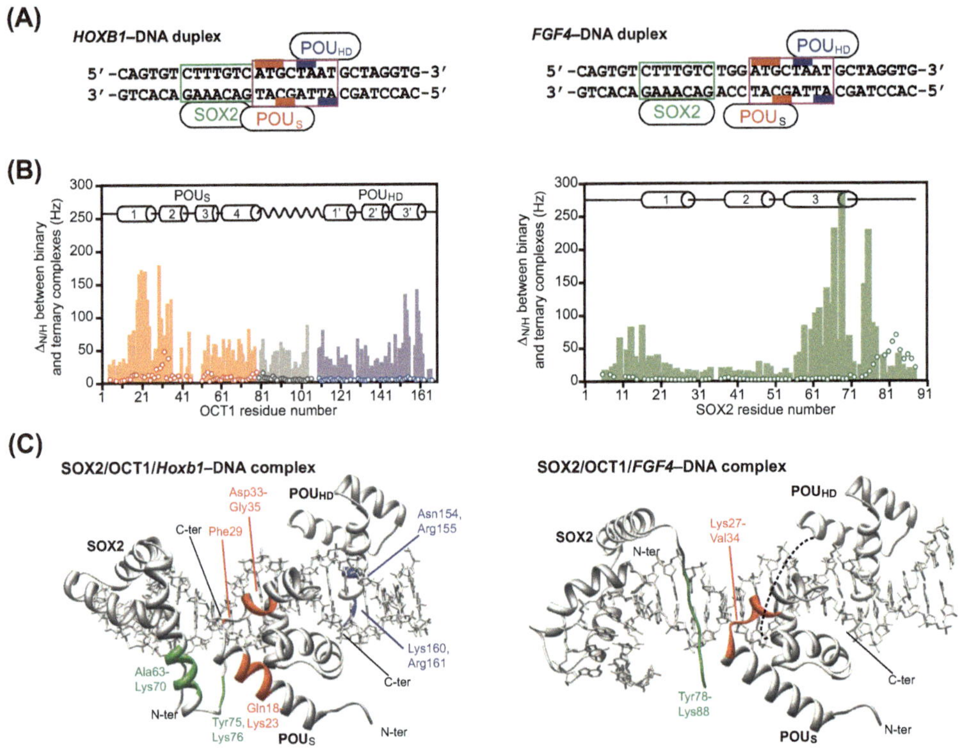

Figure 1 Structures of the ternary SOX2·OCT1·DNA complexes on the *HOXB1* promoter and *FGF4* enhancer. (A) Sequences of the *HOXB1* and *FGF4*-like DNA duplexes. The latter does not represent the actual *FGF4* enhancer sequence but adds the 3-bp insertion (TTG) between the specific binding sites for SOX2 and OCT1, thereby ensuring that differences in binding and kinetics observed for these two sequences solely reflect the 3-bp insertion. (B) $^{1}H_N/^{15}N$ backbone chemical shift perturbation mapping of OCT1 (left) and SOX2 (right) upon ternary complex formation on *HOXB1* (light bars) and *FGF4* (open circles) DNA. (C) Three-dimensional structures of the SOX2·OCT1·*Hoxb1*-DNA (PDB 1o4x) and SOX2·DNA *Fgf4*-DNA (PDB 1gt0) complexes determined by NMR (Williams et al., 2004) and X-ray crystallography (Remenyi et al., 2003), respectively. Residues showing significant $^{1}H_N/^{15}N$ chemical shift perturbations (>100 Hz for the SOX2·OCT1·*Hoxb1*-DNA complex and >20 Hz for the SOX2·OCT1·*Fgf4*-DNA complex) are mapped onto the structures. *Adapted from Takayama and Clore (2012a).*

specific target sites on the DNA within an overwhelming sea of nonspecific DNA and to study the effect of protein–protein interactions on global intermolecular translocation rates.

EXPERIMENTAL APPROACHES

The primary experimental approach we employed to study the dynamics of SOX2-DNA interactions involves the application of three NMR techniques: line-shape analysis, $^{15}N_z$-exchange spectroscopy, and PRE.

Exchange Kinetics

The strategy we developed to study the kinetics of intermolecular exchange of a transcription factor between two different DNA molecules (with or without the specific target site present) involves a simple strategy in which the sequences of two DNA molecules, A and B, are slightly different (Iwahara and Clore, 2006b; Iwahara et al., 2006). The difference may involve one or two bases at the edge of a specific target site when considering specific protein–DNA interactions, or randomly within a restricted region covering the length of a binding site (say, six to eight bases) when considering nonspecific protein–DNA interactions.The result is that the backbone chemical shifts (^1H_N and ^{15}N) for the uniformly ^{15}N-labeled transcription factor will be slightly different when bound to either DNA-A or DNA-B.

When intermolecular exchange of the transcription factor between the two DNA molecules is slow (defined as $k_{ex} << 2\pi\Delta\nu$, where $\Delta\nu$ is the difference in Hz between the chemical shifts of the two states), two sets of resonances will be observed corresponding to the two complexes with the different DNA molecules, and the rate constants for the exchange process can be readily determined by fitting the mixing time dependence of the magnetization of auto- (M_z^{AA}; M_z^{BB}) and exchange cross- (M_z^{AB}; M_z^{BA}) peaks in a two-dimensional $^{15}N_z$-exchange spectrum given by a matrix form of the McConnell equations (McConnell, 1958):

$$\frac{d}{dT_{mix}}\begin{pmatrix} M_z^{\mathrm{AA}} \\ M_z^{\mathrm{BB}} \\ M_z^{\mathrm{AB}} \\ M_z^{\mathrm{BA}} \end{pmatrix} = -\begin{bmatrix} (k_{\mathrm{AB}}^{\mathrm{app}}+R_1^{\mathrm{A}}) & 0 & -k_{\mathrm{BA}}^{\mathrm{app}} & 0 \\ 0 & (k_{\mathrm{BA}}^{\mathrm{app}}+R_1^{\mathrm{B}}) & 0 & -k_{\mathrm{AB}}^{\mathrm{app}} \\ -k_{\mathrm{AB}}^{\mathrm{app}} & 0 & (k_{\mathrm{BA}}^{\mathrm{app}}+R_1^{\mathrm{B}}) & 0 \\ 0 & -k_{\mathrm{BA}}^{\mathrm{app}} & 0 & (k_{\mathrm{AB}}^{\mathrm{app}}+R_1^{\mathrm{A}}) \end{bmatrix}\begin{pmatrix} M_z^{\mathrm{AA}} \\ M_z^{\mathrm{BB}} \\ M_z^{\mathrm{AB}} \\ M_z^{\mathrm{BA}} \end{pmatrix} \quad (1)$$

where k_{AB}^{app} and k_{BA}^{app} are the apparent rate constants for transfer of the transcription factor from DNA-A to DNA-B and vice versa, respectively, and R_1^A and R_1^B are the longitudinal relaxation rates of the protein in complex with DNA-A and DNA-B, respectively.

When exchange of the transcription factor between the two DNA molecules is intermediate ($k_{ex} \sim 2\pi\Delta\nu$) or fast ($k_{ex} << 2\pi\Delta\nu$), only a single cross-peak per residue will be observed in the ^1H-^{15}N correlation spectrum of the protein. Under these conditions, the ^1H_N transverse relaxation rate for a given protein residue, either measured directly in the ^1H dimension from the spectrum (using Lorentzian line-shape fitting) or through transverse (R_2) relaxation measurements, is related to the rate of exchange (k_{ex}) of the transcription factor between DNA-A and DNA-B. The observed transverse relaxation rate R_2^{obs} is given by:

$$R_2^{\mathrm{obs}} = p_{\mathrm{A}}R_2^{\mathrm{A}} + p_{\mathrm{B}}R_2^{\mathrm{B}} + R_{\mathrm{ex}}^{\mathrm{inter}} \quad (2)$$

where p_A and p_B are the populations of the transcription factor bound to DNA-A and DNA-B, respectively; R_2^A and R_2^B are the corresponding R_2 rates for the protein in states A and B; and R_{ex}^{inter} is the exchange contribution arising from intermolecular translocation of the transcription factor between DNA-A and DNA-B. In the intermediate exchange regime, R_2^{obs} can only be calculated numerically using the McConnell equations (McConnell, 1958), but in the fast exchange regime R_{ex}^{inter} can be calculated analytically using the equation (Iwahara et al., 2006):

$$R_{ex}^{inter} = 4\pi^2 \, \Delta \nu^2 p_A p_B / k_{ex}^{inter} \tag{3}$$

where

$$k_{ex}^{inter} = k_{AB}^{app} + k_{BA}^{app}$$

Intermolecular translocation of a transcription factor between two DNA duplexes can occur via two mechanisms that are readily dissected by examining the dependence of the apparent exchange rate constants on the concentration of free DNA (Iwahara and Clore, 2006b; Iwahara et al., 2006; Doucleff and Clore, 2008): (1) Spontaneous dissociation of the protein into free solution followed by reassociation of free protein and DNA is a first-order process that is independent of DNA concentration (under conditions where the total DNA is in excess over protein and the dissociation rate constant is significantly smaller than the pseudo–first-order association rate constant); and (2) direct intermolecular transfer from one DNA duplex to another without dissociation of the protein into free solution is a bimolecular exchange reaction whose rate is directly proportional to the concentration of free DNA. Thus, the slope of a plot of the apparent first-order rate constants versus the concentration of free DNA yields the bimolecular rate constants for direct intermolecular transfer, whereas the intercept is equal to half the dissociation rate constant (the statistical factor of 2 arising from transitions between the same species).

PRE and Detection of Short-Lived Binding Intermediates

The PRE arises from magnetic dipolar interactions between a nucleus (usual protons) and the unpaired electrons of a paramagnetic center and results in an increase in the rate of nuclear magnetization, generally measured as the transverse PRE, Γ_2, given by the difference in R_2 rates between the paramagnetic sample and a diamagnetic control (Clore and Iwahara, 2009). In studying protein-DNA complexes, the simplest strategy involves introducing (d)T-ethylenediaminetetraacetic acid (EDTA) chelated, for example, to Mn^{2+} for the paramagnetic sample and Ca^{2+} for the diamagnetic control, at a particular site synthetically, and measuring the intermolecular PRE effects on a uniformly ^{15}N-labeled protein using ^1H-^{15}N correlation-based experiments (Iwahara et al., 2003; Iwahara and Clore, 2006a). The magnitude of the PRE effect is proportional to the inverse sixth power of the ensemble average of the distance r between the nucleus of interest and the paramagnetic center. Because of the large magnetic moment of the unpaired electron, these effects can extend out to sizable distances (up to about 35 Å for Mn^{2+}). When exchange between a major species and an invisible minor species is slow on the PRE time scale, the observed Γ_2 values will simply reflect the Γ_2 values for the major species. However, when the exchange between these species is fast, the observed Γ_2 values will be population-weighted averages of the Γ_2 values for the major and minor species (Iwahara and Clore, 2006a; Clore and Iwahara, 2009). If there are distances between the paramagnetic label and several protons that are significantly shorter in the minor species than the major one, the footprint of the minor species will be apparent in the PRE profiles measured on the major species, even though the minor species may constitute less than 1% of the total population. These effects can be used to directly probe the involvement of sparsely populated states in the search process whereby transcription factors locate their specific target site within an overwhelming sea of nonspecific DNA sites by measuring the intermolecular PRE profiles observed on the transcription factor in the presence of DNA paramagnetically tagged (usually dT-EDTA–Mn^{2+}) at several locations (one at a time) (Iwahara and Clore, 2006a; Clore, 2011a,b; Takayama and Clore, 2011).

Dissecting the contributions to the intermolecular PREs from intramolecular sliding of the protein along the DNA and intermolecular translocation between DNA molecules can be accomplished by carrying out experiments in the presence of an equimixture of specific and nonspecific DNA (Iwahara and Clore, 2006a; Takayama and Clore, 2011, 2012b). In one sample, the nonspecific duplex is paramagnetically tagged so that intermolecular PREs can only occur via intermolecular translocation between the DNA bearing the specific site (whose spectrum is observed) and the paramagnetically tagged nonspecific DNA. In the second sample, the specific DNA is paramagnetically tagged and the observed PREs will therefore have contributions from both intra- and intermolecular translocation. If sliding (intramolecular translocation) and intermolecular translocation occur, the ratio of the PREs of sample 2 to sample 1 will be larger than 1.

INTERACTIONS OF SOX2 WITH NONSPECIFIC DNA

Sampling of Nonspecific DNA by SOX2

Intermolecular PRE profiles were measured for ^{15}N-labeled SOX2 in the presence of a 29–base pair (bp) DNA duplex with the (d)T-EDTA–Mn^{2+} paramagnetic label covalently attached to a thymine base at either end (Figure 2(A)) (Takayama and Clore, 2012b). Because the exchange of SOX2 between nonspecific sites is fast, the PRE profiles are a weighted average over all possible nonspecific configurations. The PRE profiles arising from the paramagnetic label at the two sites are similar both qualitatively and quantitatively (Figure 2(B) and (C)). The regions of SOX2 that exhibit significant PREs

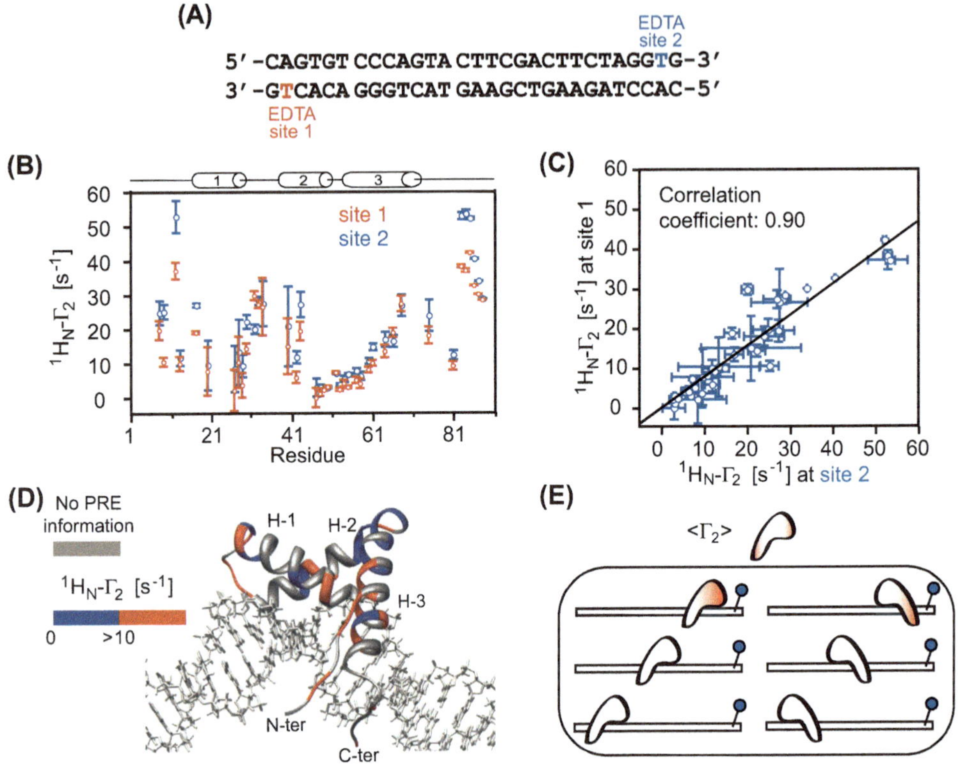

Figure 2 Sampling of binding sites upon nonspecific binding of SOX2 to DNA. (A) Sequence of nonspecific DNA duplex used for PRE measurements. The paramagnetic label (EDTA chelated to Mn^{2+}) was added to two thymines individually (red, site 1; blue, site 2) located at the two ends of the duplex. (B) PRE profiles measured on backbone amide protons (^1H_N-Γ_2) from sites 1 and 2. (C) Correlation between PREs observed from sites 1 and 2. (D) PRE profiles originating from site 2 mapped onto the structure of the specific SOX2-DNA complex (color scale shown on left). (E) Schematic representation of nonspecific DNA binding states sampled by SOX2 (with SOX2 in two orientations related by 180° rotation relative to the long axis of the DNA) with color coding representing the strength of the PRE and the paramagnetic label depicted by the blue sphere. *Adapted from Takayama and Clore (2012b).*

(^{1}H-$\Gamma_2 > 10\,s^{-1}$) are all located in structural elements that are in close proximity to the DNA (Figure 2(D)). One can therefore conclude that SOX2 binds to numerous nonspecific binding sites in two orientations, related by 180° rotation relative to the long axis of the DNA (Figure 2(E)). These results are similar to those observed for the sampling of nonspecific DNA binding sites by the related HMG-box family member HMG-B1, which only binds nonspecifically to DNA (Iwahara et al., 2004).

Kinetics of Intermolecular and Intramolecular Translocation of SOX2 on Nonspecific DNA

To examine the rates of intermolecular and intramolecular translocation of SOX2 on nonspecific DNA, we measured R_2 rates (using line-width analysis) for complexes of SOX2 with two different nonspecific DNA duplexes (Figure 3(A)), both individually and as 1:1 mixtures, as a function of DNA concentration (Takayama and Clore, 2012b). Several resonances exhibit significant ^1H_N or ^{15}N chemical shift differences between the two nonspecific complexes, and the corresponding cross-peaks in the sample comprising a 1:1 mixture of the nonspecific complexes appear halfway between those for the individual complexes and are broadened owing to the exchange contribution from intermolecular translocation (Figure 3(B)). The R_2 rates for the individual complexes are

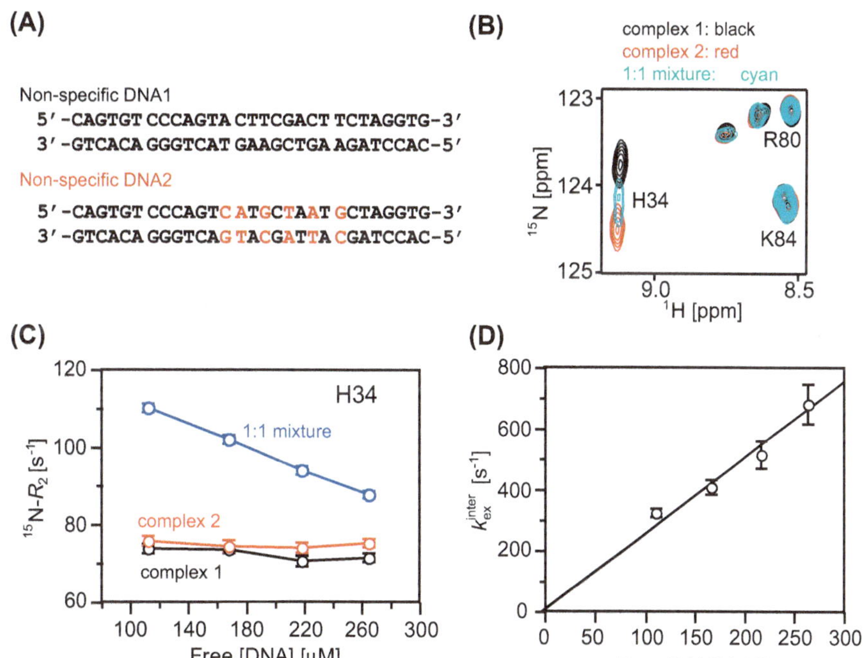

Figure 3 Intermolecular translocation of SOX2 between nonspecific DNA duplexes studied by line-width analysis. (A) Sequences of the two nonspecific DNA duplexes used with the base changes between the two duplexes indicated in red in the second DNA duplex. (B) Selected region of the ^1H-^{15}N correlation spectra of $^{15}N/^2H$-labeled SOX2 for complexes with the first (black) and second (red) DNA duplexes, and a 1:1 mixture of both DNA duplexes (cyan) at 30 °C in the presence of 25 mM added NaCl. The concentration of SOX2 is 0.4 mM; the concentration of duplexes is 0.5 mM in the individual complexes and 0.25 mM in the 1:1 mixture of the two complexes. (C) Dependence of transverse relaxation rates as a function of free DNA concentration. (D) Dependence of the overall apparent intermolecular translocation rate, k_{ex}^{inter}, on the free DNA concentration. *Adapted from Takayama and Clore (2012b).*

unaffected by the concentration of free DNA, but those for the 1:1 mixture decrease as the concentration of DNA increases (Figure 3(C)). The apparent intermolecular rate constant, k_{ex}^{inter}, calculated from Eqn (3), is linearly proportional to the free DNA concentration (Figure 3(D)), indicative of direct transfer from one DNA duplex to the other without dissociation of SOX2 into free solution, and yields a second-order association rate constant of about $10^6\,M^{-1}\,s^{-1}$.

The R_2 rates for the individual complexes are independent of DNA concentration (Figure 3(C)). Because these R_2 rates include an exchange contribution from intramolecular translocation (i.e., sliding) and because one can assume that the average chemical shift difference between nonspecific sites on a single DNA molecule will be approximately the same as those between the complexes with the two different nonspecific DNA duplexes, one can conclude that the rate for intramolecular translocation, k_{ex}^{intra}, is much faster than that for intermolecular translocation (Iwahara et al., 2006), which allows one to obtain a lower limit for k_{ex}^{intra} from the extrapolated value of $k_{ex}^{inter} \sim 900\ s^{-1}$ at the free DNA concentration (about 360 μM) when the R_2 rates for the 1:1 mixture are the same as those for the individual complexes (Figure 3(C) and (D)). This corresponds to a lower limit of about $0.1\,\mu m^2\,s^{-1}$ for the one-dimensional diffusion coefficient for sliding (Clore, 2011b).

KINETICS OF INTERMOLECULAR TRANSLOCATION OF SOX2 BETWEEN SPECIFIC DNA DUPLEXES

The Binary SOX2 DNA Complex

The rates for intermolecular translocation of SOX2 between the two specific sites, differing by a single bp mutation just 5′ of the 7-bp recognition sequence (Figure 4(A) and (B)), can be readily determined from $^{15}N_z$-exchange experiments by monitoring the auto- and exchange-peak intensities in a ^{1}H-^{15}N TROSY-based z-exchange spectrum (Sahu et al., 2007) (Figure 4(C) and (D)) as a function of mixing time for those residues that exhibit significantly different $^{1}H_N/^{15}N$ chemical shifts in the two complexes (Takayama and Clore, 2012b). In contrast to nonspecific binding in which the exchange of SOX2 between the two different nonspecific DNA duplexes was fast on the chemical shift time scale, exchange between the specific complexes is slow owing to tight binding (K_D about 10 nM at 150 mM NaCl) so that there are two sets of cross-peaks, one for each complex. Also in contrast to the nonspecific case, the apparent intermolecular exchange rate constants, k_{AB}^{app} and k_{BA}^{app}, are independent of the concentration of free DNA (Figure 4(E)), which indicates that dissociation from the DNA is rate-limiting and intermolecular transfer occurs exclusively via full dissociation followed by reassociation (Takayama and Clore, 2012b). The dissociation rate constants given by $k_A^{off} = 2k_{AB}^{app}$ and $k_B^{off} = 2k_{BA}^{app}$ have values of 5–$6\,s^{-1}$ at 30 °C, from which one can calculate corresponding association rate constants (given the measured K_D of about 10 nM) of approximately

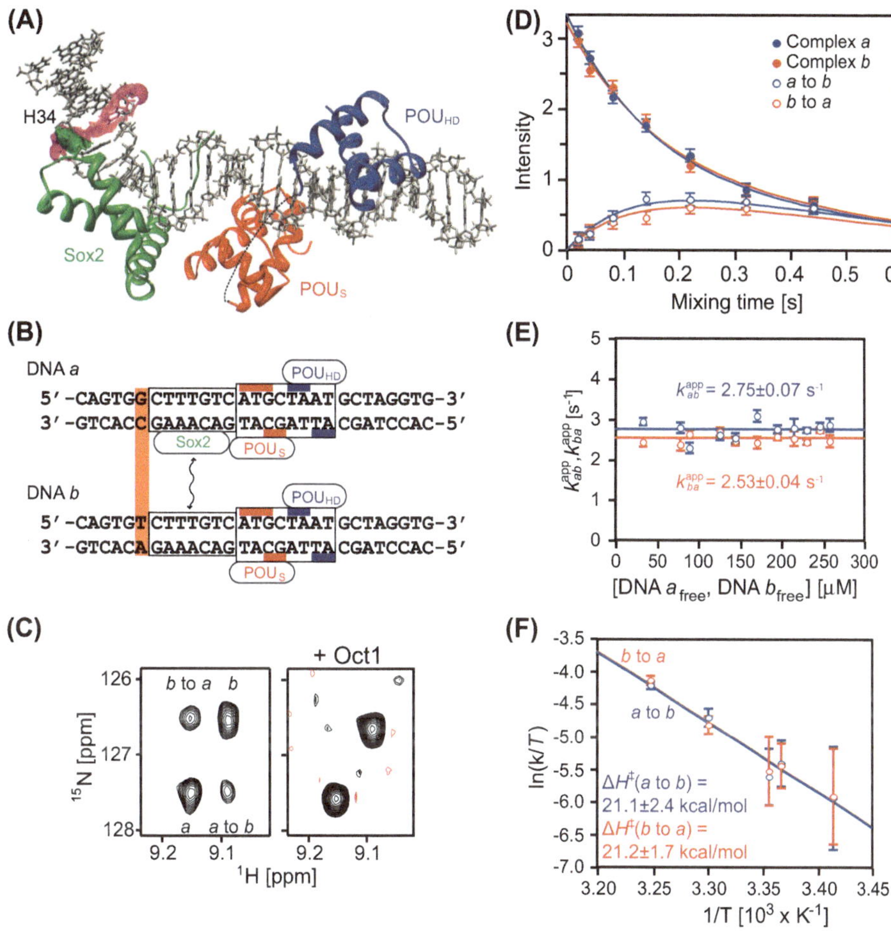

Figure 4 Intermolecular translocation of SOX2 between specific DNA complexes studied by $^{15}N_z$-exchange spectroscopy. (A) Structure of the SOX2·OCT1·*Hoxb1*-DNA complex with SOX2 in green and the POU_S and POU_{HD} domains of OCT1 in red and blue, respectively. (B) Sequences of the two DNA duplexes used for the $^{15}N_z$-exchange experiments with the bp difference between the two duplexes indicated by the purple bar (and also shown in purple in panel A). (C) Example of $^{15}N_z$-exchange data at a mixing time of 120 ms showing auto (*a* and *b*) and exchange (*a* to *b*, and *b* to *a*) peaks. Exchange is abolished upon addition of OCT1 (right panel). (D) Mixing time dependence of auto- and exchange peaks. (E) The apparent intermolecular exchange rates are independent of the concentration of free DNA, indicating that intermolecular exchange proceeds via dissociation followed by reassociation. (F) Eyring plots of the apparent intermolecular exchange rates as a function of temperature. *Adapted from Takayama and Clore (2012b).*

$5 \times 10^8\,M^{-1}\,s^{-1}$. The temperature dependence of the dissociation rate constants (Figure 4(F)) yields values of ~21 and ~4 kcal·mol^{-1} for the activation enthalpy ($\Delta H^{\ddagger}$) and entropy ($T\Delta S^{\ddagger}$), respectively, indicating that the energy barrier to dissociation is largely enthalpic.

The Ternary SOX2·CT1·*Hoxb1*-DNA Complex

In the *HOXB1* promoter, the specific DNA binding sites for SOX2 and OCT1 are immediately adjacent to one another, burying about 540 Å^2 of accessible surface at the interface between SOX2 and the POU_S domain of OCT1 (Williams et al., 2004) (cf. Figure 1 (C), left panel). Upon addition of OCT1, the rates of intermolecular translocation of both SOX2 and OCT1 are dramatically reduced and no longer detectable by z-exchange spectroscopy (Doucleff and Clore, 2008; Takayama and Clore, 2012b) (cf. Figure 4(C), right panel where no exchange peaks are observed for SOX2 upon addition of OCT1). Given a ^{15}N-R_1 relaxation rate of about 1 s^{-1}, the minimum exchange rate measurable by $^{15}N_z$-exchange spectroscopy is about 0.2 s^{-1}, so that the intermolecular exchange rates for both SOX2 and OCT1 between specific sites in the context of the ternary complex on the *HOXB1* promoter are reduced by at least an order of magnitude relative to those in the binary complexes.

The Ternary SOX2·OCT1·*Fgf4*-DNA Complex

In the *FGF-4* enhancer, the specific binding sites for SOX2 and OCT1 are separated by 3 bp, and only about 240 Å^2 of accessible surface is buried at the interface between SOX2 and OCT1 (Remenyi et al., 2003) (Figure 1(C), right panel). As a result, the reduction in intermolecular exchange rate constants is significantly less than that in the ternary complex on the *HOXB1* promoter, enabling one to probe the impact of protein–protein interactions on global exchange rates using $^{15}N_z$-exchange spectroscopy and on cooperative binding using equilibrium fluorescence anisotropy measurements (Takayama and Clore, 2012a).

The K_Ds values for the binding of SOX2 and OCT1 to their respective sites on the *FGF-4* enhancer are 5.3 and 44 nM, respectively. These are reduced by about 16-fold (to approximately 0.33 and 2.7 nM) in the context of the ternary complex (Takayama and Clore, 2012a).

The association and dissociation rate constants for the binding of SOX2 are increased (from about 1×10^9 to about $5 \times 10^9\,M^{-1}\,s^{-1}$) and decreased (from about 5.3 to 1.5 s^{-1}), respectively, in the presence of OCT1. As in the case of the binary SOX2·DNA complex, the intermolecular transfer of SOX2 occurs by jumping (i.e., dissociation and reassociation) and does not involve direct intermolecular transfer (Takayama and Clore, 2012a). For OCT1, however, both jumping and direct intermolecular transfer mechanisms are operative (Takayama and Clore, 2012a). In the presence of SOX2, the dissociation rate constant for OCT1 is reduced by about 20% (from 4.4 to 3.5 s^{-1}), whereas the association rate constant is increased by over 10-fold (from approximately 1×10^8 to $13 \times 10^8\,M^{-1}\,s^{-1}$). The direct intersegment transfer rate constant for the POU_{HD} domain is unaffected by the presence of SOX2 (about $1.8 \times 10^4\,M^{-1}\,s^{-1}$) but that for the POU_S domain is reduced by close to 40% (from 3.4×10^4 to $2.2 \times 10^4\,M^{-1}\,s^{-1}$), which is expected because SOX2 only contacts POU_S in the ternary complex.

These data lead us to propose the following model for the sequence of binding, intersegment transfer, and dissociation events involved in the combinatorial control of gene expression by SOX2 and OCT1 (Takayama and Clore, 2012a) (Figure 5). The initial step involves binding of SOX2 to its specific DNA target site, as supported by the observation that the association rate constant for the formation of the SOX2·DNA binary complex is 10-fold higher than that for OCT1 and that the intermolecular translocation rate for SOX2 is slow, involving full dissociation, whereas that for OCT1 is rapid owing to direct intermolecular transfer at high DNA concentrations. This is also consistent with the localization of SOX2 to the cell nucleus, whereas OCT1 is

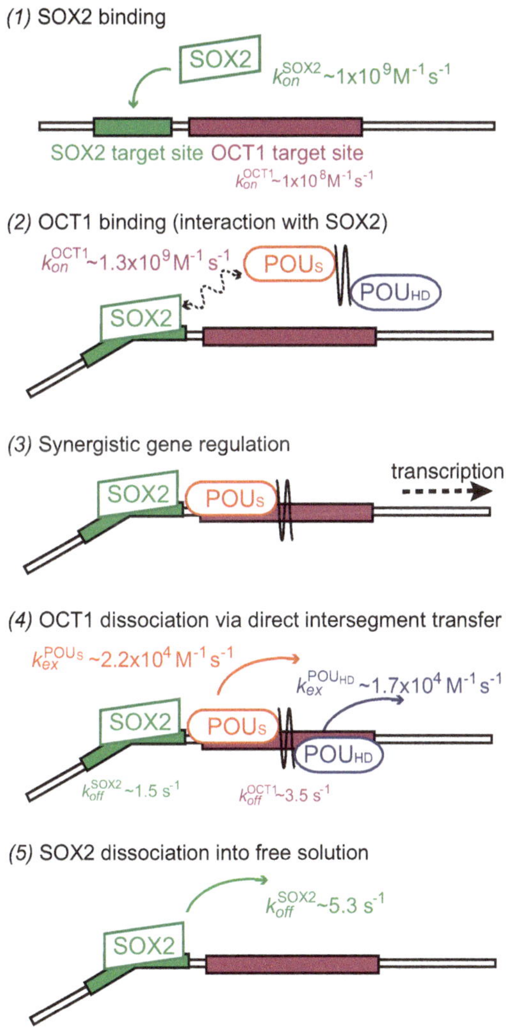

Figure 5 Model for the sequence of binding, intersegment transfer, and dissociation events involved in synergistic transcription regulation by SOX2 and OCT1 derived from *z*-exchange data on the SOX2·OCT1·*Fgf4*-DNA complex. *Adapted from Takayama and Clore (2012a).*

found in both the cytoplasm and the nucleus. The presence of specifically bound SOX2 accelerates the binding of OCT1 to its target site adjacent to that of SOX2 by just over an order of magnitude, and the ternary complex is stabilized by protein–protein interactions between SOX2 and the POU_S domain of OCT1. Once the specific ternary complex is formed, OCT1 and SOX2 activate transcription synergistically. OCT1 subsequently dissociates from the ternary complex largely by direct intersegment transfer, which, at high concentrations of DNA found in vivo, is significantly faster than dissociation into free solution. Intersegment transfer can occur to another specific site on a different promoter or to a nonspecific site. These sites may be located on different DNA molecules or at a separation of greater than about 150 bp on the same DNA, allowing transfer to occur via DNA looping. Finally, SOX2 dissociates slowly from its specific site and can subsequently re-associate to other specific or nonspecific sites.

INTERPLAY BETWEEN SOX2 AND OCT1 IN TRANSLOCATION INVOLVING SPARSELY POPULATED STATES PROBED BY PRE

Short-Lived Intermediates in Translocation of SOX2 from the Specific SOX2 DNA Binary Complex

Using a 29-bp DNA duplex with paramagnetic tags at three sites (one at a time) (Takayama and Clore, 2012b) (Figure 6(A)) reveals that at low-added salt concentrations (0 M NaCl), the PRE profiles from sites 1 and 2, located just 5′ to or within the SOX2 binding site but on the major groove, respectively, are consistent with the structure of the specific SOX2 DNA binary complex (Figure 6(B) and (C), top panels). However, the PRE profiles from site 3, located distally on the 3′ end of the DNA binding site, reveal the presence of small ($<7\,s^{-1}$) but significant PREs (Figure 6(B), top right panel). At higher salt concentrations (50 or 150 mM NaCl), additional PREs within helix 3 are

Figure 6 PRE detection of sparsely populated intermediate states involved in specific DNA target site location by SOX2 (A) 29mer *Hoxb1*-DNA duplex used with the three sites (one at a time) paramagnetically labeled with EDTA-Mn^{2+} indicated in blue. The binding sites for SOX2 and OCT1 are indicated. (B) PRE profiles observed on $^{15}N/^{2}H$-labeled SOX2 (in the absence of OCT1) originating from the three sites in the presence of 0 (top panels) or 50 and 150 mM (bottom panels) added NaCl. The blue lines in the top panels are the PREs back-calculated from the structure of the specific SOX2–DNA complex (Murphy et al., 2001). (C) PRE profiles at 0 (top) or 150 mM (bottom) added NaCl mapped onto the structure of the specific SOX2·DNA complex. The color scale for the PREs is shown at the bottom of the figure. (D) Schematic of intermolecular and intramolecular translocation pathways involving SOX2 bound specifically to DNA, as deduced from PRE measurements involving mixtures of specific and nonspecific DNA duplexes. The major translocation pathway involves intermolecular translocation initiated by sliding from the specific site to an immediately adjacent nonspecific site, followed by direct intersegment transfer to the nonspecific DNA duplex and then back to the specific DNA duplex. *Adapted from Takayama and Clore (2012b).*

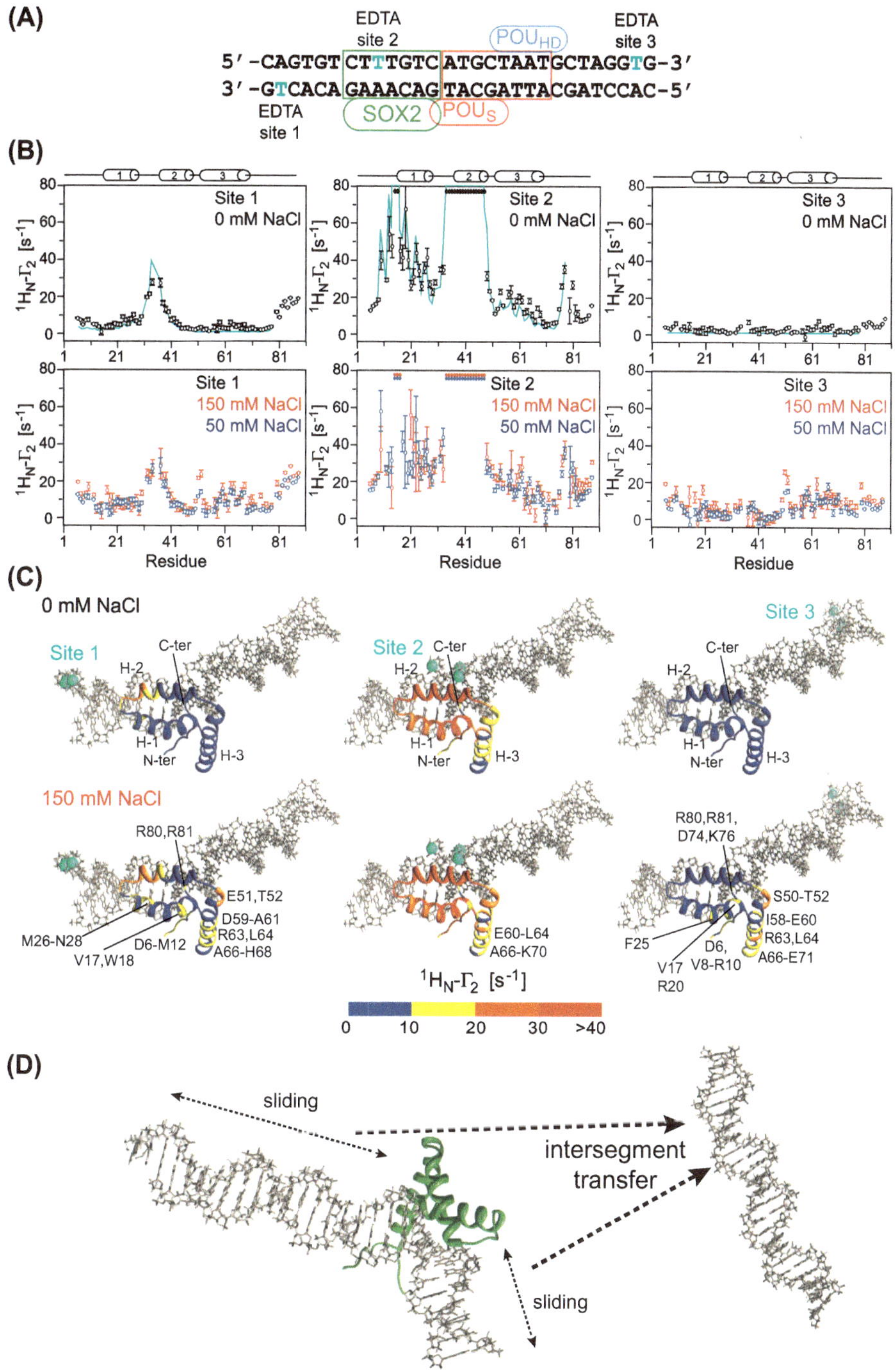

(A)
EDTA site 2
POU_HD
EDTA site 3
5′-CAGTGTCTTTGTCATGCTAATGCTAGGTG-3′
3′-GTCACAGAAACAGTACGATTACGATCCAC-5′
EDTA site 1
SOX2
POU_S
(B)
Site 1 0 mM NaCl
Site 2 0 mM NaCl
Site 3 0 mM NaCl
Site 1 150 mM NaCl 50 mM NaCl
Site 2 150 mM NaCl 50 mM NaCl
Site 3 150 mM NaCl 50 mM NaCl
1H_N-Γ2 [s-1]
Residue
(C)
0 mM NaCl
Site 1
Site 2
Site 3
C-ter
H-2
H-1
N-ter
H-3
150 mM NaCl
R80,R81
E51,T52
D59-A61
R63,L64
A66-H68
M26-N28
D6-M12
V17,W18
E60-L64
A66-K70
R80,R81, D74,K76
S50-T52
I58-E60
R63,L64
A66-E71
F25
D6, V8-R10
V17 R20
1H_N-Γ2 [s-1]
0 10 20 30 >40
(D)
sliding
intersegment transfer
sliding

observed from site 1 and all PREs from site 3 are increased in magnitude (Figure 6(B) and (C), bottom panels), although the ^{1}H–^{15}N correlation spectrum remains unchanged and represents that of the specific complex (Takayama and Clore, 2012b). These PRE data are a direct reflection of sparsely populated (<1%) transient states that come into close proximity of the paramagnetic label. The use of samples comprising an equimixture of specific and nonspecific DNA, in which the paramagnetic label is attached to either the nonspecific (intermolecular translocation only) or specific (contributions from intramolecular and intermolecular translocation) DNA duplexes indicates that the PREs are largely caused by intermolecular translocation with a small (about 20%) contribution from sliding (Takayama and Clore, 2012b) (Figure 6(D)). Given the slow dissociation of the specific SOX2·DNA complex (see Section The Binary SOX2·DNA Complex), intermolecular translocation of SOX2, once bound to its specific target site, must occur by SOX2 sliding out of the specific site to make a nonspecific interaction with the DNA, followed by intersegment transfer between nonspecific sites located on different DNA molecules (Takayama and Clore, 2012b).

Short-Lived Intermediates in Translocation of SOX2 and OCT1 from the Specific SOX2·OCT1·*Hoxb1*-DNA Complex

One can use the same 29-bp DNA duplex used to study the binary SOX2·DNA complex to examine the impact of protein–protein interactions on the sampling of sparsely populated states of SOX2 and OCT1 within the context of the specific SOX2·OCT1·*Hoxb1*-DNA complex (The OCT1 binding site is immediately 3′ of the SOX2 site (Figure 6(A)). For both SOX2 and OCT1 the presence of short-lived transient states, populated below 1%, is detected by PRE, especially at higher salt concentrations (50–150 mM NaCl) (Takayama and Clore, 2012b).

For SOX2, these PREs arise entirely from intermolecular translocation because no measurable intramolecular PRE contributions can be detected (Figure 7(A)). But just as in the case of the SOX2·DNA binary complex, this must first involve sliding to a nonspecific site immediately adjacent to the specific one because dissociation of SOX2 from its specific site is so slow (Takayama and Clore, 2012b).

In the case of OCT1, analysis of the PRE data indicates that the POU_S domain is largely fixed to its specific site through interaction with SOX2 (Figure 7(B)) (Takayama and Clore, 2012b), in direct contrast to the situation with the specific OCT1 DNA complex in which the POU_S domain largely directs intermolecular translocation (Takayama and Clore, 2011). For the POU_{HD} domain, however, the PRE data indicate that translocation in the context of the SOX2·OCT1·*Hoxb1*-DNA ternary complex is primarily intermolecular with some contribution from intramolecular translocation (Takayama and Clore, 2012b), in contrast to the situation in the binary OCT1·DNA complex in which intramolecular translocation predominates (Takayama and Clore, 2011). Furthermore, the intramolecular translocation of POU_{HD} to nonspecific sites on the 5′-side of the SOX2-specific site likely

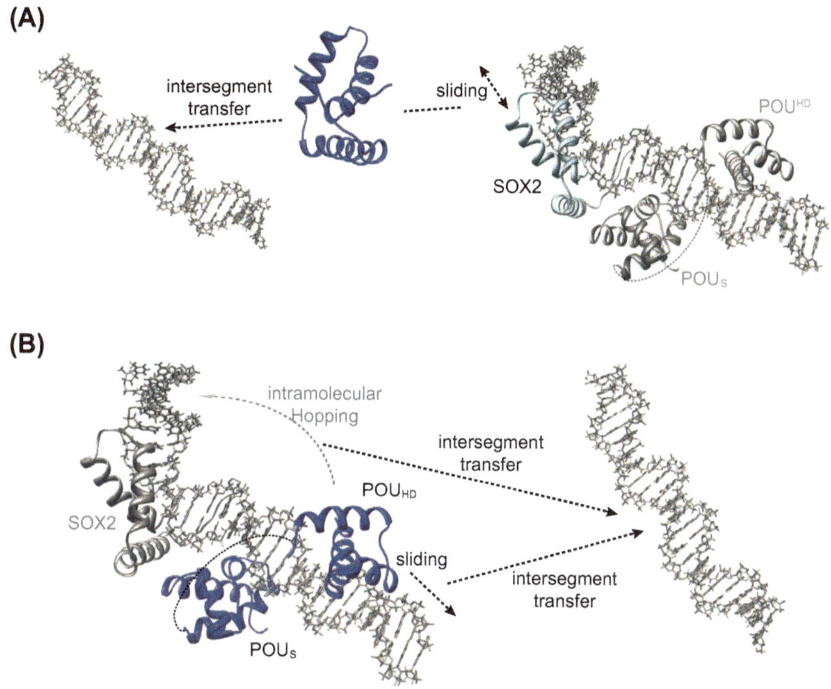

Figure 7 Schematic of intermolecular and intramolecular translocation pathways for (A) SOX2 and (B) OCT1 in the context of the specific ternary SOX2·OCT1·*Hoxb1*-DNA complex, derived from PRE measurements involving mixtures of specific and nonspecific DNA duplexes. *Adapted from Takayama and Clore (2012b).*

occurs via hopping rather than sliding because the presence of both SOX2 and the POU_S domain would block sliding of the POU_{HD} domain (Figure 7(B)).

CONCLUDING REMARKS

Data afforded by various NMR experiments including $^{15}N_z$-exchange, line-width analysis, and PRE show that SOX2 efficiently explores the DNA landscape to locate its specific target site using both direct intersegment transfer between nonspecific sites on different DNA duplexes and sliding along the nonspecific DNA (Takayama and Clore, 2012a,b). Once a specific ternary SOX2·OCT1·DNA complex is formed, further sampling of DNA sites by SOX2 is reduced but can still occur by initial sliding of SOX2 to an adjacent nonspecific site followed by intersegment transfer. OCT1 alone explores the DNA landscape by using the complementary interplay of intramolecular sliding by the POU_{HD} domain and intersegment transfer of the POU_S domain, with the latter acting as a fly cast. Once the ternary complex with SOX2 is formed, the POU_S domain is largely fixed to its specific sites through protein–protein interactions with SOX2 which are significantly stronger in the case of the complex on the *HOXB1* promoter than the *FGF4* enhancer owing to the shorter distance between the SOX2 and OCT1 sites (immediately adjacent versus separated by 3 bp, resulting in a twofold reduction for the latter in the buried accessible surface area at the interface between SOX2 and OCT1 (Williams et al., 2004; Remenyi et al., 2003)). However, the POU_{HD} domain can still sample other DNA sites by initially sliding to an immediately adjacent nonspecific site, just like SOX2, followed by either intersegment transfer to another DNA duplex or directly hopping to another nonspecific site on the same DNA molecule. The former involves the formation of a bridged intermediate complex in which POU_S and SOX2 are located on the first DNA duplex while POU_{HD} is transiently bound to the second DNA duplex, thereby promoting the complete release of OCT1 from the ternary complex by a first-order process involving dissociation of POU_S from the first DNA duplex followed by subsequent association onto the second DNA duplex. This contrasts to the mechanism used in the specific OCT1·DNA binary complex in which POU_S locates the second DNA duplex by intersegment transfer, thereby promoting the subsequent transfer of POU_{HD}. Thus, direct interaction of SOX2 and OCT1 bound to adjacent specific DNA binding sites modulates the translocation pathways employed by OCT1 to scan alternative DNA target sites via transient, sparsely populated states.

ACKNOWLEDGMENTS

This work was supported by the Intramural Program of the National Institute of Diabetes and Digestive and Kidney Diseases, National Institutes of Health and by the AIDS Targeted Antiviral Program of the Office of the Director of the National Institutes of Health.

REFERENCES

Clore, G.M., 2011a. Exploring sparsely populated states of macromolecules by diamagnetic and paramagnetic NMR relaxation. Protein Sci. 20, 229–246.

Clore, G.M., 2011b. Exploring translocation of proteins on DNA by NMR. J. Biomol. NMR 51, 209–219.

Clore, G.M., Iwahara, J., 2009. Theory, practice, and applications of paramagnetic relaxation enhancement for the characterization of transient low-population states of biological macromolecules and their complexes. Chem. Rev. 109, 4108–4139.

Dailey, L., Basilico, C., 2001. Coevolution of HMG domains and homeodomains and the generation of transcriptional regulation by Sox/POU complexes. J. Cell. Physiol. 186, 315–328.

Doucleff, M., Clore, G.M., 2008. Global jumping and domain-specific intersegment transfer between DNA cognate sites of the multidomain transcription factor Oct-1. Proc. Natl. Acad. Sci. U.S.A. 105, 13871–13876.

Iwahara, J., Anderson, D.E., Murphy, E.C., Clore, G.M., 2003. EDTA-derivatized deoxythymidine as a tool for rapid determination of protein binding polarity to DNA by intermolecular paramagnetic relaxation enhancement. J. Am. Chem. Soc. 125, 6634–6635.

Iwahara, J., Clore, G.M., 2006a. Detecting transient intermediates in macromolecular binding by paramagnetic NMR. Nature 440, 1227–1230.

Iwahara, J., Clore, G.M., 2006b. Direct observation of enhanced translocation of a homeodomain between DNA cognate sites by NMR exchange spectroscopy. J. Am. Chem. Soc. 128, 404–405.

Iwahara, J., Schwieters, C.D., Clore, G.M., 2004. Characterization of nonspecific protein-DNA interactions by ^{1}H paramagnetic relaxation enhancement. J. Am. Chem. Soc. 126, 12800–12808.

Iwahara, J., Zweckstetter, M., Clore, G.M., 2006. NMR structural and kinetic characterization of a homeodomain diffusing and hopping on nonspecific DNA. Proc. Natl. Acad. Sci. U.S.A. 103, 15062–15067.

Kamachi, Y., Uchikawa, M., Kondoh, H., 2000. Pairing SOX off: with partners in the regulation of embryonic development. Trends Genet. 16, 182–187.

Mcconnell, H.M., 1958. Reaction rates by nuclear magnetic resonance. J. Chem. Phys. 28, 430–431.

Murphy, E.C., Zhurkin, V.B., Louis, J.M., Cornilescu, G., Clore, G.M., 2001. Structural basis for SRY-dependent 46-X,Y sex reversal: modulation of DNA bending by a naturally occurring point mutation. J. Mol. Biol. 312, 481–499.

Remenyi, A., Lins, K., Nissen, L.J., Reinbold, R., Scholer, H.R., Wilmanns, M., 2003. Crystal structure of a POU/HMG/DNA ternary complex suggests differential assembly of Oct4 and SOX2 on two enhancers. Genes Dev. 17, 2048–2059.

Sahu, D., Clore, G.M., Iwahara, J., 2007. TROSY-based *z*-exchange spectroscopy: application to the determination of the activation energy for intermolecular protein translocation between specific sites on different DNA molecules. J. Am. Chem. Soc. 129, 13232–13237.

Takayama, Y., Clore, G.M., 2011. Intra- and intermolecular translocation of the bi-domain transcription factor OCT1 characterized by liquid crystal and paramagnetic NMR. Proc. Natl. Acad. Sci. U.S.A. 108, E169–E176.

Takayama, Y., Clore, G.M., 2012a. Impact of protein/protein interactions on global intermolecular translocation rates of the transcription factors SOX2 and OCT1 between DNA cognate sites analyzed by z-exchange NMR spectroscopy. J. Biol. Chem. 287, 26962–26970.

Takayama, Y., Clore, G.M., 2012b. Interplay between minor and major groove-binding transcription factors SOX2 and OCT1 in translocation on DNA studied by paramagnetic and diamagnetic NMR. J. Biol. Chem. 287, 14349–14363.

Williams Jr., D.C., Cai, M., Clore, G.M., 2004. Molecular basis for synergistic transcriptional activation by OCT1 and SOX2 revealed from the solution structure of the 42-kDa OCT1.SOX2.*Hoxb1*-DNA ternary transcription factor complex. J. Biol. Chem. 279, 1449–1457.

Wolberger, C., 1999. Multiprotein-DNA complexes in transcriptional regulation. Annu. Rev. Biophys. Biomol. Struct. 28, 29–56.

CHAPTER 4

Posttranscriptional Modulation of *Sox2* Activity by miRNAs

Nilima Prakash[1,2,3]
[1]Helmholtz Zentrum München, Deutsches Forschungszentrum für Gesundheit und Umwelt (GmbH), Institute of Developmental Genetics, Germany; [2]Technische Universität München, Lehrstuhl für Entwicklungsgenetik c/o Helmholtz Zentrum München, Germany; [3]Hamm-Lippstadt University of Applied Sciences, Germany

INTRODUCTION

SOX2, a member of the SoxB1 group of sex determining region Y–related high-mobility group-box (SOX) proteins that has emerged at the transition from unicellular to multicellular organisms (see Chapter 6, Evolution of *Sox2* and Functional Redundancy in Relation to Other *SoxB1* Genes), is a versatile transcription factor (TF) with a multitude of functions during development and adulthood (reviewed by Guth and Wegner, 2008; Kamachi and Kondoh, 2013; Sarkar and Hochedlinger, 2013). Nevertheless, SOX2 is probably best known for its essential role in the core pluripotency network (together with the TFs OCT4 (POU5F1), NANOG, and KLF4) of embryonic stem cells (ESCs) and induced pluripotent stem cells (iPSCs) (Chapter 10, Sox2-Dependent Regulation of Pluripotent Stem Cells), its fundamental function in neural development and adult neurogenesis (Chapter 11, Sox2-Dependent Regulation of Neural Stem Cells and CNS Development), and its dysregulation in all types of tumors (reviewed by Liu et al., 2013; Pevny and Placzek, 2005; Sarkar and Hochedlinger, 2013). A notable and common observation in these and other contexts is that the amount of available SOX2 protein in a cell needs to be precisely balanced: Both decreased (fewer copy numbers) and increased (more copy numbers) dosage of the *Sox2* gene have detrimental effects on pluripotent stem cell (PSC)/multipotent stem cell maintenance and tissue homeostasis (reviewed by Kamachi and Kondoh, 2013; Liu et al., 2013; Pevny and Placzek, 2005; Sarkar and Hochedlinger, 2013). The preservation of a PSC/multipotent cell state and/or the accurate transition between one cell state and the other, particularly during embryonic development, therefore requires a precise control of SOX2 protein dosage and activity in a cell, which is accomplished through transcriptional, posttranscriptional, and posttranslational mechanisms (reviewed by Kamachi and Kondoh, 2013). The *Sox2* gene itself is under tight transcriptional (including autoregulatory) control, but this mechanism may not be sufficiently dynamic to ensure a smooth transition between cell states or secure enough to warrant stable preservation of a specific cell state. Posttranscriptional mechanisms, including the posttranscriptional regulation of messenger ribonucleic acid

Sox2
http://dx.doi.org/10.1016/B978-0-12-800352-7.00004-9

(mRNA) stability and/or translation, and posttranslational protein modifications are indeed important regulators of SOX2 dosage and activity in a cell (reviewed by Kamachi and Kondoh, 2013; Kashyap et al., 2009; Sarkar and Hochedlinger, 2013). The posttranscriptional control of *Sox2* mRNA stability and/or translation by noncoding RNAs (ncRNAs) has emerged in recent years as one of the most important mechanisms to keep SOX2 protein levels in check. This mechanism is involved in establishing and maintaining a PSC or multipotent stem cell state and allows the smooth transition to a more committed and/or differentiated cell state during normal tissue homeostasis, but also prevents the dysregulation of SOX2 protein levels in an already more committed/differentiated cell that may lead to its oncogenic transformation. This chapter reviews our current knowledge about the posttranscriptional modulation of *Sox2* gene expression by ncRNAs and especially microRNAs in PSCs, multipotent neural stem cells (NSCs)/neural progenitor cells (NPCs), and cancer stem cells (CSCs)/tumor cells. Although several of these functions of SOX2 or its ortholog(s) are conserved between vertebrates and invertebrates (reviewed by Guth and Wegner, 2008) and the *Sox2* gene appears to be transcriptionally and posttranscriptionally regulated in a similar manner across species, the focus is on the ncRNA-mediated modulation of SOX2 expression in mammalian (mouse and human) tissues and cells, because all of the data presented here were derived in this context.

BIOGENESIS AND FUNCTION OF MICRORNAs AND LARGE INTERGENIC NONCODING RNAs IN ANIMALS

Noncoding RNAs are transcripts that are generally not translated into a polypeptide or protein sequence, although exceptions exist when an ncRNA contains an open-reading frame and might be translated into a protein (reviewed by Qureshi and Mehler, 2012). The huge family of ncRNAs is composed of, among many others, the small microRNAs (miRNAs or miRs) and the large intergenic ncRNAs (lincRNAs). As their name indicates, miRNAs are short (usually between 18 and 24 nucleotides (nt) long), single-stranded RNAs (ssRNAs) that are encoded in the genome as a either monocistronic or polycistronic gene with its own promoter, or as part of the intron of another gene, the so-called mirtrons (Figure 1(A) and (B)) (reviewed by Ha and Kim, 2014; Krol et al., 2010; Qureshi and Mehler, 2012; Yang and Lai, 2011). In contrast to the miRNAs, lincRNAs are long (by definition at least 200 nt long) ssRNAs that are transcribed from intergenic regions in the genome and can be modified by alternative splicing (Figure 1(C)) (reviewed by Esteller, 2011; Fatica and Bozzoni, 2014; Qureshi and Mehler, 2012). In most cases, RNA polymerase II mediates the transcription of the miRNA, protein coding, or lincRNA gene, and the primary transcripts are subsequently modified by 5′ capping, polyadenylation, and RNA editing (Figure 1) (reviewed by Ha and Kim, 2014; Qureshi and Mehler, 2012).

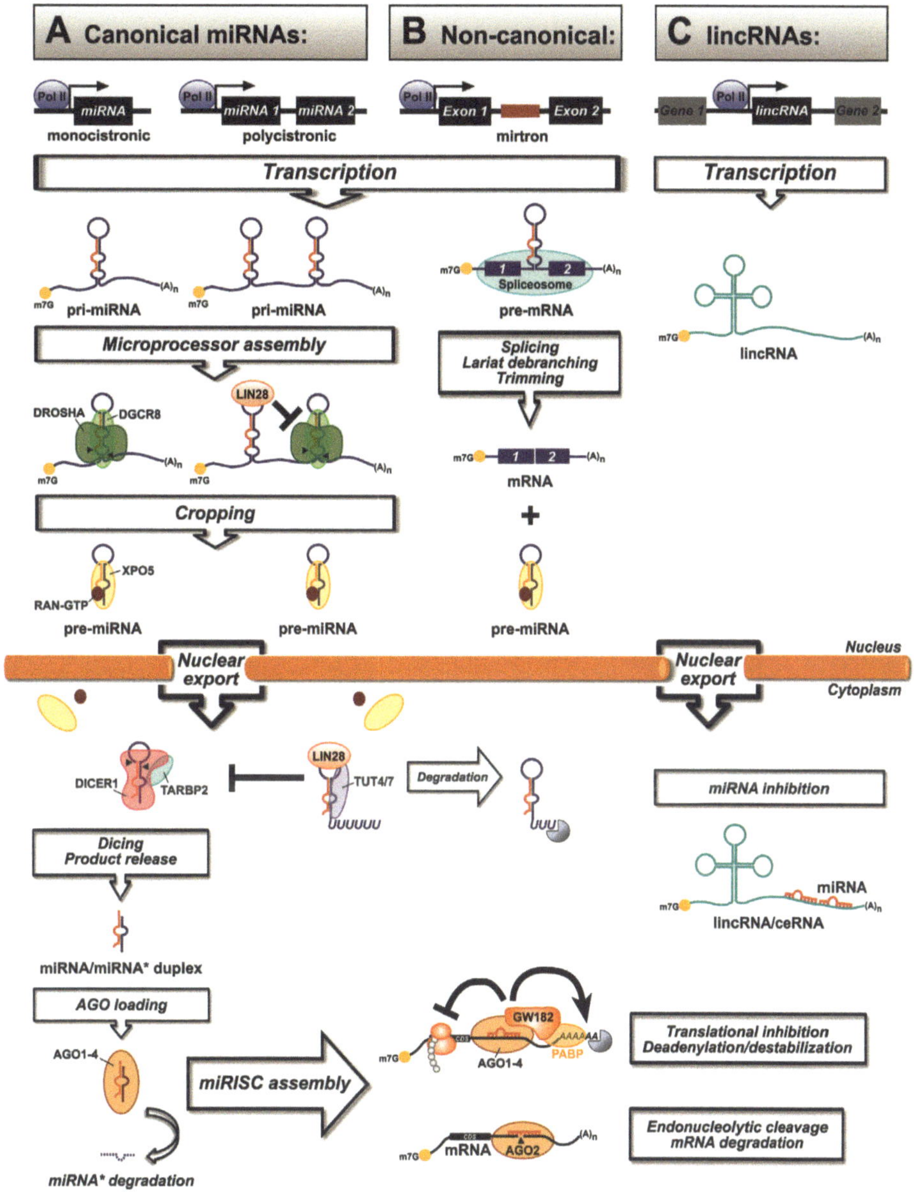

Figure 1 *Biogenesis and function of miRNAs.* Schematic depiction of the different steps during the biogenesis of miRNAs in the canonical (A) and noncanonical (B) pathways, of lincRNA synthesis (C), and of their corresponding functions as mature miRNAs or ceRNAs. See the text for details. Binding of LIN28 to the terminal loop of the pri-miRNA or pre-miRNA, respectively, inhibits the assembly of the DROSHA/DGCR8 microprocessor complex in the nucleus or the DICER1/TARBP2 complex in the cytoplasm. Oligouridylation of the pre-miRNA in the cytoplasm, through recruitment of terminal uridylyl transferases (TUT4/7, also known as ZCCHC11/ZCCHC6) by the terminal loop-bound LIN28, also blocks its processing by DICER1 and induces its degradation by 3′–5′ exonucleases. Imperfect base pairing between the miRNA and its target mRNA induces the GW182-mediated deadenylation and destabilization of the mRNA and/or the inhibition of its translation, whereas perfect base pairing between the miRNA and its target mRNA provokes the endonucleolytic cleavage by AGO2 and subsequent degradation of the mRNA. Apart from several other functions not depicted here, lincRNAs can act as ceRNAs or miRNA sponges through binding of miRNAs at complementary sites in the cytoplasm. Abbreviations: $(A)_n$, poly(A) tail; m7G, 7-methylguanosine cap; PABP (also known as PABPC1), poly(A) binding protein; Pol II, RNA polymerase II. Note that PABP is not part of the miRISC.

Because animal miRNAs need to be processed in several biocatalytic steps within the nucleus and cytoplasm before they can exert their function, a short summary of miRNA biogenesis is given here (Figure 1) (reviewed by Ha and Kim, 2014; Krol et al., 2010; Qureshi and Mehler, 2012; Yates et al., 2013). Processing of the primary transcript (pri-miRNA) containing a stem-loop structure or the pre-mRNA containing a mirtron occurs in two steps within and outside the nucleus. In the canonical pathway (Figure 1(A)), the pri-miRNA is cleaved in the nucleus by a microprocessor complex, including the endoribonuclease DROSHA and the double-stranded RNA (dsRNA)-binding protein DGCR8 (DiGeorge syndrome critical region gene 8), into an approximately 65- to 70-nt-long stem-loop precursor miRNA (pre-miRNA). In the DROSHA and DGCR8-independent noncanonical pathway (Figure 1(B)), the mirtron is excised from its host mRNA by splicing and folded into the stem-loop pre-miRNA after lariat debranching and eventually trimmed of the 5′ or 3′ ends by exonucleases. The pre-miRNA is subsequently exported from the nucleus into the cytoplasm by EXPORTIN-5 (XPO5, also known as EXP5) in a complex with RAN-GTP via the nuclear pore complex. In the cytoplasm, the pre-miRNA is bound by the dsRNA-binding protein TARBP2 (TAR (HIV) RNA-binding protein 2) and the endoribonuclease DICER1 and processed by endonucleolytic cleavage into an approximately 18- to 24-nt-long miRNA/miRNA* duplex. One strand of the miRNA/miRNA* duplex, usually the one with the lower 5′-end Watson-Crick base-pairing stability in the duplex, is then selected as the so-called "guide" (mature miRNA) strand and incorporated into a bigger protein complex, the miRNA-induced silencing complex (miRISC). The other so-called "passenger" (or miRNA*) strand of the miRNA duplex is released from this complex and subsequently degraded. In some cases, however, the miRNA* strand is retained in the miRISC complex and functions as an miRNA as well. One core component of miRISC is the Argonaute proteins (AGO1-4 in mammals), which directly interact with the mature miRNA and possess, at least in the case of AGO2, strong RNAse H-like activity mediating the endonucleolytic cleavage and degradation of the target mRNA(s). Another key functional component of miRISC is the glycine–tryptophan protein of 182 kDa (GW182, also known as trinucleotide repeat containing 6a protein (TNRC6a)), which mediates the deadenylation of the target mRNA(s). Alternative miRNA biogenic pathways include the generation of short hairpins and dsRNAs that act like miRNAs in a DROSHA/DGCR8 and DICER1-independent, noncanonical manner (reviewed by Ha and Kim, 2014; Yang and Lai, 2011).

Base pairing between the miRNA and the complementary sequence of the target mRNA, usually located within the mRNA's 3′ untranslated region (3′ UTR) but also detected within the coding sequence (CDS), induces either the degradation or the translational inhibition of the target mRNA by molecular mechanisms that are still not fully understood (reviewed by Hausser and Zavolan, 2014; Huntzinger and Izaurralde, 2011; Pasquinelli, 2012; Yates et al., 2013). Nevertheless, the degree of miRNA:mRNA sequence complementarity, in particular of a heptamer or octamer seed sequence (usually nt position 2–8) at the 5′ end of the miRNA, appears to influence this outcome: perfect base pairing between the miRNA and its target mRNA supports the endonucleolytic cleavage of the mRNA by AGO2 proteins, whereas imperfect base pairing between miRNA and target mRNA induces deadenylation and destabilization of the mRNA (and may also result in its degradation) or the inhibition of its translation (Figure 1). LincRNAs, by contrast, act in the nucleus as scaffolds for chromatin-modifying complexes or mediators of long-range chromatin interactions, and in the cytoplasm as modulators of mRNA stability and translational control. In addition, lincRNAs bind to and sequester specific miRNAs, acting as miRNA sponges or so-called competing endogenous RNAs (ceRNAs) that protect target mRNAs from miRNA-mediated repression or degradation (Figure 1) (reviewed by Fatica and Bozzoni, 2014; Hausser and Zavolan, 2014).

MODULATORY ROLES OF MICRORNAs IN THE POSTTRANSCRIPTIONAL CONTROL OF GENE EXPRESSION

MicroRNAs act together with other miRNAs that have homologous seed sequences or the same target mRNA(s), which underscores the target and functional redundancy of many miRNAs (reviewed by Hausser and Zavolan, 2014; Huntzinger and Izaurralde, 2011; Pasquinelli, 2012). The interaction of one miRNA with its target mRNA usually has only moderate effects on the translation of this mRNA (less than twofold changes in protein levels (Baek et al., 2008); reviewed by Hausser and Zavolan, 2014). The loss of one miRNA is thus frequently compensated in an organism and results in only subtle or complete absence of phenotypic defects (Ebert and Sharp, 2012; Park et al., 2012; reviewed by Bartel, 2009). However, miRNAs cooperate either as large miRNA families with identical or highly similar seed sequences and thus mRNA targets, or as several miRNAs with different seed sequences targeting different binding sites in one and the same mRNA (reviewed by Bartel, 2009; Flynt and Lai, 2008). By exerting such pleiotropic functions, miRNAs can act as genetic switches that are required, for instance, for the stable and robust commitment of a PSC/multipotent stem cell to a specific cell fate or lineage, and/or as tuning agents that facilitate, for example, the transition between different cellular of developmental stages (reviewed by Bartel, 2009; Flynt and Lai, 2008). The activity of a switch miRNA generally leads to the complete degradation or translational inhibition of the target mRNA(s) and results in a complementary expression pattern of the switch miRNA and its target(s). The activity of a tuning miRNA, by contrast, only downregulates the levels of its target mRNA(s) and/or protein(s) and results in the coexpression of the tuning miRNA and its target(s) within the same cells or tissues. Furthermore, miRNAs and their target mRNAs can engage in reciprocal (repressive–repressive) or unilateral (repressive–activating) negative feedback loops cross-regulating their own expression and/or activity. This is the case, for example, if the miRNA targets (represses) an mRNA encoding a transcriptional regulator (e.g., a TF) that in turn represses or activates the transcription of this miRNA (reviewed by Flynt and Lai, 2008; Krol et al., 2010). As a consequence of such negative feedback loops, an miRNA can initially act as a tuning miRNA for a specific target mRNA and becomes a switch miRNA once the levels of this target mRNA fall below a certain threshold (Mukherji et al., 2011; reviewed by Bartel, 2009). Apart from working as genetic switches, miRNAs have therefore been proposed to act mostly as rheostats that fine-tune protein expression levels in a cell or tissue and thereby confer robustness to biological processes (reviewed by Ebert and Sharp, 2012).

ncRNA-MEDIATED MODULATION OF SOX2 EXPRESSION IN PSCs

SOX2 is one of the core pluripotency TFs required for maintenance of the self-renewing and PSC state in human and mouse ESCs and for reprogramming of human and mouse somatic cells into iPSCs (see Chapter 10, Sox2-Dependent Regulation of Pluripotent Stem Cells) (reviewed by Kashyap et al., 2009; Sarkar and Hochedlinger, 2013). The lack of zygotic SOX2 protein impedes the derivation of pluripotent ESCs from the inner cell mass (ICM) of the *Sox2* mutant (knockout) blastocyst (Avilion et al., 2003; Keramari et al., 2010), whereas a moderate (approximately twofold) increase in SOX2 protein dosage in ESCs leads to the loss of their self-renewing ability and induces their differentiation into ectodermal and mesodermal lineages, and a larger (approximately fourfold) increase of SOX2 dosage leads to ESC death (Kopp et al., 2008), which indicates that the SOX2 dosage needs to be precisely balanced in pluripotent ESCs. Likewise, a lower SOX2 dosage combined with higher doses of the other two reprogramming factors, OCT4 and KLF4, generates more completely reprogrammed and higher-quality iPSCs (Carey et al., 2011; Yamaguchi et al., 2011; reviewed by Buganim et al., 2013). The loss of *Dgcr8* and *Dicer1* expression in mouse ESCs, or the small interfering RNA (siRNA)-mediated knockdown of *DROSHA* and *DICER1* in human ESCs results in cell-cycle progression and proliferation defects of the ESCs and their failure to differentiate properly because of the upregulated and/or persistent expression of pluripotency factors in the mutant cells (Kanellopoulou et al., 2005; Murchison et al., 2005; Qi et al., 2009; Wang et al., 2007; reviewed by Mallanna and Rizzino, 2010; Martinez and Gregory, 2010; Wright and Ciosk, 2013). This was the first indication that miRNAs have an important role in the control of ESC proliferation and differentiation, but not as much in the maintenance of their pluripotency. SOX2 is indeed engaged in a complex transcriptional and posttranscriptional network, including several miRNAs and miRNA families, with autoregulatory and feed-forward components and bivalent states in ESCs (Marson et al., 2008; reviewed by Mallanna and Rizzino, 2010; Rosa and Brivanlou, 2013; Wright and Ciosk, 2013). Concordantly, DROSHA/DICER1/AGO2-processed miRNAs are required for the reprogramming of mouse fibroblasts into iPSCs and ESC-specific miRNAs promote the efficient reprogramming of somatic cells into iPSCs, whereas differentiated somatic cell-specific miRNAs present a barrier that needs to be overcome in this process (Li et al., 2011; reviewed by Li and He, 2012; Wright and Ciosk, 2013; Yang and Rana, 2013). Although consistent with the observation that ESC pluripotency per se is not affected in the absence of DROSHA/DGCR8 and DICER1-processed miRNAs, there are surprisingly few miRNAs known to date that directly target the *Sox2* mRNA (CDS and 3′ UTR) and modulate the activity of SOX2 in this context.

Mouse miR-134 Targets the SOX2 CDS and Promotes the Differentiation of Mouse ESCs

An miRNA that is significantly upregulated upon retinoic acid-induced differentiation of murine ESCs is mouse *miR-134*, which is predicted to target the *Sox2* CDS (Tay et al., 2008a). Overexpression of a *pre-miR-134* oligomer in mouse ESCs results in reduced SOX2 protein levels but less affected *Sox2* mRNA levels, which suggests that the posttranscriptional regulation of *Sox2* by *miR-134* mostly affects its translation into SOX2 protein (Tay et al., 2008a). The silent mutation of one conserved (among mouse, rhesus monkey, and human) out of five predicted *miR-134* target sites within the *Sox2* CDS abolishes the repressive effect of *miR-134* on SOX2 protein translation, which indicates that this *miR-134* target site within the *Sox2* CDS is indeed functional (Figure 2) (Tay et al., 2008a). The overexpression of *pre-miR-134*, together with either one or two other miRNAs (*pre-miR-296* and *pre-miR-470*) that are also activated during the process, in mouse ESCs induces their differentiation and the upregulation of certain ectodermal and trophectodermal markers and suppresses the expression of SOX2 and two other pluripotency TFs (OCT4 and NANOG) in these cells (Tay et al., 2008a). Conversely, the inhibition of *miR-134*, either alone or together with *miR-296* and *miR-470* through overexpression of antisense oligomers in mouse ESCs under differentiating conditions, blocks their differentiation into ectodermal lineages and maintains the normal levels of SOX2 (and OCT4/NANOG) protein in these cells (Tay et al., 2008a). *miR-134* also targets the 3′ UTR of the pluripotency TF *Nanog* and is predicted to target the *Oct4* mRNA (Tay et al., 2008b), most likely forming part of a broader miRNA-controlled network that downregulates several pluripotency TFs during the differentiation of PSCs (reviewed by Mallanna and Rizzino, 2010; Wright and Ciosk, 2013). Together, these data demonstrate that by targeting a conserved binding site within the CDS of the *Sox2* transcript, *miR-134* miRNA inhibits the translation of this mRNA and thereby downregulates the SOX2 protein levels in PSCs, which is necessary, at least in the mouse, for the proper induction of their differentiation. *miR-134* appears to act in concert with other miRNAs also targeting the *Sox2* transcript in this context (Figure 3(A)).

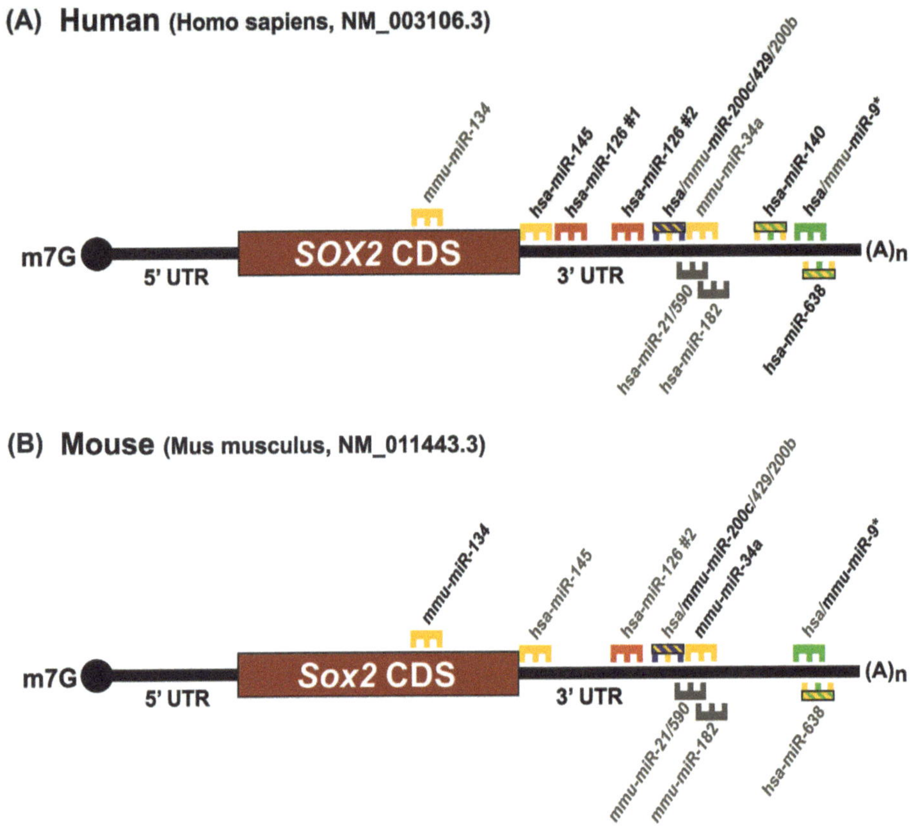

Figure 2 ***Position and conservation of miRNA binding sites within the human and mouse SOX2 transcript.*** (A) Schematic depiction of the 2520-nt-long human *SOX2* mRNA (based on NCBI RefSeq NM_003106.3) and the approximate positions of experimentally validated or predicted miRNA binding sites within this sequence. (B) Schematic depiction of the 2457-nt-long mouse *Sox2* mRNA (based on NCBI RefSeq NM_011443.3) and the approximate positions of experimentally validated or predicted miRNA binding sites within this sequence. Color coding of the bars representing the validated miRNA binding sites is the same as in Figure 3 and depicts the cells or tissues in which the corresponding miRNA functions (blue, PSCs; yellow, multipotent stem cells (e.g., NSCs/NPCs); green, differentiated somatic cells; red, CSCs/cancer cells; dashed colored bars, the corresponding miRNA acts in more than one cell/tissue type). Gray bars indicate the position of predicted (by TargetScan 6.2 (Friedman et al., 2009)) and highly conserved miRNA binding sites within the human and mouse *SOX2* mRNA that have not yet been experimentally addressed. Confirmed human or mouse miRNAs directly targeting the human or mouse *SOX2* mRNA, respectively, are depicted in black letters; gray letters denote mouse or human miRNAs that may target a conserved (by sequence and position) but experimentally not validated binding site within the human or mouse *SOX2* mRNA, respectively; # indicates more than one confirmed/predicted binding site for one and the same miRNA. Abbreviations: $(A)_n$, poly(A) tail; CDS, coding sequence; hsa, Homo sapiens (human); m7G, 7-methylguanosine cap; mmu, Mus musculus (mouse); UTR, untranslated region.

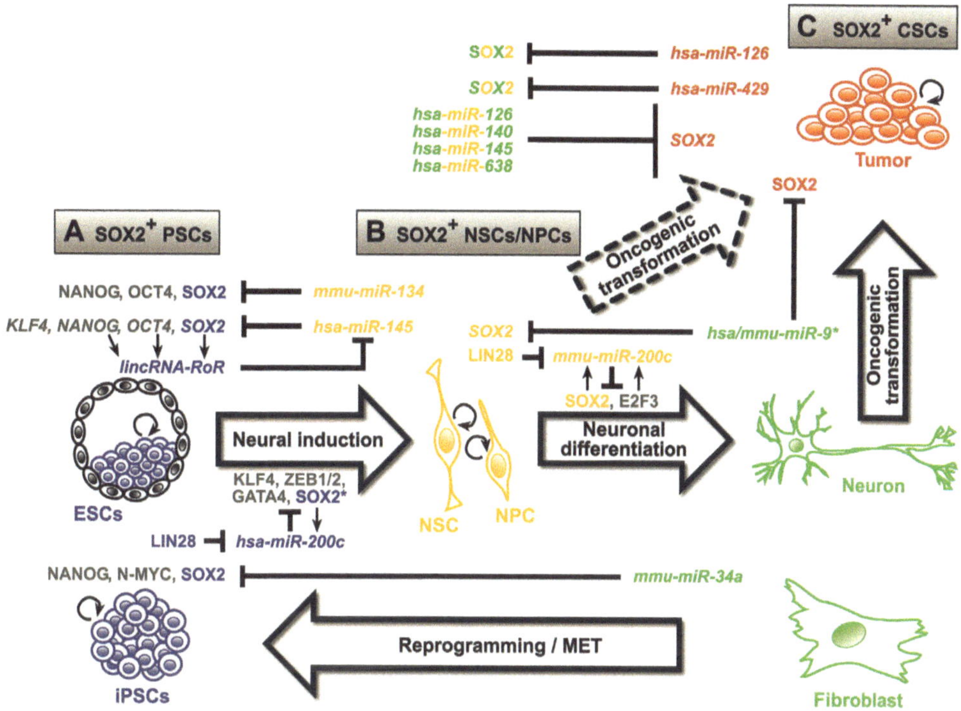

Figure 3 ***Posttranscriptional modulation of SOX2 activity by miRNAs in diverse biological contexts.*** (A) Modulation of SOX2 activity by miRNAs in SOX2-expressing (SOX2+) and self-renewing PSCs (ESCs and iPSCs) and during their differentiation into more committed stem/precursor cells (exemplified by their neural induction to NSCs/NPCs), or during their reprogramming from somatic cells. See Section ncRNA-Mediated Modulation of SOX2 Expression in PSCs for details. (B) Modulation of SOX2 activity by miRNAs in SOX2-expressing (SOX2+) and self-renewing multipotent stem cells (exemplified by NSCs/NPCs) and during their terminal differentiation into somatic cells (exemplified by a neuron). See Section miRNA-Mediated Modulation of SOX2 Expression in Neural Stem/Progenitor Cells for details. (C) Modulation of SOX2 activity by miRNAs in SOX2-expressing (SOX2+) and self-renewing CSCs or in SOX2-negative cancer cells. See Section miRNA-Mediated Modulation of SOX2 Expression in CSCs for details. Upper case letters indicate miRNA-mediated modulation of SOX2 at the protein level; upper and lower case italics indicate miRNA-mediated modulation of *SOX2/Sox2* at the mRNA level. Color coding is as follows: blue, miRNAs or factors expressed in PSCs; yellow, miRNAs or factors expressed in NSCs/NPCs or other types of multipotent stem cells; green, miRNAs or factors expressed in differentiated somatic cells (e.g., neurons/glial cells, fibroblasts); red, miRNAs or factors expressed in CSCs and tumor cells; green/yellow letters, miRNAs or factors that may be expressed in differentiated somatic as well as multipotent stem/precursor cells in a given tissue; gray, other pluripotency, self-renewal and/or EMT TFs that are also modulated by the corresponding miRNA within a broader regulatory network. SOX2*: the direct targeting (binding) of *SOX2* by *miR-200c* has not been demonstrated in this context. Arrows indicate activation, bars indicate inhibition. Abbreviations: hsa, *Homo sapiens* (human); mmu, *Mus musculus* (mouse).

Human miR-145 Targets the SOX2 3′ UTR, Inhibits Self-Renewal, and Promotes the Differentiation of Human ESCs and iPSCs

Another miRNA that is expressed at relatively low levels in self-renewing (undifferentiated) human ESCs but is strongly upregulated upon their differentiation is human *miR-145* (Xu et al., 2009). The *SOX2* 3′ UTR contains a conserved (among mouse, rat and human) and functional target site for *miR-145* (Figure 2) (Liu et al., 2012; Xu et al., 2009). This *miR-145* target site is probably coopted with other miRNA binding sites within the *SOX2* transcript to downregulate the expression of SOX2 in differentiating human ESCs, because mutation of the *miR-145* target site in the presence of endogenous *miR-145* is not sufficient to completely rescue the activation of a *SOX2* 3′ UTR reporter construct (Xu et al., 2009). In contrast to *miR-134*, the overexpression of a *pre-miR-145* oligomer in human ESCs leads to reduced *SOX2* mRNA but not SOX2 protein levels through a mechanism that was not further investigated by Xu et al. (2009). The overexpression of *pre-miR-145* in human ESCs or iPSCs under self-renewing culture conditions strongly reduces their proliferative capacity (Liu et al., 2012; Xu et al., 2009), increases the apoptotic cell death rate of human ESCs, and induces their differentiation into mesodermal and ectodermal lineages (Xu et al., 2009). The antisense oligomer-mediated inhibition of *miR-145* in human ESCs, by contrast, increases both the *SOX2* mRNA and the SOX2 protein levels in these cells (Xu et al., 2009). Accordingly, the inhibition of *miR-145* activity in human ESCs under differentiating conditions leads to a reduced differentiation rate, the upregulation of the pluripotency TFs OCT4 and KLF4 and self-renewal markers, and the dysregulation of endodermal, mesodermal, and trophectodermal lineage markers in these cells (Xu et al., 2009). The overexpression or inhibition of *miR-145* has pleiotropic effects on human ESC self-renewal and differentiation because *miR-145* targets not only the *SOX2* but also the *OCT4* and *KLF4* 3′ UTRs and *NANOG* CDS, three other TFs required for maintenance of the pluripotent and self-renewing ESC state (Wang et al., 2013b; Xu et al., 2009). Notably, OCT4 directly represses the promoter region of the human *hsa-miR-145* gene, forming a reciprocal negative feedback loop with this miRNA (Xu et al., 2009). Moreover, SOX2, OCT4, and NANOG bind to the promoter region and activate the expression of a lincRNA that facilitates the reprogramming of human somatic cells into iPSCs and was therefore dubbed "lincRNA-regulator of reprogramming" (*lincRNA-RoR*) (Loewer et al., 2010; Wang et al., 2013b). *Linc-RoR* is highly enriched in the cytoplasm of self-renewing (undifferentiated) human ESCs and its expression rapidly declines even before that of the pluripotency TFs upon induction of their differentiation (Wang et al., 2013b). *LincRNA-RoR* acts like an miRNA sponge (ceRNA) in human ESCs and maintains their self-renewing and PSC state through direct binding of mature *miR-145* miRNAs at complementary sites (Figure 1(C)) and protecting *miR-145*-targeted *SOX2, OCT4, KLF4,* and *NANOG* mRNAs from degradation and translational inhibition (Wang et al., 2013b; Xu et al., 2009). Upon induction of human ESC differentiation, the rapid decline of *lincRNA-RoR* levels in the differentiating cells releases this stoichiometric inhibition of *miR-145* activity and allows the sustained translational repression of the pluripotency TFs by increasing *miR-145* levels in these cells. Together, these data uncovered another type of miRNA-mediated and partly reciprocal negative feedback loop between *miR-145*, *SOX2*, and the other three core pluripotency TFs, which is required for the proper differentiation of human ESCs especially into ectodermal and mesodermal lineages. This miRNA regulatory circuitry is in turn modulated by a reciprocal positive feedback loop between the core pluripotency TFs and *lincRNA-RoR*, which is necessary for maintenance of the self-renewing and pluripotent human ESC state. As noted previously, *miR-145* most likely acts in concert with other miRNAs (including *miR-134*) targeting the *SOX2* transcript in this context (Figure 3(A)).

Mouse miR-34a Targets the SOX2 3′ UTR in PSCs and Inhibits the Reprogramming of Mouse Embryonic Fibroblasts into iPSCs

The murine *miR-34* family, which is composed of the three members with identical seed sequences, *miR-34a*, *miR-34b*, and *miR-34c*, is a third miRNA family that is upregulated in a p53-dependent manner during the direct conversion of mouse embryonic fibroblasts (MEFs) into iPSCs using the four reprogramming TFs, OCT4, SOX2, KLF4, and C-MYC (Choi et al., 2011). Transformation-related protein 53 (p53, also known as TRP53) is a strong inhibitor of somatic cell reprogramming (Hong et al., 2009; Kawamura et al., 2009; Marion et al., 2009). Ablation of one or all three *miR-34* miRNAs in MEFs from *miR-34a*$^{-/-}$ and/or *miR-34b/c*$^{-/-}$ mutant mice, or depletion of *miR-34a* after transfection of an antisense oligomer into wild-type MEFs significantly increases their reprogramming efficiency into iPSCs (Choi et al., 2011). The *Sox2* 3′ UTR contains one functional *miR-34a* binding site (Figure 2) and another predicted but not yet experimentally validated *miR-34b/c* binding site (Choi et al., 2011). Accordingly, overexpression of *miR-34a*, *miR-34b*, or *miR-34c* mimics (oligomers) in mouse ESCs leads to a significant reduction of SOX2 (and NANOG/N-MYC, which are also targeted by *miR-34*) protein but not *Sox2* (or *Nanog/N-Myc*) mRNA levels, whereas the lack of these miRNAs in *miR-34a*$^{-/-}$ single and *miR-34a*$^{-/-}$;*miR-34b/c*$^{-/-}$ triple-mutant iPSCs results in elevated SOX2 (and NANOG/N-MYC) protein levels (Choi et al., 2011). Because of the elevated SOX2, NANOG, and N-MYC levels, *miR-34a*$^{-/-}$ iPSCs display a delayed downregulation of these pluripotency TFs upon withdrawal of the self-renewing conditions and induction of differentiation, and differentiate more slowly than wild-type iPSCs under these conditions (Choi et al., 2011). This report revealed that the p53-mediated activation of *miR-34* miRNAs during the reprogramming of somatic cells (MEFs) into iPSCs imposes a barrier to this process by downregulating the levels of SOX2 and the two other pluripotency/reprogramming TFs, NANOG and N-MYC, in these cells, and disrupting the establishment of the self-renewal and pluripotency network required for efficient reprogramming. Conversely, the upregulation of *miR-34* miRNAs may promote the differentiation of PSCs, an issue that was not investigated by Choi et al. (2011).

Human and Mouse miR-200c May Target the *SOX2* 3′ UTR in Human and Mouse PSCs and Promote the Self-Renewing and Pluripotent Stem Cell State and the Reprogramming of MEFs into iPSCs

In contrast to the previous cases, an miRNA that is predominantly downregulated (but in some cases also upregulated) during the differentiation of mouse and human ESCs is *miR-200c*, a member of the conserved vertebrate *miR-200* family composed of four other miRNAs: *miR-200a, miR-200b, miR-141*, and *miR-429* (Bar et al., 2008; Du et al., 2013; Gill et al., 2011; Huang et al., 2013; Lakshmipathy et al., 2007; Li et al., 2009; Wellner et al., 2009; reviewed by Peter, 2009). The expression levels of this miRNA family, and in particular those of *miR-200c*, are highest in pluripotent (undifferentiated) human ESCs and strongly upregulated during the reprogramming of MEFs into iPSCs (Du et al., 2013; Gill et al., 2011; Huang et al., 2013; Lakshmipathy et al., 2007; Li et al., 2009; Samavarchi-Tehrani et al., 2010; Suh et al., 2004; Wang et al., 2013a). The five *miR-200* members are transcribed from two polycistronic gene clusters located on different chromosomes: the tricistronic *miR-200b/a/429* cluster on human chromosome 1 or mouse chromosome 4, and the bicistronic *miR-200c/141* cluster on human chromosome 12 or mouse chromosome 6 (reviewed by Trumbach and Prakash, 2014). SOX2 and OCT4 bind to and activate the promoter of the murine *mmu-miR-200b/a/429* and *mmu-miR-200c/141* gene cluster, respectively, during reprogramming of MEFs into iPSCs (Wang et al., 2013a). These two gene clusters encode mature *miR-200* miRNAs whose seed sequences differ by just one nt and are therefore classified into two seed sequence subgroups: Subgroup I consists of *miR-200a* and *miR-141* (seed sequence 5′-AACACUG-3′), and subgroup II consists of *miR-200b, miR-200c*, and *miR-429* (seed sequence 5′-AAUACUG-3′). *Sox2* and *Klf4* 3′ UTRs each harbor a conserved (among chick, dog, mouse, rat, and human) *miR-200b/c/429* binding site (Figure 2) (Wellner et al., 2009). Although the regulatory impact (targeting) of *miR-200* miRNAs on the expression and/or activity of SOX2 and KLF4 in PSCs has not been studied to date, a general requirement of these miRNAs for the maintenance (in PSCs) or acquisition (during somatic cell reprogramming) of the self-renewing and PSC state has been well established by several reports. The depletion of *miR-200c* by antisense oligomers in human ESCs induces their differentiation even under self-renewing conditions, downregulates the *SOX2* (and *NANOG/C-MYC*) mRNA and protein levels (an effect that appears to be specific for a particular human ESC line), and leads to cell-cycle defects and the apoptotic death of these cells (Huang et al., 2013). Conversely, the overexpression of *miR-200c* mimics (oligomers) in human ESCs inhibits their differentiation, an effect that can be rescued by the simultaneous overexpression of the zinc-finger TF GATA4, a direct target of *miR-200c* (Huang et al., 2013). The downregulation of *SOX2* levels after *miR-200c* depletion in human ESCs thus appears to be an indirect consequence of the upregulation of GATA4 in these cells (Huang et al., 2013). The overexpression of *miR-200* miRNAs (oligomers) also facilitates, whereas *miR-200* inhibition with antisense oligomers attenuates, the reprogramming of MEFs into iPSCs with the TFs OCT4, SOX2, KLF4, and C-MYC (Samavarchi-Tehrani et al., 2010; Wang et al., 2013a). The transfection of mature *miR-200c* together with mature *miR-302* and *miR-369* miRNAs into murine and human somatic cells (adult mouse and human adipose stromal cells and human dermal fibroblasts) is in fact sufficient to reprogram these cells into iPSCs (Miyoshi et al., 2011). Through direct repression of the homeobox TFs ZEB1/2, *miR-200* miRNAs indirectly promote the acquisition of stem cell–specific epithelial characteristics (including the expression of E-cadherin) by the MEFs and suppress the expression of fibroblast-specific mesenchymal genes in these cells in a process known as mesenchymal-to-epithelial transition (MET) during iPSC reprogramming (Samavarchi-Tehrani et al., 2010; Wang et al., 2013a; reviewed by Brabletz and Brabletz, 2010; Lamouille et al., 2013). The *miR-200* family, and in particular *miR-200c*, therefore appears, on the one hand, to participate in the miRNA-mediated regulatory network required to balance the SOX2 dosage in PSCs and to maintain these cells in a self-renewing and pluripotent state. On the other hand, this miRNA family, and in particular *miR-200c*, appears to be crucial for the reprogramming of somatic cells into iPSCs by promoting the MET and safeguarding the epithelial characteristics of PSCs (Figure 3(A)).

miRNA-MEDIATED MODULATION OF SOX2 EXPRESSION IN NEURAL STEM/PROGENITOR CELLS

SOX2 is not only expressed in PSCs of the ICM and in the trophectoderm, giving rise to extraembryonic tissues, it continues to be expressed in more committed (multipotent) stem/progenitor cells of the ectodermal, mesodermal, and endodermal lineages during and after mammalian gastrulation (Avilion et al., 2003; Wood and Episkopou, 1999; reviewed by Guth and Wegner, 2008; Kamachi and Kondoh, 2013; Sarkar and Hochedlinger, 2013). A particularly prominent site of SOX2 expression and activity at these later developmental stages is the neuroectoderm, which will give rise to the central nervous system (see Chapter 11, SOX2-Dependent Regulation of Neural Stem Cells and CNS Development). SOX2 is expressed in multipotent NSCs/NPCs residing within the ventricular and subventricular zone (VZ/SVZ) of the neuroepithelium that are capable of self-renewal and differentiation into neurons and glial cells (astrocytes and oligodendrocytes) (Ellis et al., 2004; Ferri et al., 2004; reviewed by Pevny and Placzek, 2005; Pevny and Nicolis, 2010). In addition, SOX2 is expressed in NSCs within the neurogenic niches of the adult mouse brain, thus marking the proliferating and multipotent NSC/NPC identity throughout the lifetime (Ellis et al., 2004; Ferri et al., 2004; reviewed by Pevny and Placzek, 2005; Pevny and Nicolis, 2010). Again, the SOX2 dosage needs to be precisely balanced in these contexts: the overexpression of SOX2 in differentiating mouse ESCs promotes their acquisition of a neuroectodermal cell fate at the expense of mesodermal and endodermal identities (Zhao et al., 2004), but the overexpression of SOX2 in neuroectodermal tissues maintains the proliferating NSC/NPC identity and blocks their neuronal differentiation (Graham et al., 2003). Conversely, the inhibition of SOX2-mediated promoter transactivation by overexpression of a SOX2-repressor domain fusion protein in NSCs/NPCs leads to their premature cell-cycle exit, delamination from the VZ, and initiation of neuronal differentiation (Graham et al., 2003). The consequences of *Sox2* ablation (by conventional or conditional mutagenesis of the mouse *Sox2* gene) in neural tissues are highly dose- and context-dependent, probably owing to redundant functions of the other two SOXB1 members, SOX1 and SOX3, which are also expressed in these tissues (Graham et al., 2003; Wood and Episkopou, 1999; reviewed by Kamachi and Kondoh, 2013; Pevny and Nicolis, 2010; Sarkar and Hochedlinger, 2013). Nevertheless, a strong reduction of SOX2 levels or a complete lack of SOX2 protein in neuroepithelial cells leads to the postnatal loss of self-renewing (proliferating) NSCs/NPCs and results in specific neurogenesis defects (Favaro et al., 2009; Ferri et al., 2004; Taranova et al., 2006). The discovery in recent years that the expression and activity of SOX2 in proliferating NSCs/NPCs are also regulated by miRNA-mediated modulatory mechanisms was therefore expected.

Mouse miR-200c Targets the SOX2 3′ UTR in Murine NSCs/NPCs and Promotes Their Cell-Cycle Exit and Neuronal Differentiation

The depletion of *Dicer1* and of DICER1-processed mature miRNAs by conditional mutagenesis in the murine mid-/hindbrain region leads to the accumulation of cycling (proliferating) NSCs/NPCs and inhibits their proper differentiation into the specific neuronal subtypes arising in this region, before the apoptotic loss of the entire mid-/hindbrain neuroepithelial tissue in late gestation (Peng et al., 2012). This *Dicer1* mutant phenotype partly resembles the consequences of SOX2 overexpression in NSCs/NPCs described above (Graham et al., 2003), and indeed, the numbers of SOX2-expressing NSCs/NPCs are increased within the VZ/SVZ of the *Dicer1* mutant mid-/hindbrain region (Peng et al., 2012). Profiling of downregulated miRNAs in the mid-/hindbrain region of midgestational *Dicer1* mutant embryos (when this tissue is not yet lost) revealed all five *miR-200* family members among the most strongly depleted (1–7% residual levels relative to wild-type embryos) miRNAs in this region. Other miRNAs such as *miR-9* and *miR-124*, which have previously been associated with NSC/NPC maintenance and/or neuronal differentiation (reviewed by Coolen and Bally-Cuif, 2009; Coolen et al., 2013; McNeill and Van Vactor, 2012), are still expressed at about 40–45% of their normal levels in the *Dicer1* mutant mid-/hindbrain tissue (Trumbach and Prakash, 2014). One member of the *miR-200* family, *miR-200c*, is expressed at particularly high levels in developing neural tissues and pluripotent ESCs differentiating into neurons (Du et al., 2013; Gill et al., 2011; Nielsen et al., 2009; Peng et al., 2012). Regarding the latter observation, available data on a potential role of this miRNA family in neural induction and differentiation of mouse and human PSCs are inconclusive.

The lentiviral overexpression of *miR-200* miRNAs during neural induction of differentiating human ESCs suppresses their neuroectodermal identity and promotes the mesendodermal and epidermal cell fate choice (Du et al., 2013), whereas the gene-targeted overexpression of *mir-200* miRNAs in differentiating mouse ESCs suppresses the mesodermal fate in these cells and promotes the acquisition of a still pluripotent but already more committed (primed) epiblast stem cell identity that will also give rise to the neuroectoderm in the gastrulating mouse embryo (Gill et al., 2011). These conflicting data most likely result from intrinsic differences between mouse and human ESCs (Tesar et al., 2007) and the incommensurable timing of *miR-200* overexpression in these cells. In SOX2-expressing ventral mid-/hindbrain neural tissues, overexpression of the *pre-miR-200c* precursor or the murine *mmu-miR-200c-141* gene cluster targets the conserved (among vertebrates) *miR-200b/c/429* binding site within the *Sox2* 3′ UTR (Figure 2) and downregulates the expression of SOX2 in NSCs/NPCs, facilitating their cell-cycle exit and neuronal differentiation (Peng et al., 2012). Conversely, the *miR-200* sponge vector-mediated inhibition of *mir-200* miRNAs in primary ventral mid-/hindbrain neural cells suppresses their differentiation into neurons and increases the numbers of SOX2-expressing NSCs/NPCs, thus phenocopying the *Dicer1* mutant mid-/hindbrain region (Peng et al., 2012). *Mir-200c* also targets the 3′ UTR of *E2f3* (Peng et al., 2012), a TF required for the proper control of cell-cycle progression/DNA synthesis and cell survival/apoptosis (reviewed by Chen et al., 2009; DeGregori, 2002). In addition to SOX2-mediated expression of the murine *mmu-miR-200b/a/429* gene cluster (Wang et al., 2013a), SOX2 and E2F3 activate the transcription of the mouse *mmu-miR-200c/141* gene cluster through conserved (among mouse, rat, and human) binding sites in the distal and proximal promoter region of this gene cluster (Peng et al., 2012). This mutual activation and repression between SOX2/E2F3 and *miR-200* miRNAs is predicted to result in a unilateral negative feedback loop, which leads to the gradual reduction of SOX2/E2F3 protein and *mir-200* miRNA levels in NSCs/NPCs and facilitates their cell-cycle exit and differentiation into postmitotic neurons (Figure 3(B)) (Peng et al., 2012; reviewed by Trumbach and Prakash, 2014). Notably, SOX2 also binds to and activates the promoter of the RNA-binding protein and pluripotency factor LIN28, which is highly expressed in NSCs/NPCs but is downregulated as soon as these cells differentiate into neural lineages (Balzer et al., 2010; Cimadamore et al., 2013), as discussed in Chapter 11. Binding of LIN28 to the stem loop of the *pre-miR-200c* miRNA inhibits its processing into the mature *miR-200c* miRNA and induces its degradation (Figure 1) (Heo et al., 2009). Positive co-regulation of SOX2 and LIN28 may in fact ensure that unilateral negative feedback modulation of SOX2 by *miR-200c* does not set in prematurely in self-renewing NSCs/NPCs (and PSCs) during development and/or adult neurogenesis, which may threaten the maintenance of the NSC/NPC (and PSC) pool and the integrity of pre- and postnatal (neural) tissues (Figure 3(A) and (B)).

Mouse miR-9* Targets SOX2 3′ UTR during Neuronal Differentiation of Murine P19 Teratocarcinoma Cells

In contrast to the *miR-200* miRNAs, an miRNA that is strongly upregulated in NSCs/NPCs at the onset of neurogenesis and neuronal differentiation is *miR-9* and its complementary *miR-9** strand (Jo et al., 2012; Krichevsky et al., 2006; Peng et al., 2012; Shibata et al., 2011; Zhao et al., 2009). Although conclusive evidence is still missing in the in vivo context, *miR-9** was shown to target a conserved (between mouse and human) binding site directly within the *Sox2* 3′ UTR (Figure 2), leading to a complementary expression pattern of *miR-9** miRNA (upregulated) and *Sox2* mRNA (downregulated) during the induced neuronal differentiation of P19 teratocarcinoma cells (Jo et al., 2012). This finding suggests that *miR-9** may cooperate with *miR-200c* for the downregulation of SOX2 levels in multipotent NSCs/NPCs during the induction of neuronal differentiation, but *miR-9** alone may be responsible for the complete shutdown of SOX2 expression in these cells, which is required for their definitive differentiation into neurons and other neural (glial) cells (Figure 3(B)).

miRNA-MEDIATED MODULATION OF SOX2 EXPRESSION IN CSCs

Apart from its activity in PSCs, NSCs/NPCs, and other multipotent stem cells of the healthy organism, SOX2 expression and function are dysregulated in all kinds of benign and malign tumors and particularly in the CSCs with tumor-initiating and -propagating potentials (reviewed by Dong et al., 2004; Liu et al., 2013; Reya et al., 2001; Sarkar and Hochedlinger, 2013). This is probably not surprising in light of the similarities between normal stem cells and cancer cells, especially with regard to their capacity for self-renewal, rapid growth (proliferation), and differentiation into lineage-related cells (reviewed by Reya et al., 2001), and the fact that human tumors, in particular those that are poorly differentiated and aggressive (highly proliferative and metastatic), exhibit a similar gene expression signature as human ESCs (Ben-Porath et al., 2008). Genetic amplification (increased copy numbers) of the human *SOX2* locus was first detected in human prostate cancer (Sattler et al., 2000), but occurs as well in lung and esophageal squamous-cell carcinomas and small-cell lung cancer (Bass et al., 2009; Hussenet et al., 2010; McCaughan et al., 2010; Rudin et al., 2012). These cancer types, except for prostate cancer, are derived from tissues and/or cells expressing SOX2 during development or even adulthood (Arnold et al., 2011; reviewed by Sarkar and Hochedlinger, 2013), leading to the initial definition of *SOX2* as a lineage-survival oncogene (Bass et al., 2009; reviewed by Garraway and Sellers, 2006). The upregulation of SOX2 expression has been reported for many other kinds of lineage-related or -unrelated human tumors and cancer cell lines, including glioblastoma and other brain tumors of glial origin (Alonso et al., 2011; Phi et al., 2008), osteosarcoma (Basu-Roy et al., 2012), breast carcinoma (Lengerke et al., 2011), Ewing sarcoma (Riggi et al., 2010), non-small-cell lung cancers and lung adenocarcinoma cells (Chou et al., 2013), colorectal carcinoma (Ma et al., 2014), prostate cancer (Jia et al., 2011), pancreatic carcinoma (Sanada et al., 2006), hepatocellular carcinoma (Sun et al., 2013), and skin squamous-cell carcinoma (Boumahdi et al., 2014), which indicates that SOX2 has a more general oncogenic function in this context. The increased expression of SOX2 is associated in most if not all of these cases with increased proliferation, survival, invasiveness, and/or tumorigenicity of the cancer cells, and thus with malignancy and poor prognosis of the corresponding tumors. Notably, the siRNA-mediated knockdown of *SOX2* expression suppresses the tumorigenic phenotype of the cancer cells in most cases, thus highlighting the essential role of *SOX2* activation in tumor formation. Indeed, *SOX2*-expressing cells have been identified as the tumor-initiating and -propagating CSCs in benign skin papilloma/malign skin squamous-cell carcinoma (Boumahdi et al., 2014) and high-grade oligodendroglioma (Favaro et al., 2014). In contrast to these cancer (cell) types, SOX2 expression is downregulated in gastric cancer (Otsubo et al., 2008). In this particular case, SOX2 appears to act as a tumor suppressor by inhibiting the proliferation and promoting the apoptotic cell death of gastric epithelial cells (Otsubo et al., 2008).

Apart from the genetic amplification of the *SOX2* locus in lung/esophageal squamous cell carcinomas, prostate and small-cell lung cancer, SOX2 expression appears to be aberrantly activated or suppressed through transcriptional, posttranscriptional, or epigenetic (e.g., hypomethylation (Alonso et al., 2011) or hypermethylation (Otsubo et al., 2008) of *SOX2* promoter regions) mechanisms in all other tumors mentioned above. In fact, several miRNAs targeting the *SOX2* 3′ UTR have been identified over past years that act either as tumor-suppressor miRNAs (downregulated in transformed/cancerous tissues and/or cells) or as oncogenic miRNAs/oncomiRs (upregulated in transformed/cancerous tissues and/or cells (reviewed by Cho, 2007)) to control SOX2 activity levels in normal or transformed tissues.

Tumor-Suppressive Human miR-9*, miR-126, miR-140, miR-145, and miR-638 Target the *SOX2* 3′ UTR and Are Downregulated in a Variety of Human Tumor Tissues and Cancer Cells

Overexpression of the inhibitor of DNA binding 4 (ID4), a helix–loop–helix protein without DNA binding capacity that acts as a dominant-negative transcriptional repressor on E-box containing promoters (Perk et al., 2005), leads to the repression of the intronic human *hsa-miR-9/9*-1* locus and the increased expression of SOX2 protein in human glioblastoma cell lines and glioma stem cells (Jeon et al., 2011). The *miR-9** miRNA directly targets the *SOX2* 3′ UTR in these cells (Figure 2) (see also Section Mouse miR-9* Targets SOX2 3′ UTR during Neuronal Differentiation of Murine P19 Teratocarcinoma Cells), causing the up- or downregulation of SOX2 protein but not mRNA when repressed by ID4 overexpression or activated by siRNA-mediated depletion of ID4, respectively (Jeon et al., 2011). The increased expression of SOX2 in human glioblastoma cell lines and glioma stem cells promotes their self-renewal and tumorigenicity and is responsible for the chemoresistance of these cells, because SOX2 directly activates the expression of drug resistance genes (Jeon et al., 2011). The expression levels of *miR-126* are reduced in human osteosarcoma tumors and cell lines relative to normal bone tissues and osteoblast cells (Yang et al., 2014). The human *miR-126* miRNA targets two binding sites within the human *SOX2* 3′ UTR (Figure 2(A)), leading to the downregulation of *SOX2* mRNA and protein levels in osteosarcoma cells (Yang et al., 2014). Overexpression of *miR-126* oligomer mimics in osteosarcoma cell lines reduces their proliferative and invasive (tumorigenic) capacities and increases their apoptotic cell death, an effect that is abolished by the simultaneous overexpression of an *miR-126*-insensitive *SOX2* cDNA construct (*SOX2* open-reading frame lacking the 5′ and 3′ UTRs) in these cells (Yang et al., 2014). Breast tumors and breast cancer cell lines exhibit reduced *miR-140* expression levels compared with normal mammary epithelial tissues (Volinia et al., 2012; Zhang et al., 2012). The human *miR-140* miRNA directly targets a binding site within the human *SOX2* 3′ UTR (Figure 2(A)), leading to the down- or upregulation of *SOX2* mRNA and protein levels if overexpressed or depleted, respectively, in breast cancer and mammary epithelial cell lines (Zhang et al., 2012). Expression of the intronic human *hsa-miR-140* locus in mammary epithelial and breast cancer cells is inhibited in an estrogen receptor signaling-dependent manner through direct repression of an estrogen response element in the promoter region of the *miR-140* host gene (*Wwp2*) after estrogen receptor alpha (ERα) binding to this response element (Zhang et al., 2012). Overexpression of human *miR-140* or siRNA-mediated knockdown of *SOX2* in breast cancer cells reduces their self-renewal capacities and the proportion of tumor-initiating and -propagating CSCs in these cultures, an effect that is abolished by the simultaneous overexpression of an *miR-140*-insensitive *SOX2* cDNA lacking the 3′ UTR, or by estrogen treatment of ERα-responsive breast cancer cells (Zhang et al., 2012). The expression of *miR-145* is strongly reduced in Ewing sarcoma cell lines and completely abolished in Ewing sarcoma CSCs (Riggi et al., 2010). Human pediatric mesenchymal stem cells isolated from healthy donors are transformed into Ewing sarcoma CSC-like cells (among others) by the constitutive expression of an Ewing sarcoma breakpoint region 1 (EWSR1) transactivator-Friend leukemia integration 1 (FLI1) DNA-binding domain fusion protein that results from a unique chromosomal translocation observed in Ewing sarcoma family tumors (Riggi et al., 2010). The EWSR1-FLI1 fusion protein strongly represses the *hsa-miR-145* promoter and induces the expression of *SOX2* mRNA in Ewing sarcoma CSCs (Riggi et al., 2010). The *miR-145* miRNA, in turn, targets the *FLI1* 3′ UTR containing three highly and one poorly conserved *miR-145* binding sites, and may also target the *SOX2* 3′ UTR in Ewing sarcoma cells (Figure 2) (see also Section Human miR-145 Targets the SOX2 3′ UTR, Inhibits Self-Renewal, and Promotes the Differentiation of Human ESCs and iPSCs), although this was not assessed by the authors in this context

(Riggi et al., 2010). Accordingly, the retroviral overexpression of *miR-145* leads to the downregulation of EWSR1-FLI1 fusion protein but not mRNA levels in Ewing sarcoma cell lines (Riggi et al., 2010). Moreover, the retroviral overexpression of *miR-145* or siRNA-mediated depletion of SOX2 in Ewing sarcoma cells decreases their proliferative and tumorigenic potential and downregulates the mRNA levels of several stemness markers including *SOX2* (Riggi et al., 2010). Lower *miR-638* miRNA and higher *SOX2* mRNA levels are detected in non-small-cell lung cancer relative to the adjacent healthy lung tissues (Xia et al., 2014). The human *SOX2* 3′ UTR contains a conserved binding site that is directly targeted by the human *miR-638* miRNA (Figure 2) (Xia et al., 2014). Overexpression of *miR-638* oligomer mimics in lung cancer cell lines decreases the SOX2 protein levels in these cells and their proliferative and invasive capacities (Xia et al., 2014). Inhibition of *miR-638* activity in lung cancer cells using antisense oligomers has the opposite effect (increased SOX2 protein levels and proliferation/invasiveness), which can be partially reverted by the simultaneous siRNA-mediated depletion of SOX2 in these cells (Xia et al., 2014). *Mir-638* levels are also downregulated and *SOX2* mRNA and protein levels are upregulated in colorectal carcinoma tissues and colorectal cancer cell lines relative to noncancerous colorectal tissues (Ma et al., 2014). This study confirmed the presence of a direct target site for human *miR-638* in the human *SOX2* 3′ UTR (Figure 2) and showed that *SOX2* mRNA and protein levels as well as migration/invasiveness of colorectal cancer cell lines are reduced after overexpression of *miR-638* oligomer mimics in these cells, whereas *SOX2* mRNA and protein levels and migration/invasiveness of colorectal cancer cells are increased after inhibition of *miR-638* activity with antisense oligomers (Ma et al., 2014). Notably, the overexpression of *miR-638* in colorectal cancer cells induces cell morphology and protein expression changes reminiscent of MET, whereas the opposite effect (cell morphology and protein expression changes suggestive of epithelial-to-mesenchymal transition or EMT) is seen after *miR-638* inhibition (Ma et al., 2014). The migration and invasiveness of colorectal cancer cells are also inhibited by the siRNA-mediated depletion of SOX2 in these cells, and overexpression of SOX2 in the presence of *miR-638* mimics in colorectal cancer cells abolishes the *miR-638*-mediated reduction of cell migration/invasiveness and induces a mesenchymal (migratory) phenotype of these cells (Ma et al., 2014), providing a link between the increased levels of SOX2 expression in colorectal carcinoma and the enhanced metastatic potential of these tumors.

Together, the previous findings indicate that the downregulated or even abolished expression, either by a direct oncoprotein-mediated transcriptional repression of their promoters or by so far unknown mechanisms, of several tumor-suppressive miRNAs targeting the human *SOX2* 3′ UTR in a diverse array of tissues/cells leads to an upregulation of SOX2 levels in these tissues/cells and their tumorigenic/CSC transformation (Figure 3(C)), whose malignant potential directly correlates with the expression levels of SOX2 in these tumors/CSCs.

Oncogenic Human miR-126 and miR-429 Target the *SOX2* 3′ UTR and Are Upregulated in Human Gastric Cancer and Colorectal Carcinoma Cells

In contrast to the tumor-suppressor miRNAs, *miR-126* is upregulated in some gastric cancer tissues and cell lines with low SOX2 expression levels that do not result from hypermethylation of the *SOX2* promoter region (Otsubo et al., 2011). This study was the first to show that human *miR-126* miRNA targets the human *SOX2* 3′ UTR at two different binding sites (Figure 2(A)) (see also Section Tumor-Suppressive Human miR-9*, miR-126, miR-140, miR-145, and miR-638 Target the *SOX2* 3′ UTR and Are Downregulated in a Variety of Human Tumor Tissues and Cancer Cells) (Otsubo et al., 2011). Overexpression of a *pre-miR-126* oligomer in SOX2-expressing human gastric cancer cell lines reduces, as expected, the SOX2 protein levels in these cells and increases their proliferative and tumorigenic potential, an effect that can be mimicked by the siRNA-mediated knockdown of *SOX2* expression in these cells (Otsubo et al., 2011). Conversely, inhibition of *miR-126* activity with an antisense oligomer in *miR-126*-expressing gastric cancer cells increases the amount of SOX2 protein in these cells (Otsubo et al., 2011). In contrast to osteosarcomas, in which this miRNA acts like a tumor-suppressor (see Section Tumor-Suppressive Human miR-9*, miR-126, miR-140, miR-145, and miR-638 Target the *SOX2* 3′ UTR and Are Downregulated in a Variety of Human Tumor Tissues and Cancer Cells), *miR-126* appears to be an oncogene in the context of gastric tumorigenesis, in line with its proangiogenic and antiapoptotic effects in other tissues (see Otsubo et al., 2011 and references therein). The expression of human *miR-429*, a member of the *miR-200* miRNA family, is also increased and inversely correlated with *SOX2* mRNA levels in human colorectal carcinoma tissues (Li et al., 2013). The human *miR-429* miRNA directly targets a conserved *miR-200b/c/429* binding site within the human *SOX2* 3′ UTR (Figure 2) (see also Section Human and Mouse miR-200c May Target the *SOX2* 3′ UTR in Human and Mouse PSCs and Promote the Self-Renewing and Pluripotent Stem Cell State and the Reprogramming of MEFs into iPSCs and Section Mouse miR-200c Targets the SOX2 3′ UTR in Murine NSCs/NPCs and Promotes Their Cell-Cycle Exit and Neuronal Differentiation) (Li et al., 2013). Overexpression of *miR-429* oligomer mimics in human colorectal carcinoma cell lines decreases the *SOX2* mRNA and protein levels in these cells and inhibits their apoptotic cell death, an effect that is partially abrogated by the simultaneous overexpression of *SOX2* in these cells (Li et al., 2013). Notably, overexpression of *SOX2* alone also increases the apoptotic death of human colorectal carcinoma cells (Li et al., 2013). It remains to be clarified how these findings can be reconciled with the reported upregulation of *SOX2* mRNA and protein levels in colorectal carcinoma tissues and cells, and their reduced migration/invasiveness (metastatic potential) after SOX2 depletion (Ma et al., 2014) (see also Section Tumor-Suppressive Human miR-9*, miR-126, miR-140, miR-145, and miR-638 Target the *SOX2* 3′ UTR and Are Downregulated in a Variety of Human Tumor Tissues and Cancer Cells).

Together, the previous reports indicate that the upregulated expression, by as yet unknown mechanisms, of two oncomiRs targeting the human *SOX2* 3′ UTR in gastric cancer or colorectal carcinoma cells leads to the tumor-promoting downregulation of SOX2 levels in these tissues/cells (Figure 3(C)).

KEEPING SOX2 ACTIVITY IN CHECK: SOME CONCLUDING REMARKS ON POSTTRANSCRIPTIONAL MODULATION OF SOX2 EXPRESSION

The data summarized in the previous sections indicate that human and mouse *SOX2* mRNA is targeted by various miRNAs to modulate the activity of the SOX2 TF in different cellular and biological contexts (Figure 2). Several conclusions can be drawn from these findings:

1. MiRNA targeting specificity: the interaction of a particular miRNA and the *SOX2* mRNA in a given cell or tissue is determined by an at least partial complementarity (especially of the seed sequence) between the mature miRNA and the *SOX2* mRNA and their coexpression in this particular cell or tissue during at least a certain period of time, which may be either a transitional (in the case of a switch miRNA) or a homeostatic state (in the case of a tuning miRNA). One and the same miRNA (e.g., *miR-9**, *miR-126*, *miR-145*, and *miR-638*) therefore targets the *SOX2* transcript in different cellular and/or biological contexts. The functional outcome of this interaction, however, is context-dependent: *miR-9** and *miR-145*, for example, are required, on the one hand, in differentiating PSCs or NSCs/NPCs to promote the transition from the PSC or multipotent stem cell state to the differentiated cell state, and, on the other hand, to inhibit the oncogenic transformation (dedifferentiation) of the differentiated cells (Figure 3).
2. miRNA targeting redundancy: several miRNAs appear to act in a redundant manner to modulate the expression and activity of SOX2 in a given cell or tissue. For example, *miR-34a*, *miR-134*, and *miR-145* most likely cooperate in differentiating PSCs to silence the expression of SOX2 in these cells, which is necessary for their proper differentiation into the germ layer–specific lineages (Figures 2 and 3).
3. miRNA functional redundancy: several miRNAs target not only the *Sox2* transcript, but also other pluripotency or self-renewal factors within a broader regulatory network required for the correct transition between different states in a given cell or tissue. For example, *miR-34a*, *miR-134*, and *miR-145* each target other core pluripotency TFs (such as OCT4, NANOG, KLF4, and N-MYC) apart from SOX2 to ensure the proper differentiation of PSCs or to inhibit the reprogramming of somatic cells into iPSCs. Similarly, *miR-200c* targets both SOX2 and E2F3, another regulator of cell-cycle progression, in NSCs/NPCs to promote their cell-cycle exit and neuronal differentiation (Figure 3).

4. miRNA function at the *Sox2* transcript or translational level: Some miRNAs, such as *miR-9**, *miR-126*, *miR-140*, *miR-145*, *miR-429*, and *miR-638*, act on the *Sox2* transcript level, most likely through mRNA destabilization and degradation, and downregulate *Sox2* mRNA (and in most cases, also SOX2 protein) levels in the corresponding cells or tissues. By contrast, *miR-34a*, *miR-134*, and possibly *miR-200c* (although the latter has not been investigated so far) act as translational inhibitors and downregulate SOX2 protein but not *Sox2* mRNA levels in the respective cells or tissues (Figure 3).
5. Regulation of *Sox2*-targeting miRNAs by other factors: Base pairing between a mature miRNA and the *Sox2* mRNA is only limited by unfavorable biochemical constraints such as imperfect Watson-Crick base pairing within the seven- or octamer seed sequence or inaccessibility of the target site in the secondary structure (reviewed by Bartel, 2009; Pasquinelli, 2012). However, miRNA:*Sox2* mRNA interactions in a given cell or tissue can also be controlled or modulated by the coexpression of inhibitory ceRNAs, as is the case for *miR-145* and *lincRNA-RoR* in PSCs, or inhibitory RNA-binding proteins, as may be the case for *miR-200c* and LIN28 in PSCs and NSCs/NPCs (Figures 1 and 3).
6. Switch or tuning miRNAs: Those miRNAs (e.g., *miR-134*, *miR-145*, *miR-9**, and eventually also *miR-34*) that are expressed in a complementary pattern to SOX2 in differentiating PSCs or NSCs/NPCs most likely act as switch miRNAs required for the complete shutdown of SOX2 expression in these cells, which is necessary for the completion of PSC and NSC/NPC differentiation (Figure 3). Dysregulation (overexpression or silencing) of these miRNAs therefore has detrimental consequences for the maintenance of the PSC/multipotent stem cell state or the differentiation capacity, respectively, of these cells. By contrast, miRNAs (e.g., *miR-200c* and most of the tumor-suppressor/oncogenic miRNAs, such as *miR-126*, *miR-140*, *miR-145*, *miR-429*, and *miR-638*) that are expressed in an overlapping pattern with SOX2 in PSCs, NSCs/NPCs, or tumor cells/CSCs, most likely act as tuning miRNAs required for the modulation of optimal SOX2 expression levels in these cells (Figure 3). Dysregulation (overexpression or silencing) of these miRNAs has variable or less severe consequences for the maintenance of the PSC/multipotent stem cell state or the differentiation potential of these cells, but usually has a significant impact on the tumorigenic and malignant potential of the corresponding CSCs and tumor cells.

Despite these advancements in our knowledge about the posttranscriptional modulation of SOX2 expression and activity by miRNAs, several aspects remain to be addressed. First, most work on miRNA-mediated posttranscriptional modulation of SOX2 expression has been performed in cells and tissues of either mouse or human origin, although significant species-specific differences may exist between these cells or tissues. For instance, mouse and human ESCs and iPSCs represent a different stage of mammalian development (Brons et al., 2007; Tesar et al., 2007; reviewed by Mallanna and Rizzino, 2010). The precise function(s) of *miR-200* miRNAs in the neural induction and differentiation into specific lineages of human and mouse ESCs/iPSCs (see Section Mouse miR-200c Targets the Sox2 3′ UTR in Murine NSCs/NPCs and Promotes Their Cell-Cycle Exit and Neuronal Differentiation) thus remains to be clarified. Second, some of the miRNA target sites (such as the *miR-126* binding site 1 and the *miR-140* binding site) that have been identified within the human *SOX2* transcript are in fact not conserved (neither by sequence nor by position) in the mouse *Sox2* mRNA (Figure 2(A) and (B)), and orthologs or syntenic correlates of human lincRNAs, such as *lincRNA-RoR*, have not been found in mouse ESCs (Loewer et al., 2010). Therefore, caution should be taken in directly translating results from one species to another, and potential species-specific differences in the posttranscriptional modulation of SOX2 activity need to be explored in more detail. Third, the *Sox2* mRNA and especially its 3′ UTR may be targeted by many more miRNAs. For example, TargetScan 6.2 (Friedman et al., 2009) predicts the targeting of the human and mouse *SOX2* 3′ UTR by three highly conserved miRNAs: *miR-21* and *miR-590* (which have identical seed sequences), and *miR-182* (Figure 2). The identity of the tissues or cells and biological contexts in which these miRNAs may function to modulate the activity of SOX2 remain to be established. Future research is likely to render new insights into these aspects.

ACKNOWLEDGMENTS

This work was supported by the Systems Biology of Stem Cells and Reprogramming (SyBoSS) project, which has received funding from the European Union's Seventh Framework Program for research, technological development, and demonstration under Grant Agreement No. [FP7-HEALTH-F4-2010-242129]. This work was funded (in part) by the Helmholtz Portfolio Theme "Supercomputing and Modeling for the Human Brain" (SMHB).

REFERENCES

Alonso, M.M., Diez-Valle, R., Manterola, L., Rubio, A., Liu, D., Cortes-Santiago, N., Urquiza, L., Jauregi, P., Lopez De Munain, A., Sampron, N., Aramburu, A., Tejada-Solis, S., Vicente, C., Odero, M.D., Bandres, E., Garcia-Foncillas, J., Idoate, M.A., Lang, F.F., Fueyo, J., Gomez-Manzano, C., 2011. Genetic and epigenetic modifications of Sox2 contribute to the invasive phenotype of malignant gliomas. PLoS One 6, e26740.

Arnold, K., Sarkar, A., Yram, M.A., Polo, J.M., Bronson, R., Sengupta, S., Seandel, M., Geijsen, N., Hochedlinger, K., 2011. Sox2(+) adult stem and progenitor cells are important for tissue regeneration and survival of mice. Cell Stem Cell 9, 317–329.

Avilion, A.A., Nicolis, S.K., Pevny, L.H., Perez, L., Vivian, N., Lovell-Badge, R., 2003. Multipotent cell lineages in early mouse development depend on SOX2 function. Genes Dev. 17, 126–140.

Baek, D., Villen, J., Shin, C., Camargo, F.D., Gygi, S.P., Bartel, D.P., 2008. The impact of microRNAs on protein output. Nature 455, 64–71.

Balzer, E., Heine, C., Jiang, Q., Lee, V.M., Moss, E.G., 2010. LIN28 alters cell fate succession and acts independently of the let-7 microRNA during neurogliogenesis in vitro. Development 137, 891–900.

Bar, M., Wyman, S.K., Fritz, B.R., Qi, J., Garg, K.S., Parkin, R.K., Kroh, E.M., Bendoraite, A., Mitchell, P.S., Nelson, A.M., Ruzzo, W.L., Ware, C., Radich, J.P., Gentleman, R., Ruohola-Baker, H., Tewari, M., 2008. MicroRNA discovery and profiling in human embryonic stem cells by deep sequencing of small RNA libraries. Stem Cells 26, 2496–2505.

Bartel, D.P., 2009. MicroRNAs: target recognition and regulatory functions. Cell 136, 215–233.

Bass, A.J., Watanabe, H., Mermel, C.H., Yu, S., Perner, S., Verhaak, R.G., Kim, S.Y., Wardwell, L., Tamayo, P., Gat-Viks, I., Ramos, A.H., Woo, M.S., Weir, B.A., Getz, G., Beroukhim, R., O'kelly, M., Dutt, A., Rozenblatt-Rosen, O., Dziunycz, P., Komisarof, J., Chirieac, L.R., Lafargue, C.J., Scheble, V., Wilbertz, T., Ma, C., Rao, S., Nakagawa, H., Stairs, D.B., Lin, L., Giordano, T.J., Wagner, P., Minna, J.D., Gazdar, A.F., Zhu, C.Q., Brose, M.S., Cecconello, I., Ribeiro Jr., U., Marie, S.K., Dahl, O., Shivdasani, R.A., Tsao, M.S., Rubin, M.A., Wong, K.K., Regev, A., Hahn, W.C., Beer, D.G., Rustgi, A.K., Meyerson, M., 2009. SOX2 is an amplified lineage-survival oncogene in lung and esophageal squamous cell carcinomas. Nat. Genet. 41, 1238–1242.

Basu-Roy, U., Seo, E., Ramanathapuram, L., Rapp, T.B., Perry, J.A., Orkin, S.H., Mansukhani, A., Basilico, C., 2012. Sox2 maintains self renewal of tumor-initiating cells in osteosarcomas. Oncogene 31, 2270–2282.

Ben-Porath, I., Thomson, M.W., Carey, V.J., Ge, R., Bell, G.W., Regev, A., Weinberg, R.A., 2008. An embryonic stem cell-like gene expression signature in poorly differentiated aggressive human tumors. Nat. Genet. 40, 499–507.

Boumahdi, S., Driessens, G., Lapouge, G., Rorive, S., Nassar, D., Le Mercier, M., Delatte, B., Caauwe, A., Lenglez, S., Nkusi, E., Brohee, S., Salmon, I., Dubois, C., Del Marmol, V., Fuks, F., Beck, B., Blanpain, C., 2014. SOX2 controls tumour initiation and cancer stem-cell functions in squamous-cell carcinoma. Nature 511, 246–250.

Brabletz, S., Brabletz, T., 2010. The ZEB/miR-200 feedback loop–a motor of cellular plasticity in development and cancer? EMBO Rep. 11, 670–677.

Brons, I.G., Smithers, L.E., Trotter, M.W., Rugg-Gunn, P., Sun, B., Chuva de Sousa Lopes, S.M., Howlett, S.K., Clarkson, A., Ahrlund-Richter, L., Pedersen, R.A., Vallier, L., 2007. Derivation of pluripotent epiblast stem cells from mammalian embryos. Nature 448, 191–195.

Buganim, Y., Faddah, D.A., Jaenisch, R., 2013. Mechanisms and models of somatic cell reprogramming. Nat. Rev. Genet. 14, 427–439.

Carey, B.W., Markoulaki, S., Hanna, J.H., Faddah, D.A., Buganim, Y., Kim, J., Ganz, K., Steine, E.J., Cassady, J.P., Creyghton, M.P., Welstead, G.G., Gao, Q., Jaenisch, R., 2011. Reprogramming factor stoichiometry influences the epigenetic state and biological properties of induced pluripotent stem cells. Cell Stem Cell 9, 588–598.

Chen, H.Z., Tsai, S.Y., Leone, G., 2009. Emerging roles of E2Fs in cancer: an exit from cell cycle control. Nat. Rev. Cancer 9, 785–797.

Cho, W.C., 2007. OncomiRs: the discovery and progress of microRNAs in cancers. Mol. Cancer 6, 60.

Choi, Y.J., Lin, C.P., Ho, J.J., He, X., Okada, N., Bu, P., Zhong, Y., Kim, S.Y., Bennett, M.J., Chen, C., Ozturk, A., Hicks, G.G., Hannon, G.J., He, L., 2011. miR-34 miRNAs provide a barrier for somatic cell reprogramming. Nat. Cell Biol. 13, 1353–1360.

Chou, Y.T., Lee, C.C., Hsiao, S.H., Lin, S.E., Lin, S.C., Chung, C.H., Chung, C.H., Kao, Y.R., Wang, Y.H., Chen, C.T., Wei, Y.H., Wu, C.W., 2013. The emerging role of SOX2 in cell proliferation and survival and its crosstalk with oncogenic signaling in lung cancer. Stem Cells 31, 2607–2619.

Cimadamore, F., Amador-Arjona, A., Chen, C., Huang, C.T., Terskikh, A.V., 2013. SOX2-LIN28/let-7 pathway regulates proliferation and neurogenesis in neural precursors. Proc. Natl Acad. Sci. U.S.A. 110, E3017–E3026.

Coolen, M., Bally-Cuif, L., 2009. MicroRNAs in brain development and physiology. Curr. Opin. Neurobiol. 19, 461–470.

Coolen, M., Katz, S., Bally-Cuif, L., 2013. miR-9: a versatile regulator of neurogenesis. Front. Cell. Neurosci. 7, 220.

DeGregori, J., 2002. The genetics of the E2F family of transcription factors: shared functions and unique roles. Biochim. Biophys. Acta 1602, 131–150.

Dong, C., Wilhelm, D., Koopman, P., 2004. Sox genes and cancer. Cytogenet. Genome Res. 105, 442–447.

Du, Z.W., Ma, L.X., Phillips, C., Zhang, S.C., 2013. Mir-200 and miR-96 families repress neural induction from human embryonic stem cells. Development 140, 2611–2618.

Ebert, M.S., Sharp, P.A., 2012. Roles for microRNAs in conferring robustness to biological processes. Cell 149, 515–524.
Ellis, P., Fagan, B.M., Magness, S.T., Hutton, S., Taranova, O., Hayashi, S., Mcmahon, A., Rao, M., Pevny, L., 2004. SOX2, a persistent marker for multipotential neural stem cells derived from embryonic stem cells, the embryo or the adult. Dev. Neurosci. 26, 148–165.
Esteller, M., 2011. Non-coding RNAs in human disease. Nat. Rev. Genet. 12, 861–874.
Fatica, A., Bozzoni, I., 2014. Long non-coding RNAs: new players in cell differentiation and development. Nat. Rev. Genet. 15, 7–21.
Favaro, R., Appolloni, I., Pellegatta, S., Sanga, A.B., Pagella, P., Gambini, E., Pisati, F., Ottolenghi, S., Foti, M., Finocchiaro, G., Malatesta, P., Nicolis, S.K., 2014. Sox2 is required to maintain cancer stem cells in a mouse model of high-grade oligodendroglioma. Cancer Res. 74, 1833–1844.
Favaro, R., Valotta, M., Ferri, A.L., Latorre, E., Mariani, J., Giachino, C., Lancini, C., Tosetti, V., Ottolenghi, S., Taylor, V., Nicolis, S.K., 2009. Hippocampal development and neural stem cell maintenance require Sox2-dependent regulation of Shh. Nat. Neurosci. 12, 1248–1256.
Ferri, A.L., Cavallaro, M., Braida, D., Di Cristofano, A., Canta, A., Vezzani, A., Ottolenghi, S., Pandolfi, P.P., Sala, M., Debiasi, S., Nicolis, S.K., 2004. Sox2 deficiency causes neurodegeneration and impaired neurogenesis in the adult mouse brain. Development 131, 3805–3819.
Flynt, A.S., Lai, E.C., 2008. Biological principles of microRNA-mediated regulation: shared themes amid diversity. Nat. Rev. Genet. 9, 831–842.
Friedman, R.C., Farh, K.K., Burge, C.B., Bartel, D.P., 2009. Most mammalian mRNAs are conserved targets of microRNAs. Genome Res. 19, 92–105.
Garraway, L.A., Sellers, W.R., 2006. Lineage dependency and lineage-survival oncogenes in human cancer. Nat. Rev. Cancer 6, 593–602.
Gill, J.G., Langer, E.M., Lindsley, R.C., Cai, M., Murphy, T.L., Kyba, M., Murphy, K.M., 2011. Snail and the microRNA-200 family act in opposition to regulate epithelial-to-mesenchymal transition and germ layer fate restriction in differentiating ESCs. Stem Cells 29, 764–776.
Graham, V., Khudyakov, J., Ellis, P., Pevny, L., 2003. SOX2 functions to maintain neural progenitor identity. Neuron 39, 749–765.
Guth, S.I., Wegner, M., 2008. Having it both ways: sox protein function between conservation and innovation. Cell. Mol. Life Sci. 65, 3000–3018.
Ha, M., Kim, V.N., 2014. Regulation of microRNA biogenesis. Nat. Rev. Mol. Cell Biol. 15, 509–524.
Hausser, J., Zavolan, M., 2014. Identification and consequences of miRNA-target interactions - beyond repression of gene expression. Nat. Rev. Genet. 5, 599–612.
Heo, I., Joo, C., Kim, Y.K., Ha, M., Yoon, M.J., Cho, J., Yeom, K.H., Han, J., Kim, V.N., 2009. TUT4 in concert with Lin28 suppresses microRNA biogenesis through pre-microRNA uridylation. Cell 138, 696–708.
Hong, H., Takahashi, K., Ichisaka, T., Aoi, T., Kanagawa, O., Nakagawa, M., Okita, K., Yamanaka, S., 2009. Suppression of induced pluripotent stem cell generation by the p53-p21 pathway. Nature 460, 1132–1135.
Huang, H.N., Chen, S.Y., Hwang, S.M., Yu, C.C., Su, M.W., Mai, W., Wang, H.W., Cheng, W.C., Schuyler, S.C., Ma, N., Lu, F.L., Lu, J., 2013. miR-200c and GATA binding protein 4 regulate human embryonic stem cell renewal and differentiation. Stem Cell Res. 12, 338–353.
Huntzinger, E., Izaurralde, E., 2011. Gene silencing by microRNAs: contributions of translational repression and mRNA decay. Nat. Rev. Genet. 12, 99–110.
Hussenet, T., Dali, S., Exinger, J., Monga, B., Jost, B., Dembele, D., Martinet, N., Thibault, C., Huelsken, J., Brambilla, E., Du Manoir, S., 2010. SOX2 is an oncogene activated by recurrent 3q26.3 amplifications in human lung squamous cell carcinomas. PLoS One 5, e8960.
Jeon, H.M., Sohn, Y.W., Oh, S.Y., Kim, S.H., Beck, S., Kim, S., Kim, H., 2011. ID4 imparts chemoresistance and cancer stemness to glioma cells by derepressing miR-9*-mediated suppression of SOX2. Cancer Res. 71, 3410–3421.
Jia, X., Li, X., Xu, Y., Zhang, S., Mou, W., Liu, Y., Liu, Y., Lv, D., Liu, C.H., Tan, X., Xiang, R., Li, N., 2011. SOX2 promotes tumorigenesis and increases the anti-apoptotic property of human prostate cancer cell. J. Mol. Cell Biol. 3, 230–238.
Jo, M.-H., Lee, C., Ali, B., Alarifi, S., Al-Khedhairy, A., Kim, S., 2012. A bioinformatics approach for in vivo imaging of endogenous MicroRNA targets during neurogenesis. Tissue Eng. Regener. Med. 9, 157–169.

Kamachi, Y., Kondoh, H., 2013. Sox proteins: regulators of cell fate specification and differentiation. Development 140, 4129–4144.

Kanellopoulou, C., Muljo, S.A., Kung, A.L., Ganesan, S., Drapkin, R., Jenuwein, T., Livingston, D.M., Rajewsky, K., 2005. Dicer-deficient mouse embryonic stem cells are defective in differentiation and centromeric silencing. Genes Dev. 19, 489–501.

Kashyap, V., Rezende, N.C., Scotland, K.B., Shaffer, S.M., Persson, J.L., Gudas, L.J., Mongan, N.P., 2009. Regulation of stem cell pluripotency and differentiation involves a mutual regulatory circuit of the NANOG, OCT4, and SOX2 pluripotency transcription factors with polycomb repressive complexes and stem cell microRNAs. Stem Cells Dev. 18, 1093–1108.

Kawamura, T., Suzuki, J., Wang, Y.V., Menendez, S., Morera, L.B., Raya, A., Wahl, G.M., Izpisua Belmonte, J.C., 2009. Linking the p53 tumour suppressor pathway to somatic cell reprogramming. Nature 460, 1140–1144.

Keramari, M., Razavi, J., Ingman, K.A., Patsch, C., Edenhofer, F., Ward, C.M., Kimber, S.J., 2010. Sox2 is essential for formation of trophectoderm in the preimplantation embryo. PLoS One 5, e13952.

Kopp, J.L., Ormsbee, B.D., Desler, M., Rizzino, A., 2008. Small increases in the level of Sox2 trigger the differentiation of mouse embryonic stem cells. Stem Cells 26, 903–911.

Krichevsky, A.M., Sonntag, K.C., Isacson, O., Kosik, K.S., 2006. Specific microRNAs modulate embryonic stem cell-derived neurogenesis. Stem Cells 24, 857–864.

Krol, J., Loedige, I., Filipowicz, W., 2010. The widespread regulation of microRNA biogenesis, function and decay. Nat. Rev. Genet. 11, 597–610.

Lakshmipathy, U., Love, B., Goff, L.A., Jornsten, R., Graichen, R., Hart, R.P., Chesnut, J.D., 2007. MicroRNA expression pattern of undifferentiated and differentiated human embryonic stem cells. Stem Cells Dev. 16, 1003–1016.

Lamouille, S., Subramanyam, D., Blelloch, R., Derynck, R., 2013. Regulation of epithelial-mesenchymal and mesenchymal-epithelial transitions by microRNAs. Curr. Opin. Cell Biol. 25, 200–207.

Lengerke, C., Fehm, T., Kurth, R., Neubauer, H., Scheble, V., Muller, F., Schneider, F., Petersen, K., Wallwiener, D., Kanz, L., Fend, F., Perner, S., Bareiss, P.M., Staebler, A., 2011. Expression of the embryonic stem cell marker SOX2 in early-stage breast carcinoma. BMC Cancer 11, 42.

Li, J., Du, L., Yang, Y., Wang, C., Liu, H., Wang, L., Zhang, X., Li, W., Zheng, G., Dong, Z., 2013. MiR-429 is an independent prognostic factor in colorectal cancer and exerts its anti-apoptotic function by targeting SOX2. Cancer Lett. 329, 84–90.

Li, M.A., He, L., 2012. microRNAs as novel regulators of stem cell pluripotency and somatic cell reprogramming. BioEssays 34, 670–680.

Li, S.S., Yu, S.L., Kao, L.P., Tsai, Z.Y., Singh, S., Chen, B.Z., Ho, B.C., Liu, Y.H., Yang, P.C., 2009. Target identification of microRNAs expressed highly in human embryonic stem cells. J. Cell. Biochem. 106, 1020–1030.

Li, Z., Yang, C.S., Nakashima, K., Rana, T.M., 2011. Small RNA-mediated regulation of iPS cell generation. EMBO J. 30, 823–834.

Liu, K., Lin, B., Zhao, M., Yang, X., Chen, M., Gao, A., Liu, F., Que, J., Lan, X., 2013. The multiple roles for Sox2 in stem cell maintenance and tumorigenesis. Cell. Signal. 25, 1264–1271.

Liu, T., Cheng, W., Huang, Y., Huang, Q., Jiang, L., Guo, L., 2012. Human amniotic epithelial cell feeder layers maintain human iPS cell pluripotency via inhibited endogenous microRNA-145 and increased Sox2 expression. Exp. Cell Res. 318, 424–434.

Loewer, S., Cabili, M.N., Guttman, M., Loh, Y.H., Thomas, K., Park, I.H., Garber, M., Curran, M., Onder, T., Agarwal, S., Manos, P.D., Datta, S., Lander, E.S., Schlaeger, T.M., Daley, G.Q., Rinn, J.L., 2010. Large intergenic non-coding RNA-RoR modulates reprogramming of human induced pluripotent stem cells. Nat. Genet. 42, 1113–1117.

Ma, K., Pan, X., Fan, P., He, Y., Gu, J., Wang, W., Zhang, T., Li, Z., Luo, X., 2014. Loss of miR-638 in vitro promotes cell invasion and a mesenchymal-like transition by influencing SOX2 expression in colorectal carcinoma cells. Mol. Cancer 13, 118.

Mallanna, S.K., Rizzino, A., 2010. Emerging roles of microRNAs in the control of embryonic stem cells and the generation of induced pluripotent stem cells. Dev. Biol. 344, 16–25.

Marion, R.M., Strati, K., Li, H., Murga, M., Blanco, R., Ortega, S., Fernandez-Capetillo, O., Serrano, M., Blasco, M.A., 2009. A p53-mediated DNA damage response limits reprogramming to ensure iPS cell genomic integrity. Nature 460, 1149–1153.

Marson, A., Levine, S.S., Cole, M.F., Frampton, G.M., Brambrink, T., Johnstone, S., Guenther, M.G., Johnston, W.K., Wernig, M., Newman, J., Calabrese, J.M., Dennis, L.M., Volkert, T.L., Gupta, S., Love, J., Hannett, N., Sharp, P.A., Bartel, D.P., Jaenisch, R., Young, R.A., 2008. Connecting microRNA genes to the core transcriptional regulatory circuitry of embryonic stem cells. Cell 134, 521–533.

Martinez, N.J., Gregory, R.I., 2010. MicroRNA gene regulatory pathways in the establishment and maintenance of ESC identity. Cell Stem Cell 7, 31–35.

McCaughan, F., Pole, J.C., Bankier, A.T., Konfortov, B.A., Carroll, B., Falzon, M., Rabbitts, T.H., George, P.J., Dear, P.H., Rabbitts, P.H., 2010. Progressive 3q amplification consistently targets SOX2 in preinvasive squamous lung cancer. Am. J. Respir. Crit. Care Med. 182, 83–91.

McNeill, E., Van Vactor, D., 2012. MicroRNAs shape the neuronal landscape. Neuron 75, 363–379.

Miyoshi, N., Ishii, H., Nagano, H., Haraguchi, N., Dewi, D.L., Kano, Y., Nishikawa, S., Tanemura, M., Mimori, K., Tanaka, F., Saito, T., Nishimura, J., Takemasa, I., Mizushima, T., Ikeda, M., Yamamoto, H., Sekimoto, M., Doki, Y., Mori, M., 2011. Reprogramming of mouse and human cells to pluripotency using mature microRNAs. Cell Stem Cell 8, 633–638.

Mukherji, S., Ebert, M.S., Zheng, G.X., Tsang, J.S., Sharp, P.A., Van Oudenaarden, A., 2011. MicroRNAs can generate thresholds in target gene expression. Nat. Genet. 43, 854–859.

Murchison, E.P., Partridge, J.F., Tam, O.H., Cheloufi, S., Hannon, G.J., 2005. Characterization of dicer-deficient murine embryonic stem cells. Proc. Natl Acad. Sci. U.S.A. 102, 12135–12140.

Nielsen, J.A., Lau, P., Maric, D., Barker, J.L., Hudson, L.D., 2009. Integrating microRNA and mRNA expression profiles of neuronal progenitors to identify regulatory networks underlying the onset of cortical neurogenesis. BMC Neurosci. 10, 98.

Otsubo, T., Akiyama, Y., Hashimoto, Y., Shimada, S., Goto, K., Yuasa, Y., 2011. MicroRNA-126 inhibits SOX2 expression and contributes to gastric carcinogenesis. PLoS One 6, e16617.

Otsubo, T., Akiyama, Y., Yanagihara, K., Yuasa, Y., 2008. SOX2 is frequently downregulated in gastric cancers and inhibits cell growth through cell-cycle arrest and apoptosis. Br. J. Cancer 98, 824–831.

Park, C.Y., Jeker, L.T., Carver-Moore, K., Oh, A., Liu, H.J., Cameron, R., Richards, H., Li, Z., Adler, D., Yoshinaga, Y., Martinez, M., Nefadov, M., Abbas, A.K., Weiss, A., Lanier, L.L., De Jong, P.J., Bluestone, J.A., Srivastava, D., Mcmanus, M.T., 2012. A resource for the conditional ablation of microRNAs in the mouse. Cell Rep. 1, 385–391.

Pasquinelli, A.E., 2012. MicroRNAs and their targets: recognition, regulation and an emerging reciprocal relationship. Nat. Rev. Genet. 13, 271–282.

Peng, C., Li, N., Ng, Y.K., Zhang, J., Meier, F., Theis, F.J., Merkenschlager, M., Chen, W., Wurst, W., Prakash, N., 2012. A unilateral negative feedback loop between miR-200 microRNAs and Sox2/E2F3 controls neural progenitor cell-cycle exit and differentiation. J. Neurosci. 32, 13292–13308.

Perk, J., Iavarone, A., Benezra, R., 2005. Id family of helix-loop-helix proteins in cancer. Nat. Rev. Cancer 5, 603–614.

Peter, M.E., 2009. Let-7 and miR-200 microRNAs: guardians against pluripotency and cancer progression. Cell Cycle 8, 843–852.

Pevny, L., Placzek, M., 2005. SOX genes and neural progenitor identity. Curr. Opin. Neurobiol. 15, 7–13.

Pevny, L.H., Nicolis, S.K., 2010. Sox2 roles in neural stem cells. Int. J. Biochem. Cell Biol. 42, 421–424.

Phi, J.H., Park, S.H., Kim, S.K., Paek, S.H., Kim, J.H., Lee, Y.J., Cho, B.K., Park, C.K., Lee, D.H., Wang, K.C., 2008. Sox2 expression in brain tumors: a reflection of the neuroglial differentiation pathway. Am. J. Surg. Pathol. 32, 103–112.

Qi, J., Yu, J.Y., Shcherbata, H.R., Mathieu, J., Wang, A.J., Seal, S., Zhou, W., Stadler, B.M., Bourgin, D., Wang, L., Nelson, A., Ware, C., Raymond, C., Lim, L.P., Magnus, J., Ivanovska, I., Diaz, R., Ball, A., Cleary, M.A., Ruohola-Baker, H., 2009. microRNAs regulate human embryonic stem cell division. Cell Cycle 8, 3729–3741.

Qureshi, I.A., Mehler, M.F., 2012. Emerging roles of non-coding RNAs in brain evolution, development, plasticity and disease. Nat. Rev. Neurosci. 13, 528–541.

Reya, T., Morrison, S.J., Clarke, M.F., Weissman, I.L., 2001. Stem cells, cancer, and cancer stem cells. Nature 414, 105–111.

Riggi, N., Suva, M.L., De Vito, C., Provero, P., Stehle, J.C., Baumer, K., Cironi, L., Janiszewska, M., Petricevic, T., Suva, D., Tercier, S., Joseph, J.M., Guillou, L., Stamenkovic, I., 2010. EWS-FLI-1 modulates miRNA145 and SOX2 expression to initiate mesenchymal stem cell reprogramming toward Ewing sarcoma cancer stem cells. Genes Dev. 24, 916–932.

Rosa, A., Brivanlou, A.H., 2013. Regulatory non-coding RNAs in pluripotent stem cells. Int. J. Mol. Sci. 14, 14346–14373.

Rudin, C.M., Durinck, S., Stawiski, E.W., Poirier, J.T., Modrusan, Z., Shames, D.S., Bergbower, E.A., Guan, Y., Shin, J., Guillory, J., Rivers, C.S., Foo, C.K., Bhatt, D., Stinson, J., Gnad, F., Haverty, P.M., Gentleman, R., Chaudhuri, S., Janakiraman, V., Jaiswal, B.S., Parikh, C., Yuan, W., Zhang, Z., Koeppen, H., Wu, T.D., Stern, H.M., Yauch, R.L., Huffman, K.E., Paskulin, D.D., Illei, P.B., Varella-Garcia, M., Gazdar, A.F., De Sauvage, F.J., Bourgon, R., Minna, J.D., Brock, M.V., Seshagiri, S., 2012. Comprehensive genomic analysis identifies SOX2 as a frequently amplified gene in small-cell lung cancer. Nat. Genet. 44, 1111–1116.

Samavarchi-Tehrani, P., Golipour, A., David, L., Sung, H.K., Beyer, T.A., Datti, A., Woltjen, K., Nagy, A., Wrana, J.L., 2010. Functional genomics reveals a BMP-driven mesenchymal-to-epithelial transition in the initiation of somatic cell reprogramming. Cell Stem Cell 7, 64–77.

Sanada, Y., Yoshida, K., Ohara, M., Oeda, M., Konishi, K., Tsutani, Y., 2006. Histopathologic evaluation of stepwise progression of pancreatic carcinoma with immunohistochemical analysis of gastric epithelial transcription factor SOX2: comparison of expression patterns between invasive components and cancerous or nonneoplastic intraductal components. Pancreas 32, 164–170.

Sarkar, A., Hochedlinger, K., 2013. The sox family of transcription factors: versatile regulators of stem and progenitor cell fate. Cell Stem Cell 12, 15–30.

Sattler, H.P., Lensch, R., Rohde, V., Zimmer, E., Meese, E., Bonkhoff, H., Retz, M., Zwergel, T., Bex, A., Stoeckle, M., Wullich, B., 2000. Novel amplification unit at chromosome 3q25-q27 in human prostate cancer. Prostate 45, 207–215.

Shibata, M., Nakao, H., Kiyonari, H., Abe, T., Aizawa, S., 2011. MicroRNA-9 regulates neurogenesis in mouse telencephalon by targeting multiple transcription factors. J. Neurosci. 31, 3407–3422.

Suh, M.R., Lee, Y., Kim, J.Y., Kim, S.K., Moon, S.H., Lee, J.Y., Cha, K.Y., Chung, H.M., Yoon, H.S., Moon, S.Y., Kim, V.N., Kim, K.S., 2004. Human embryonic stem cells express a unique set of microRNAs. Dev. Biol. 270, 488–498.

Sun, C., Sun, L., Li, Y., Kang, X., Zhang, S., Liu, Y., 2013. Sox2 expression predicts poor survival of hepatocellular carcinoma patients and it promotes liver cancer cell invasion by activating Slug. Med. Oncol. 30, 503.

Taranova, O.V., Magness, S.T., Fagan, B.M., Wu, Y., Surzenko, N., Hutton, S.R., Pevny, L.H., 2006. SOX2 is a dose-dependent regulator of retinal neural progenitor competence. Genes Dev. 20, 1187–1202.

Tay, Y., Zhang, J., Thomson, A.M., Lim, B., Rigoutsos, I., 2008a. MicroRNAs to Nanog, Oct4 and Sox2 coding regions modulate embryonic stem cell differentiation. Nature 455, 1124–1128.

Tay, Y.M., Tam, W.L., Ang, Y.S., Gaughwin, P.M., Yang, H., Wang, W., Liu, R., George, J., Ng, H.H., Perera, R.J., Lufkin, T., Rigoutsos, I., Thomson, A.M., Lim, B., 2008b. MicroRNA-134 modulates the differentiation of mouse embryonic stem cells, where it causes post-transcriptional attenuation of Nanog and LRH1. Stem Cells 26, 17–29.

Tesar, P.J., Chenoweth, J.G., Brook, F.A., Davies, T.J., Evans, E.P., Mack, D.L., Gardner, R.L., Mckay, R.D., 2007. New cell lines from mouse epiblast share defining features with human embryonic stem cells. Nature 448, 196–199.

Trumbach, D., Prakash, N., 2014. The conserved miR-8/miR-200 microRNA family and their role in invertebrate and vertebrate neurogenesis. Cell Tissue Res. 359, 161–177.

Volinia, S., Galasso, M., Sana, M.E., Wise, T.F., Palatini, J., Huebner, K., Croce, C.M., 2012. Breast cancer signatures for invasiveness and prognosis defined by deep sequencing of microRNA. Proc. Natl Acad. Sci. U.S.A. 109, 3024–3029.

Wang, G., Guo, X., Hong, W., Liu, Q., Wei, T., Lu, C., Gao, L., Ye, D., Zhou, Y., Chen, J., Wang, J., Wu, M., Liu, H., Kang, J., 2013a. Critical regulation of miR-200/ZEB2 pathway in Oct4/Sox2-induced mesenchymal-to-epithelial transition and induced pluripotent stem cell generation. Proc. Natl Acad. Sci. U.S.A. 110, 2858–2863.

Wang, Y., Medvid, R., Melton, C., Jaenisch, R., Blelloch, R., 2007. DGCR8 is essential for microRNA biogenesis and silencing of embryonic stem cell self-renewal. Nat. Genet. 39, 380–385.

Wang, Y., Xu, Z., Jiang, J., Xu, C., Kang, J., Xiao, L., Wu, M., Xiong, J., Guo, X., Liu, H., 2013b. Endogenous miRNA sponge lincRNA-RoR regulates Oct4, Nanog, and Sox2 in human embryonic stem cell self-renewal. Dev. Cell 25, 69–80.

Wellner, U., Schubert, J., Burk, U.C., Schmalhofer, O., Zhu, F., Sonntag, A., Waldvogel, B., Vannier, C., Darling, D., Zur Hausen, A., Brunton, V.G., Morton, J., Sansom, O., Schuler, J., Stemmler, M.P., Herzberger, C., Hopt, U., Keck, T., Brabletz, S., Brabletz, T., 2009. The EMT-activator ZEB1 promotes tumorigenicity by repressing stemness-inhibiting microRNAs. Nat. Cell Biol. 11, 1487–1495.

Wood, H.B., Episkopou, V., 1999. Comparative expression of the mouse Sox1, Sox2 and Sox3 genes from pre-gastrulation to early somite stages. Mech. Dev. 86, 197–201.

Wright, J.E., Ciosk, R., 2013. RNA-based regulation of pluripotency. Trends Genet. 29, 99–107.

Xia, Y., Wu, Y., Liu, B., Wang, P., Chen, Y., 2014. Downregulation of miR-638 promotes invasion and proliferation by regulating SOX2 and induces EMT in NSCLC. FEBS Lett. 588, 2238–2245.

Xu, N., Papagiannakopoulos, T., Pan, G., Thomson, J.A., Kosik, K.S., 2009. MicroRNA-145 regulates OCT4, SOX2, and KLF4 and represses pluripotency in human embryonic stem cells. Cell 137, 647–658.

Yamaguchi, S., Hirano, K., Nagata, S., Tada, T., 2011. Sox2 expression effects on direct reprogramming efficiency as determined by alternative somatic cell fate. Stem Cell Res. 6, 177–186.

Yang, C., Hou, C., Zhang, H., Wang, D., Ma, Y., Zhang, Y., Xu, X., Bi, Z., Geng, S., 2014. miR-126 functions as a tumor suppressor in osteosarcoma by targeting Sox2. Int. J. Mol. Sci. 15, 423–437.

Yang, C.S., Rana, T.M., 2013. Learning the molecular mechanisms of the reprogramming factors: let's start from microRNAs. Mol. Biosyst. 9, 10–17.

Yang, J.S., Lai, E.C., 2011. Alternative miRNA biogenesis pathways and the interpretation of core miRNA pathway mutants. Mol. Cell 43, 892–903.

Yates, L.A., Norbury, C.J., Gilbert, R.J., 2013. The long and short of microRNA. Cell 153, 516–519.

Zhang, Y., Eades, G., Yao, Y., Li, Q., Zhou, Q., 2012. Estrogen receptor alpha signaling regulates breast tumor-initiating cells by down-regulating miR-140 which targets the transcription factor SOX2. J. Biol. Chem. 287, 41514–41522.

Zhao, C., Sun, G., Li, S., Shi, Y., 2009. A feedback regulatory loop involving microRNA-9 and nuclear receptor TLX in neural stem cell fate determination. Nat. Struct. Mol. Biol. 16, 365–371.

Zhao, S., Nichols, J., Smith, A.G., Li, M., 2004. SoxB transcription factors specify neuroectodermal lineage choice in ES cells. Mol. Cell. Neurosci. 27, 332–342.

CHAPTER 5

The Role of SOX2-Interacting Proteins in Gene Regulation by SOX2

Raymond A. Poot
Department of Cell Biology, Erasmus MC, Rotterdam, Netherlands

INTRODUCTION

The family of Sox transcription factors is defined as a family of proteins all of which harbor an HMG box DNA-binding domain first identified in the male sex-determining gene SRY (Gubbay et al., 1990). Currently 20 Sox genes are known to fulfill important roles in development (Lefebvre et al., 2007). Among the first badge of Sox genes to be identified was a gene termed a2 (Gubbay et al., 1990), whose name was later changed to SOX2 (Collignon et al., 1996; Denny et al., 1992). Mouse knockout studies showed that SOX2 is required to maintain the epiblast (Avilion et al., 2003). SOX2 may function even earlier in development, because ribonucleic acid (RNA) interference studies showed that depleting *Sox2* messenger RNA arrests the mouse embryo at the morula stage (Keramari et al., 2010). However, the action of SOX2 is not restricted to early development. The first functional role identified for SOX2 was in lens development, where it induces crystallin genes (Kamachi et al., 1995). This highly tissue-specific activity of SOX2 was an early indication of things to come. Subsequently, SOX2 was shown to be expressed and to act in the development of the trophectoderm (Avilion et al., 2003), retina (Taranova et al., 2006), foregut (Que et al., 2007), taste buds (Okubo et al., 2006), inner ear (Kiernan et al., 2005), lungs (Que et al., 2009), hair follicles (Clavel et al., 2012), pituitary (Kelberman et al., 2006), and brain (Ferri et al., 2004). Cells, often with stem cell characteristics, were derived from some of these tissues and were shown to depend on SOX2 activity for their viability and/or self-renewal: embryonic stem cells (ESCs) (Avilion et al., 2003) (from the inner cell mass), trophectoderm stem cells (TSCs) (Adachi et al., 2013), retinal progenitors (Taranova et al., 2006), and neural stem cells (NSCs) (Graham et al., 2003). This versatility of SOX2 to act in different tissues is striking and to my knowledge unequaled by any other transcription factor. Two scenarios could explain this activity of SOX2: (1) It performs the same activity in all cells where it is expressed, and (2) its activity depends on the cell type. There is far more evidence for the latter option, as will be discussed subsequently in Chapter 8. This suggests that the activity of SOX2 is modified by cell-specific factors; or, to put it more concretely, its genome-binding sites to activate genes likely change depending on the available cell type-specific factors. These cell type–specific factors may interact with SOX2

Sox2
http://dx.doi.org/10.1016/B978-0-12-800352-7.00005-0

in a DNA-independent manner, which brings us to the topic of this chapter: SOX2-interacting proteins and their influence on SOX2 activity in gene regulation. I will briefly discuss SOX2 purification methodologies to arm the reader against publications that report SOX2-interacting proteins that in my opinion are false owing to poor purification methodologies. Subsequently, I will discuss examples in which interacting proteins affect SOX2 activity in gene regulation. Because the level of knowledge varies according to different cell types, I decided to discuss all information about a particular cell type and, when applicable, draw parallels with other cell types.

UNBIASED IDENTIFICATION OF SOX2 INTERACTION PARTNERS

Established SOX2 interaction partners such as PAX6, OCT4, BRN2, and OTX2 were primarily identified by their role as essential co-activators of SOX2-dependent enhancers and not by their protein–protein interaction with SOX2 (Kamachi et al., 2001; Yuan et al., 1995; Catena et al., 2004; Danno et al., 2008), although SOX2 does form (often weak) protein–protein interactions with those proteins (Kamachi et al., 2001; Ambrosetti et al., 1997; Danno et al., 2008; van den Berg et al., 2010; Yuan et al., 1995). However, unbiased biochemical screens for SOX2-interacting proteins have been performed and identified other putative SOX2 interaction partners. Here, I will give my opinion about the quality of the used purification methods and consequently the reliability of the identifications. We previously purified SOX2 from ESCs (Gontan et al., 2009) and NSCs (Engelen et al., 2011). For our SOX2 purifications, we prepared nuclear extracts according to the Dignam protocol (Dignam et al., 1983) and treated the nuclear extracts with the RNA–deoxyribonucleic acid (DNA) digesting nuclease benzonase (van den Berg et al., 2010). Both steps were implemented to minimize the amount of DNA in the extracts. Because SOX2 is a transcription factor, it will bind such DNA, leading to false identifications of DNA-mediated interactions, often with abundant chromatin-binding factors or DNA-end binding (repair) proteins. That such concerns are warranted was shown 20 years ago (Lai and Herr, 1992). Recently, SOX2 was purified from ESCs and brain tumor cells without controlling for DNA in the extract (Mallanna et al., 2010; Gao et al., 2012; Cox et al., 2013). Indeed, proteins such as XRCC5, XRCC6 (repair proteins that bind DNA ends), and histones (chromatin) were prominently observed in the analysis of their SOX2 purifications (Gao et al., 2012; Mallanna et al., 2010), casting doubt on the validity of other identified SOX2 interaction partners. Fang et al. (2011) performed SOX2 purification without DNase and reported the prominent presence of XRCC5 and XRCC6 and other repair proteins in their SOX2 interactome. Groucho repressor proteins were reported in purifications of SOX2 from NSCs after crosslinking of the cells and no DNase addition to the resulting extracts (Liu et al., 2014). We did not observe Groucho proteins in SOX2 purifications from NSCs (Engelen et al., 2011). This does not mean that all identified interaction partners in these purifications are necessarily false, but the identification of likely contaminants as SOX2 interaction partners shows that one has to be careful with some of the published biochemical data on SOX2.

SOX2 IN ESCs

OCT4

The role of SOX2 in gene regulation in ESCs was immediately linked to what was to become its most studied interaction partner in ESCs, a Pou-homeodomain protein called OCT4 (Yuan et al., 1995). SOX2 was shown to bind next to OCT4 on adjacent binding sites for Sox and Pou proteins in the enhancer of the *fgf4* gene. The spacing between the OCT4 and SOX2 binding sites can vary between different enhancers. Whereas in the *fgf4* enhancer there are three nucleotides between the two motifs (Yuan et al., 1995), in the enhancer of the *utf1* gene, the SOX2 and Pou sites are immediately adjacent (Miyagi et al., 2006). Crystal structures of the ternary complex of the Pou and HMG box DNA-binding domains on the *fgf4* and *utf1* enhancers suggested that SOX2 and OCT4 can indeed form different conformations on different enhancers (Remenyi et al., 2003). SOX2 and OCT4 were also suggested to have DNA-independent contacts (Remenyi et al., 2003), which explains the observed (weak) interactions between OCT4 and SOX2 in solution (Yuan et al., 1995; van den Berg et al., 2010). The importance of the OCT4–SOX2 pair in regulating gene expression and ESC identity became clear from the groundbreaking work of Shinya Yamanaka, who reprogrammed fibroblasts into ESC-like induced pluripotent cells (iPSCs) by the ectopic expression of OCT4, SOX2, KLF4, and c-myc (Takahashi and Yamanaka, 2006). Whereas c-Myc and KLF4 could be omitted (Huangfu et al., 2008) or replaced by factors such as ESRRB (Feng et al., 2009) or Nanog and Lin28 (Yu et al., 2007), for a long time OCT4 and SOX2 were essential for the generation of iPSCs. Genome-wide binding site studies showed the abundant teaming up of OCT4 and SOX2 on ESC enhancers (Chen et al., 2008), to the extent that an unbiased identification of ESC enhancers by chromatin immunoprecipitation (ChIP) for enhancer binder p300 yielded a Pou–Sox motif (Chen et al., 2008). The observation by Niwa and coworkers that SOX2 was dispensable for ESC pluripotency if OCT4 expression were artificially maintained was therefore surprising (Masui et al., 2007). Moreover, they showed that the expression of several hallmark OCT4–SOX2 target genes such as *fgf4* and *utf1* was only moderately affected by SOX2 depletion (Masui et al., 2007). It was shown that other Sox factors such as SOX4, SOX5, and SOX15 could bind to Oct–Sox enhancers and thereby may complement the loss of SOX2 (Masui et al., 2007).

An elegant single-molecule imaging study revealed the order of events of SOX2 and OCT4 finding and binding to a target enhancer (Chen et al., 2014). First, SOX2 engages with the DNA and performs a trial-and-error sampling of DNA sequences that takes several minutes before it finds a specific binding site, after which OCT4 joins and together remains bound for 12–14 s (Chen et al., 2014). This suggests that SOX2 has a primary role in target site selection and its binding partners only assist in stabilizing the specific SOX2–DNA complex. Consequently, SOX2 interactions with OCT4 were suggested to occur only on the DNA. This implies that differences in SOX2-binding sites between different cell types (see below) do not result primarily because of different binding preferences of a SOX2–partner complex. It appears more likely that in different cell types, different SOX2-binding sites are accessible in the chromatin and the different partner proteins are needed only to stabilize SOX2 binding to those sites. However, this leads to the chicken and egg problem of why different SOX2-binding sites are available in different cell types in the first place. It also appears to contradict the conclusions of a study on the initial engagement of iPSC reprogramming factors OCT4, SOX2, KLF4, and c-myc (OSK) with the fibroblast genome (Soufi et al., 2012). It was shown that OSK act as pioneering factors and bind to closed and unmarked fibroblast chromatin and recruit c-Myc for further opening of the chromatin (Soufi et al., 2012) (see also Chapter 11). How SOX2 recognizes the correct OCT4–SOX2 sites in fibroblast chromatin is unclear. Maybe under these conditions, SOX2 binds to Sox sites randomly, but with only OCT4 as the binding partner, only bound Oct–Sox sites will be stabilized and will be productive in stimulating transcription of the accompanying gene. Indeed, SOX2 with neurogenic factor MASH1 supports the generation of neurons from pericytes, which suggests that stabilization of different SOX2–DNA complexes by MASH1 promotes a different cell fate (Karow et al., 2012). What may happen during reprogramming is the highly inefficient process of binding of SOX2 to all of its often nucleosome-occupied potential binding sites. Some of these sites will be stabilized by available SOX2 cofactors (OCT4 in iPSCs and MASH1 in neuronal reprogramming) and lead to gene activation, which coincides with the recruitment of histone acetylation complexes and chromatin remodeling complexes, which further open and keep open productive SOX2 sites (often in enhancers). In this way, an epigenetic landscape is created in which only the correct, previously productive SOX2 sites are open to subsequent SOX2 binding, which is much more efficient. This may explain why reprogramming of fibroblasts to iPSCs is a highly inefficient and slow process. In (early) development, cell fate changes and the accompanying change of SOX2-binding sites occurs on the order of days, that is, more rapidly than during reprogramming. One reason why this is feasible may be the open chromatin structure that has been suggested for cells during early development, for example, ESCs (Meshorer et al., 2006). With such open chromatin, the sampling of binding sites by SOX2 and the subsequent stabilization of correct SOX2 sites by partner proteins would be faster and allow for relatively fast changes in the SOX2-binding pattern and, in parallel, the epigenetic landscape during subsequent developmental stages.

NANOG

Nanog colocalizes with OCT4 and SOX2 to many binding sites in ESCs (Chen et al., 2008) and could be expected to be an interaction partner of SOX2. Earlier Nanog purifications had not identified SOX2 (Wang et al., 2006; Costa et al., 2013), but a recent publication showed a robust and direct interaction between SOX2 and Nanog (Gagliardi et al., 2013) (see also Chapter 10). The strong interaction between SOX2 and Nanog was illustrated by the observation that a SOX2–Nanog complex could be purified to homogeneity from bacteria expressing both proteins (Gagliardi et al., 2013). Because of this stable SOX2–Nanog interaction, it would be interesting to see whether the SOX2–Nanog complex has a role in primary binding site selection by SOX2, as described in the earlier discussed single-molecule study (Chen et al., 2014). Amino acid mutations showed that SOX2 interacts with Nanog via a triple (S–X–T/S–Y) motif in the serine-rich regions between the two activation domains at its C-terminal half, whereas Nanog interacts with SOX2 with tryptophans in its WD domain (Gagliardi et al., 2013). Stacking of the tyrosine rings in the SOX2 interaction domain with the tryptophan rings in the Nanog WD domain was suggested as the mechanism behind the SOX2–Nanog interaction. This mode of interaction was supported by the observation that the exchange of SOX2 tyrosines for phenylalanines still supported interaction with Nanog (Gagliardi et al., 2013). The ability of SOX2 to interact with Nanog was essential for its role in maintaining ESC pluripotency (Gagliardi et al., 2013). Intriguingly, the S–X–S/T–Y motif is not present in other Sox proteins, and indeed SOX2 was the only Sox protein identified as a Nanog-interacting protein. SELEX experiments showed that the SOX2–Nanog complex binds to a composite Nanog site (with AAT as its prominent nucleotides) and Sox site. The expression of a number of genes harboring such a SOX2–Nanog binding site was indeed altered upon abrogating the SOX2–Nanog interaction, which suggests that the SOX2–Nanog complex regulates its own set of target genes (Gagliardi et al., 2013).

EXPORTIN4

Our purification of SOX2 from ESCs yielded Exportin4 (EXP4), a protein known as a nuclear exporter (Lipowsky et al., 2000), as its top hit interaction partner (Gontan et al., 2009). However, the addition of RanGTP to in vitro binding assays disrupted SOX2–EXP4 interaction, which suggests that EXP4 acts as a nuclear import factor for SOX2 (Gontan et al., 2009). Indeed, in vitro nuclear import assays showed the robust nuclear import of SOX2 by EXP4 (Gontan et al., 2009). Interestingly, single amino acid mutations in the HMG box domain of SRY in XY females were shown to impair its nuclear localization (Harley et al., 2003). We showed that these SRY mutants no longer bind EXP4 and have impaired EXP4-dependent in vitro nuclear import. Amino acids in the HMG box of SRY that are needed for EXP4 binding and for the binding of SRY to DNA are conserved in all Sox proteins, which suggests that EXP4 is a nuclear import factor for all Sox proteins (Harley et al., 2003). Exportin4 was also shown to bind SOX9 but not SOX11 (Tsuchiya et al., 2011). It is intriguing that universally conserved amino acids in the Sox HMG box are required for both EXP4 binding and DNA binding. Accordingly, Tsuchiya et al. (2011) convincingly showed that by binding to the HMG box EXP4 interferes with the binding of SOX9 to its target enhancers and the subsequent activation of its target genes. This suggests that EXP4 is not just an import factor for Sox proteins, but also a dynamic regulator of their transcriptional activity.

SOX2 IN TSCs

SOX2 is essential for the maintenance of not only the inner cell mass, but also the trophectoderm lineage (Avilion et al., 2003). Maybe not surprisingly, it was observed that SOX2 is also essential for maintaining stem cells that can be derived from the trophectoderm (TSCs) (Adachi et al., 2013). SOX2-binding sites in TSCs revealed enrichment of the AP-2 transcription factor motif. Because AP-2 factor TFAP2C is highly expressed in TSCs, it was tested and found to co-bind many SOX2 sites in TSCs. TFAP2C expression can induce TSC fate in ESCs and recruits SOX2 to its binding sites in TSCs. TFAP2C and SOX2 co-purify in immunoprecipitations, but it is unclear whether this reflects their adjacent binding to common DNA elements or whether they have a stable protein–protein interaction in solution (Adachi et al., 2013). The TSC-fate inducing action of TFAP2C and its global cooperativity with SOX2 in gene regulation suggests a central role for the SOX2-TFAP2C couple in maintaining TSC identity (Adachi et al., 2013).

SOX2 IN NSCs

The role of SOX2 in neural development has been intensely studied over the years (Pevny and Nicolis, 2010). SOX2 is a marker for NCSs both during development and in the adult brain (Ellis et al., 2004; Suh et al., 2007) and is essential for the viability of NSCs both in vivo and in vitro (Favaro et al., 2009) (our unpublished observations). We purified SOX2 from NSCs and identified its co-purifying partners. The list of 50 interaction partners contained common chromatin-modifying complexes such as NuRD, TRRAP and SWI-SNF, which bind to many transcription factors (van den Berg et al., 2010). The robust interaction of SOX2 with chromatin remodeling factor CHD7 appeared more SOX2-specific, because CHD7 did not interact with a number of other transcription factors in ESCs (van den Berg et al., 2010), where it is also expressed (Schnetz et al., 2009). Mutations in SOX2 and CHD7 cause the SOX2 anophthalmia syndrome (Fantes et al., 2003) and CHARGE syndrome (Vissers et al., 2004), respectively, which have a number of phenotypic features in common that lead occasionally to misdiagnosis of SOX2 patients in CHARGE patients (Williamson et al., 2006). SOX2 and CHD7 were shown to overlap significantly in their binding sites and to have a large set of common target genes in NSCs (Engelen et al., 2011). Intriguingly, mutations in a subset of these target genes cause human syndromes with a phenotypic overlap with SOX2 syndrome and CHARGE syndrome (Engelen et al., 2011). This suggests an overlap in the gene-regulatory network that is affected in these syndromes

(Engelen et al., 2011). We tested this hypothesis for the semicircular canal defects that are nearly 100% penetrant for CHARGE syndrome as well as in Alagille syndrome, a syndrome caused by mutations in the Notch ligand Jagged1 (JAG1). *Jagged1* was identified as a target gene of SOX2 and CHD7 in NSCs, because it was bound and activated by both proteins. Strikingly, in CHD7+/− mice, *Jag1* expression was severely decreased in the developing inner ear at a stage (embryonic day 10.5) that is critical for semicircular canal development (Engelen et al., 2011). We concluded that the identical inner ear phenotypes in CHARGE and Alagille syndromes were likely caused by the same molecular event; decreased expression of JAG1 during inner ear development, either as a result of haploid-insufficient expression or mutation of upstream regulator CHD7. SOX2 and CHD7 have a highly overlapping expression pattern in mouse development (Engelen et al., 2011). Together with their strong interaction and their propensity to cooperate in gene regulation, one may wonder why SOX2 syndrome and CHARGE syndrome do not have more features in common, because SOX2-CHD7 target genes should be widespread in development. In my opinion, this is because of gene dosage effects; having 50% of SOX2 protein or CHD7 protein may be sufficient for normal expression of many SOX–CHD7 target genes. In support of this hypothesis, lowering the expression level of SOX2 in mouse below 50%, by combining a *Sox2* KO allele with a *Sox2* allele in which specific enhancers are mutated, leads to eye and inner ear phenotypes similar to malformations caused by CHD7 mutation (Taranova et al., 2006; Langer et al., 2012; Kiernan et al., 2005). Moreover, analyses of the molecular defects in malformed eyes of SOX2 hypomorphic mice showed a near absence of the expression of Notch pathway genes, including *Hes5* (Taranova et al., 2006). We showed that SOX2 and CHD7 regulate Notch genes in NSCs (Engelen et al., 2011), which suggests that the eye malformations in CHD7 mutants may have the same molecular cause as in SOX2 hypomorphic mice.

An analysis of the genome-wide binding sites of SOX2 in NSCs showed enrichment of the SOX motif, as expected, but also enrichment of a bHLH motif, whereas no enrichment of the POU motif was observed (Engelen et al., 2011). This suggested that SOX2 had switched from a POU factor such as OCT4 in ESCs to a bHLH factor in NSCs as its major interaction partner on the genome. Among the SOX2 interaction partners in NSCs is the bHLH factor Twist1, but it is unclear whether this is the major bHLH partner of SOX2 on the NSC genome (Engelen et al., 2011). The different motifs in SOX2-binding sites in ESCs versus NSCs suggest that SOX2 occupies different binding sites in these cell types. Indeed, SOX2-binding sites in ESCs and NSCs show little genome-wide overlap (data not shown). Even on genes that are regulated by SOX2 in ESCs and NSCs, such as *Jag1* (Masui et al., 2007; Engelen et al., 2011) (Figure 1(A)), or the *Sox2* gene itself (Figure 1(B)), SOX2-binding sites are at different positions. Possible mechanisms underlying the recognition of different binding sites by SOX2 in different cell types have been discussed above.

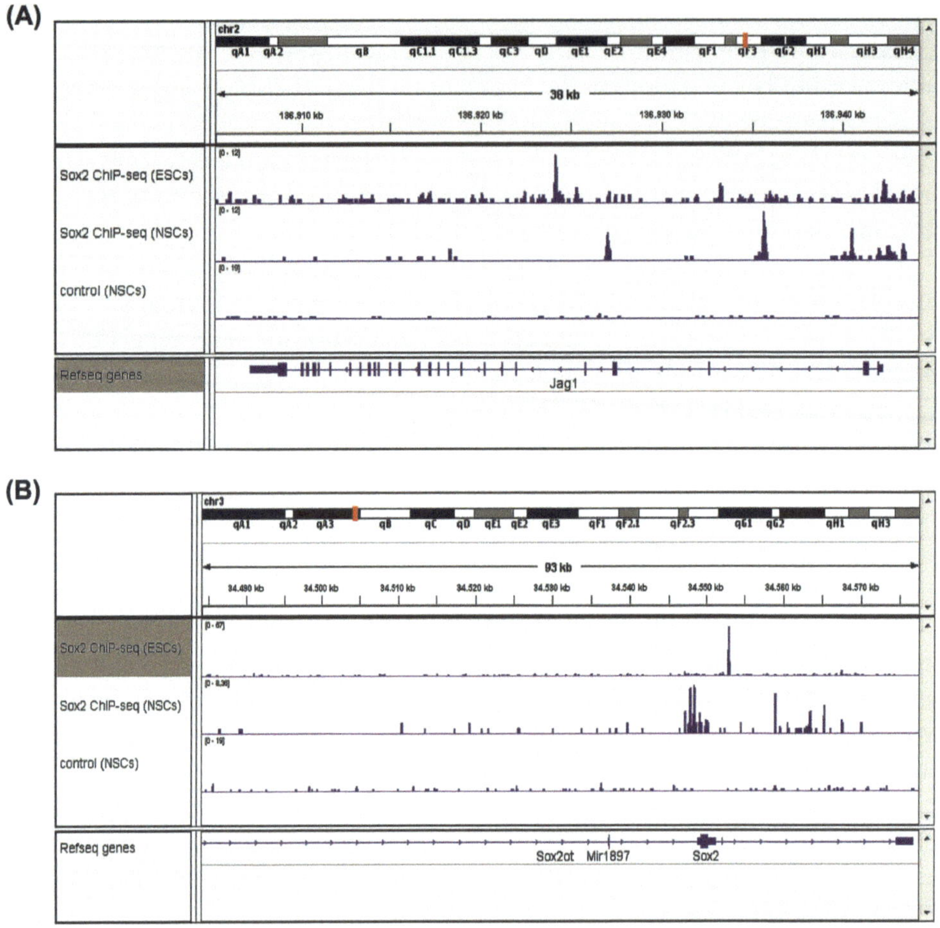

Figure 1 Sox2 has different binding sites in ESCs and NSCs. (A) Sox2-binding profile (by ChIP-seq) on the *Jag1* gene in ESCs and NSCs. *Jag1* is regulated by Sox2 in both ESCs and NSCs. (B) Sox2-bindings profile on the *Sox2* gene in ESCs and NCSs. Sox2 ChIPseq reads in ESCs and NSCS are plotted on the mouse genome around the *Jag1* gene (A) or around the *Sox2* gene (B). Control ChIP in NSCs (using immunoglobulin G) is indicated; gene position is indicated; span width of the view is indicated in kilobytes.

BRN1/2

Although POU domain factors do not appear to be major SOX2 partners in NSCs (Engelen et al., 2011), some examples of gene regulation by these factors are known. The upstream enhancer of the mouse SOX2 gene contains a *SOX–POU* site (Catena et al., 2004). Not surprisingly, in ESCs, expression from this enhancer is driven by OCT4 and SOX2 (Catena et al., 2004). In NSCs, neural POU factors BRN1 and BRN2 replace OCT4 as the SOX2 partner and together they drive expression in SOX2 in neural progenitor cells (Catena et al., 2004). The Nestin gene is another example of a gene that contains an upstream Pou-Sox enhancer that is activated in neural progenitors by the combined action of SOX2 and BRN2 (Tanaka et al., 2004). Whether there is a principal difference in expression pattern or regulatory mechanism between enhancers driven by SOX2 and a bHLH factor and enhancers dependent on SOX2 and BRN2 is currently unclear. We did not observe BRN1 or BRN2 as an interaction partner of SOX2 in NSCs (Engelen et al., 2011), which suggests that a SOX2–BRN interaction in solution, if any, is weak.

EYE

Anophthalmia, microphthalmia, and coloboma are structural eye defects of different severity caused by mutation of a selected set of genes including *Sox2, Otx2, Pax6*, and *Rax* (Williamson and FitzPatrick, 2014). These mutations are mechanistically related and are part of two transcriptional networks centered on SOX2 in the eye, as will be discussed subsequently.

PAX6

PAX6 was the first transcriptional cofactor of SOX2 to be identified (Kamachi et al., 2001). In a two hybrid–like screen on the DC5 enhancer of the δ-crystallin gene in yeast to find the activation partner of SOX2, the cDNA of paired domain transcription factor PAX6 was identified (Kamachi et al., 2001). SOX2 and PAX6 were shown to bind to their adjacent Sox and a degenerate paired domain binding site. Both the Sox site and the PAX6 site showed some binding of their respective proteins in *electrophoretic mobility shift assay (EMSA)* experiments but the combined binding of SOX2 and PAX6 was more efficient, illustrating the symbiotic binding of the two proteins to the DC5 enhancer (Kamachi et al., 2001). SOX2 and PAX6 expressed in bacteria were shown to have a weak DNA-independent interaction, which was surprisingly stabilized by bacterial DNA (Kamachi et al., 2001). We did not observe PAX6 in our purifications of SOX2 from NSCs, although PAX6 is expressed in these cells, which suggests that the protein–protein interaction between SOX2 and PAX6 is weak (Engelen et al., 2011). The binding of SOX2 and PAX6 to the DC5 enhancer drives the transcription of the δ-Crystallin gene (Kamachi et al., 2001). Ectopic expression of SOX2 and PAX6 in chicken ectodermic tissue is sufficient to induce δ-Crystallin expression. Moreover, ectopic expression of SOX2 and PAX6 induced morphological changes reminiscent of the lens placode, which suggests that other key genes in lens induction are also regulated by SOX2 and PAX6 (Kamachi et al., 2001). The creation of compound SOX2–PAX6 heterozygotes suggests that SOX2 and PAX6 cooperation in gene regulation encompasses all aspects of eye development (Smith et al., 2009). SOX2–PAX6 compound heterozygotes often have severe eye defects, similar to full PAX6 KO mice (Smith et al., 2009).

OTX2

A separate regulatory circuit in eye development is controlled by SOX2 and OTX2. SOX2 and OTX2 were found to bind the upstream enhancer of the Rax gene and activate its expression (Danno et al., 2008). RAX1 is a paired-type homeobox transcription factor that regulates genes involved in specifying the eye field or the proliferation of retinal progenitors. *Rax1*$^{-/-}$ mice have no eyes and Rax mutations cause anophthalmia in humans (Voronina et al., 2004). It was observed that a region upstream of the Rax gene, CNS1, is conserved in vertebrates and contains adjacent Sox and Otx sites (Voronina et al., 2004). SOX2 and OTX2 mutations cause anophthalmia in humans (Fantes et al., 2003; Ragge et al., 2005) and *Otx2* mutations cause anophthalmia in mice (Simeone and Acampora, 2001). Consequently, the binding of these proteins to CNS1 was tested and confirmed in Xenopus both in vitro (EMSA) and in vivo (ChIP). SOX2 and OTX2 bind synergistically to the Sox and Otx sites in CNS1, which are always separated by six nucleotides. SOX2 and OTX2 have a direct, DNA-independent protein–protein interaction that facilitates the synergy in binding the CNS1 enhancer of the Rax gene. Indeed, activation of the CNS1 enhancer of Rax requires both SOX2 and OTX2 in Xenopus and human cells (Danno et al., 2008). A recent article suggests that the teaming up of SOX2 and OTX2 is not limited to eye development but also occurs in epiblast stem cells and perhaps in its in vivo equivalent, the epiblast (Buecker et al., 2014). It was shown that upon differentiation of ESCs to epiblast stem cells (epiSCs), OCT4 changes position on the genome from ESC sites to epiSC sites directed by the upregulated factor OTX2 (Buecker et al., 2014). These epiSC sites harbor a Pou, Sox and Otx motif, which suggests that they bind OCT4, SOX2, and OTX2, although actual relocalization of SOX2 with OCT4 and OTX2 was not shown (Buecker et al., 2014). This observation has the intriguing suggestion that OCT4 and OTX2 interact with nonoverlapping parts of the DNA-bound SOX2 protein and SOX2 can stabilize OCT4 and OTX2 binding simultaneously on triple binding sites. Binding of a POU protein to (some of the) SOX2–Otx sites in the eye has not been reported and SOX2 sites in the eye have not been analyzed genome-wide to search for co-occurring motifs, but the observations in epiblast stem cells make triple binding of SOX2, OTX2, and an eye-expressed POU protein a realistic possibility.

CONCLUDING REMARKS

In this review I have illustrated the remarkable flexibility of SOX2 to bind the genome together with one or more interaction partners, which are different in different cell types, and activate the appropriate cell type–specific expression program. What is even more remarkable is that the SOX2-binding partners are diverse in their DNA-binding domain and other features. To date, only the structure of SOX2 with a POU domain protein has been resolved, so it is unclear whether common features exist in the DNA complexes of SOX2 with its various interaction partners. For me, two remaining questions stand out: (1) Why has nature deemed SOX2 so suitable to maintain the cell fate of many different tissues and their associated stem cells? (2) How do SOX2 DNA complexes activate transcription? One can only speculate about answers to these questions. Regarding question 1, perhaps the HMG box domain is highly suited to combine the binding to a consensus sequence with high flexibility to interact with binding partners. The strong DNA-bending activity of the HMG box (Pontiggia et al., 1994), which is essential for SOX2 activity (Scaffidi and Bianchi, 2001), may have a role in this feature or the overall appeal of SOX2 as a molecule to be used for regulatory purposes. Question 2 is equally unclear. SOX2 does not appear to bind to the mediator complex or RNA polymerase 2 itself to bridge the distance between the enhancer and the promoter (Engelen et al., 2011). In fact, none of the interaction partners of SOX2 has a documented interaction with the basal transcription machinery. SOX2 was shown to bind next to the estrogen receptor–like transcription factor ESRRB on a minority of its binding sites in ESCs (Hutchins et al., 2013; Adachi et al., 2013). ESRRB has a strong association with the mediator complex and RNA polymerase 2 (van den Berg et al., 2010). Nevertheless, how SOX2 would activate transcription from its many sites devoid of ESRRB (Adachi et al., 2013) is unclear. These questions highlight that there is still much to be learned about SOX2 in years to come.

REFERENCES

Adachi, K., Nikaido, I., Ohta, H., Ohtsuka, S., Ura, H., Kadota, M., Wakayama, T., Ueda, H.R., Niwa, H., 2013. Context-dependent wiring of Sox2 regulatory networks for self-renewal of embryonic and trophoblast stem cells. Mol. Cell 52, 380–392.

Ambrosetti, D.C., Basilico, C., Dailey, L., 1997. Synergistic activation of the fibroblast growth factor 4 enhancer by Sox2 and Oct-3 depends on protein-protein interactions facilitated by a specific spatial arrangement of factor binding sites. Mol. Cell Biol. 17, 6321–6329.

Avilion, A.A., Nicolis, S.K., Pevny, L.H., Perez, L., Vivian, N., Lovell-Badge, R., 2003. Multipotent cell lineages in early mouse development depend on SOX2 function. Genes Dev. 17, 126–140.

van den Berg, D.L., Snoek, T., Mullin, N.P., Yates, A., Bezstarosti, K., Demmers, J., Chambers, I., Poot, R.A., 2010. An Oct4-centered protein interaction network in embryonic stem cells. Cell Stem Cell 6, 369–381.

Buecker, C., Srinivasan, R., Wu, Z., Calo, E., Acampora, D., Faial, T., Simeone, A., Tan, M., Swigut, T., Wysocka, J., 2014. Reorganization of enhancer patterns in transition from naive to primed pluripotency. Cell Stem Cell 14, 838–853.

Catena, R., Tiveron, C., Ronchi, A., Porta, S., Ferri, A., Tatangelo, L., Cavallaro, M., Favaro, R., Ottolenghi, S., Reinbold, R., Scholer, H., Nicolis, S.K., 2004. Conserved POU binding DNA sites in the Sox2 upstream enhancer regulate gene expression in embryonic and neural stem cells. J. Biol. Chem. 279, 41846–41857.

Chen, J., Zhang, Z., Li, L., Chen, B.C., Revyakin, A., Hajj, B., Legant, W., Dahan, M., Lionnet, T., Betzig, E., Tjian, R., Liu, Z., 2014. Single-molecule dynamics of enhanceosome assembly in embryonic stem cells. Cell 156, 1274–1285.

Chen, X., Xu, H., Yuan, P., Fang, F., Huss, M., Vega, V.B., Wong, E., Orlov, Y.L., Zhang, W., Jiang, J., Loh, Y.H., Yeo, H.C., Yeo, Z.X., Narang, V., Govindarajan, K.R., Leong, B., Shahab, A., Ruan, Y., Bourque, G., Sung, W.K., Clarke, N.D., Wei, C.L., Ng, H.H., 2008. Integration of external signaling pathways with the core transcriptional network in embryonic stem cells. Cell 133, 1106–1117.

Clavel, C., Grisanti, L., Zemla, R., Rezza, A., Barros, R., Sennett, R., Mazloom, A.R., Chung, C.Y., Cai, X., Cai, C.L., Pevny, L., Nicolis, S., Ma'ayan, A., Rendl, M., 2012. Sox2 in the dermal papilla niche controls hair growth by fine-tuning BMP signaling in differentiating hair shaft progenitors. Dev. Cell 23, 981–994.

Collignon, J., Sockanathan, S., Hacker, A., Cohen-Tannoudji, M., Norris, D., Rastan, S., Stevanovic, M., Goodfellow, P.N., Lovell-Badge, R., 1996. A comparison of the properties of Sox-3 with Sry and two related genes, Sox-1 and Sox-2. Development 122, 509–520.

Costa, Y., Ding, J., Theunissen, T.W., Faiola, F., Hore, T.A., Shliaha, P.V., Fidalgo, M., Saunders, A., Lawrence, M., Dietmann, S., Das, S., Levasseur, D.N., Li, Z., Xu, M., Reik, W., Silva, J.C., Wang, J., 2013. NANOG-dependent function of TET1 and TET2 in establishment of pluripotency. Nature 495, 370–374.

Cox, J.L., Wilder, P.J., Gilmore, J.M., Wuebben, E.L., Washburn, M.P., Rizzino, A., 2013. The SOX2-interactome in brain cancer cells identifies the requirement of MSI2 and USP9X for the growth of brain tumor cells. PLoS One 8, e62857.

Danno, H., Michiue, T., Hitachi, K., Yukita, A., Ishiura, S., Asashima, M., 2008. Molecular links among the causative genes for ocular malformation: Otx2 and Sox2 coregulate Rax expression. Proc. Natl Acad. Sci. U.S.A. 105, 5408–5413.

Denny, P., Swift, S., Brand, N., Dabhade, N., Barton, P., Ashworth, A., 1992. A conserved family of genes related to the testis determining gene, SRY. Nucleic Acids Res. 20, 2887.

Dignam, J.D., Lebovitz, R.M., Roeder, R.G., 1983. Accurate transcription initiation by RNA polymerase II in a soluble extract from isolated mammalian nuclei. Nucleic Acids Res. 11, 1475–1489.

Ellis, P., Fagan, B.M., Magness, S.T., Hutton, S., Taranova, O., Hayashi, S., Mcmahon, A., Rao, M., Pevny, L., 2004. SOX2, a persistent marker for multipotential neural stem cells derived from embryonic stem cells, the embryo or the adult. Dev. Neurosci. 26, 148–165.

Engelen, E., Akinci, U., Bryne, J.C., Hou, J., Gontan, C., Moen, M., Szumska, D., Kockx, C., Van Ijcken, W., Dekkers, D.H., Demmers, J., Rijkers, E.J., Bhattacharya, S., Philipsen, S., Pevny, L.H., Grosveld, F.G., Rottier, R.J., Lenhard, B., Poot, R.A., 2011. Sox2 cooperates with Chd7 to regulate genes that are mutated in human syndromes. Nat. Genet. 43, 607–611.

Fang, X., Yoon, J.G., Li, L., Tsai, Y.S., Zheng, S., Hood, L., Goodlett, D.R., Foltz, G., Lin, B., 2011. Landscape of the SOX2 protein-protein interactome. Proteomics 11, 921–934.

Fantes, J., Ragge, N.K., Lynch, S.A., Mcgill, N.I., Collin, J.R., Howard-Peebles, P.N., Hayward, C., Vivian, A.J., Williamson, K., Van Heyningen, V., Fitzpatrick, D.R., 2003. Mutations in SOX2 cause anophthalmia. Nat. Genet. 33, 461–463.

Favaro, R., Valotta, M., Ferri, A.L., Latorre, E., Mariani, J., Giachino, C., Lancini, C., Tosetti, V., Ottolenghi, S., Taylor, V., Nicolis, S.K., 2009. Hippocampal development and neural stem cell maintenance require Sox2-dependent regulation of Shh. Nat. Neurosci. 12, 1248–1256.

Feng, B., Jiang, J., Kraus, P., Ng, J.H., Heng, J.C., Chan, Y.S., Yaw, L.P., Zhang, W., Loh, Y.H., Han, J., Vega, V.B., Cacheux-Rataboul, V., Lim, B., Lufkin, T., Ng, H.H., 2009. Reprogramming of fibroblasts into induced pluripotent stem cells with orphan nuclear receptor Esrrb. Nat. Cell Biol. 11, 197–203.

Ferri, A.L., Cavallaro, M., Braida, D., Di Cristofano, A., Canta, A., Vezzani, A., Ottolenghi, S., Pandolfi, P.P., Sala, M., Debiasi, S., Nicolis, S.K., 2004. Sox2 deficiency causes neurodegeneration and impaired neurogenesis in the adult mouse brain. Development 131, 3805–3819.

Gagliardi, A., Mullin, N.P., Ying Tan, Z., Colby, D., Kousa, A.I., Halbritter, F., Weiss, J.T., Felker, A., Bezstarosti, K., Favaro, R., Demmers, J., Nicolis, S.K., Tomlinson, S.R., Poot, R.A., Chambers, I., 2013. A direct physical interaction between Nanog and Sox2 regulates embryonic stem cell self-renewal. EMBO J. 32, 2231–2247.

Gao, Z., Cox, J.L., Gilmore, J.M., Ormsbee, B.D., Mallanna, S.K., Washburn, M.P., Rizzino, A., 2012. Determination of protein interactome of transcription factor Sox2 in embryonic stem cells engineered for inducible expression of four reprogramming factors. J. Biol. Chem. 287, 11384–11397.

Gontan, C., Guttler, T., Engelen, E., Demmers, J., Fornerod, M., Grosveld, F.G., Tibboel, D., Gorlich, D., Poot, R.A., Rottier, R.J., 2009. Exportin 4 mediates a novel nuclear import pathway for Sox family transcription factors. J. Cell Biol. 185, 27–34.

Graham, V., Khudyakov, J., Ellis, P., Pevny, L., 2003. SOX2 functions to maintain neural progenitor identity. Neuron 39, 749–765.

Gubbay, J., Collignon, J., Koopman, P., Capel, B., Economou, A., Munsterberg, A., Vivian, N., Goodfellow, P., Lovell-Badge, R., 1990. A gene mapping to the sex-determining region of the mouse Y chromosome is a member of a novel family of embryonically expressed genes. Nature 346, 245–250.

Harley, V.R., Layfield, S., Mitchell, C.L., Forwood, J.K., John, A.P., Briggs, L.J., Mcdowall, S.G., Jans, D.A., 2003. Defective importin beta recognition and nuclear import of the sex-determining factor SRY are associated with XY sex-reversing mutations. Proc. Natl. Acad. Sci. U.S.A. 100, 7045–7050.

Huangfu, D., Maehr, R., Guo, W., Eijkelenboom, A., Snitow, M., Chen, A.E., Melton, D.A., 2008. Induction of pluripotent stem cells by defined factors is greatly improved by small-molecule compounds. Nat. Biotechnol. 26, 795–797.

Hutchins, A.P., Choo, S.H., Mistri, T.K., Rahmani, M., Woon, C.T., Ng, C.K., Jauch, R., Robson, P., 2013. Co-motif discovery identifies an Esrrb-Sox2-DNA ternary complex as a mediator of transcriptional differences between mouse embryonic and epiblast stem cells. Stem Cells 31, 269–281.

Kamachi, Y., Sockanathan, S., Liu, Q., Breitman, M., Lovell-Badge, R., Kondoh, H., 1995. Involvement of SOX proteins in lens-specific activation of crystallin genes. EMBO J. 14, 3510–3519.

Kamachi, Y., Uchikawa, M., Tanouchi, A., Sekido, R., Kondoh, H., 2001. Pax6 and SOX2 form a co-DNA-binding partner complex that regulates initiation of lens development. Genes Dev. 15, 1272–1286.

Karow, M., Sanchez, R., Schichor, C., Masserdotti, G., Ortega, F., Heinrich, C., Gascon, S., Khan, M.A., Lie, D.C., Dellavalle, A., Cossu, G., Goldbrunner, R., Gotz, M., Berninger, B., 2012. Reprogramming of pericyte-derived cells of the adult human brain into induced neuronal cells. Cell Stem Cell 11, 471–476.

Kelberman, D., Rizzoti, K., Avilion, A., Bitner-Glindzicz, M., Cianfarani, S., Collins, J., Chong, W.K., Kirk, J.M., Achermann, J.C., Ross, R., Carmignac, D., Lovell-Badge, R., Robinson, I.C., Dattani, M.T., 2006. Mutations within Sox2/SOX2 are associated with abnormalities in the hypothalamo-pituitary-gonadal axis in mice and humans. J. Clin. Invest. 116, 2442–2455.

Keramari, M., Razavi, J., Ingman, K.A., Patsch, C., Edenhofer, F., Ward, C.M., Kimber, S.J., 2010. Sox2 is essential for formation of trophectoderm in the preimplantation embryo. PLoS One 5, e13952.

Kiernan, A.E., Pelling, A.L., Leung, K.K., Tang, A.S., Bell, D.M., Tease, C., Lovell-Badge, R., Steel, K.P., Cheah, K.S., 2005. Sox2 is required for sensory organ development in the mammalian inner ear. Nature 434, 1031–1035.

Lai, J.S., Herr, W., 1992. Ethidium bromide provides a simple tool for identifying genuine DNA-independent protein associations. Proc. Natl. Acad. Sci. U.S.A. 89, 6958–6962.

Langer, L., Taranova, O., Sulik, K., Pevny, L., 2012. SOX2 hypomorphism disrupts development of the prechordal floor and optic cup. Mech. Dev. 129, 1–12.

Lefebvre, V., Dumitriu, B., Penzo-Mendez, A., Han, Y., Pallavi, B., 2007. Control of cell fate and differentiation by Sry-related high-mobility-group box (Sox) transcription factors. Int. J. Biochem. Cell Biol. 39, 2195–2214.

Lipowsky, G., Bischoff, F.R., Schwarzmaier, P., Kraft, R., Kostka, S., Hartmann, E., Kutay, U., Gorlich, D., 2000. Exportin 4: a mediator of a novel nuclear export pathway in higher eukaryotes. EMBO J. 19, 4362–4371.

Liu, Y.R., Laghari, Z.A., Novoa, C.A., Hughes, J., Webster, J.R., Goodwin, P.E., Wheatley, S.P., Scotting, P.J., 2014. Sox2 acts as a transcriptional repressor in neural stem cells. BMC Neurosci. 15, 95.

Mallanna, S.K., Ormsbee, B.D., Iacovino, M., Gilmore, J.M., Cox, J.L., Kyba, M., Washburn, M.P., Rizzino, A., 2010. Proteomic analysis of Sox2-associated proteins during early stages of mouse embryonic stem cell differentiation identifies Sox21 as a novel regulator of stem cell fate. Stem Cells 28, 1715–1727.

Masui, S., Nakatake, Y., Toyooka, Y., Shimosato, D., Yagi, R., Takahashi, K., Okochi, H., Okuda, A., Matoba, R., Sharov, A.A., Ko, M.S., Niwa, H., 2007. Pluripotency governed by Sox2 via regulation of Oct3/4 expression in mouse embryonic stem cells. Nat. Cell Biol. 9, 625–635.

Meshorer, E., Yellajoshula, D., George, E., Scambler, P.J., Brown, D.T., Misteli, T., 2006. Hyperdynamic plasticity of chromatin proteins in pluripotent embryonic stem cells. Dev. Cell 10, 105–116.

Miyagi, S., Nishimoto, M., Saito, T., Ninomiya, M., Sawamoto, K., Okano, H., Muramatsu, M., Oguro, H., Iwama, A., Okuda, A., 2006. The Sox2 regulatory region 2 functions as a neural stem cell-specific enhancer in the telencephalon. J. Biol. Chem. 281, 13374–13381.

Okubo, T., Pevny, L.H., Hogan, B.L., 2006. Sox2 is required for development of taste bud sensory cells. Genes Dev. 20, 2654–2659.

Pevny, L.H., Nicolis, S.K., 2010. Sox2 roles in neural stem cells. Int. J. Biochem. Cell Biol. 42, 421–424.

Pontiggia, A., Rimini, R., Harley, V.R., Goodfellow, P.N., Lovell-Badge, R., Bianchi, M.E., 1994. Sex-reversing mutations affect the architecture of SRY-DNA complexes. EMBO J. 13, 6115–6124.

Que, J., Luo, X., Schwartz, R.J., Hogan, B.L., 2009. Multiple roles for Sox2 in the developing and adult mouse trachea. Development 136, 1899–1907.

Que, J., Okubo, T., Goldenring, J.R., Nam, K.T., Kurotani, R., Morrisey, E.E., Taranova, O., Pevny, L.H., Hogan, B.L., 2007. Multiple dose-dependent roles for Sox2 in the patterning and differentiation of anterior foregut endoderm. Development 134, 2521–2531.

Ragge, N.K., Brown, A.G., Poloschek, C.M., Lorenz, B., Henderson, R.A., Clarke, M.P., Russell-Eggitt, I., Fielder, A., Gerrelli, D., Martinez-Barbera, J.P., Ruddle, P., Hurst, J., Collin, J.R., Salt, A., Cooper, S.T., Thompson, P.J., Sisodiya, S.M., Williamson, K.A., Fitzpatrick, D.R., Van Heyningen, V., Hanson, I.M., 2005. Heterozygous mutations of OTX2 cause severe ocular malformations. Am. J. Hum. Genet. 76, 1008–1022.

Remenyi, A., Lins, K., Nissen, L.J., Reinbold, R., Scholer, H.R., Wilmanns, M., 2003. Crystal structure of a POU/HMG/DNA ternary complex suggests differential assembly of Oct4 and Sox2 on two enhancers. Genes Dev. 17, 2048–2059.

Scaffidi, P., Bianchi, M.E., 2001. Spatially precise DNA bending is an essential activity of the sox2 transcription factor. J. Biol. Chem. 276, 47296–47302.

Schnetz, M.P., Bartels, C.F., Shastri, K., Balasubramanian, D., Zentner, G.E., Balaji, R., Zhang, X., Song, L., Wang, Z., Laframboise, T., Crawford, G.E., Scacheri, P.C., 2009. Genomic distribution of CHD7 on chromatin tracks H3K4 methylation patterns. Genome Res. 19, 590–601.

Simeone, A., Acampora, D., 2001. The role of Otx2 in organizing the anterior patterning in mouse. Int. J. Dev. Biol. 45, 337–345.

Smith, A.N., Miller, L.A., Radice, G., Ashery-Padan, R., Lang, R.A., 2009. Stage-dependent modes of Pax6-Sox2 epistasis regulate lens development and eye morphogenesis. Development 136, 2977–2985.

Soufi, A., Donahue, G., Zaret, K.S., 2012. Facilitators and impediments of the pluripotency reprogramming factors' initial engagement with the genome. Cell 151, 994–1004.

Suh, H., Consiglio, A., Ray, J., Sawai, T., D'amour, K.A., Gage, F.H., 2007. In vivo fate analysis reveals the multipotent and self-renewal capacities of Sox2+ neural stem cells in the adult hippocampus. Cell Stem Cell 1, 515–528.

Takahashi, K., Yamanaka, S., 2006. Induction of pluripotent stem cells from mouse embryonic and adult fibroblast cultures by defined factors. Cell 126, 663–676.

Tanaka, S., Kamachi, Y., Tanouchi, A., Hamada, H., Jing, N., Kondoh, H., 2004. Interplay of SOX and POU factors in regulation of the Nestin gene in neural primordial cells. Mol. Cell. Biol. 24, 8834–8846.

Taranova, O.V., Magness, S.T., Fagan, B.M., Wu, Y., Surzenko, N., Hutton, S.R., Pevny, L.H., 2006. SOX2 is a dose-dependent regulator of retinal neural progenitor competence. Genes Dev. 20, 1187–1202.

Tsuchiya, M., Ogawa, H., Suzuki, T., Sugiyama, N., Haraguchi, T., Hiraoka, Y., 2011. Exportin 4 interacts with Sox9 through the HMG Box and inhibits the DNA binding of Sox9. PLoS One 6, e25694.

Vissers, L.E., Van Ravenswaaij, C.M., Admiraal, R., Hurst, J.A., De Vries, B.B., Janssen, I.M., Van Der Vliet, W.A., Huys, E.H., De Jong, P.J., Hamel, B.C., Schoenmakers, E.F., Brunner, H.G., Veltman, J.A., Van Kessel, A.G., 2004. Mutations in a new member of the chromodomain gene family cause CHARGE syndrome. Nat. Genet. 36, 955–957.

Voronina, V.A., Kozhemyakina, E.A., O'kernick, C.M., Kahn, N.D., Wenger, S.L., Linberg, J.V., Schneider, A.S., Mathers, P.H., 2004. Mutations in the human RAX homeobox gene in a patient with anophthalmia and sclerocornea. Hum. Mol. Genet. 13, 315–322.

Wang, J., Rao, S., Chu, J., Shen, X., Levasseur, D.N., Theunissen, T.W., Orkin, S.H., 2006. A protein interaction network for pluripotency of embryonic stem cells. Nature 444, 364–368.

Williamson, K.A., Fitzpatrick, D.R., 2014. The genetic architecture of microphthalmia, anophthalmia and coloboma. Eur. J. Med. Genet. 57, 369–380.

Williamson, K.A., Hever, A.M., Rainger, J., Rogers, R.C., Magee, A., Fiedler, Z., Keng, W.T., Sharkey, F.H., Mcgill, N., Hill, C.J., Schneider, A., Messina, M., Turnpenny, P.D., Fantes, J.A., Van Heyningen, V., Fitzpatrick, D.R., 2006. Mutations in SOX2 cause anophthalmia-esophageal-genital (AEG) syndrome. Hum. Mol. Genet. 15, 1413–1422.

Yu, J., Vodyanik, M.A., Smuga-Otto, K., Antosiewicz-Bourget, J., Frane, J.L., Tian, S., Nie, J., Jonsdottir, G.A., Ruotti, V., Stewart, R., Slukvin, I.I., Thomson, J.A., 2007. Induced pluripotent stem cell lines derived from human somatic cells. Science 318, 1917–1920.

Yuan, H., Corbi, N., Basilico, C., Dailey, L., 1995. Developmental-specific activity of the FGF-4 enhancer requires the synergistic action of Sox2 and Oct-3. Genes Dev. 9, 2635–2645.

PART 2

Gene Regulatory Networks Centered by Sox2

CHAPTER 6

Evolution of *Sox2* and Functional Redundancy in Relation to Other *SoxB1* Genes

Yusuke Kamachi
Graduate School of Frontier Biosciences, Osaka University, Osaka, Japan
Present address: School of Environmental Science and Engineering, Kochi University of Technology, Japan

INTRODUCTION

In the *Sox* gene family studies, the mammalian *Sox2* gene and the other *SoxB1* genes, *Sox1* and *Sox3*, were first identified as a group of genes with high sequence similarity to the sex-determining gene *Sry* within the high mobility group (HMG) box, which encodes the deoxyribonucleic acid (DNA)-binding HMG domain (Gubbay et al., 1990). From an evolutionary perspective, however, cumulative evidence indicates that the *Sry* genes are unique to the mammalian sex-determining system and that *Sry* on the Y chromosome originated from *Sox3* on the X chromosome (Foster et al., 1994).

In this chapter, we will present an overview of the evolutionary aspects of *Sox* genes, with a focus on genes in the *SoxB1* and *SoxB2* groups, during metazoan (animal) evolution. We will also discuss the origin and expansion of the *SoxB* genes during metazoan evolution, a process that ultimately yielded both functional redundancy and functional diversification among the *SoxB1* and *SoxB2* genes.

THE SOX FAMILY AND HMG DOMAIN SUPERFAMILY

Since the discovery of the human and mouse *SRY*/*Sry* genes (Gubbay et al., 1990; Koopman et al., 1991; Sinclair et al., 1990), a number of *Sry*-related HMG-box (*Sox*) genes have been identified in various vertebrate and invertebrate species on the basis of the homology within the HMG box (Gubbay et al., 1990; Pevny and Lovell-Badge, 1997; Wright et al., 1993). For instance, 20 different *Sox* genes have been discovered in mice and humans (Schepers et al., 2002). These *Sox* genes were initially classified into five major groups (*SoxB, SoxC, SoxD, SoxE*, and *SoxF*) and additional minor groups (*SoxA, SoxG*, and *SoxH*) according to their HMG domain amino acid sequences (Table 1; Figure 1) (Bowles et al., 2000; Pevny and Lovell-Badge, 1997; Wright et al., 1993). The Sox HMG domain amino acid sequences are highly similar within each group (approximately 90% identity) but only approximately 50% identical between groups (Wright et al., 1993). A comparison of the amino acid sequences

Sox2
http://dx.doi.org/10.1016/B978-0-12-800352-7.00006-2

Table 1 Sox groups and their members

Group	Mammal *Homo sapiens*	Fish *Danio rerio*	Ascidian *Ciona intestinalis*	Sponge *Amphimedon queenslandica*
SoxA	SRY			
SoxB1	SOX1	Sox1a/b	SoxB1	SoxB1
	SOX2	Sox2		
	SOX3	Sox3		
SoxB1/SoxG	SOX15	Sox19a/b		
SoxB2	SOX14	Sox14a/b	SoxB2	SoxB2
	SOX21	Sox21a/b		
SoxC	SOX4	Sox4a/b	SoxC	SoxC
	SOX11	Sox11a/b		
	SOX12	Sox12		
SoxD	SOX5	Sox5	SoxD	
	SOX6	Sox6		
	SOX13	Sox13		
SoxE	SOX8	Sox8a/b	SoxE	
	SOX9	Sox9a/b		
	SOX10	Sox10		
SoxF	SOX7	Sox7	SoxF	SoxF
	SOX17	Sox17		
	SOX18	Sox18		
		Sox32		
SoxH	SOX30		SoxH	

of the full-length Sox proteins supported this classification and indicated that the amino acid sequences outside the HMG domain are also similar within each group but dissimilar between groups (Bowles et al., 2000; Wegner, 1999). The SoxB group members can be further divided into the SoxB1 and SoxB2 subgroups because amino acid sequences outside the HMG domain are dissimilar between the subgroups (Bowles et al., 2000; Kamachi et al., 2000; Uchikawa et al., 1999). The amino acid sequence similarity of each Sox group outside the HMG domain is kept high among vertebrate species but decreases between vertebrate and invertebrate Sox proteins. Amino acid sequences of SoxB1 proteins of humans, zebrafish, ascidians, and sponges are aligned in Figure 2 for comparison.

The Sox proteins belong to the HMG domain superfamily and constitute a large proportion of this superfamily. In the human genome, 20 *SOX* genes are found among total of 47 HMG domain protein-coding genes (Stros et al., 2007). The HMG domain superfamily proteins are grouped into two main classes: HMGB/UBF and Sox/Tcf/MATA (Soullier et al., 1999). Members of the former class contain one or more HMG domains that bind DNA without sequence specificities. Nonhistone proteins such as HMG box 1/2 (HMGB1/2, previously called HMG-1/2) belong to this class. However, members of the latter class contain a single HMG domain that binds specific DNA sequences.

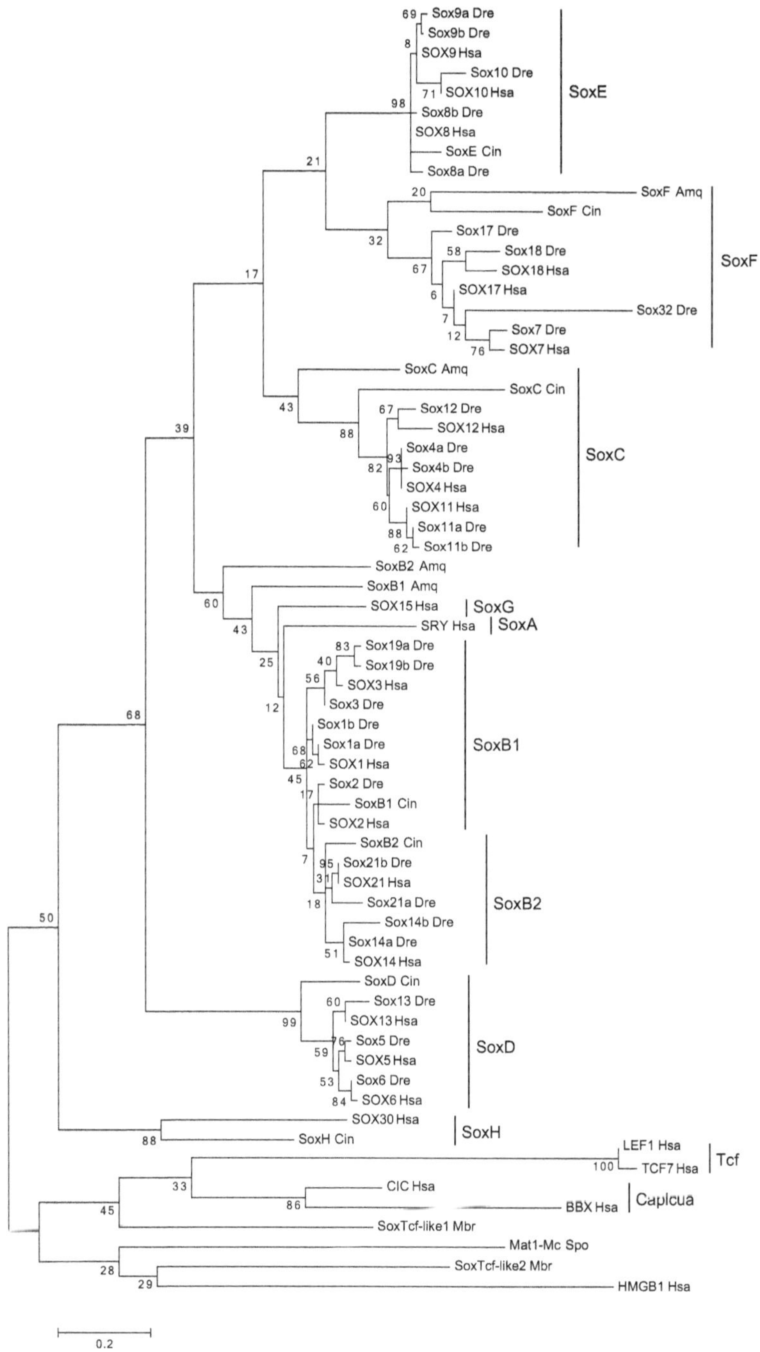

Figure 1 Phylogenetic tree of Sox family proteins. The phylogenetic tree of Sox family proteins and related HMG proteins was constructed using the maximum likelihood method, based on the JTT matrix-based model implemented in MEGA6 (Tamura et al., 2013). Seventy-nine amino acid sequences in the HMG domain were used for this analysis. The tree with the highest log likelihood (−3858.4857) is shown. Values in bootstrap analysis percentages with 500 replicate runs are shown next to the branches. The tree is drawn to scale, with branch lengths indicating the number of substitutions per site. In addition to the Sox proteins of *Homo sapiens* (Has, human), *Danio rerio* (Dre, zebrafish), *Ciona intestinalis* (Cin), and *Amphimedon queenslandica* (Amq), human TCF/LEF, Capicua, and HMGB1 proteins, *Monosiga brevicollis* (Mbr) Sox/Tcf-like proteins, and *Schizosaccharonyces pombe* (Spo) Mat1-Mc are included in this analysis.

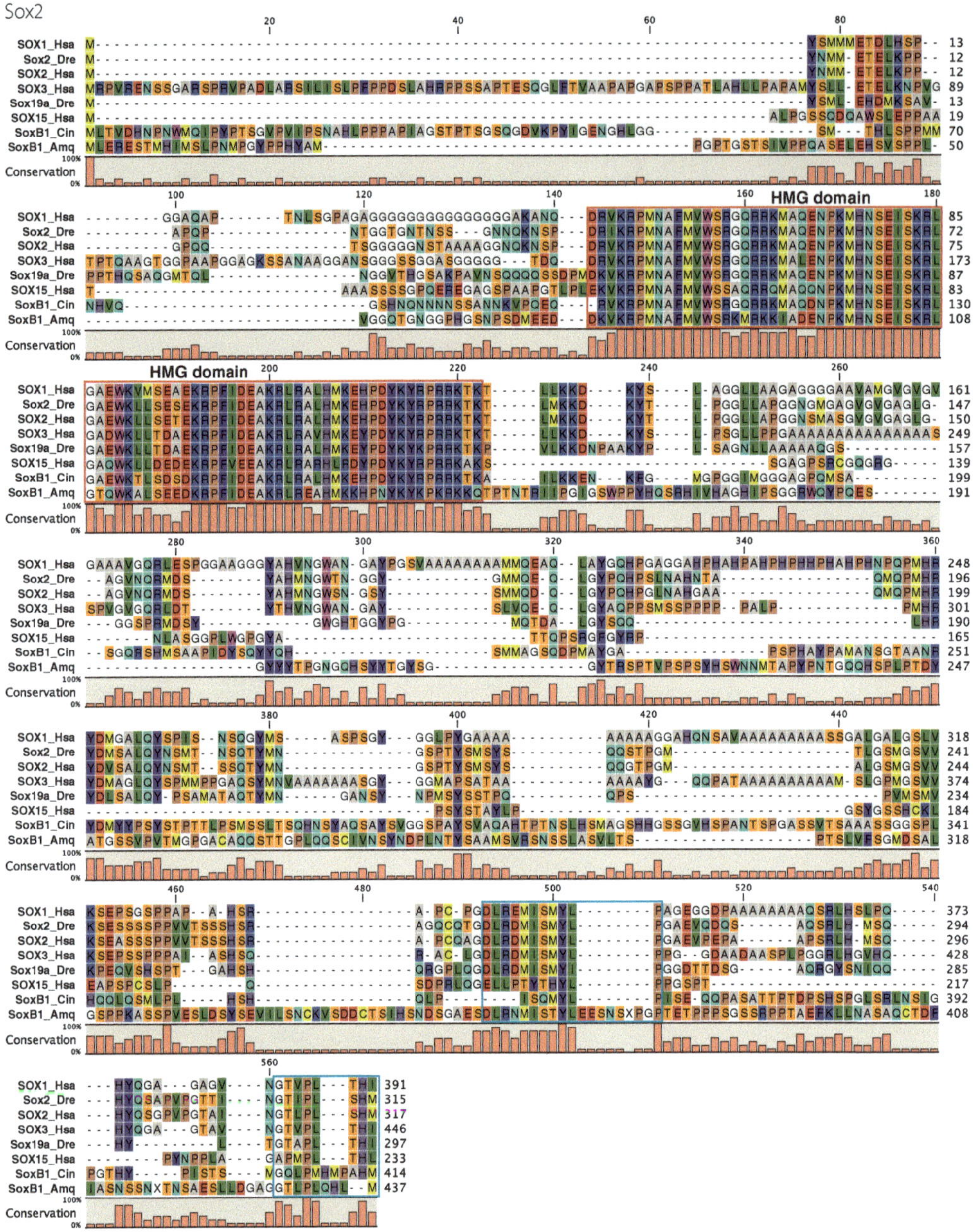

Figure 2 Alignment of SoxB1 proteins. Alignment of human (Has) SOXB1 proteins, SOX1, SOX2, and SOX3, and SOXG protein SOX15; zebrafish (Dre) Sox2 and Sox19a proteins; ascidian *Ciona intestinalis* (Cin) SoxB1 protein; and sponge *Amphimedon queenslandica* (Amq) SoxB1 protein. Red boxes indicate 79–amino acid sequences in the HMG domain. Blue boxes indicate conserved SoxB1-specific C-terminal short sequence motifs.

Sox family, Tcf/Lef family, and MATA_HMG domain proteins (*Schizosaccharonyces pombe* Mat1-Mc and other yeast mating–type gene products) belong to this class. The HMG domains of both classes contain approximately 80 amino acid residues in a similar twisted L shape that contains three α-helices and an N-terminal β strand (Weiss, 2001).

The term "HMG" was derived from a high mobility group of polyacrylamide gel electrophoresis-sorted nonhistone chromosomal proteins (Goodwin et al., 1973) and has subsequently caused some confusion regarding their relationships. These proteins have been found to comprise three families of DNA-binding proteins: HMGB1/2 (HMG domain superfamily), HMG AT-hook 1/2 (HMGA1/2, previously called HMG-I(Y/C), AT-hook family), and HMG nucleosomal binding domain 1/2 (HMGN1/2, previously called HMG-14/17, nucleosome binding protein family) (Hock et al., 2007). These three families are totally unrelated with respect to amino acid sequences.

Although HMG domain proteins were initially identified in vertebrate cells, the HMG superfamily appears to have an ancient evolutionary origin as a nucleic acid binding domain, because HMGB/UBF-class HMG proteins have been found in all eukaryotic organisms examined to date (Sebe-Pedros et al., 2011; Sessa and Bianchi, 2007). These organisms include a broad range of species both from Unikonta (including Opisthokonta and Amoebozoa) and Bikonta (including Archaeplastida [Plantae sensu lato] and Excavata). In contrast, the Sox/Tcf/MATA-class HMG proteins have been found only in Opisthokonta lineages, including Metazoa (animals), Choanoflagellatea, and Fungi. Despite this diversity, these HMG domains for sequence-specific DNA binding are considered to share a common ancestral proto-HMG domain without sequence specificity (Sessa and Bianchi, 2007; Soullier et al., 1999). Among these, true Sox/Tcf family members appear to be specific to metazoans, whereas MATA_HMG family members are only found in Fungi (Larroux et al., 2008; Sebe-Pedros et al., 2011). Together, these findings suggest that the HMGB/UBF-type HMG domain proteins arose early in eukaryotic evolution and that the Sox/Tcf and MATA_HMG families may have arisen independently in the metazoan and fungal lineages, respectively.

EVOLUTION OF THE *SOX* FAMILY GENES

When did the ancestral *Sox* genes arise during evolution? Two *Sox/Tcf-like* genes were identified in the genome of the unicellular choanoflagellate *Monosiga brevicollis* (King et al., 2008). However, the HMG domain sequences of these Sox/Tcf-like proteins exhibit low levels of identity (<40%) with metazoan Sox proteins (Zhong et al., 2011), which is significantly lower than the levels of identity observed among the metazoan Sox proteins (≥46%) (Lefebvre et al., 2007). The Sox/Tcf-like proteins also do not include the signature Sox family sequence (RPMNFMVW) in the N-terminal part of the Sox HMG domain (Bowles et al., 2000) and more strongly resemble Tcf and Capicua families (Figure 1). These data suggest that the choanoflagellate *Sox/Tcf-like* genes may share ancestry with the *Sox/Tcf* family but cannot be categorized within the current *Sox* family. Interestingly,

Sox/Tcf-like genes are not found in the genome of the unicellular holozoan *Capsaspora owczarzaki*, which has been putatively assigned to the sister group of metazoans and choanoflagellates (Sebe-Pedros et al., 2011), which suggests that a *proto-Sox/Tcf* gene may have appeared in the common ancestor of metazoans and choanoflagellates (Figure 3).

In the past decade, the genomes of species in early-branching metazoan phyla, including poriferans (sponges), placozoans, ctenophores (comb jellies), and cnidarians (anemones and jellyfish), have been sequenced. Analyses of these genomes have revealed that the genome organization and contents are similar among metazoans including more complex bilaterians (Degnan et al., 2009; Larroux et al., 2008). Remarkably, a vast majority of transcription factor families, including Sox, T-box, Ets and nuclear receptors, and signaling pathway components, including Wnt, transforming growth factor-β and Hedgehog, exist in poriferans, placozoans, ctenophores, and cnidarians, although these gene families were considered in association with complex bilaterian development. Analogous to the case of Sox, these regulatory proteins lack direct orthologs in choanoflagellates (King et al., 2008). This finding suggests that a set of these regulatory proteins was established after the divergence of the choanoflagellate and metazoan lineages but before the divergence among metazoan lineages. However, further analysis will be required to confirm these evolutionary scenarios because a more recent analysis of the *C. owczarzaki* genome revealed that some transcription factors formerly

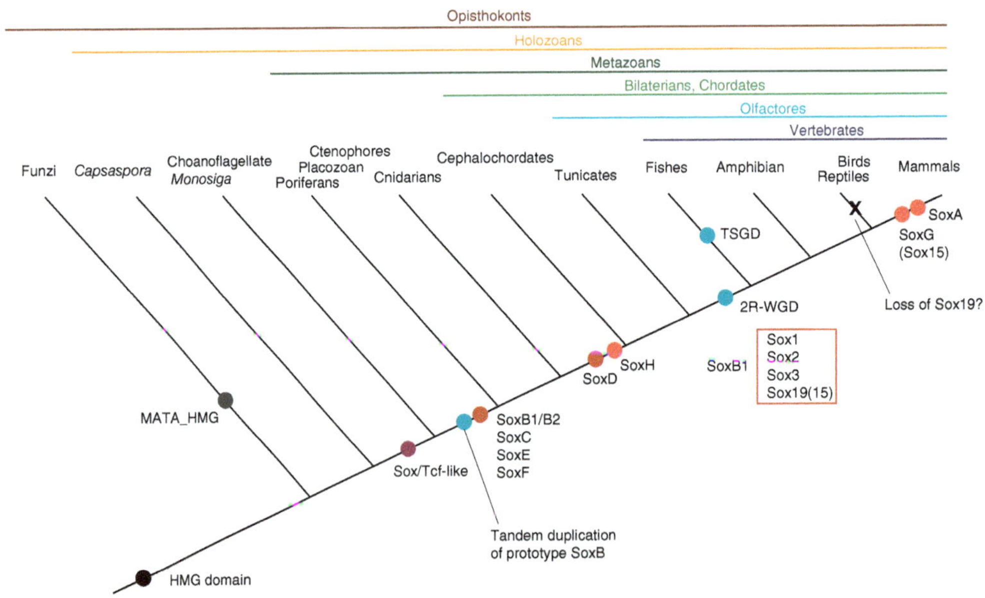

Figure 3 Evolution of HMG domain and Sox family. Hypothetical origins of major Sox groups (red dots), minor Sox groups (pink dots), Sox/Tcf-like (purple dot), MATA_HMG domain (gray dot) and HMG domain (black dot) are plotted on a simplified cladogram showing the phylogenetic relationship of representative lineages of Opisthokonta. Duplication events (blue dots) are also plotted. 2R-WGD, two rounds of whole-genome duplication; TSGD, teleost-specific genome duplication.

thought to be metazoan specific, such as T-box and Runx, have a more ancient origin and were specifically lost in the choanoflagellate lineage (Sebe-Pedros et al., 2011).

The phylogenetic placement of poriferans, placozoans, ctenophores, and cnidarians has not been fully resolved (Marlow and Arendt, 2014). Ctenophore genome analyses have suggested the placement of ctenophores as the most basal animal lineage (Moroz et al., 2014; Ryan et al., 2013), whereas other studies have suggested the placement of either poriferans or placozoans at the earliest-branching animal lineage (Nosenko et al., 2013; Schierwater et al., 2009). In any case, genome analyses of these potentially basal animal species from the poriferan, placozoan, and ctenophore lineages suggest that a basic set of *Sox* family genes was already established in Urmetazoa, the hypothetical first multicellular animals or the last common ancestor of Metazoa (Jager et al., 2006). Among poriferan species, four and seven *Sox* genes were identified in the demosponge *Amphimedon queenslandica* (Larroux et al., 2008) and the calcareous sponge *Sycon ciliatum* (Fortunato et al., 2012), respectively, whereas five *Sox* genes were identified in the placozoan *Trichoplax adhaerens* (Srivastava et al., 2008), and six *Sox* genes in the ctenophore *Mnemiopsis leidyi* (Schnitzler et al., 2014). These *Sox* genes have been classified into three major groups (B, C, and F) in the demosponge *A. queenslandica* (Larroux et al., 2008), four groups (B, C, E, and F) in the calcareous sponge *S. ciliatum* (Fortunato et al., 2012), three groups (B, C, and E) in *Trichoplax* (Schnitzler et al., 2014; Srivastava et al., 2008), and four groups (B, C, E, and F) in the ctenophore *M. leidyi* (Schnitzler et al., 2014). Some *Sox* genes from these species have not been classified into the major Sox groups and are likely lineage specific. The variable numbers of *Sox* genes among these species also suggest the occurrence of lineage-specific gene duplication and/or loss during evolution.

From these analyses, it appears that urmetazoans harbored at least a prototype set of genes composed of *SoxB, SoxC, SoxE,* and *SoxF* (Figure 3). Interestingly, because the genomes of Amphimedon and Trichoplax include both *SoxB1-like* and *SoxB2-like* genes, it appears that the duplication of a hypothetical prototype *SoxB* gene and subsequent divergence into prototype *SoxB1* and *SoxB2* genes occurred before the establishment of Urmetazoa (see also Section 6.5). Genes that can be clearly classified into the *SoxD* group have not been found in poriferans, placozoans, ctenophores, or cnidarians but have been found in most bilaterian metazoans (Larroux et al., 2008). Taken together, this suggests that the full complement of the major six *Sox* groups (B1, B2, C, D, E, and F) was established in Urbilateria, the hypothetical last common ancestor of bilaterians (Figure 3).

EXPANSION OF THE *SOX* GENES AND EMERGENCE OF *SOX2* IN THE VERTEBRATE LINEAGE

The set of *Sox* genes was apparently stably maintained after the establishment of the major *Sox* groups in Urbilateria. The basal chordate *Ciona intestinalis* (tunicate) harbors six *Sox* genes, each of which belongs to one of the major groups (B1, B2, C, D, E, and F), and an additional *SoxH-like* gene (Table 1; Figure 1) (Leveugle et al., 2004; Yamada et al., 2003). *SoxH* genes, including human *SOX30*, are most similar to *SoxD* genes, as evidenced by the presence of a remnant coiled-coil domain that is characteristic of SoxD (Kamachi and Kondoh, 2013) and may have been produced from a *SoxD* duplicate just after the emergence of *SoxD* (Figure 3).

In contrast, mammalian species, including the human and mouse, generally have 20 *Sox* genes, as mentioned above (Schepers et al., 2002). Fish species harbor more *Sox* genes; for instance, zebrafish harbor 27 *Sox* genes (Table 1; Figure 1) (http://zfin.org/). How did the number of *Sox* genes increase in the vertebrate lineage? It is widely accepted that two rounds of whole-genome duplication (2R-WGD) occurred early in the vertebrate lineage after separation from the basal chordates (Figure 3) (Hokamp et al., 2003; Leveugle et al., 2004). Therefore, each single invertebrate gene corresponds to up to four equally related genes (paralogs) that are located on different chromosomes, which is dubbed as the "one-to-four rule" (Spring, 1997). Whole-genome duplication (WGD) during vertebrate evolution was initially proposed by Susumu Ohno, who suggested that gnathostome vertebrates underwent two successive rounds of tetraploidization (genome duplication) (Ohno, 1970). Ohno also emphasized the importance of gene duplication in supplying raw genetic material to biological evolution, in which duplicated genes could potentially acquire novel functions. In honor of Ohno's contribution, paralogs that arise through genome duplication are also called "ohnologs" (Wolfe, 2000), distinguishing those from paralogs generated by tandem duplication. In agreement with the one-to-four rule, each of the mammalian major *Sox* groups (*SoxB1, SoxC, SoxD, SoxE*, and *SoxF*) has three closely related genes and the *SoxB2* group has two similar genes.

The genomes of basal chordates can provide a good reference for understanding vertebrate genome organization. Recent molecular phylogenetic analyses support the basal position of cephalochordates within chordates and a sister relationship between tunicates and vertebrates, which together form the clade of Olfactores (Figure 3) (Delsuc et al., 2006; Putnam et al., 2008). Interestingly, a draft genome of the amphioxus *Branchiostoma floridae*, a representative cephalochordate, revealed that chromosome-scale linkage (macro-synteny) between the amphioxus and vertebrate genomes has been highly conserved, whereas conservation of local gene order (micro-synteny) was limited (Putnam et al., 2008). Through this comparative genomic analysis, Putnam et al. (2008) reconstructed 17 ancient chordate linkage groups that may correspond to the proto-chromosomes of the last common chordate ancestor. Furthermore, when these proto-chromosomes were compared with the mammalian genomes, global fourfold-conserved syntenies were often observed, thus confirming the occurrence of 2R-WGD on the vertebrate stem. When the human *SOX* genes were plotted onto human chromosomes that are regrouped according to the 17 ancestral chordate linkage groups, the genes within each major *Sox* group were placed on the same ancestral chordate linkage group with few exceptions (Figure 4(A)). For instance, the *SoxB1* genes, *SOX1* (on chromosome 13), *SOX2* (on chromosome 3), and *SOX3* (on chromosome X), were all placed on the

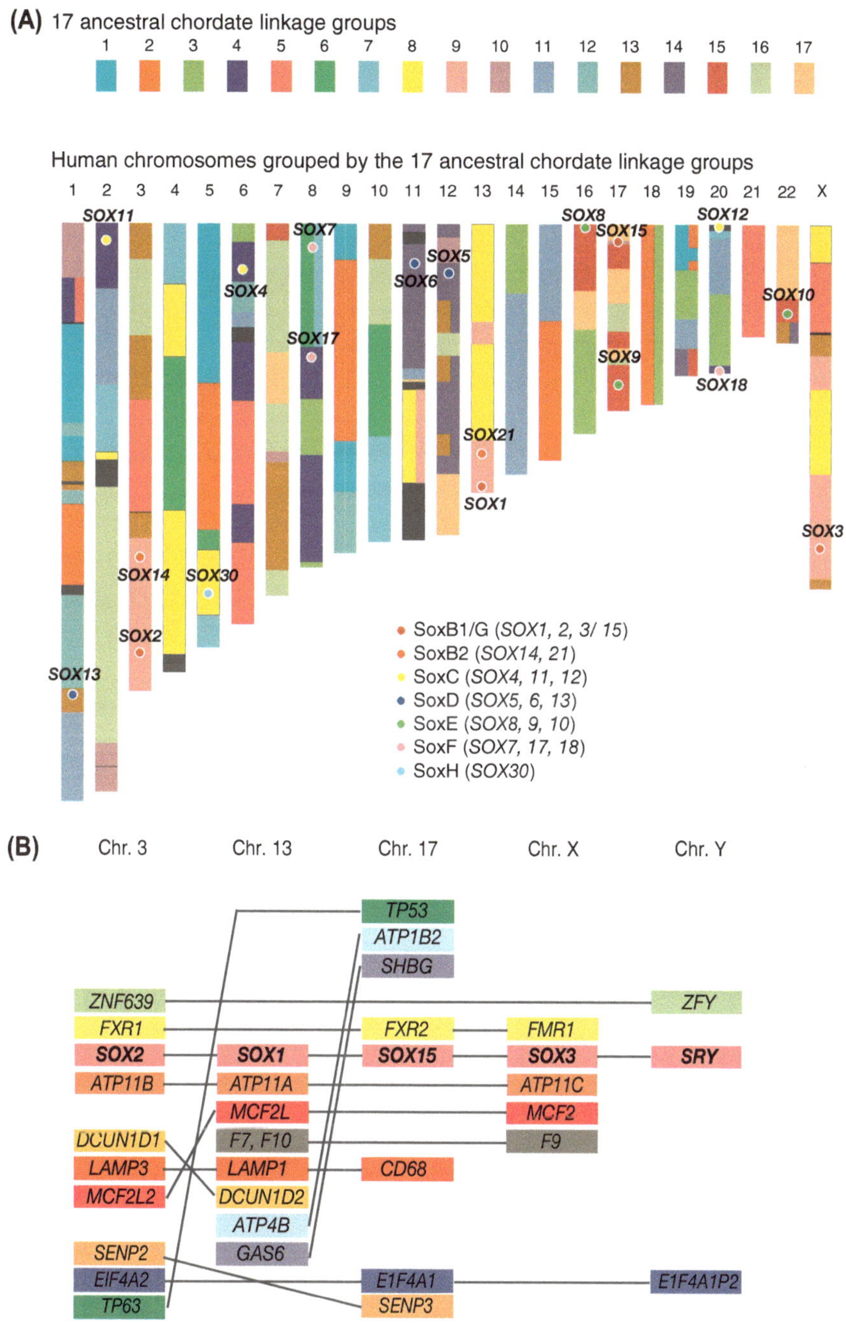

Figure 4 Expansion of *Sox* family genes through two rounds of whole-genome duplication. (A) Plotting of human *SOX* genes onto human chromosomes that have been regrouped according to the 17 ancestral chordate linkage groups. *(Modified from Putnam et al. (2008).)* (B) Conserved local synteny of human chromosomal segments containing the *SOX1*, *SOX2*, *SOX3*, *SOX15* (SoxG), and *SRY* (SoxA) genes.

ninth ancestral chordate linkage group (colored in pink in Figure 4(A)). This finding confirms that the vertebrate *Sox* genes indeed expanded through 2R-WGD, and thus the members belonging to each major *Sox* group are ohnologs.

Furthermore, when human chromosomal segments containing the *SoxB1* genes were compared with each other, significant conservation of local synteny was also observed (Figure 4(B)). Consistent with the model that *SRY* is originated from *SOX3*, limited local synteny was also observed for the *SRY*-containing Y chromosomal segment. Although *SOX15* was initially classified into the *SoxG* group because of its limited similarity with the *SoxB1* group members, the chromosomal segment containing *SOX15* also exhibits conserved local synteny. This ohnologous relationship is further supported by the fact that the zebrafish *sox19* gene, which is classified into the *SoxB1* group, was found to be the ortholog of *SOX15*, as revealed in a comparative analysis of synteny and genomic organization, thus suggesting that *Sox15* is derived from a hypothetical *proto-Sox19* gene (Kamachi et al., 2009; Okuda et al., 2010). Although the amino acid sequences of teleost Sox19 proteins are more divergent from the authentic SoxB1 members (Sox1, Sox2, and Sox3), they exhibit higher similarity with the authentic SoxB1 members than do the mammalian Sox15 proteins (Figure 2) (Okuda et al., 2010). These findings further support the generation of *Sox1, Sox2, Sox3/Sry*, and *Sox19/Sox15* from a single ancestral *SoxB1* gene through 2R-WGD in the vertebrate lineage.

An additional WGD is considered to have occurred in the ray-finned (actinopterygian) fish lineage, known as teleost-specific genome duplication (TSGD) (Figure 3) (Meyer and Schartl, 1999; Meyer and Van de Peer, 2005). This extended the one-to-four rule to a one-to-eight rule and further expanded the numbers of *Sox* genes in fish species (Cresko et al., 2003; Hoegg et al., 2004; Koopman et al., 2004). Recent studies have indicated that the fish-specific genome duplication occurred in the teleost stem lineage after the divergence of early branching ray-finned fish lineages, such as sturgeons and gars. Consistently, two copies of *Sox* genes often exist in teleost fish genomes for each cognate *Sox* in other vertebrate species. Zebrafish *SoxB1* genes include two *sox1* genes, *sox1a* and *sox1b*, and two *Sox15* orthologs, *sox19a* and *sox19b*, and single copies of *sox2* and *sox3* (Okuda et al., 2010).

FUNCTIONAL REDUNDANCY AND DIVERSIFICATION AMONG THE *SOXB1* GENES

Ohno also suggested that WGDs were likely key mechanisms in evolution, because WGD would enable duplicated genes to become functionally diversified (Ohno, 1970). After duplication, the evolutionary fates of the duplicated copies would be: (1) to maintain the original functions redundantly (functional redundancy), (2) to be lost (nonfunctionalization), (3) to partition the original functions between the duplicated copies (subfunctionalization), and/or (4) to acquire novel functions (neofunctionalization)

(Canestro et al., 2013). Although subfunctionalization may be advantageous for the initial maintenance of the duplicated copies, it is compatible with subsequent neofunctionalization. Neofunctionalization and subfunctionalization can occur by changes in protein-coding sequences and/or *cis*-regulatory sequences.

As expected from the amino acid sequence similarities among the vertebrate authentic SoxB1 proteins (Sox1, Sox2, and Sox3), these proteins exhibit similar transcriptional characteristics. For instance, these SoxB1 proteins activate the δ-crystallin DC5 and nestin NES30 enhancers in cooperation with target-specific partner factors Pax6 and Brn1/2, respectively, in cultured cells (Kamachi et al., 1998, 2001; Tanaka et al., 2004). Zebrafish Sox19a and Sox19b, two orthologs of mammalian Sox15 (SoxG), also activate these enhancers in a similar fashion, although Sox15 is less active in these assays in a manner consistent with its highly diverged amino acid sequences (Okuda et al., 2006). Analogously, induced pluripotent stem cells can be generated using either Sox1 or Sox3 instead of Sox2, along with the other three Yamanaka factors, although Sox15 exhibits negligible activity in this assay (Nakagawa et al., 2008).

The similar transcriptional regulatory characteristics of the SoxB1 proteins suggest that these proteins function redundantly in tissues wherein they are coexpressed. Indeed, single *Sox1* or *Sox3* knockout mice display only mild abnormalities in the central nervous system (CNS), presumably reflecting extensive coexpression of *Sox1, Sox2*, and *Sox3* (Nishiguchi et al., 1998; Rizzoti et al., 2004; Weiss et al., 2003). In contrast, *Sox2*-deficient embryos die approximately at the time of implantation, reflecting the exclusive expression of *Sox2* in the inner cell mass of mouse embryos (Avilion et al., 2003). Consistently, a knockdown of either of *sox2, sox3, sox19a*, or *sox19b* in zebrafish embryos caused only mild developmental abnormalities, whereas simultaneous knockdown of all four *SoxB1* genes resulted in severe developmental abnormalities (Dee et al., 2008; Okuda et al., 2010). Furthermore, the *sox2, sox3, sox19a*, and *sox19b* genes were equally able to rescue *SoxB1*-depleted zebrafish embryos, which suggests that the functions of SoxB1 proteins are interchangeable at least during early zebrafish development (Okuda et al., 2010). However, a study indicated that some SoxB1 target genes, including *Dbx1*, may be preferentially activated by Sox3 during neurogenesis, thus suggesting that functional differences among the SoxB1 proteins also exist (Rogers et al., 2014) (see also Chapter 11). These observations altogether suggest that at the protein function level, neofunctionalization and subfunctionalization of *SoxB1* genes appear to be minimal except in the case of *Sox15* gene.

Neofunctionalization and subfunctionalization of the *SoxB1* genes are more clearly manifested in their regulation. *Sox1, Sox2*, and *Sox3* often share expression domains during embryogenesis, although their precise spatiotemporal expression patterns differ. *Sox1, Sox2*, and *Sox3* are all expressed in the CNS from the time of formation of the embryonic neural ectoderm and continue to be expressed throughout adulthood, but the timing of their activation and tissue areas of their expression vary among these three genes (Okuda et al., 2006; Rogers et al., 2013; Uchikawa et al., 1999, 2011). Furthermore, the *SoxB1* genes exhibit function shuffling, in which the expression patterns of a paralog are shared with a nonorthologous paralog. *Sox2* expression precedes *Sox3* expression in the mouse neural ectoderm development, whereas *Sox3* expression precedes *Sox2* expression in chicken neuroectoderm development (Uchikawa et al., 2011). In zebrafish embryos, expression of *sox19a* covers the entire area of the neural ectoderm, whereas *sox2* and *sox3* are expressed in localized domains (Okuda et al., 2006). This differential expression and function shuffling among the *SoxB1* genes may be explained by lineage-specific subfunctionalization of the *SoxB1* genes, in which original *cis*-regulatory sequences may have been differentially maintained or modified during vertebrate evolution.

In the mammalian lineage, the derivation of the *Sox15* gene from the *proto-Sox19* gene appears to have involved neofunctionalization in both protein-coding sequences and *cis*-regulatory sequences, because in mice *Sox15* has a unique function in skeletal muscle regeneration, in which the *SoxB1* genes appear to have no roles (Lee et al., 2004; Meeson et al., 2007). *Sox15* may also have a unique function in placental development (Yamada et al., 2006). *Sox15* genes have been found only in eutherians (placental mammals), and a *Sox19/Sox15*-like gene, originally called *SoxD*, is found in *Xenopus* (Mizuseki et al., 1998) but not in marsupials, birds, or reptiles (Ito, 2010) (www.ensembl.org).

Studies have suggested that 2R-WDG may not have been directly linked to the emergence of vertebrate characteristics such as a complex brain and placodes, but instead contributed to their subsequent diversification (Canestro et al., 2013). The *SoxB* genes in the sponge *S. ciliatum* and ctenophores *M. leidyi* and *Pleurobrachia pileus* are expressed in the ectodermal cells that contribute to sensory cells/organs with photoreception functions (Fortunato et al., 2012; Jager et al., 2008; Schnitzler et al., 2014). Therefore, the function of SoxB proteins in light-sensing system development can be regarded as the ancestral form of vertebrate SoxB1 functions in the nervous system development, including development of the CNS and sensory organs.

DIVERSIFICATION OF *SOXB* INTO *SOXB1* AND *SOXB2* THROUGH TANDEM DUPLICATION

Subdivision of the SoxB group into SoxB1 and SoxB2 was first proposed on the basis of a comparison of the full-length protein sequences of chicken Sox1/Sox2/Sox3 and Sox14/Sox21 (Uchikawa et al., 1999). The amino acid sequences of the HMG domain and group B homology domain, which immediately follows the HMG domain, are well conserved but the remaining portions are dissimilar between SoxB1 and SoxB2 proteins (Uchikawa et al., 1999). Furthermore, the SoxB1 and SoxB2 proteins have been shown to act primarily as a transcriptional activator and repressor, respectively (Sandberg et al., 2005; Uchikawa et al., 1999).

A study indicated that the SoxB1/SoxB2 subdivision occurred before divergence of the earliest-branching animal lineages including sponges (Zhong et al., 2011). The authors of that study demonstrated that the subgroup-specific C-terminal short sequence motifs existed in most metazoan SoxB1 and SoxB2 proteins (Figure 1). The authors further showed that the discrimination of SoxB1/SoxB2 based on the two diagnostic residues—amino acid positions 2/78, which are R/T in SoxB1 and H/P in SoxB2—in the HMG domain was widely applicable to almost all the metazoan SoxB1/SoxB2 proteins. Therefore, it appears that a set of proto-SoxB1 and proto-SoxB2 had already been created before the establishment of Urmetazoa (the last common ancestor of Metazoa) (Figure 3).

SOX2/SOX14 and *SOX1/SOX21* are linked on human chromosomes 3 and 13, respectively (Malas et al., 1999). Similar *SoxB1/SoxB2* linkages have been conserved in a number of vertebrates, including the chimpanzee, chicken, platypus, and cichlid fishes, although these linkages have been disrupted in the remaining half of vertebrate species, including the mouse and zebrafish (Kirby et al., 2002; Kuroiwa et al., 2002; Mazzuchelli et al., 2011; Okuda et al., 2006; Popovic and Stevanovic, 2009). In the dog, only the *SOX1/SOX21* linkage has been preserved. Therefore, the ancestral set of *proto-SoxB1* and *proto-SoxB2* appears to have been arranged in tandem, and this linkage has been independently broken through chromosome rearrangements during evolution (Figure 5). As first suggested by Kirby (2002), the hypothetical ancestral prototype *SoxB* appears to have experienced tandem duplication of a large chromosomal segment. This is further supported by the fact that in humans, similar sets of genes (e.g., *CLDN, ZIC, TFDP, MBNL*) are located around both *SoxB1* (*SOX2/SOX1*) and *SoxB2* (*SOX14/SOX21*) (Okuda et al., 2006).

Did the hypothetical ancestral SoxB resemble either SoxB1 or SoxB2? The predominance of SoxB1 versus SoxB2 functional roles in current organisms suggests that SoxB1 more strongly resembles the ancestral SoxB with respect to molecular functions. In this case, it appears that SoxB2 acquired a divergent molecular function as a transcriptional repressor through neofunctionalization after tandem duplication.

EVOLUTIONARY PATHWAY OF *SOXB* GENES

A possible evolutionary pathway of *SoxB* genes leading to the mammalian lineage is summarized in Figure 5. A pair of *proto-SoxB1* and *proto-SoxB2* genes appears to have arisen by tandem duplication of a genomic segment that contained the hypothetical ancestral *SoxB*. Subsequently, 2R-WGD occurred on the vertebrate stem, during which two *SoxB2* genes appear to have been lost, resulting in establishment of the prototypic vertebrate *SoxB1* and *SoxB2* gene composition: four *SoxB1* ohnologs (*Sox1, Sox2, Sox3,* and *Sox19*) and two *SoxB2* ohnologs (*Sox14* and *Sox21*). The *Sox19*-type genes are characterized by the possession of an intron at a conserved position in the coding sequence, in contrast to the other intron-less *SoxB1/B2* genes (Okuda et al., 2006). In placental mammals, *Sox19*-type genes appear as *Sox15*, the derivation of which represents an example of neofunctionalization of an ohnolog. In birds and reptiles, *Sox19*-type

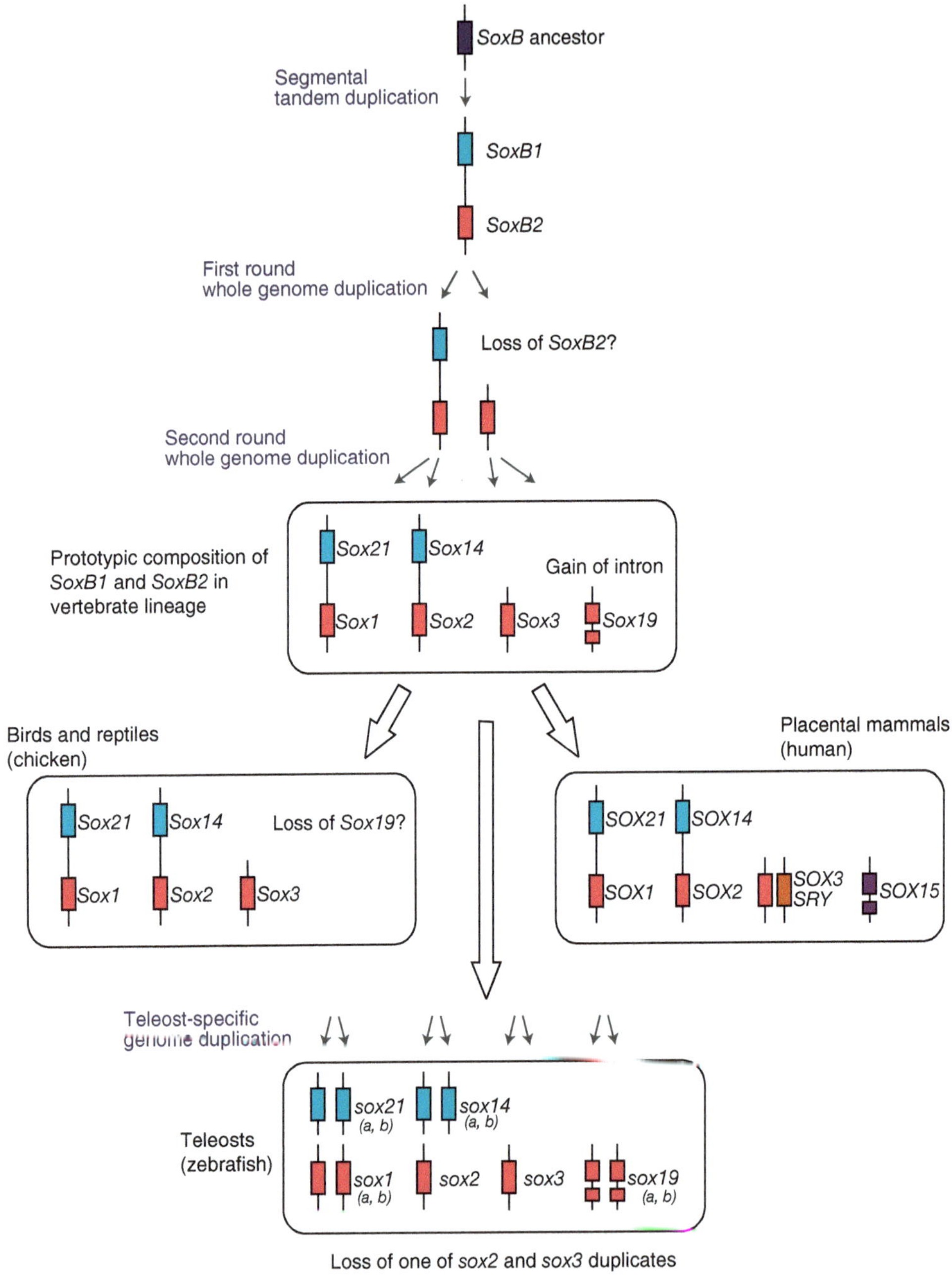

Figure 5 Model of evolutionary pathway of *SoxB* genes.

genes appear to have been lost. In teleost fish, an additional round of gene duplication produced pairs of *sox1, sox19, sox14,* and *sox21*, whereas single copies of *sox2* and *sox3* appear to have been lost. Thus, the function of *Sox2* should be understood in the context of functional diversity of the *SoxB* genes that developed during the evolutionary process.

REFERENCES

Avilion, A.A., Nicolis, S.K., Pevny, L.H., Perez, L., Vivian, N., Lovell-Badge, R., 2003. Multipotent cell lineages in early mouse development depend on SOX2 function. Genes Dev. 17, 126–140.

Bowles, J., Schepers, G., Koopman, P., 2000. Phylogeny of the SOX family of developmental transcription factors based on sequence and structural indicators. Dev. Biol. 227, 239–255.

Canestro, C., Albalat, R., Irimia, M., Garcia-Fernandez, J., 2013. Impact of gene gains, losses and duplication modes on the origin and diversification of vertebrates. Semin. Cell Dev. Biol. 24, 83–94.

Cresko, W.A., Yan, Y.L., Baltrus, D.A., Amores, A., Singer, A., Rodriguez-Mari, A., Postlethwait, J.H., 2003. Genome duplication, subfunction partitioning, and lineage divergence: sox9 in stickleback and zebrafish. Dev. Dyn. 228, 480–489.

Dee, C.T., Hirst, C.S., Shih, Y.H., Tripathi, V.B., Patient, R.K., Scotting, P.J., 2008. Sox3 regulates both neural fate and differentiation in the zebrafish ectoderm. Dev. Biol. 320, 289–301.

Degnan, B.M., Vervoort, M., Larroux, C., Richards, G.S., 2009. Early evolution of metazoan transcription factors. Curr. Opin. Genet. Dev. 19, 591–599.

Delsuc, F., Brinkmann, H., Chourrout, D., Philippe, H., 2006. Tunicates and not cephalochordates are the closest living relatives of vertebrates. Nature 439, 965–968.

Fortunato, S., Adamski, M., Bergum, B., Guder, C., Jordal, S., Leininger, S., Zwafink, C., Rapp, H.T., Adamska, M., 2012. Genome-wide analysis of the sox family in the calcareous sponge *Sycon ciliatum*: multiple genes with unique expression patterns. EvoDevo 3, 14.

Foster, J.W., Dominguez-Steglich, M.A., Guioli, S., Kwok, C., Weller, P.A., Stevanović, M., Weissenbach, J., Mansour, S., Young, I.D., Goodfellow, P.N., 1994. Campomelic dysplasia and autosomal sex reversal caused by mutations in an SRY-related gene. Nature 372, 525–530.

Goodwin, G.H., Sanders, C., Johns, E.W., 1973. A new group of chromatin-associated proteins with a high content of acidic and basic amino acids. Eur. J. Biochem. 38, 14–19.

Gubbay, J., Collignon, J., Koopman, P., Capel, B., Economou, A., Munsterberg, A., Vivian, N., Goodfellow, P., Lovell-Badge, R., 1990. A gene mapping to the sex-determining region of the mouse Y chromosome is a member of a novel family of embryonically expressed genes. Nature 346, 245–250.

Hock, R., Furusawa, T., Ueda, T., Bustin, M., 2007. HMG chromosomal proteins in development and disease. Trends Cell Biol. 17, 72–79.

Hoegg, S., Brinkmann, H., Taylor, J.S., Meyer, A., 2004. Phylogenetic timing of the fish-specific genome duplication correlates with the diversification of teleost fish. J. Mol. Evol. 59, 190–203.

Hokamp, K., McLysaght, A., Wolfe, K.H., 2003. The 2R hypothesis and the human genome sequence. J. Struct. Funct. Genomics 3, 95–110.

Ito, M., 2010. Function and molecular evolution of mammalian Sox15, a singleton in the SoxG group of transcription factors. Int. J. Biochem. Cell Biol. 42, 449–452.

Jager, M., Queinnec, E., Chiori, R., Le Guyader, H., Manuel, M., 2008. Insights into the early evolution of SOX genes from expression analyses in a ctenophore. J. Exp. Zool. B Mol. Dev. Evol. 310, 650–667.

Jager, M., Queinnec, E., Houliston, E., Manuel, M., 2006. Expansion of the SOX gene family predated the emergence of the Bilateria. Mol. Phylogenet. Evol. 39, 468–477.

Kamachi, Y., Iwafuchi, M., Okuda, Y., Takemoto, T., Uchikawa, M., Kondoh, H., 2009. Evolution of non-coding regulatory sequences involved in the developmental process: reflection of differential employment of paralogous genes as highlighted by Sox2 and group B1 Sox genes. Proc. Jpn. Acad. Ser. B Phys. Biol. Sci. 85, 55–68.

Kamachi, Y., Kondoh, H., 2013. Sox proteins: regulators of cell fate specification and differentiation. Development 140, 4129–4144.

Kamachi, Y., Uchikawa, M., Collignon, J., Lovell-Badge, R., Kondoh, H., 1998. Involvement of Sox1, 2 and 3 in the early and subsequent molecular events of lens induction. Development 125, 2521–2532.

Kamachi, Y., Uchikawa, M., Kondoh, H., 2000. Pairing SOX off with partners in the regulation of embryonic development. Trends Genet. 16, 182–187.

Kamachi, Y., Uchikawa, M., Tanouchi, A., Sekido, R., Kondoh, H., 2001. Pax6 and SOX2 form a co-DNA-binding partner complex that regulates initiation of lens development. Genes Dev. 15, 1272–1286.

King, N., Westbrook, M.J., Young, S.L., Kuo, A., Abedin, M., Chapman, J., Fairclough, S., Hellsten, U., Isogai, Y., Letunic, I., Marr, M., Pincus, D., Putnam, N., Rokas, A., Wright, K.J., Zuzow, R., Dirks, W., Good, M., Goodstein, D., Lemons, D., Li, W., Lyons, J.B., Morris, A., Nichols, S., Richter, D.J., Salamov, A., Sequencing, J.G., Bork, P., Lim, W.A., Manning, G., Miller, W.T., McGinnis, W., Shapiro, H., Tjian, R., Grigoriev, I.V., Rokhsar, D., 2008. The genome of the choanoflagellate *Monosiga brevicollis* and the origin of metazoans. Nature 451, 783–788.

Kirby, P.J., Waters, P.D., Delbridge, M., Svartman, M., Stewart, A.N., Nagai, K., Graves, J.A., 2002. Cloning and mapping of platypus SOX2 and SOX14: insights into SOX group B evolution. Cytogenet. Genome Res. 98, 96–100.

Koopman, P., Gubbay, J., Vivian, N., Goodfellow, P., Lovell-Badge, R., 1991. Male development of chromosomally female mice transgenic for Sry. Nature 351, 117–121.

Koopman, P., Schepers, G., Brenner, S., Venkatesh, B., 2004. Origin and diversity of the SOX transcription factor gene family: genome-wide analysis in *Fugu rubripes*. Gene 328, 177–186.

Kuroiwa, A., Uchikawa, M., Kamachi, Y., Kondoh, H., Nishida-Umehara, C., Masabanda, J., Griffin, D.K., Matsuda, Y., 2002. Chromosome assignment of eight SOX family genes in chicken. Cytogenet. Genome Res. 98, 189–193.

Larroux, C., Luke, G.N., Koopman, P., Rokhsar, D.S., Shimeld, S.M., Degnan, B.M., 2008. Genesis and expansion of metazoan transcription factor gene classes. Mol. Biol. Evol. 25, 980–996.

Lee, H.J., Goring, W., Ochs, M., Muhlfeld, C., Steding, G., Paprotta, I., Engel, W., Adham, I.M., 2004. Sox15 is required for skeletal muscle regeneration. Mol. Cell. Biol. 24, 8428–8436.

Lefebvre, V., Dumitriu, B., Penzo-Méndez, A., Han, Y., Pallavi, B., 2007. Control of cell fate and differentiation by Sry-related high-mobility-group box (Sox) transcription factors. Int. J. Biochem. Cell. Biol. 39, 2195–2214.

Leveugle, M., Prat, K., Popovici, C., Birnbaum, D., Coulier, F., 2004. Phylogenetic analysis of *Ciona intestinalis* gene superfamilies supports the hypothesis of successive gene expansions. J. Mol. Evol. 58, 168–181.

Malas, S., Duthie, S., Deloukas, P., Episkopou, V., 1999. The isolation and high-resolution chromosomal mapping of human SOX14 and SOX21; two members of the SOX gene family related to SOX1, SOX2, and SOX3. Mamm. Genome 10, 934–937.

Marlow, H., Arendt, D., 2014. Evolution: ctenophore genomes and the origin of neurons. Curr. Biol. 24, R757–R761.

Mazzuchelli, J., Yang, F., Kocher, T.D., Martins, C., 2011. Comparative cytogenetic mapping of Sox2 and Sox14 in cichlid fishes and inferences on the genomic organization of both genes in vertebrates. Chromosome Res. 19, 657–667.

Meeson, A.P., Shi, X., Alexander, M.S., Williams, R.S., Allen, R.E., Jiang, N., Adham, I.M., Goetsch, S.C., Hammer, R.E., Garry, D.J., 2007. Sox15 and Fhl3 transcriptionally coactivate Foxk1 and regulate myogenic progenitor cells. EMBO J. 26, 1902–1912.

Meyer, A., Schartl, M., 1999. Gene and genome duplications in vertebrates: the one-to-four (-to-eight in fish) rule and the evolution of novel gene functions. Curr. Opin. Cell Biol. 11, 699–704.

Meyer, A., Van de Peer, Y., 2005. From 2R to 3R: evidence for a fish-specific genome duplication (FSGD). Bioessays 27, 937–945.

Mizuseki, K., Kishi, M., Shiota, K., Nakatani, S., Sasai, Y., 1998. SoxD: an essential mediator of induction of anterior neural tissues in Xenopus embryos. Neuron 21, 77–85.

Moroz, L.L., Kocot, K.M., Citarella, M.R., Dosung, S., Norekian, T.P., Povolotskaya, I.S., Grigorenko, A.P., Dailey, C., Berezikov, E., Buckley, K.M., Ptitsyn, A., Reshetov, D., Mukherjee, K., Moroz, T.P., Bobkova, Y., Yu, F., Kapitonov, V.V., Jurka, J., Bobkov, Y.V., Swore, J.J., Girardo, D.O., Fodor, A., Gusev, F., Sanford, R., Bruders, R., Kittler, E., Mills, C.E., Rast, J.P., Derelle, R., Solovyev, V.V., Kondrashov, F.A., Swalla, B.J., Sweedler, J.V., Rogaev, E.I., Halanych, K.M., Kohn, A.B., 2014. The ctenophore genome and the evolutionary origins of neural systems. Nature 510, 109–114.

Nakagawa, M., Koyanagi, M., Tanabe, K., Takahashi, K., Ichisaka, T., Aoi, T., Okita, K., Mochiduki, Y., Takizawa, N., Yamanaka, S., 2008. Generation of induced pluripotent stem cells without Myc from mouse and human fibroblasts. Nat. Biotechnol. 26, 101–106.

Nishiguchi, S., Wood, H., Kondoh, H., Lovell-Badge, R., Episkopou, V., 1998. Sox1 directly regulates the gamma-crystallin genes and is essential for lens development in mice. Genes Dev. 12, 776–781.

Nosenko, T., Schreiber, F., Adamska, M., Adamski, M., Eitel, M., Hammel, J., Maldonado, M., Muller, W.E., Nickel, M., Schierwater, B., Vacelet, J., Wiens, M., Worheide, G., 2013. Deep metazoan phylogeny: when different genes tell different stories. Mol. Phylogenet. Evol. 67, 223–233.

Ohno, S., 1970. Evolution by Gene Duplication. Springer-Verlag.

Okuda, Y., Ogura, E., Kondoh, H., Kamachi, Y., 2010. B1 SOX coordinate cell specification with patterning and morphogenesis in the early zebrafish embryo. PLoS Genet. 6, e1000936.

Okuda, Y., Yoda, H., Uchikawa, M., Furutani-Seiki, M., Takeda, H., Kondoh, H., Kamachi, Y., 2006. Comparative genomic and expression analysis of group B1 sox genes in zebrafish indicates their diversification during vertebrate evolution. Dev. Dyn. 235, 811–825.

Pevny, L.H., Lovell-Badge, R., 1997. Sox genes find their feet. Curr. Opin. Genet. Dev. 7, 338–344.

Popovic, J., Stevanovic, M., 2009. Remarkable evolutionary conservation of SOX14 orthologues. J. Genet. 88, 15–24.

Putnam, N.H., Butts, T., Ferrier, D.E., Furlong, R.F., Hellsten, U., Kawashima, T., Robinson-Rechavi, M., Shoguchi, E., Terry, A., Yu, J.K., Benito-Gutierrez, E.L., Dubchak, I., Garcia-Fernandez, J., Gibson-Brown, J.J., Grigoriev, I.V., Horton, A.C., de Jong, P.J., Jurka, J., Kapitonov, V.V., Kohara, Y., Kuroki, Y., Lindquist, E., Lucas, S., Osoegawa, K., Pennacchio, L.A., Salamov, A.A., Satou, Y., Sauka-Spengler, T., Schmutz, J., Shin, I.T., Toyoda, A., Bronner-Fraser, M., Fujiyama, A., Holland, L.Z., Holland, P.W., Satoh, N., Rokhsar, D.S., 2008. The amphioxus genome and the evolution of the chordate karyotype. Nature 453, 1064–1071.

Rizzoti, K., Brunelli, S., Carmignac, D., Thomas, P.Q., Robinson, I.C., Lovell-Badge, R., 2004. SOX3 is required during the formation of the hypothalamo-pituitary axis. Nat. Genet. 36, 247–255.

Rogers, N., Cheah, P.S., Szarek, E., Banerjee, K., Schwartz, J., Thomas, P., 2013. Expression of the murine transcription factor SOX3 during embryonic and adult neurogenesis. Gene Expression Patterns 13, 240–248.

Rogers, N., McAninch, D., Thomas, P., 2014. Dbx1 is a direct target of SOX3 in the spinal cord. PLoS One 9, e95356.

Ryan, J.F., Pang, K., Schnitzler, C.E., Nguyen, A.D., Moreland, R.T., Simmons, D.K., Koch, B.J., Francis, W.R., Havlak, P., Program, N.C.S., Smith, S.A., Putnam, N.H., Haddock, S.H., Dunn, C.W., Wolfsberg, T.G., Mullikin, J.C., Martindale, M.Q., Baxevanis, A.D., 2013. The genome of the ctenophore *Mnemiopsis leidyi* and its implications for cell type evolution. Science 342, 1242592.

Sandberg, M., Kallstrom, M., Muhr, J., 2005. Sox21 promotes the progression of vertebrate neurogenesis. Nat. Neurosci. 8, 995–1001.

Schepers, G.E., Teasdale, R.D., Koopman, P., 2002. Twenty pairs of sox: extent, homology, and nomenclature of the mouse and human sox transcription factor gene families. Dev. Cell 3, 167–170.

Schierwater, B., Eitel, M., Jakob, W., Osigus, H.J., Hadrys, H., Dellaporta, S.L., Kolokotronis, S.O., Desalle, R., 2009. Concatenated analysis sheds light on early metazoan evolution and fuels a modern "urmetazoon" hypothesis. PLoS Biol. 7, e20.

Schnitzler, C.E., Simmons, D.K., Pang, K., Martindale, M.Q., Baxevanis, A.D., 2014. Expression of multiple Sox genes through embryonic development in the ctenophore *Mnemiopsis leidyi* is spatially restricted to zones of cell proliferation. EvoDevo 5, 15.

Sebe-Pedros, A., de Mendoza, A., Lang, B.F., Degnan, B.M., Ruiz-Trillo, I., 2011. Unexpected repertoire of metazoan transcription factors in the unicellular holozoan *Capsaspora owczarzaki*. Mol. Biol. Evol. 28, 1241–1254.

Sessa, L., Bianchi, M.E., 2007. The evolution of High Mobility Group Box (HMGB) chromatin proteins in multicellular animals. Gene 387, 133–140.

Sinclair, A.H., Berta, P., Palmer, M.S., Hawkins, J.R., Griffiths, B.L., Smith, M.J., Foster, J.W., Frischauf, A.M., Lovell-Badge, R., Goodfellow, P.N., 1990. A gene from the human sex-determining region encodes a protein with homology to a conserved DNA-binding motif. Nature 346, 240–244.

Soullier, S., Jay, P., Poulat, F., Vanacker, J.M., Berta, P., Laudet, V., 1999. Diversification pattern of the HMG and SOX family members during evolution. J. Mol. Evol. 48, 517–527.

Spring, J., 1997. Vertebrate evolution by interspecific hybridisation–are we polyploid? FEBS Lett. 400, 2–8.

Srivastava, M., Begovic, E., Chapman, J., Putnam, N.H., Hellsten, U., Kawashima, T., Kuo, A., Mitros, T., Salamov, A., Carpenter, M.L., Signorovitch, A.Y., Moreno, M.A., Kamm, K., Grimwood, J., Schmutz, J., Shapiro, H., Grigoriev, I.V., Buss, L.W., Schierwater, B., Dellaporta, S.L., Rokhsar, D.S., 2008. The Trichoplax genome and the nature of placozoans. Nature 454, 955–960.

Stros, M., Launholt, D., Grasser, K.D., 2007. The HMG-box: a versatile protein domain occurring in a wide variety of DNA-binding proteins. Cell. Mol. Life Sci. 64, 2590–2606.

Tamura, K., Stecher, G., Peterson, D., Filipski, A., Kumar, S., 2013. MEGA6: molecular evolutionary genetics analysis version 6.0. Mol. Biol. Evol. 30, 2725–2729.

Tanaka, S., Kamachi, Y., Tanouchi, A., Hamada, H., Jing, N., Kondoh, H., 2004. Interplay of SOX and POU factors in regulation of the Nestin gene in neural primordial cells. Mol. Cell. Biol. 24, 8834–8846.

Uchikawa, M., Kamachi, Y., Kondoh, H., 1999. Two distinct subgroups of Group B Sox genes for transcriptional activators and repressors: their expression during embryonic organogenesis of the chicken. Mech. Dev. 84, 103–120.

Uchikawa, M., Yoshida, M., Iwafuchi-Doi, M., Matsuda, K., Ishida, Y., Takemoto, T., Kondoh, H., 2011. B1 and B2 Sox gene expression during neural plate development in chicken and mouse embryos: universal versus species-dependent features. Dev. Growth Differ. 53, 761–771.

Wegner, M., 1999. From head to toes: the multiple facets of Sox proteins. Nucleic Acids Res. 27, 1409–1420.

Weiss, J., Meeks, J.J., Hurley, L., Raverot, G., Frassetto, A., Jameson, J.L., 2003. Sox3 is required for gonadal function, but not sex determination, in males and females. Mol. Cell. Biol. 23, 8084–8091.

Weiss, M.A., 2001. Floppy SOX: mutual induced fit in hmg (high-mobility group) box-DNA recognition. Mol. Endocrinol. 15, 353–362.

Wolfe, K., 2000. Robustness–it's not where you think it is. Nat. Genet. 25, 3–4.

Wright, E.M., Snopek, B., Koopman, P., 1993. Seven new members of the Sox gene family expressed during mouse development. Nucleic Acids Res. 21, 744.

Yamada, K., Kanda, H., Tanaka, S., Takamatsu, N., Shiba, T., Ito, M., 2006. Sox15 enhances trophoblast giant cell differentiation induced by Hand1 in mouse placenta. Differentiation 74, 212–221.

Yamada, L., Kobayashi, K., Degnan, B., Satoh, N., Satou, Y., 2003. A genomewide survey of developmentally relevant genes in *Ciona intestinalis*. IV. Genes for HMG transcriptional regulators, bZip and GATA/Gli/Zic/Snail. Dev. Genes Evol. 213, 245–253.

Zhong, L., Wang, D., Gan, X., Yang, T., He, S., 2011. Parallel expansions of Sox transcription factor group B predating the diversifications of the arthropods and jawed vertebrates. PLoS One 6, e16570.

CHAPTER 7

Regulation of *Sox2* via Many Enhancers of Distinct Specificities

Masanori Uchikawa[1], Hisato Kondoh[2]
[1]Graduate School of Frontier Biosciences, Osaka University, Suita, Osaka, Japan; [2]Faculty of Life Sciences, Kyoto Sangyo University, Kyoto, Japan

INTRODUCTION

The transcription factor SOX2 has pivotal roles during embryonic processes (Kamachi et al., 1998; Avilion et al., 2003; Kiernan et al., 2005; Taranova et al., 2006; Kamachi and Kondoh, 2013) and the regulation of stem cells including those in adult tissues (Ellis et al., 2004; Takahashi and Yamanaka, 2006; Suh et al., 2007; Pevny and Nicolis, 2010; Arnold et al., 2011). In addition, investigations have highlighted the involvement of SOX2 in cancer regulation, as mentioned in Chapter 1. Therefore, clarification of *Sox2* gene regulation is important to evaluate the role of SOX2 in individual regulatory processes.

We have investigated *Sox2* gene-associated enhancers employing a functional assay using chicken embryo electroporation (Uchikawa et al., 2003). In chicken embryos, *Sox2* expression is first activated in the epiblast area surrounding Hensen's node around the time when the definitive primitive streak is formed and then continues in the neural plate and then in the sensory placodes. We thought it unlikely that *Sox2* expression in the wide area of neural and sensory primordial tissues is regulated by a simple set of enhancers. Therefore, we adopted the following approach. We isolated and determined the 50-kb chicken genomic sequence surrounding the *Sox2* gene, produced various deoxyribonucleic acid (DNA) fragments, and placed each of them in a *tkEGFP* vector for chicken embryo electroporation. The *tkEGFP* vector alone does not express a detectable level of enhanced green fluorescent protein (EGFP) in electroporated chicken embryos; however, it expressed EGFP in specific tissues when an enhancer sequence was inserted, reflecting the spatiotemporal specificity of the enhancer (Uchikawa et al., 2003; Kondoh and Uchikawa, 2008) (Figure 1).

This initial survey successfully identified 11 neural and placodal enhancers with distinct spatial and temporal specificities. Neural enhancers N1 to N5 were active in different yet overlapping domains of the embryonic central nervous system (CNS) and their ensemble covered the entire *Sox2* expression domain in early chicken embryos. These enhancers were conserved in the nucleotide sequences, genomic arrangements, and regional activity specificity in mammalian genomes (Uchikawa et al., 2003; Kamachi et al., 2009). This cross-species comparison also highlighted the advantage of using the compactly organized chicken genome for an enhancer survey.

Sox2
http://dx.doi.org/10.1016/B978-0-12-800352-7.00007-4

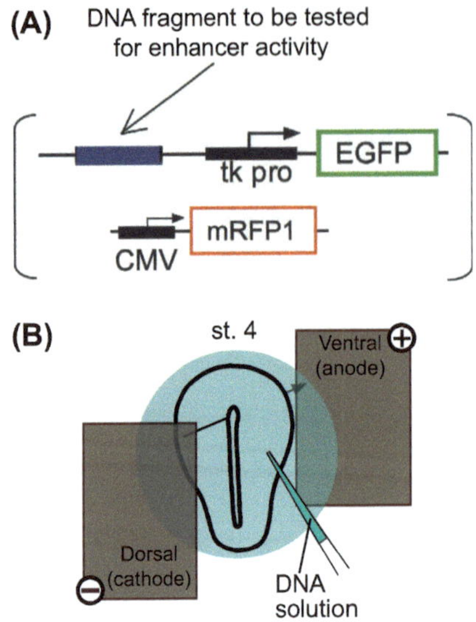

Figure 1 Strategy to identify enhancers using chicken embryo electroporation. (A) Deoxyribonucleic acid fragments tested for enhancer activity were inserted into the tkEGFP vector and mixed with CMV-mRFP1 (electroporation control). (B) Stage 4 chicken embryos were electroporated with tkEGFP vector carrying a genomic fragment from the epiblast side and cultured using a modified New's method. *Adapted from Uchikawa (2008).*

Deleting individual enhancers in the mouse genome reduced *Sox2* expression in the embryonic domains corresponding to enhancer activity (Iwafuchi-Doi et al., 2011; Takemoto et al., 2011; Yoshida et al., 2014, and our unpublished results). However, the initial reduction of *Sox2* expression in an enhancer-associated domain was subsequently rescued, suggesting the existence of additional enhancers that overlap in specificity with one of the *Sox2* enhancers. This notion prompted us to extend the enhancer survey from the original 50 kb to a wider 205-kb region in the chicken genome. The location of the *Sox2* gene in a gene-sparse region facilitated the analysis. This second analysis identified 16 additional enhancers with various new specificities. Presumably, a larger number of enhancers exist that were not identified using chicken embryo electroporation. The sequence blocks conserved in chicken and human genomes are possible hypothetical enhancers.

The analysis also revealed the correlation among the enhancers, the conserved sequence blocks, and the exons of the noncoding *Sox2ot* ribonucleic acid (RNA) genes, which have been suggested to modulate *Sox2* functions (Askarian-Amiri et al., 2014).

We also compared the compositions of the enhancers that regulate *Sox2* and *Sox3* genes. As discussed in Chapter 6, *SoxB1* genes arose because of gene duplication events, and gene expression patterns are most similar between *Sox2* and *Sox3* (Okuda et al., 2006; Uchikawa et al., 2011; Nishimura et al., 2012). Despite the overall similarity in the *Sox2* and *Sox3* expression patterns, their enhancers do not show much similarity in sequence or spatiotemporal specificity (Nishimura et al., 2012). One exception is similarity in the regulation of *Sox2* N1 and *Sox3* NP1 enhancers, which are both involved in regulating axial stem cells, bipotential precursors for neural and mesodermal development (Nishimura et al., 2012).

CHARACTERIZATION OF NEURAL ENHANCERS DISTRIBUTED IN THE 50-KB *SOX2*-PROXIMAL REGION OF THE CHICKEN GENOME

Various overlapping fragments derived from the 50-kb sequence were tested for enhancer activity by electroporating stage (st.) 4 (approximately 18h incubation) chicken embryo from the epiblast side and placing them in New's culture (Figure 1). Once a fragment showed enhancer activity, its subfragments were tested again. In this way, we identified five regions showing neural enhancer activity, which we named N1 to N5 (Figure 2).

Interestingly, these enhancer activity-bearing sequences corresponded to a fraction of the sequence blocks that are conspicuously conserved between the chicken and human genomes, as characterized by the criterion of more than 60% identity in a stretch longer than 100base pairs (bp) (Uchikawa et al., 2003). However, the arrays of sequences conserved between human and mouse were much longer than enhancers, and the enhancers were embedded in such arrays. By contrast, only a portion of the enhancer sequences is conserved between chicken and lower vertebrates (Uchikawa et al., 2004), reflecting species-dependent variations in the details of *Sox2* expression pattern (Okuda et al., 2006; Uchikawa et al., 2011). Thus, the chicken and human genome sequences have a phylogenetic distance only appropriate for indicating candidate enhancer sequences at the *Sox2* locus as conserved sequence blocks.

The conserved sequence blocks are arranged co-linearly among chicken, mouse, and human genomes and larger (about 60-kb) genomic spans occurred in the latter two species. A fraction of the conserved sequence blocks in the *Sox2* locus is missing from the mouse genome.

When the mouse sequences that corresponded to the chicken *Sox2* enhancers were tested in transgenic mice using LacZ as reporters, they also exhibited enhancer activities with nearly identical specificities in the CNS domains (Figure 2).

However, *Sox2* expression patterns are subtly different between mouse and chicken embryos at the early developmental stages. In chicken embryos, *Sox2* expression starts only at the epiblast stage (ES) (st. approximately 4), whereas in mouse embryos, zygotic *Sox2* expression is initiated before the implantation stage, presumably at the inner cell mass stage from which ES cells are derived. In addition, the maternal SOX2 protein is carried to mouse fertilized eggs. The N2 and SRR2 enhancers, as determined by chicken embryo electroporation (Uchikawa et al., 2003), and activity in ES cells (Tomioka et al., 2002), respectively, are both active in the ES cells, whereas only the N2 enhancer is active in the epiblast (Iwafuchi-Doi et al., 2011).

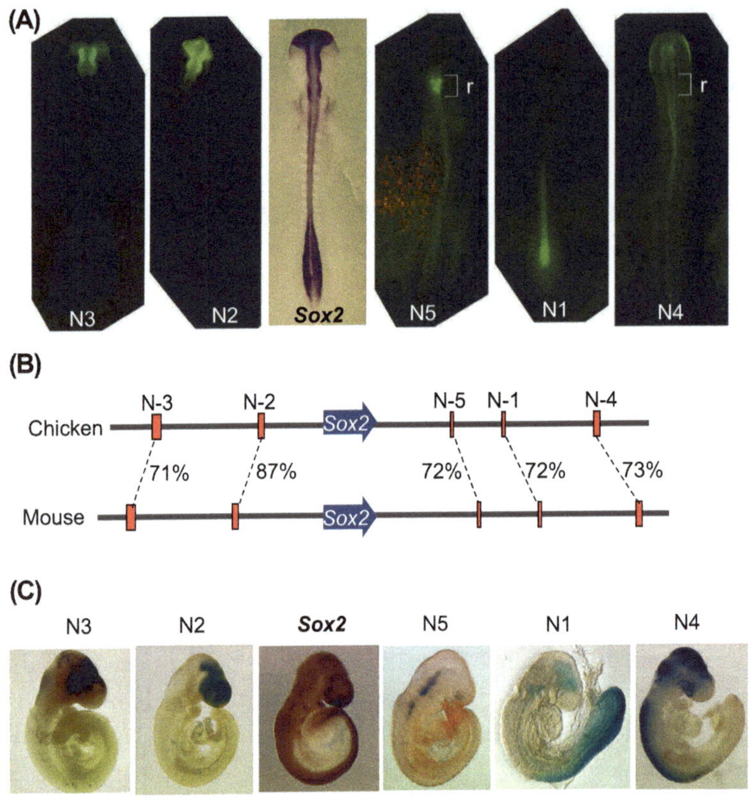

Figure 2 Genomic arrangements and activities of N1–N5 enhancers in chicken and mouse embryos. (A) Enhancer activities in electroporated chicken embryos at st. 11 as indicated by EGFP expression. Note that the addition of these enhancer activities roughly reconstructs the *Sox2* expression pattern indicated by in situ hybridization. (B) Comparison of genomic positions and sequence similarities shown as percent sequence identity in chicken and mouse enhancer sequences. (C) Mouse enhancer activities tested using primary transgenic mouse embryos at E9.5 using LacZ as the transgene and X-gal staining. *Adapted from Kamachi et al. (2009).*

CHARACTERIZATION OF INDIVIDUAL ENHANCERS THAT EXHIBIT ACTIVITIES IN THE NEURAL PRIMORDIA

The N2 Enhancer Activates *Sox2* in the Epiblast and Anterior Neural Plate

The N2 enhancer is 2.5–3 kb upstream of the *Sox2* coding sequence in the chicken genome and 3.9–4.4 kb upstream in the mouse genome. N2 is activated at the earliest stage among the neural enhancers, at the epiblast stage in chicken embryos, and at the preimplantation stage in mouse embryos. Once the neural plate is formed, the N2 enhancer activity is confined to the prosencephalon (telencephalon and diencephalon) region, and its activity continues to the progeny of the prosencephalon at later stages of CNS development. This enhancer has also been described as the *Sox2*–5′ enhancer (Zappone et al., 2000;

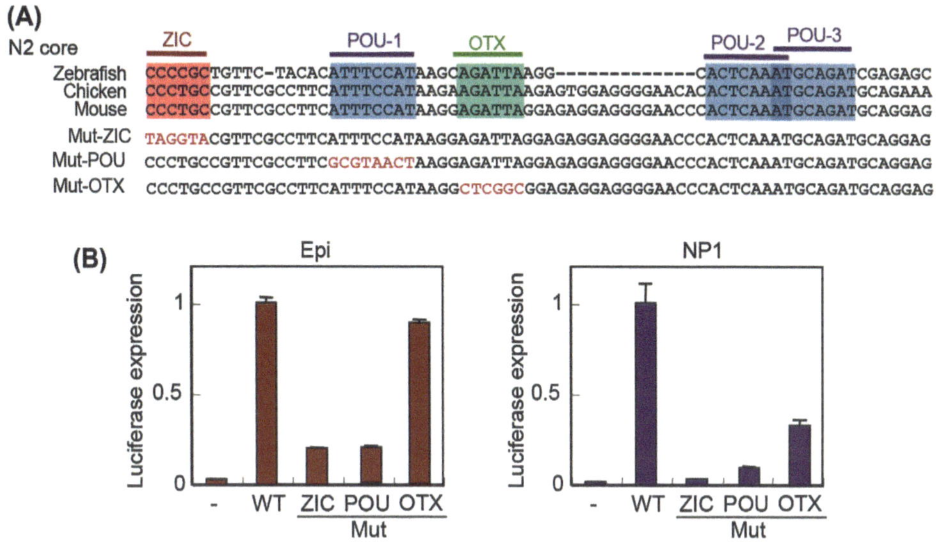

Figure 3 Arrangement of transcription factor binding sites in the N2 core sequence. (A) Phylogenetic conservation of the transcription factor binding sites and mutant versions of the mouse N2 core. (B) Effect of mutations on N2 core enhancer activity tested in epiblast stem cells (Epi) and derived neural precursors (NP1). Zic and Pou site mutations inactivated the enhancer in both cell types, whereas the OTX binding site mutation had a greater effect in NP1 cells. *Adapted from Iwafuchi-Doi et al. (2012).*

Catena et al., 2004; Ferri et al., 2004) and as SRR1 because of its activity in ES cells (Tomioka et al., 2002). The N2 enhancer spans a 530-base region for full activity. However, its core sequence is 73 bp long, as determined by two criteria: that is, (1) the minimal length of the enhancer that exhibits activity in the form of a multimer analogous to the original full-length enhancer; and (2) its deletion from the original full-length sequence inactivates the enhancer.

Detailed mutational analysis using epiblast stem cells and transactivation by exogenous transcription factors in fibroblasts indicated that the N2 core enhancer is activated by a combination of three transcription factors, such as OTX2, ZIC2/3, and POU5F1 or POU3Fs (Iwafuchi-Doi et al., 2011, 2012) (Figure 3). This is in contrast to regulation of the SRR2 enhancer, which depends on the cooperative action of SOX2 and POU5F1 (Tomioka et al., 2002) and is presumed to be involved in cross-activation of SOX2 and POU5F1 in ES cells (Kamachi and Kondoh, 2013). *Sox2* is presumably activated in ES cells by two parallel pathways: SRR2-mediated activation initiated by the maternally derived SOX2 and POU5F1 in embryos, which is present at the preimplantation stages (Avilion et al., 2003), and N2-mediated de novo activation that depends on OTX2, ZIC2/3, and POU5F1. Indeed, the important contribution of OTX2 in preimplantation embryos has been reported (Acampora et al., 2013).

OTX2, ZIC2, POU5F1, and POU3F1, which are involved in activating the N2 enhancer in egg cylinder postimplantation-stage embryos, are expressed uniformly in the epiblast. However, at stages past embryonic day (E)6.5 and when anteroposterior differences of the embryos are generated, OTX2, ZIC2, and POU3F1 expression together with SOX2 increases in the anterior region of embryos that gives rise to the anterior neural plate (ANP), whereas POU5F1 is strongly downregulated in the same anterior region. These changes are initiated by inhibition of Nodal signaling in the anterior region of the epiblast. Downregulation of POU5F1 follows the anterior decrease of Nodal signaling, because of activation of NR6A1, which represses *Pou5f1* (Fuhrmann et al., 2001), leading to the development of the ANP directly from the epiblast (Iwafuchi-Doi et al., 2012). This process can be recapitulated using epiblast stem cells, which are established from the epiblast (Brons et al., 2007; Tesar et al., 2007) and maintain the major features of the epiblast (Iwafuchi-Doi et al., 2012). Epiblast stem cells are maintained in the presence of activin (Nodal replacement), whereas activin removal from the culture medium quickly activates the ANP developmental process.

During this transition phase from the epiblast to the ANP, the major-acting POU5F1 factor is replaced by POU3F1, but the N2 enhancer activity and *Sox2* expression persist, because the transcription factor combination of OTX1 and ZIC2 cooperates with POU5F1 or POU3F1 to activate the N2 enhancer (Iwafuchi-Doi et al., 2012).

Homozygous inactivation of the N2 enhancer in the mouse genome *Sox2*$^{\Delta N2/\Delta N2}$ causes a large decrease in *Sox2* expression from the ES to the early ANP stage, confirming that the N2 enhancer is the major regulator of *Sox2* expression during these stages (Iwafuchi-Doi et al., 2011). Despite this initial decrease in *Sox2* expression after implantation, *Sox2* expression level in the ANP recovers after E8, and homozygous mice develop without morphological abnormalities or fertility defects. Normal development of *Sox2*$^{\Delta N2/\Delta N2}$ embryos is ascribed, at least partly, to the compensatory function of *Sox3* during these stages, despite low *Sox2* expression in the ES and early ANP. The recovery of *Sox2* expression in *Sox2*$^{\Delta N2/\Delta N2}$ embryos after E8 strongly suggests the existence of additional enhancers that activate *Sox2* and ANP and compensate for loss of the N2 enhancer. This prompted us to investigate a wider genomic region for *Sox2*-regulating enhancers, as will be described below. Enhancer C[27], with regional specificity similar but not identical to the N2 enhancer, has been discovered 35 kb downstream of *Sox2* (Okamoto et al., 2015).

Even at later stages of forebrain development, a contribution by multiple enhancers including N2 in the regulation of neural stem/progenitor cells is indicated. Mouse mutants homozygous for a deletion in the N2 enhancer region develop normally, but heterozygotes for the N2 deletion with a *Sox2*-null allele show defects during forebrain development (Zappone et al., 2000) (see also Chapter 11). This observation indicates that although the *Sox2* expression level decreased in the forebrain region by homozygous N2 enhancer deletion, the level remained above the threshold to cause developmental defects, and the *Sox2* expression level decreased below the threshold by deletion of the N2 enhancer only after the removal of an active *Sox2* gene (see Chapter 12 for such a threshold).

The N1 Enhancer Is Involved in the Regulation of Axial Stem Cells

The N1 enhancer is located approximately 13 kb downstream (of the *Sox2* 5′ end) and the full-length enhancer spans 300 bp. In contrast to the N2 enhancer, which shows activity in the anterior-most region of the developing CNS, the N1 enhancer shows activity in the posterior end of the growing CNS, that is, in the spinal cord precursor. More precisely, the N1 enhancer shows activity in the region of the epiblast immediately lateral to the primitive streak, called the caudal lateral epiblast (Takemoto et al., 2006; Wilson et al., 2009).

The core sequence determined by the criteria described above was 55 bp long and was activated by the combined action of Wnt and fibroblast growth factor (FGF) signals (Takemoto et al., 2006) (Figure 4). A mutational analysis indicated that the N1 enhancer sequence consisted of five regulatory blocks (A to E). Blocks A and B included the LEF1 binding sequences that mediate Wnt signal-dependent activation of the N1 enhancer (Figure 4). Block C is also involved in activating the enhancer, although the mechanism has not been clarified. Block D mediates FGF signal-dependent enhancer activation through a motif resembling an ETS factor binding sequence.

Block E is unique and represses N1 activation in cells that migrate through the primitive streak into the mesodermal compartment. The wild-type N1 enhancer is active only in the epiblast layer of embryonic tissue abutting the primitive streak. However, activation of N1 continues to the paraxial mesoderm compartment in the N1 enhancer block E mutant, as indicated by *tkEGFP* expression. This observation indicates that the wild-type N1 enhancer is activated in the caudal lateral epiblast but is then turned off in cells that migrate into the mesodermal compartment, and this inhibition of the N1 enhancer is mediated by block E (Takemoto et al., 2006). This observation also suggests a model in which cells residing in the caudal lateral epiblast are bipotential for neural and mesodermal development, but only the cell population that remains at the epiblastic site maintains N1 enhancer activity, activates *Sox2*, and develops into the neural plate (Figure 5(A)). A prediction from this model is that mesodermal compartment cells would develop into neural tissues rather than into mesodermal tissues under conditions in which N1 enhancer activity is maintained.

Such a case was reported by Chapman and Papaioannou (1998). Paraxial mesoderm (somites) failed to develop in mutant embryos deficient in the T-box transcription factor TBX6 and were replaced by bilateral ectopic neural tubes. We examined development of *Tbx6* mutant embryos in detail (Takemoto et al., 2011) and found that in *Tbx6* mutant embryos, the paraxial mesoderm precursors at E8.5 activated the N1 enhancer

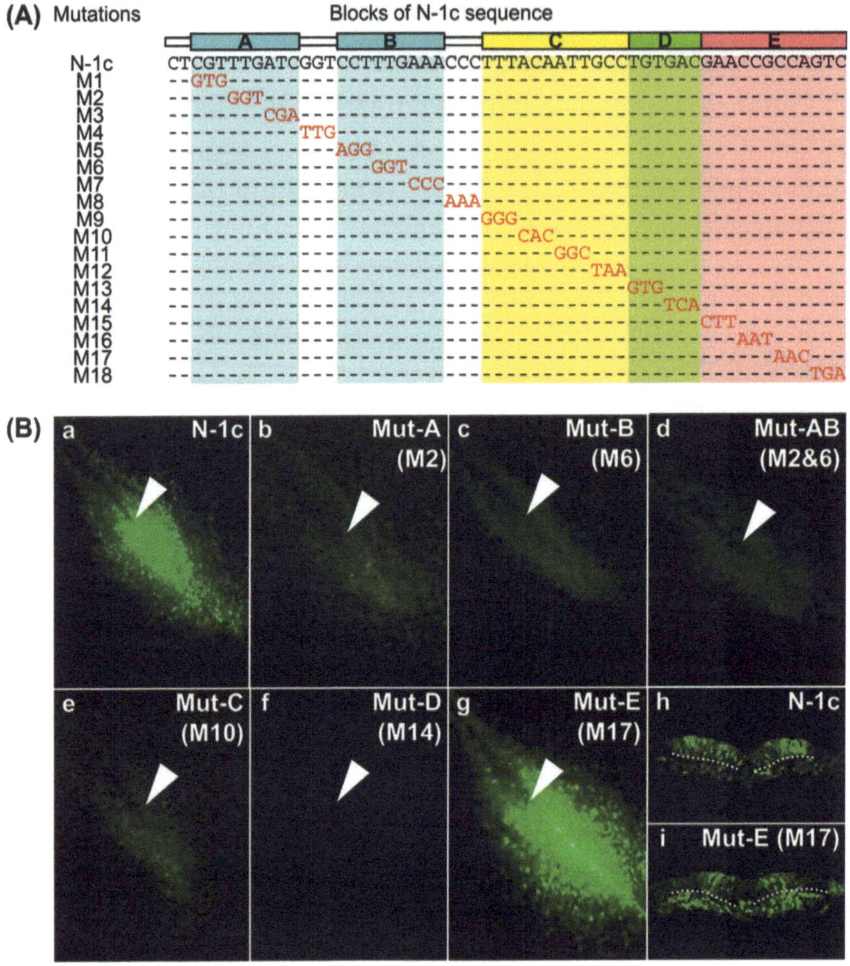

Figure 4 Functional blocks of the N1 enhancer core sequence. (A) Mutations in the N1 core enhancer (N-1c) that determined functional blocks A–E. Blocks A and B contain a canonical LEF1 binding sequence TTTG(A/T) and function as Wnt-responsive elements, and block D carrying the ETS family binding motif functions as FGF-responsive elements. (B) Mutations in blocks A–D resulted in N1 core inactivation, whereas mutations in block E strongly activated the N1 core enhancer (g) in the mesodermal compartment, as shown in cross-section (i). *Adapted from Takemoto et al. (2006).*

and expressed *Sox2*, and *Sox2* expressing cells developed into bilateral neural tubes at E9.5 at the expense of the paraxial mesoderm. We deleted the N1 enhancer sequence from the genome, which suppressed ectopic *Sox2* expression and ectopic neural tube development, to demonstrate involvement of the N1 enhancer in the *Tbx6* mutant phenotype (Figure 5(B)). Furthermore, when *Sox2* was forced to be expressed in the normal (wild-type) unsegmented paraxial mesoderm precursors, these cells also developed into neural tubes (Takemoto et al., 2011). These observations established that

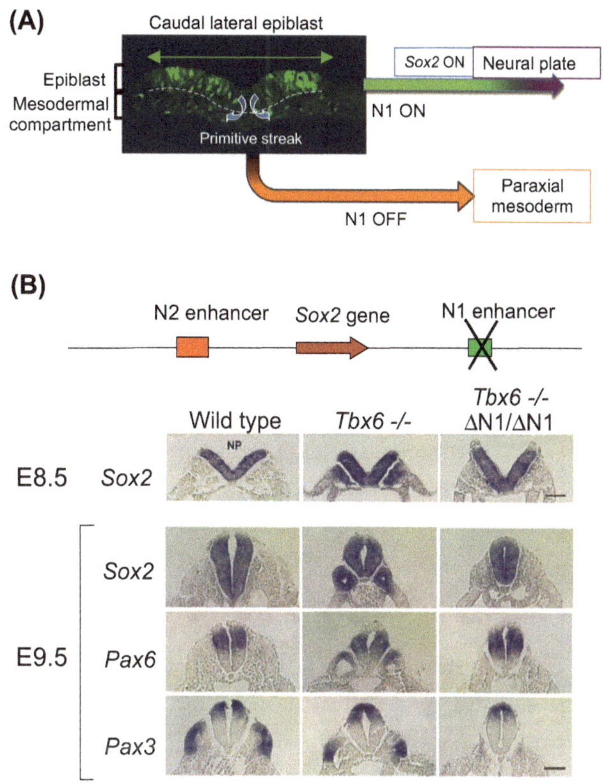

Figure 5 Regulation of axial stem cells. (A) Regulation of axial stem cells with neural and mesodermal bipotentiality, which are regulated by activation and repression, respectively, of the *Sox2* N1 enhancer. (B) The *Tbx6* mutant phenotype and its rescue by deleting the N1 enhancer. *Tbx6* mutant embryos expressed *Sox2* in the paraxial mesoderm precursors at E8.5, and developed ectopic neural tubes that express *Sox2*, *Pax6*, and *Pax3* at the paraxial mesoderm position at E9.5. These Tbx6 mutant-characteristic phenotypes were suppressed by homozygous deletion of the *Sox2* N1 enhancer (ΔN1/ΔN1). *Adapted from Takemoto et al. (2011).*

paraxial mesoderm precursors, a derivative of the caudal lateral epiblast, remain competent to develop into neural tubes; N1 enhancer activity must be turned off for normal paraxial mesodermal development to occur.

The action of TBX6 on the N1 enhancer is not direct because TBX6 fails to bind the N1 DNA sequence (Takemoto et al., 2011), but it occurs indirectly regulating *Wnt3a* expression in the paraxial mesoderm compartment. As described above, activation of the N1 enhancer relies on Wnt and FGF signaling (Takemoto et al., 2006, 2011). *Wnt3a* expression in the paraxial mesoderm precursor of normal embryos is low whereas *Wnt3a* is strongly expressed in the paraxial mesoderm of *Tbx6* mutant embryos, resulting in ectopic activation of the N1 enhancer and subsequent activation of *Sox2*. The inhibition of the chicken N1 enhancer in the mesodermal compartment depends on the E segment (Takemoto et al., 2006), but the N1 inhibition presumably relies on a different mechanism in mice, because the E segment sequence is not conserved in mouse N1. This may be attributed to a difference in the acting T-box transcription factors, that is, TBX6 in mouse and TBX6L in chicken embryos (Knezevic et al., 1997). TBX6 and TBX6L are similarly expressed in the caudal lateral epiblast, but they share similarity as transcription factors only in the DNA-binding T-box sequence.

Sox2 is only weakly activated in the caudal lateral epiblast, the tissue area for the axial stem cells, despite activation of the N1 enhancer. This is because of the strong inhibitory action of bone morphogenetic protein (BMP) signals against *Sox2* expression emanating from primitive streak-abutting tissues. The *Sox2* gene in the caudal lateral epiblast is released from BMP inhibition only by the posterior migration of the node that secretes BMP antagonists (e.g., Chordin and Noggin), and strong *Sox2* expression starts in cells that will develop into the posterior neural plate. During the time window for the repression of *Sox2* expression by BMPs, a significant fraction of cells migrates into the mesodermal compartment via the primitive streak, turns off the N1 enhancer, and proceeds into the mesodermal developmental pathway (Figure 5(A)).

Deleting the N1 enhancer sequence ($Sox2^{\Delta N1/\Delta N1}$) in the wild-type embryo cleared *Sox2* expression in the caudal lateral epiblast but did not cause a significant defect in the morphological development. A possible cause for the lack of a significant morphological defect was functional compensation by *Sox3*, which is also weakly expressed in the caudal lateral epiblast (Uchikawa et al., 2011; Yoshida et al., 2014). $Sox2^{\Delta N1/\Delta N1}$; *Sox3*-null double mutants produce more abundant mesodermal precursors and more enlarged somites than normal, but neural development proceeds normally (Yoshida et al., 2014). This observation suggests that the unidentified enhancer(s) that activate *Sox2* at the axial level anterior to the node backed up the loss in the N1 enhancer activity and sustained neural plate development.

The existence of neuro-mesodermal bipotential cells positioned close to the node and primitive streak was first indicated by single cell tracing studies (Wilson et al., 2009). The most compelling evidence for the common origin of the trunk neural tube and paraxial mesoderm was provided by tracing clones in mouse embryos using activation of the LacZ marker gene that occurs infrequently via intragenic recombination events (Tzouanacou et al., 2009). However, these earlier studies did not clarify the exact distribution of such cells or the mechanisms underlying the bipotentiality of the cells. Combining these cell tracing results and our analysis of the N1 enhancer regulation established that the neural tube and paraxial mesoderm in the trunk of embryos are derived from bipotential axial stem cells and that the N1 enhancer-dependent regulation of *Sox2* in the axial stem cells is crucial for the dichotomous fate decision.

The N3 Enhancer in Visual System Development

The N3 enhancer is located approximately 15 kb upstream of the *Sox2* gene, and the full-length enhancer spans about a 580-bp sequence (Uchikawa et al., 2003; Inoue et al., 2007).

The N3 enhancer in chicken embryos is initially activated in the diencephalon and mesencephalon at st. approximately 8, and activation extends to the optic vesicle (future retina) that evaginates from the diencephalon. When the optic vesicle contacts the surface ectoderm, N3 is also activated in the contact area of surface ectoderm, which elicits a local surge in *Sox2* expression leading to formation of the lens placode (Inoue et al., 2007) (see also Chapter 12 for eye tissue development).

Through a series of deletion analysis, a 36-bp core sequence was determined that contains SOX2 and PAX6 binding sequences in spacing and orientation identical to those of the DC5 *delta-crystallin* enhancer (Kamachi et al., 2001; Inoue et al., 2007) (see also Chapters 8 and 12), in which functional and molecular cooperation of SOX2 and PAX6 was demonstrated. Moreover, similar to the DC5 enhancer, the PAX6 binding sequence deviates from the consensus determined by PAX6 single molecule binding (Epstein et al., 1994), and PAX6 binding depends on simultaneous binding of the N3 enhance sequence by SOX2 (Inoue et al., 2007).

The N3 enhancer is considered part of the *Sox2*–*Pax6* coactivation loop (see Chapters 8 and 12). Surface ectoderm expresses a low level of *Sox2* because of the activity of N4 enhancer activity before activating the N3 enhancer, and also a low level of *Pax6*. Presumably, transcription factor levels at this stage are below the threshold of driving the *Sox2*–*Pax6* coactivation loop (Figure 6). Contact of the head ectoderm by the optic vesicle presumably activates a molecular cue to activate the N3 enhancer, which strongly activates *Sox2* and promotes operation of the *Sox2*–*Pax6* coactivation loop. This molecular cue has not been identified, but it must be the key player in the lens induction event proposed more than a century ago by H. Spemann (Spemann, 1901), based on his classic optic vesicle ablation experiments.

The N4 Enhancer Is Active in the Preplacodal Surface Ectoderm and the CNS

The N4 enhancer is located 22 kb downstream of the *Sox2* gene and the full-length sequence is 480 bp long (Uchikawa et al., 2003). The N4 enhancer is activated in two groups of embryonic tissues: the head surface ectoderm, which gives rise to various placodal tissues, and two CNS subdomains, the midbrain and spinal cord. An essential sequence stretch with specificity analogous to the original N4 was shortened to a 110-bp region called mini-N4, but no further division into core-like minimum short sequences was possible (Saigou et al., 2010). A mutational analysis of the 110-bp sequence showed multiple and heterologous elements distributed widely in the mini-N4 sequence. Some of the elements are essential for the mini-N4 enhancer activity in surface ectoderm and neural tissues, whereas others are important for neural tissue activity only, and the rest are limited to the surface ectoderm (Figure 7(A)).

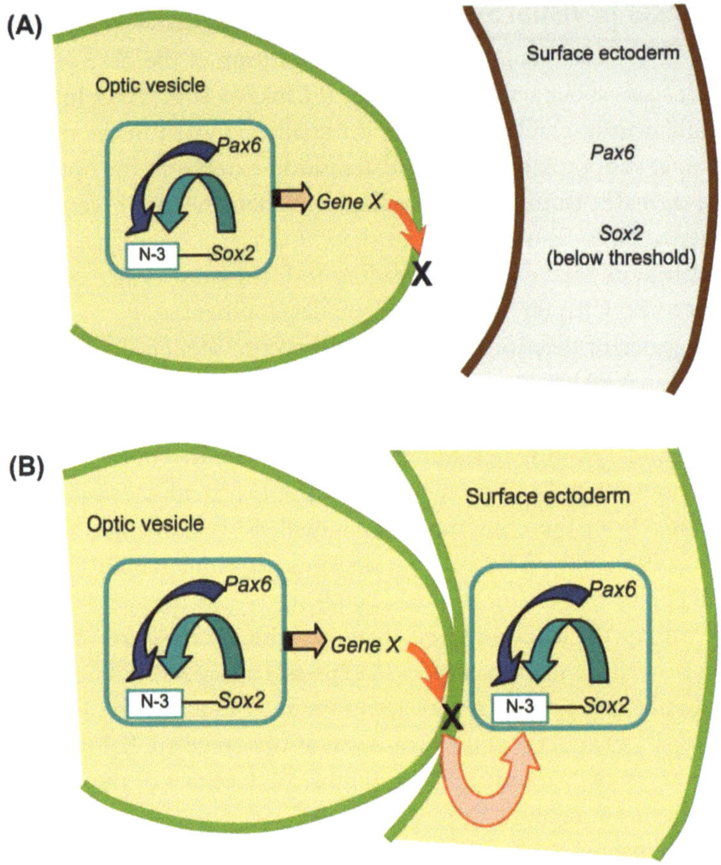

Figure 6 Activation of the *Sox2–Pax6* coactivation loop in the lens placode by contact of surface ectoderm with the optic vesicle. (A) The *Sox2–Pax6* coactivation loop is only in operation in the optic vesicle before contact with the cephalic ectoderm. (B) However, contact activates the *Sox2–Pax6* coactivation loop in the cephalic ectoderm via the action of hypothetical cueing molecule X, which elicits development of the lens placode. *Adapted from Inoue et al. (2007).*

This analysis suggests complex interactions between transcription factors that bind to the N4 sequence and generate the activation pattern of the N4 enhancer. The 3′ half of the mini-N4 enhancer (mini-N4[211–260]) includes major elements for surface ectoderm enhancer activity, and octamerization of the mini-N4[211–260] sequence results in artificial enhancer activity in the surface ectoderm except for the most anterior region (Saigou et al., 2010) (Figure 7(B)). This mini-N4[211–260] octamer enhancer is useful for gene manipulation targeted to the cephalic surface ectoderm (Sugahara et al., unpublished).

ADDITIONAL ENHANCERS OUTSIDE THE CENTRAL 50-KB REGION

The limited impact of deleting the N1 or N2 enhancer sequences on *Sox2* expression and overall embryogenesis strongly suggests the existence of additional enhancers that at

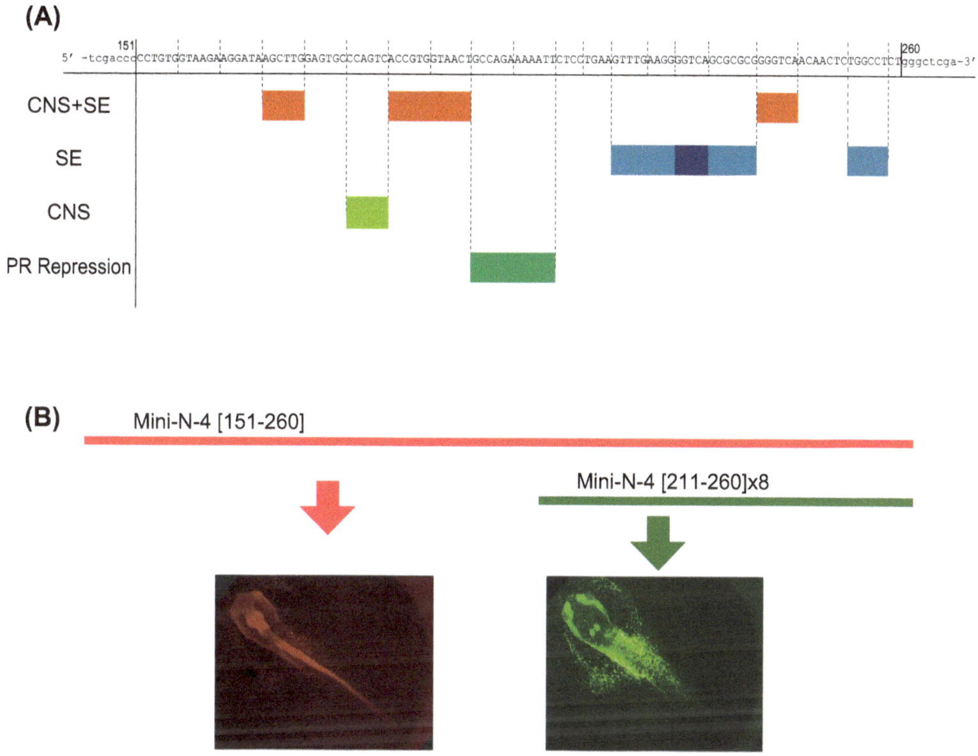

Figure 7 Composition of mini-N4 enhancer regulatory elements and the mini-N4[211–260] octamer activity. (A) Arrangement of the mini-N4 regulatory elements. CNS + SE indicates activation in the CNS and surface ectoderm, SE indicates activation in the surface ectoderm only, CNS indicates activation in the CNS only, and "PR repression" is a repressive element effective in the prosencephalon. (B) Mini-N4[151–260] and Mini-N4[211–260] octamer enhancer activities in the same electroporated embryo shown as red (mRFP1) and green (EGFP) fluorescence. *Adapted from Saigou et al. (2010).*

least partially compensate for the loss of these enhancers. Thus, we widened the area of enhancer analysis to the 205-kb region of the chicken genome, which corresponds to the 460- and 581-kb regions in the mouse and human genomes, respectively. We prepared a series of 5-kb fragments of the chicken genomic sequences with 1 kb terminal overlaps in the regions outside the central 50 kb of the initial survey. These DNA fragments were tested individually by chicken embryo electroporation (Figure 1) using the *tk-EGFP* reporter vector for enhancer activity up to developmental st. 18 (Okamoto et al., 2015).

This analysis detected 16 additional enhancers activated in neural and/or sensory organ primordial that are distributed widely in the *Sox2*-encompassing genomic region (Okamoto et al., 2015) (Figure 8(A)). The specificities of these and earlier identified neural enhancers were individually unique in the enhancer-active CNS domains and the timing of activation; however, significant overlap was observed among neural-specific enhancer activities (Figure 8(B)). In most cases, enhancer activities associated with an

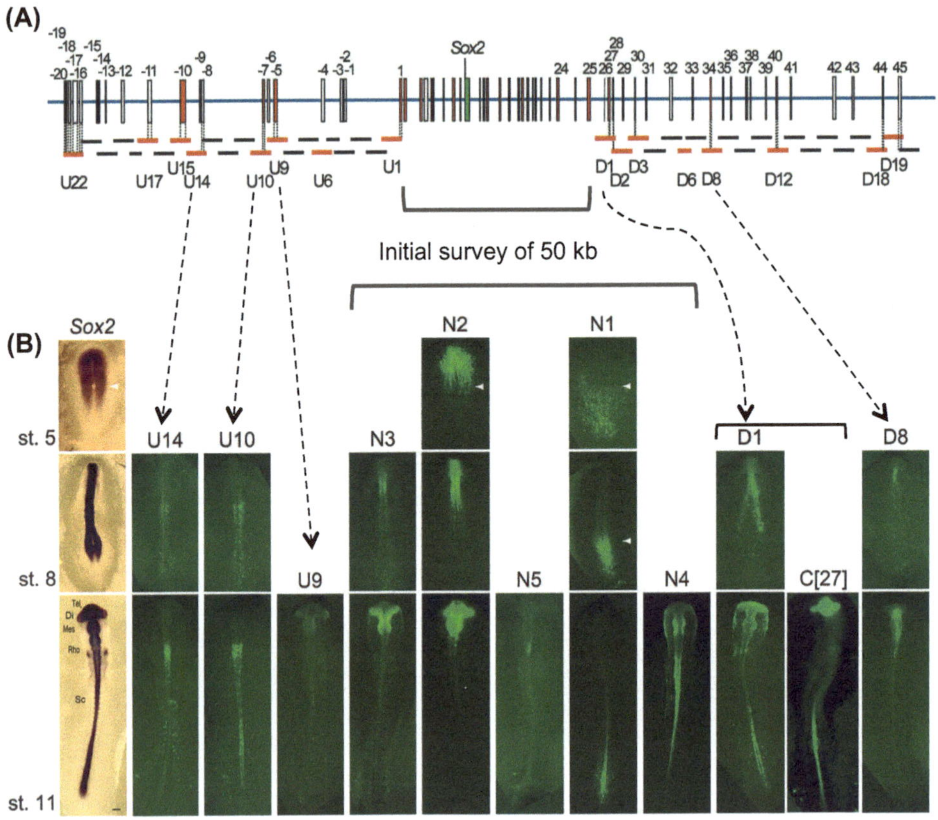

Figure 8 Neural enhancers in the 205-kb genomic span of the chicken *Sox2* locus. (A) Distribution of the enhancer-bearing sequences, highlighted in red, compared with the conserved sequence blocks shown as thin boxes. (B) Comparison of neural enhancer activities in the temporal sequence of activation and tissue coverage. The N2 and N1 enhancers are activated first in the anterior and posterior CNS, respectively. Enhancer activity borne by the D1 sequence represents the sum of two enhancer activities; C[27], showing activity in the forebrain and posterior spinal cord, as shown parallel to D1 in the figure, and C[28], with activity in the cephalic ectoderm and otic placode. *Adapted from Okamoto et al. (2015).*

approximately 5-kb fragment represented a single enhancer. However, D1 enhancer activity was further divided into C[27] (brain and posterior spinal cord) and C[28] (head surface ectoderm and otic placode) enhancer activities (Figure 8(B)).

C[27] enhancer activity in the forebrain was similar to that of N2 enhancer, although it was activated later than N2. C[27] enhancer may be responsible for the late onset of *Sox2* activation in the forebrain in *Sox2*$^{\Delta N2/\Delta N2}$ mutant embryos (Iwafuchi-Doi et al., 2011). C[28] enhancer is active in the cephalic ectoderm analogous to the N4 enhancer, which suggests that C[28] enhancer also contributes to low-level *Sox2* expression before placodal development. C[27] enhancer activity in the posterior spinal cord also suggests the possibility that C[27] contributes to *Sox2* activation after N1.

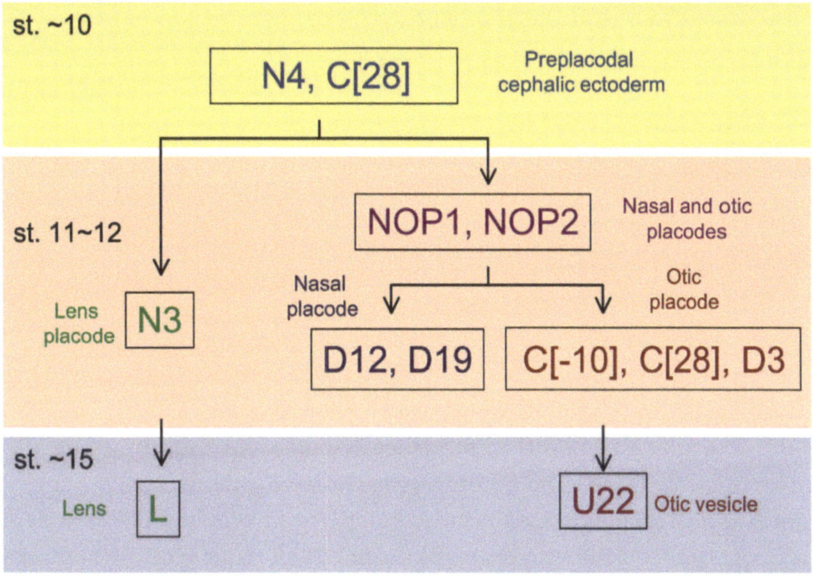

Figure 9 Hierarchy of *Sox2* enhancer action during sensory development. *Sox2* enhancers showing activities in the cephalic ectoderm and sensory placodes in greater than st. 11 embryos can be ordered according to their coverage of more defined or developmentally advanced tissues. *Adapted from Okamoto et al. (2015).*

HIERARCHY IN THE ACTION OF SENSORY PLACODE-SPECIFIC ENHANCERS

Two enhancers in the *Sox2*-proximal 50-kb region of the chicken genome involved in lens development were identified: N3 in the lens placode to lens vesicle and L in lens fiber; the latter is absent in mammals (see Chapter 12). By contrast, the NOP1 and NOP2 enhancers, which have specificity to both nasal and otic placodes, have been identified (Uchikawa et al., 2003). This finding suggests that nasal and otic placodes are under analogous regulatory mechanisms at the initial developmental stage. Interestingly, both NOP1 and NOP2 enhancers are located to the 5′ and 3′ sides of the *Sox2* gene, respectively, and share DNA sequence motifs for SOX2-SALL4 heterodimer binding. NOP1 is activated by cooperation between SOX2 and SALL4 (Sugahara et al., unpublished). This SOX2-SALL4 interaction represents a new example of a SOX2–partner factor interaction (see Chapter 8).

Extending the genomic area to the 205-kb region for the enhancer search identified several more enhancers with either otic or nasal placodal specificity. Considering the temporal order of activation of these enhancers, the hierarchy of enhancer actions involved in development of the sensory placodes is ordered as shown in Figure 9.

Although the NOP1 enhancer is activated earliest during development of the nasal and otic placodes, its activity continues to later stages of inner ear development when sensory patches and acoustic ganglia develop (Neves et al., 2012; Evsen et al., 2013). The NOP1 enhancer is repressed by basic helix–loop–helix (bHLH) transcription factors, and this inhibition is responsible for inhibiting *Sox2* expression by bHLH transcription factors. NEUROGENIN and NEUROD1 are required for acoustic neuron differentiation (Evsen et al., 2013).

CORRELATIONS BETWEEN THE ENHANCERS AND THE CONSERVED SEQUENCE BLOCKS

The 11 enhancers identified in the *Sox2*-proximal 50-kb region corresponded to a subset of conserved sequence blocks, which are defined as sequences that show more than 60% identity over a 100-bp length between the chicken and human genomes (Uchikawa et al., 2003). Forty more conserved sequence blocks were found outside the central 50-kb region in the 200-kb region encompassing the *Sox2* gene. Most of the approximately 5-kb sequences that demonstrated enhancer activity in our functional assay included conserved sequence blocks. Some of approximately 5-kb sequences were compared with constituent conserved sequence blocks for enhancer activity. This analysis indicated that the enhancer activities of U15, U10, U9, U1, and D8 are borne by the conserved sequence blocks C[−10], C[−7], C[−5], C[1], and C[34], respectively, whereas D1 sequence activities in the CNS and surface ectoderm are split into enhancer activities of C[27] and C[28]. These observations confirmed that the conserved sequence blocks often act as enhancers.

Additional conserved sequence blocks that were not assigned enhancers in our assay presumably act as enhancers under other developmental contexts. Enhancers in chromatin structures are associated with local histone H3 modifications that reflect the status of individual enhancers. H3K27 acetylation (H3K27Ac) marks active enhancers, whereas H3K4 mono-methylation (H3K4me1) is associated with primed enhancers (Creyghton et al., 2010; Rada-Iglesias et al., 2011; Zentner et al., 2011). A comparison of mouse conserved sequence blocks with the H3 modification patterns in ES cells, and the ES-derived neural progenitors showed some interesting correlations (Figure 10). The conserved sequence blocks included in the U17 and U22 sequences were enriched with H3K4me1 in ES cells, and the U22 sequence was enriched for H3K27Ac in the neural progenitors, which was consistent with the observation that U17 and U22 act as enhancers. Moreover, the conserved sequence blocks C[−12] and C[−15] were strongly enriched for H3K27Ac in neural progenitor cells, which suggests that they are active enhancers. These observations support the view that many other conserved sequence blocks also function as enhancers under conditions different from our assay using chicken embryo electroporation.

Notably, additional enhancers possibly exist outside the 205-kb region in the chicken genome (460-kb region in the mouse genome). For example, hypothetical *Sox2* regulatory

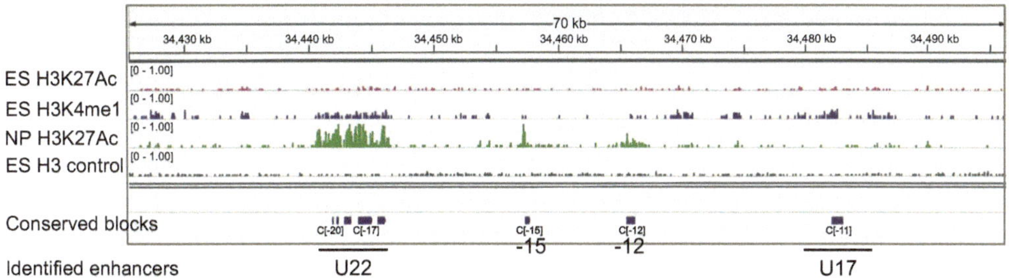

Figure 10 Correlation between histone H3 modification patterns and potential enhancers. The chromatin-immunoprecipitation-sequencing (ChIP-seq) data published by Creyghton et al. (2010) were reanalyzed by Okamoto et al. (2015). The H3K4me1 distribution indicative of primed enhancers in ES cells are compared with active (H3K27Ac) states in ES cells and ES-derived neural progenitor (NP) cells. The U22 enhancer sequence including the conserved sequence blocks C[−16] to C[−20] is enriched for the H3K4me1 modification on ES cells and densely marked by H3K27Ac in NP cells. The conserved sequence blocks C[−15] and C[−12] are also enriched for H3K27Ac in neural progenitor cells. *Adapted from Okamoto et al. (2015).*

sequences involved in the *Ysb* and *Lcc* mouse mutations causing inner ear defects have been mapped outside this region (Dong et al., 2002; Kiernan et al., 2005) (see also Chapter 15).

COMPARISON WITH *SOX3* ENHANCERS

Although SOX2 is the major player among SOXB1 transcription factors in amniotes including mammals and avians, this has evolved relatively recently in vertebrate phylogeny; *Sox2* is a minor player in amphibian and fishes, and its expression does not cover the entire CNS or all sensory placodes (see Chapter 6). Among the *SoxB1* genes, *Sox3* conserves its expression in the major part of neural primordia and sensory placodes and is considered the founder *SoxB1* gene. Even in higher vertebrates in which *Sox2* expression is more prevalent, *Sox2* and *Sox3* expression patterns have remarkable similarities. Therefore, it is interesting to compare the organization of the *Sox2* and *Sox3* enhancers.

The 231-kb region encompassing the *Sox3* gene was analyzed for enhancers in the same manner as *Sox2* enhancers, and 13 enhancers associated with embryonic neural and placodal tissues were identified (Nishimura et al., 2012). Earlier studies on the *Sox3*-proximal regions using mouse and *Xenopus* embryos identified three enhancers (Brunelli et al., 2003; Rogers et al., 2009). Taken together, the *Sox3* gene is associated with at least 13 enhancers. These *Sox3* enhancers are substantially different from *Sox2* enhancers in individual sequence and spatiotemporal specificity despite overall similarity of the *Sox2* and *Sox3* expression pattern (Nishimura et al., 2012). Most *Sox3* enhancers are distributed downstream of the gene, whereas the *Sox2* enhancers are distributed widely from upstream to downstream of the gene (Nishimura et al., 2012).

Thus, *Sox2* and *Sox3* enhancers diverge in sequence, organization, and specificity, but they are regulated similarly in a few limited cases (Nishimura et al., 2012). Most intriguing is the *Sox3* NP1 enhancer, which is activated in the caudal lateral epiblast, similar to the *Sox2* N1 enhancer. An analysis of the mutational effects and the effects of inhibitors against signaling systems indicated that the *Sox3* NP1 enhancer is regulated by Wnt and FGF signals, similar to the *Sox2* N1 enhancer, although the arrangement of the individual regulatory element is different. Thus, Wnt and FGF-dependent regulation would allow for coordinated regulation of *Sox3* and *Sox2* in axial stem cell-derived cell populations that develop into neural and mesodermal tissues.

RELATIONSHIPS AMONG *SOX2*-ASSOCIATED ENHANCERS, CONSERVED SEQUENCE BLOCKS, AND *SOX2OT* EXONS

Although the *Sox2* gene is located in a genomic region remote from other protein-encoding genes, a variety of long noncoding transcripts collectively called *Sox2ot*s (*Sox2* overlapping transcripts) are also transcribed from the same DNA strand of the same genomic region (Fantes et al., 2003; Amaral et al., 2009). In addition, transcription of a noncoding RNA approximately 200 kb downstream of the mouse *Sox2* gene has been reported (Guttman et al., 2009). Therefore, it is appropriate to discuss relationships between exons of these noncoding RNAs and the conserved sequence blocks, as well as the possible involvement of *Sox2*-associated enhancers in the regulation of these transcripts.

As many as 30 *Sox2OT* variants have been reported in humans (GRCh37, Ensembl: http://www.ensembl.org/), and 16 *Sox2ot* in mice (GRCm38, Ensembl). These mammalian *Sox2ot*s have similar exon organization (Figure 11). The major group of *Sox2ot* sequences uses C[−6], C[4] and the region including C[23] and C[24]. Modified versions of this *Sox2ot* group include additional exons or a missing C[4] exon. The second *Sox2ot* group starts from far upstream and is called *Sox2dot*s (*Sox2* distal overlapping transcripts) (Amaral et al., 2009). *Sox2dot*s mostly use C[−9] as an internal exon and frequently employ C[4] as the last exon. The third *Sox2ot* group uses C[4] as the first exon. Therefore, C[4] is a *Sox2ot* hub exon, acting as the first, intermediate, or last exon. Thus, C[4] and C[−6], which were not assigned to an enhancer, represent *Sox2ot* exons in the mouse and human. Exon organization in chicken and zebrafish *Sox2ot*s is simpler, in which C[−6] and C[4] comprise the first two exons in most transcripts (Amaral et al., 2009).

However, some enhancer sequences serve as *Sox2ot* exons. The N4 enhancer sequence C[24] is often employed as the last *Sox2ot* exon. Moreover, C[14] of the N5 enhancer sequence is included as an internal exon in a form of human *Sox2OT* (Figure 11(A)), and C[17], which corresponds to the N1 enhancer sequence, is the first exon of a form of mouse *Sox2ot* (Figure 11(B)). The major *Sox2dot* exon C[−9] sequence may show U14 enhancer activity. In addition, the *hs192* conserved region (VISTA Enhancer

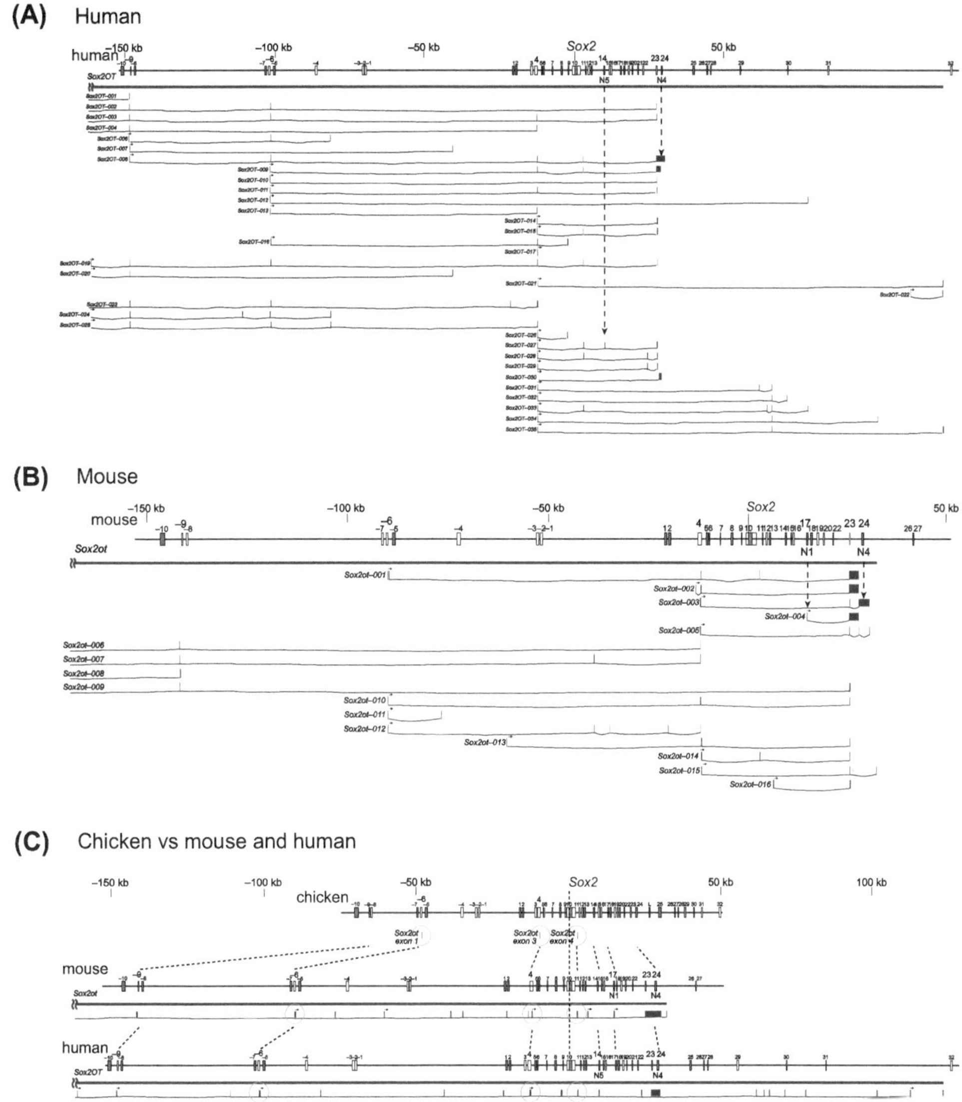

Figure 11 Organization of *Sox2ot* exons compared with the conserved sequence blocks and *Sox2* enhancers. (A) Representative human *Sox2OT*s in which the conserved sequence blocks C[−9], C[−6], and C[4] are the common exons. The conserved sequence blocks are indicated by numbers on the map shown on top, the enhancer sequences are indicated by red rectangles, and the corresponding enhancers with *Sox2OT* exons are indicated by broken lines with arrowheads. (B) The same comparison using mouse *Sox2ots*. Note that exon composition of mouse *Sox2ot* is similar but not identical to that of human *Sox2OT*, and the *Sox2ot* gene defined by most distal exons terminates more proximal to *Sox2* in mice. (C) Comparison of chicken *Sox2ots* with an ensemble of mouse and human *Sox2ots*.

Browser: http://enhancer.lbl.gov/), corresponding to one of the first *Sox2dot* exons and located about 520 kb upstream of the mouse *Sox2* gene, shows enhancer activity in a region of the E11.5 mouse embryo telencephalon (Pennacchio et al., 2006; Amaral et al., 2009; Visel et al., 2013). Exons of long noncoding RNAs occasionally overlap with enhancer sequences (Ulitsky and Bartel, 2013). Overlaps in the *Sox2ot* exons and *Sox2*-associated enhancers represent such examples.

The *Sox2ot* expression pattern resembles that of *Sox2* during early embryogenesis: for example, strong expression during formation of neural tissues in chicken and zebrafish embryos and during mouse ES cell-derived cell differentiation (Amaral et al., 2009). Therefore, it is possible that many *Sox2*-associated enhancers are involved in regulating both *Sox2* and *Sox2ot*. However, *Sox2ot* expression is strongly activated without accompanying *Sox2* during mesodermal development. Notably, the U14 enhancer has activity in both the CNS and mesodermal somites (Okamoto et al., 2015). This enhancer may at least partially account for mesodermal activation of *Sox2ot*.

CONCLUSIONS

As described in this chapter, many enhancer elements are associated with the *Sox2* locus and differ in tissue specificity and activation stages. Possible involvement of *Sox2ot* in the overall regulation of *Sox2* adds further complexity. This situation illustrates how a single transcription factor gene with important functions is furnished with numerous regulatory elements in the genome to accomplish many roles in different biological contexts during the life of an animal.

In-depth analysis of the regulation of individual enhancers provides a unique approach to mechanisms underlying embryonic regulation. Analysis of the N1 enhancer action typically indicates the nature and regulation of axial stem cells, which have neuro-mesodermal bipotentiality of development.

REFERENCES

Acampora, D., Di Giovannantonio, L.G., Simeone, A., 2013. Otx2 is an intrinsic determinant of the embryonic stem cell state and is required for transition to a stable epiblast stem cell condition. Development 140, 43–55.

Amaral, P.P., Neyt, C., Wilkins, S.J., Askarian-Amiri, M.E., Sunkin, S.M., Perkins, A.C., Mattick, J.S., 2009. Complex architecture and regulated expression of the Sox2ot locus during vertebrate development. RNA 15, 2013–2027.

Arnold, K., Sarkar, A., Yram, M.A., Polo, J.M., Bronson, R., Sengupta, S., Seandel, M., Geijsen, N., Hochedlinger, K., 2011. Sox2(+) adult stem and progenitor cells are important for tissue regeneration and survival of mice. Cell Stem Cell 9, 317–329.

Askarian-Amiri, M.E., Seyfoddin, V., Smart, C.E., Wang, J., Kim, J.E., Hansji, H., Baguley, B.C., Finlay, G.J., Leung, E.Y., 2014. Emerging role of long non-coding RNA SOX2OT in SOX2 regulation in breast cancer. PLoS One 9, e102140.

Avilion, A.A., Nicolis, S.K., Pevny, L.H., Perez, L., Vivian, N., Lovell-Badge, R., 2003. Multipotent cell lineages in early mouse development depend on SOX2 function. Genes Dev. 17, 126–140.

Brons, I.G., Smithers, L.E., Trotter, M.W., Rugg-Gunn, P., Sun, B., Chuva de Sousa Lopes, S.M., Howlett, S.K., Clarkson, A., Ahrlund-Richter, L., Pedersen, R.A., Vallier, L., 2007. Derivation of pluripotent epiblast stem cells from mammalian embryos. Nature 448, 191–195.

Brunelli, S., Silva Casey, E., Bell, D., Harland, R., Lovell-Badge, R., 2003. Expression of Sox3 throughout the developing central nervous system is dependent on the combined action of discrete, evolutionarily conserved regulatory elements. Genesis 36, 12–24.

Catena, R., Tiveron, C., Ronchi, A., Porta, S., Ferri, A., Tatangelo, L., Cavallaro, M., Favaro, R., Ottolenghi, S., Reinbold, R., Scholer, H., Nicolis, S.K., 2004. Conserved POU binding DNA sites in the Sox2 upstream enhancer regulate gene expression in embryonic and neural stem cells. J. Biol. Chem. 279, 41846–41857.

Chapman, D.L., Papaioannou, V.E., 1998. Three neural tubes in mouse embryos with mutations in the T-box gene Tbx6. Nature 391, 695–697.

Creyghton, M.P., Cheng, A.W., Welstead, G.G., Kooistra, T., Carey, B.W., Steine, E.J., Hanna, J., Lodato, M.A., Frampton, G.M., Sharp, P.A., Boyer, L.A., Young, R.A., Jaenisch, R., 2010. Histone H3K27ac separates active from poised enhancers and predicts developmental state. Proc. Natl. Acad. Sci. U.S.A. 107, 21931–21936.

Dong, S., Leung, K.K., Pelling, A.L., Lee, P.Y., Tang, A.S., Heng, H.H., Tsui, L.C., Tease, C., Fisher, G., Steel, K.P., Cheah, K.S., 2002. Circling, deafness, and yellow coat displayed by yellow submarine (ysb) and light coat and circling (lcc) mice with mutations on chromosome 3. Genomics 79, 777–784.

Ellis, P., Fagan, B.M., Magness, S.T., Hutton, S., Taranova, O., Hayashi, S., McMahon, A., Rao, M., Pevny, L., 2004. SOX2, a persistent marker for multipotential neural stem cells derived from embryonic stem cells, the embryo or the adult. Dev. Neurosci. 26, 148–165.

Epstein, J., Cai, J., Glaser, T., Jepeal, L., Maas, R., 1994. Identification of a Pax paired domain recognition sequence and evidence for DNA-dependent conformational changes. J. Biol. Chem. 269, 8355–8361.

Evsen, L., Sugahara, S., Uchikawa, M., Kondoh, H., Wu, D.K., 2013. Progression of neurogenesis in the inner ear requires inhibition of Sox2 transcription by neurogenin1 and neurod1. J. Neurosci. 33, 3879–3890.

Fantes, J., Ragge, N.K., Lynch, S.A., McGill, N.I., Collin, J.R., Howard-Peebles, P.N., Hayward, C., Vivian, A.J., Williamson, K., van Heyningen, V., FitzPatrick, D.R., 2003. Mutations in SOX2 cause anophthalmia. Nat. Genet. 33, 461–463.

Ferri, A.L., Cavallaro, M., Braida, D., Di Cristofano, A., Canta, A., Vezzani, A., Ottolenghi, S., Pandolfi, P.P., Sala, M., DeBiasi, S., Nicolis, S.K., 2004. Sox2 deficiency causes neurodegeneration and impaired neurogenesis in the adult mouse brain. Development 131, 3805–3819.

Fuhrmann, G., Chung, A.C., Jackson, K.J., Hummelke, G., Baniahmad, A., Sutter, J., Sylvester, I., Scholer, H.R., Cooney, A.J., 2001. Mouse germline restriction of Oct4 expression by germ cell nuclear factor. Dev. Cell 1, 377–387.

Guttman, M., Amit, I., Garber, M., et al., 2009. Chromatin signature reveals over a thousand highly conserved large non-coding RNAs in mammals. Nature 458, 223–227.

Inoue, M., Kamachi, Y., Matsunami, H., Imada, K., Uchikawa, M., Kondoh, H., 2007. PAX6 and SOX2-dependent regulation of the Sox2 enhancer N-3 involved in embryonic visual system development. Genes Cells 12, 1049–1061.

Iwafuchi-Doi, M., Matsuda, K., Murakami, K., Niwa, H., Tesar, P.J., Aruga, J., Matsuo, I., Kondoh, H., 2012. Transcriptional regulatory networks in epiblast cells and during anterior neural plate development as modeled in epiblast stem cells. Development 139, 3926–3937.

Iwafuchi-Doi, M., Yoshida, Y., Onichtchouk, D., Leichsenring, M., Driever, W., Takemoto, T., Uchikawa, M., Kamachi, Y., Kondoh, H., 2011. The Pou5f1/Pou3f-dependent but SoxB-independent regulation of conserved enhancer N2 initiates Sox2 expression during epiblast to neural plate stages in vertebrates. Dev. Biol. 352, 354–366.

Kamachi, Y., Iwafuchi, M., Okuda, Y., Takemoto, T., Uchikawa, M., Kondoh, H., 2009. Evolution of non-coding regulatory sequences involved in the developmental process: reflection of differential employment of paralogous genes as highlighted by Sox2 and group B1 Sox genes. Proc. Jpn. Acad. Ser. B Phys. Biol. Sci. 85, 55–68.

Kamachi, Y., Kondoh, H., 2013. Sox proteins: regulators of cell fate specification and differentiation. Development 140, 4129–4144.

Kamachi, Y., Uchikawa, M., Collignon, J., Lovell-Badge, R., Kondoh, H., 1998. Involvement of Sox1, 2 and 3 in the early and subsequent molecular events of lens induction. Development 125, 2521–2532.

Kamachi, Y., Uchikawa, M., Tanouchi, A., Sekido, R., Kondoh, H., 2001. Pax6 and SOX2 form a co-DNA-binding partner complex that regulates initiation of lens development. Genes Dev. 15, 1272–1286.

Kiernan, A.E., Pelling, A.L., Leung, K.K., Tang, A.S., Bell, D.M., Tease, C., Lovell-Badge, R., Steel, K.P., Cheah, K.S., 2005. Sox2 is required for sensory organ development in the mammalian inner ear. Nature 434, 1031–1035.

Knezevic, V., De Santo, R., Mackem, S., 1997. Two novel chick T-box genes related to mouse brachyury are expressed in different, non-overlapping mesodermal domains during gastrulation. Development 124, 411–419.

Kondoh, H., Uchikawa, M., 2008. Dissection of chick genomic regulatory regions. Methods Cell Biol. 87, 313–336.

Neves, J., Uchikawa, M., Bigas, A., Giraldez, F., 2012. The prosensory function of Sox2 in the chicken inner ear relies on the direct regulation of Atoh1. PLoS One 7, e30871.

Nishimura, N., Kamimura, Y., Ishida, Y., Takemoto, T., Kondoh, H., Uchikawa, M., 2012. A systematic survey and characterization of enhancers that regulate Sox3 in neuro-sensory development in comparison with Sox2 enhancers. Biology (Basel) 1, 714–735.

Okamoto, R., Uchikawa, M., Kondoh, H., 2015. Sixteen additional enhancers associated with the chicken Sox2 locus outside the central 50-kb region. Dev. Growth Differ. 57, 24–39.

Okuda, Y., Yoda, H., Uchikawa, M., Furutani-Seiki, M., Takeda, H., Kondoh, H., Kamachi, Y., 2006. Comparative genomic and expression analysis of group B1 sox genes in zebrafish indicates their diversification during vertebrate evolution. Dev. Dyn. 235, 811–825.

Pennacchio, L.A., Ahituv, N., Moses, A.M., et al., 2006. In vivo enhancer analysis of human conserved non-coding sequences. Nature 444, 499–502.

Pevny, L.H., Nicolis, S.K., 2010. Sox2 roles in neural stem cells. Int. J. Biochem. Cell. Biol. 42, 421–424.

Rada-Iglesias, A., Bajpai, R., Swigut, T., Brugmann, S.A., Flynn, R.A., Wysocka, J., 2011. A unique chromatin signature uncovers early developmental enhancers in humans. Nature 470, 279–283.

Rogers, C.D., Harafuji, N., Archer, T., Cunningham, D.D., Casey, E.S., 2009. *Xenopus* Sox3 activates sox2 and geminin and indirectly represses Xvent2 expression to induce neural progenitor formation at the expense of non-neural ectodermal derivatives. Mech. Dev. 126, 42–55.

Saigou, Y., Kamimura, Y., Inoue, M., Kondoh, H., Uchikawa, M., 2010. Regulation of Sox2 in the pre-placodal cephalic ectoderm and central nervous system by enhancer N-4. Dev. Growth Differ. 52, 397–408.

Spemann, H., 1901. Ueber Correlationen in der Entwickelung des Auges. Verh. Anat. Geselschaft 61–79.

Suh, H., Consiglio, A., Ray, J., Sawai, T., D'Amour, K.A., Gage, F.H., 2007. In vivo fate analysis reveals the multipotent and self-renewal capacities of Sox2+ neural stem cells in the adult hippocampus. Cell Stem Cell 1, 515–528.

Takahashi, K., Yamanaka, S., 2006. Induction of pluripotent stem cells from mouse embryonic and adult fibroblast cultures by defined factors. Cell 126, 663–676.

Takemoto, T., Uchikawa, M., Kamachi, Y., Kondoh, H., 2006. Convergence of Wnt and FGF signals in the genesis of posterior neural plate through activation of the Sox2 enhancer N-1. Development 133, 297–306.

Takemoto, T., Uchikawa, M., Yoshida, M., Bell, D.M., Lovell-Badge, R., Papaioannou, V.E., Kondoh, H., 2011. Tbx6-dependent Sox2 regulation determines neural or mesodermal fate in axial stem cells. Nature 470, 394–398.

Taranova, O.V., Magness, S.T., Fagan, B.M., Wu, Y., Surzenko, N., Hutton, S.R., Pevny, L.H., 2006. SOX2 is a dose-dependent regulator of retinal neural progenitor competence. Genes Dev. 20, 1187–1202.

Tesar, P.J., Chenoweth, J.G., Brook, F.A., Davies, T.J., Evans, E.P., Mack, D.L., Gardner, R.L., McKay, R.D., 2007. New cell lines from mouse epiblast share defining features with human embryonic stem cells. Nature 448, 196–199.

Tomioka, M., Nishimoto, M., Miyagi, S., Katayanagi, T., Fukui, N., Niwa, H., Muramatsu, M., Okuda, A., 2002. Identification of Sox-2 regulatory region which is under the control of Oct-3/4-Sox-2 complex. Nucleic Acids Res. 30, 3202–3213.

Tzouanacou, E., Wegener, A., Wymeersch, F.J., Wilson, V., Nicolas, J.F., 2009. Redefining the progression of lineage segregations during mammalian embryogenesis by clonal analysis. Dev. Cell 17, 365–376.

Uchikawa, M., 2008. Enhancer analysis by chicken embryo electroporation with aid of genome comparison. Dev. Growth Differ. 50, 467–474.

Uchikawa, M., Ishida, Y., Takemoto, T., Kamachi, Y., Kondoh, H., 2003. Functional analysis of chicken Sox2 enhancers highlights an array of diverse regulatory elements that are conserved in mammals. Dev. Cell 4, 509–519.

Uchikawa, M., Takemoto, T., Kamachi, Y., Kondoh, H., 2004. Efficient identification of regulatory sequences in the chicken genome by a powerful combination of embryo electroporation and genome comparison. Mech. Dev. 121, 1145–1158.

Uchikawa, M., Yoshida, M., Iwafuchi-Doi, M., Matsuda, K., Ishida, Y., Takemoto, T., Kondoh, H., 2011. B1 and B2 Sox gene expression during neural plate development in chicken and mouse embryos: universal versus species-dependent features. Dev. Growth Differ. 53, 761–771.

Ulitsky, I., Bartel, D.P., 2013. lincRNAs: genomics, evolution, and mechanisms. Cell 154, 26–46.

Visel, A., Taher, L., Girgis, H., et al., 2013. A high-resolution enhancer atlas of the developing telencephalon. Cell 152, 895–908.

Wilson, V., Olivera-Martinez, I., Storey, K.G., 2009. Stem cells, signals and vertebrate body axis extension. Development 136, 1591–1604.

Yoshida, M., Uchikawa, M., Rizzoti, K., Lovell-Badge, R., Takemoto, T., Kondoh, H., 2014. Regulation of mesodermal precursor production by low-level expression of B1 Sox genes in the caudal lateral epiblast. Mech. Dev. 132, 59–68.

Zappone, M.V., Galli, R., Catena, R., Meani, N., De Biasi, S., Mattei, E., Tiveron, C., Vescovi, A.L., Lovell-Badge, R., Ottolenghi, S., Nicolis, S.K., 2000. Sox2 regulatory sequences direct expression of a (beta)-geo transgene to telencephalic neural stem cells and precursors of the mouse embryo, revealing regionalization of gene expression in CNS stem cells. Development 127, 2367–2382.

Zentner, G.E., Tesar, P.J., Scacheri, P.C., 2011. Epigenetic signatures distinguish multiple classes of enhancers with distinct cellular functions. Genome Res. 21, 1273–1283.

CHAPTER 8

SOX2–Partner Factor Interactions and Enhancer Regulation

Hisato Kondoh[1], Yusuke Kamachi[2]
[1]Faculty of Life Sciences, Kyoto Sangyo University, Kyoto, Japan; [2]Graduate School of Frontier Biosciences, Osaka University, Suita, Osaka, Japan

INTRODUCTION

Following the discovery of SOX2 as a member of SRY-related high mobility group (HMG) domain-containing deoxyribonucleic acid (DNA)-binding proteins (SOX proteins), analysis of its expression pattern in developing embryos suggested its involvement in early-stage embryogenesis and the development of the nervous system and sensory organs (Uwanogho et al., 1995; Collignon et al., 1996). In 1995, two groups reported identification of the regulatory target genes of SOX2 and that binding of the SOX2 protein alone to an enhancer sequence does not elicit gene activation, but does so only in cooperation with a partner transcription factor that binds to an adjacent DNA site (Kamachi et al., 1995; Yuan et al., 1995) (Figure 1(A)). This requirement of a partner factor to regulate transcription appears to hold true for virtually all SOX2 family transcription factors; a particular combination of SOX2 and a partner factor determines their regulatory target and changes in the partnership alter an ensemble of activated genes and elicit the cell state transition (Kamachi et al., 2000). This mechanism is called the SOX–partner code, as reviewed previously (Kondoh and Kamachi, 2010; Kamachi and Kondoh, 2013). For example, SOX2 alters its partner from PO5F1 to POU3F factors during development from early stem cells (embryonic stem cells (ESCs) and epiblast stem cells) to neural progenitors (Iwafuchi-Doi et al., 2011, 2012; Lodato et al., 2013), whereas POU5F1 changes its partner from SOX2 to SOX17 during endodermal development from ESCs (Aksoy et al., 2013).

SOX2–POU5F1 INTERACTION

Lisa Dailey et al. investigated *Fgf4* gene regulation in teratocarcinoma and ESC lines, identified an essential enhancer, and found that the enhancer is bound by SOX2 and PO5F1 (OCT3/4) at adjacent DNA sites (Yuan et al., 1995). In addition, simultaneous binding of SOX2 and POU5F1 to appropriately arranged binding sites was essential for activating the *fibroblast growth factor 4* (*Fgf4*) enhancer; a mutational alteration in one of the binding sites of insertion of more than three bases inactivated the enhancer (Dailey et al., 1994; Yuan et al., 1995; Ambrosetti et al., 1997). This discovery was the prelude to identifying many

Sox2
http://dx.doi.org/10.1016/B978-0-12-800352-7.00008-6

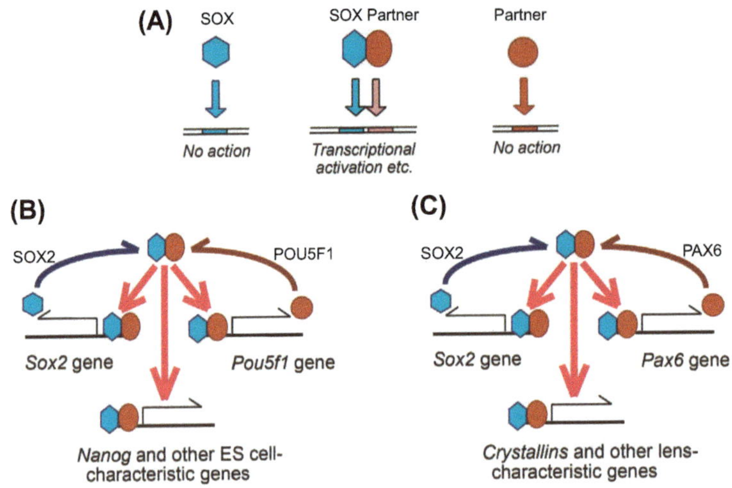

Figure 1 Cooperation between SOX2 and a partner transcription factor. (A) Cooperation of SOX2 with a partner factor binding to an adjacent DNA sites elicits transcriptional regulation by their complex. This mechanism is involved in the co-activation loops of *Sox2–Pou5f1* genes by SOX2 and POU5F1 in ESCs (B) and of *Sox2–Pax6* genes by SOX2 and PAX6 during development of the visual system (C).

enhancers associated with genes involved in the maintenance of the ESC state (Boyer et al., 2005; Chen et al., 2008; Kondoh and Kamachi, 2010). Of particular importance is the finding that *Sox2* and *Pou5f1* genes themselves are activated by the combination of SOX2 and POU5F1 (Tomioka et al., 2002; Chew et al., 2005). The ESC state is stabilized because of this co-activation loop involving SOX2 and PO5F1 (Figure 1(B)). It is one of the reasons for including SOX2 and PO5F1 as essential players in the transcription factor quartet to produce induced pluripotent cells (iPS) (Takahashi and Yamanaka, 2006).

Some representative enhancer core sequences bound by SOX2 and POU5F1 are shown in Figure 2. At least three different classes of SOX2-POU5F1 co-binding modes to the core enhancer are recognized. In the class 1 mode, which represents most cases, the SOX2 binding sequence [A/T]TTGT(T) is immediately followed by the octamer binding sequence ATGCAAAT or its variations shared by many POU transcription factors as their binding sequence. However, a few bases are inserted in the SOX2 and POU5F1 binding sequences with the same orientation in the class 2 mode, such as a single base insertion to *developmental pluripotency-associated 4* (*Dppa4*) enhancer (Chakravarthy et al., 2008) and a three-base insertion to the *Fgf4* enhancer. The class 3 mode is represented by the *Nestin* enhancer, and the orientation of the SOX2 binding and octamer sequences is completely altered (Tanaka et al., 2004).

A crystallographic study was performed using a SOX2-POU factor complex made with SOX2 and POU2F1 (OCT1), a POU5F1 replacement, on the *Fgf4* enhancer sequence (class 2) (Remenyi et al., 2003). Based on the crystallographic data, a model was built for the SOX2-POU5F1 complex made on the *Undifferentiated Transcription Factor 1*

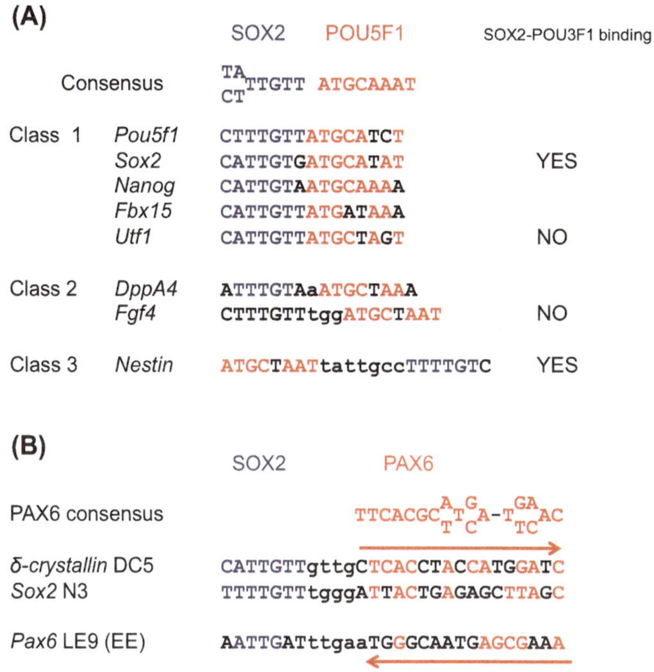

Figure 2 SOX2-POU5F1 and SOX2-PAX6 co-binding sequences. (A) Three classes of SOX2-POU5F1 binding sequences compared with consensus sequences for SOX2 (blue) and POU factor (red) binding. Note that the naturally occurring SOX2-POU5F1 binding sequence contains an octamer (ATGCAAAT)-related sequence that deviates from the canonical octamer in a few base positions. (B) Three well-characterized SOX2-PAX6 co-binding sequences. Note that the PAX6-interacting sequence considerably deviates from the PAX6 standalone binding consensus sequence. These deviations are essential for eliciting the transactivation potential after SOX2 and PAX6 co-binding.

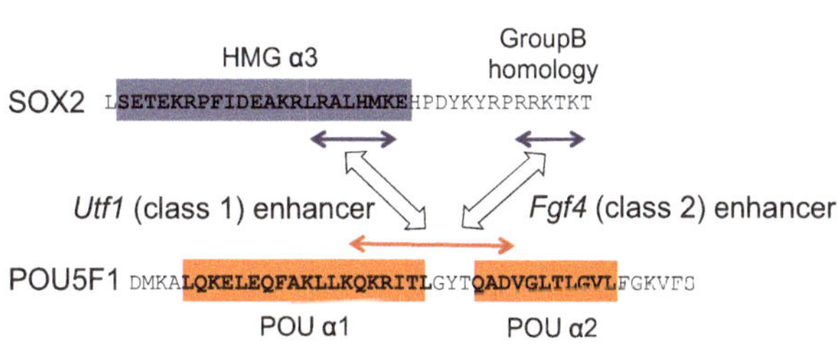

Figure 3 Amino acid residues of SOX2 and POU5F1 that interact in the SOX2-POU5F1–DNA complexes depend on the enhancer sequence they bind (Remenyi et al., 2003). POU5F1 (bottom) interacts with SOX2 at the loop region between POU domain α-helices 1 and 2. The SOX2 interaction site (top) depends on spacing of the SOX2-interacting and POU-interacting sequences. The C-proximal portion of the SOX2 α-helix 3 in the case of the *Utf1* (class 1) enhancer and the Group B homology region of the *Fgf4* (class 2) enhancer interacts with POU5F1.

(*UTF1*) enhancer (class 1). That study revealed the following features of the interaction between the SOX2-POU5F1 factors on the bound DNA (Figure 3): (1) The C-terminal end region of the SOX2 HMG domain third α-helix and the loop at the junction of the POU5F1 POU domain α1-α2 helices are in direct contact on bound DNA during class 1 binding. (2) Inserting three bases between the SOX2 and POU5F1 binding sequences in the *Fgf4* enhancer adds rotational and axial shifts in the HMG–POU interaction interface; however, the interaction is maintained by a change in the SOX2 domain from the HMG third α-helix to the C-proximal extension of the HMG domain, which interacts with the POU domain α1-α2 loop (Remenyi et al., 2003). The C-proximal extension region corresponds to the group B homology region rich in basic amino acid residues and is shared by SOXB1 and SOXB2 factors (Uchikawa et al., 1999). This group B-homology region has also been suggested to be the SOX2 interface in the SOX2-PAX6–DNA complex (Remenyi et al., 2003). Therefore, SOX2 is flexible in its interactions with various transcription factors at the C-proximal region of the HMG domain by possessing various facets of protein–protein interaction.

The octamer motif ATGCAAAT is bound by various POU factors and is not unique to POU5F1 binding. However, the occurrence of the consensus ATGCAAAT is rare in the natural enhancer sequence and deviations from the consensus by one or two bases are common (Figure 2). This deviation from the consensus sequence presumably determines the specificity of the POU factors that co-bind to DNA with SOX2.

Simultaneous binding of POU3F factors (POU3F1 [OCT6], or POU3F2 [Brn2]) with SOX2 to the SOX2-POU5F1 co-binding sequences shown in Figure 2 depends on the POU factor binding motifs. The *Sox2* SRR2 enhancer with the POU binding sequence ATGCATAT permits co-binding of the POU3F factor with SOX2 (Tomioka et al., 2002), whereas enhancers with ATGCTAAT (*Fgf4*) (Ambrosetti et al., 1997) or ATGCTAGT (*UTF1*) (Nishimoto et al., 1999) exclude co-binding with SOX2. ATGCTAAT is also found in the *Nestin* enhancer, and POU3F factors and POU5F1 equally cooperate with SOX2 during the enhancer activation in this class (Tanaka et al., 2004). Therefore, a particular deviation from the octamer consensus sequence has a discriminative effect under the condition of tightly associated SOX2-POU5F1 co-binding (classes 1 and 2).

In addition, co-binding of SOX2 and the POU factor to adjacent sites does not guarantee their functional cooperation. For example, although SOX2 and PO2F1 co-bind to the *Fgf4* enhancer sequence and have been used for crystallographic analysis of SOX2-POU–DNA ternary complexes, the complexes fail to activate the enhancer (Remenyi et al., 2003). However, the SOX2-POU2F1 pair activates the *Pax6* EE enhancer containing the uniquely arranged POU-SOX2 binding site sequence ATTCAAATACAAT (the POU binding sequence is underlined) (Donner et al., 2007) and *Hoxb1* enhancer with class 1 arrangement of the SOX2 and POU factor binding sequences, CTTTGTCATGCTAAT (Di Rocco et al., 2001). Therefore, the compatibility of the SOX2-POU factor pair with a combination of SOX2 and POU binding sequences is determined by their arrangement, spacing, and the particular way in which they deviate from the canonical octamer sequence.

The *Nestin* enhancer Nes30 sequence, which represents class 3, has larger spacing in the unique arrangement of the SOX2 and POU binding sequences. It binds to either POU5F1 or POU3F factors (POU3F1 [OCT6], POU3F2 [BRN2], POU3F3 [BRN1], and POU3F4 [BRN4]) without discrimination with SOX2 but without a notable cooperative effect (Tanaka et al., 2004). It has not been clarified how the SOX2 and POU factors interact on DNA in this binding sequence configuration. In vitro studies are in favor of the model in which two factors independently glide on DNA and meet at the co-binding sequences, as discussed in Chapter 3. Alternatively, it will be significant if the SOX2 and POU factors interact without DNA, as indicated by pull-down experiments (Ambrosetti et al., 1997; Tanaka et al., 2004). Preformed SOX2-POU factor complexes presumably facilitate paired binding of these factors to nearby DNA sites, suggesting a model in which the SOX2-POU heterodimer approaches target DNA sites, and binding of SOX2 and a POU factor to nearby sites is established when the DNA sequence is compatible with binding of the particular heterodimer. However, some reservation to this model is discussed in Chapter 5.

CASES FOR OVERLAP OF HIGH-DENSITY CLUSTERS OF SOX2 BINDING AND POU BINDING SEQUENCES

Both ESCs and ESC-derived neural progenitor cells (NPCs) abundantly express SOX2 and Pou factors: POU5F1 in the former and POU3F2 (BRN2) in the latter. Chromatin immunoprecipitation-sequencing (ChIP-seq) analyses of both types of cells indicate that a limited fraction of the ChIP-seq peaks bound by both SOX2 and POU5F1 are co-bound by SOX2 and POU3F2 (Lodato et al., 2013). Our analysis of their data deposited in the GEO database indicated that many of these ChIP-seq peaks shared by SOX2-POU5F1 in ESCs and SOX2-POU3F2 in NPCs are characterized by a high score of fold-enrichment relative to background in ESCs. One example is a prominent ChIP-seq peak downstream of the *Sox21* gene, previously designated as *Sox21*-19 in analysis of highly conserved non-coding sequences (Woolfe et al., 2005). The *Sox21*-19 region has been tested for enhancer activity using transgenic zebrafish embryos and was demonstrated to have strong enhancer activity in the embryonic central nervous system (Woolfe et al., 2005). An inspection of the DNA sequences in this region indicated that dense clusters of SOX2 and POU factor binding sequences overlapped (Figure 4); however, no tightly associated SOX2-POU binding motifs corresponding to these heterodimer binding classes were found. This finding suggests that the overlap of high-density clusters SOX2 and POU factor binding sequences, not necessarily at close proximity, is sufficient to build a strong enhancer and may represent a new fourth class of SOX2-POU factor co-binding.

Binding of the SOX2 and POU factors will not depend on the preformed SOX2-POU complex in the class 4 arrangements of SOX2 and POU factor co-binding sequences, because locally enriched SOX2 and POU binding sequences will be found more efficiently by randomly scanning single transcription factors. Because SOX2 recruits the chromatin remodeling factor CHD7, as discussed below and in Chapter 5, the local accumulation of SOX2 will help access other transcription factors, such as partner factors. These combined effects may result in a greater occupancy by SOX2 and POU factors than by the class 1–3 cases, which was reflected by the greater enrichment of ChIP-seq reads.

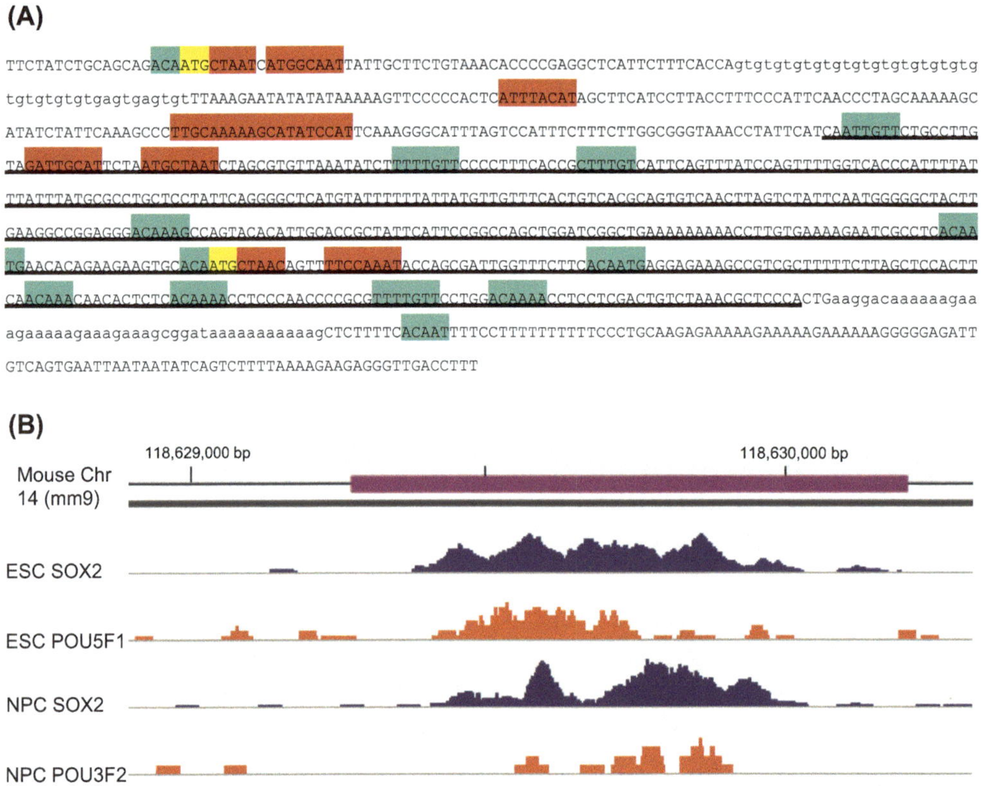

Figure 4 Organization of the SOX2 and POU binding sequences in the mouse *Sox21*-19 enhancer (Woolfe et al., 2005). (A) The *Sox21*-19 enhancer sequence (underlined) and flanking sequence are shown. SOX2 binding sequences including the core sequence [A/T]TTGT or its reverse ACAA[T/A], which often appear as TTTTGTT (AACAAAA) or C[A/T]TTGT (ACAA[T/A]G), are highlighted in blue; the Pou binding sequences with not more than a two-base deviation from the canonical octamer sequence (see Figure 2) are highlighted in red; and their overlap is indicated by yellow. (B) SOX2 and POU factor chromatin immunoprecipitation-sequencing (ChIP-seq) fragment distributions showing in vivo binding of factors in ESCs and neural progenitor cells (NPCs) aligned with a region of mouse chromosome 14 downstream of *Sox21*. The sequence shown in (A) includes the *Sox21*-19 enhancer and is indicated by the purple bar. The data indicate that the *Sox21*-19 region is bound by SOX2 and POU5F1/POU3F2 in ESCs and NPCs. However, the details of binding profiles differ between ESCs and NPCs. Original data are those reported by Lodato et al. (2013) and deposited in the GEO database with accession number GSE38850. The data were analyzed using SraTailor (Oki et al., 2014). The full vertical scales are two reads per million for ESC SOX2 and POU5F1 ChIP-seq, four reads per million for NPC SOX2, and one read per million for NPC POU3F2.

One study (Chen et al., 2014) investigated single-molecule kinetics of fluorescent protein–tagged SOX2 and POU5F1 in ESCs. Using single molecule tracking, the authors observed a number of intranuclear loci in which the molecules displayed a significant residence time of approximately 10s. This residence time was not affected so greatly by introducing mutations at the SOX2–partner factor interface (Figure 3), which suggests that the residence to SOX2 binding site is not stabilized by interaction with partner factors. However, it is likely that their observations were strongly biased on class 4 targets in which residence by transcription factors is more frequent. Indeed, the overall intranuclear diffusion rate of SOX2 reflecting the entire ensemble of SOX2–target interactions was significantly augmented by a mutation at the interface of SOX2–partner factor interaction in our unpublished study. Therefore, it is likely that target hunting by SOX2 and/or stability of SOX2 binding to specific sites depends on the nature of the target sites: pairing with a partner for class 1–3 targets or singly for class 4 targets. Indeed, a nuclear magnetic resonance study on the interaction of SOX2 with *Utf1* and *Fgf4* enhancer sequences (Takayama and Clore, 2012) demonstrated that the SOX2 exchange rate was significantly reduced by interaction with POU2F1 (also discussed in Chapter 3).

SOX2–PAX6 INTERACTION

As described above, the DC5 enhancer, which regulates the *δ-crystallin* gene, was the first identified genetic element to depend on SOX2-PAX6 interaction (Kamachi et al., 1995, 2001). Further survey of SOX2-PAX6 targets indicates their association with the development of the visual system, including the *Sox2–Pax6* co-activation system (Kamachi et al., 1998; Aota et al., 2003; Inoue et al., 2007) (Figure 1(C)). The biological significance of *Sox2–Pax6* co-activation loops is discussed in Chapter 12; here, we focus on ternary molecular interactions among SOX2, PAX6, and target DNAs.

The DC5 sequence contained two adjacent genetic elements bound by different nuclear factors when it was initially defined. Simultaneous binding by these two factors is required to elicit enhancer activity (Kamachi and Kondoh, 1993). One of the factors was identified as SOX2 by expression cloning of cDNAs that code for a DC5 DNA-binding protein (Kamachi et al., 1995); however, cloning the second factor required a more sophisticated functional assay, which identified it as PAX6 (Kamachi et al., 2001). The DC5 PAX6 binding sequence deviates considerably from the PAX6 standalone consensus binding sequence (Epstein et al., 1994) and fails to achieve strong PAX6 binding (Kamachi et al., 2001). However, PAX6 binds cooperatively to the DC5 sequence when SOX2 is present and forms a strong SOX2-PAX6–DNA ternary complex (Figure 5). Therefore, the DC5 sequence serves as a SOX2-PAX6 heterodimer binding sequence rather than as the sum of the SOX2 and PAX6 binding sequences. The second SOX2-PAX6 heterodimer binding sequence was found in the *Sox2* N3 enhancer core. It had the same spatial organization as the SOX2-interacting and PAX6-interacting DNA sequences and exhibited SOX2-dependent PAX6 binding analogs to DC5 (Figure 5).

Modeling a three-dimensional structure of the SOX2-PAX6 complex on DC5 DNA using the crystallographic data of SOX2-POU2F1–DNA complex (Remenyi et al., 2003) and PAX6–DNA complex crystallographic data (Xu et al., 1999) indicated a molecular interaction between SOX2 and PAX6 analogous to the SOX2-POU5F1 interaction: namely

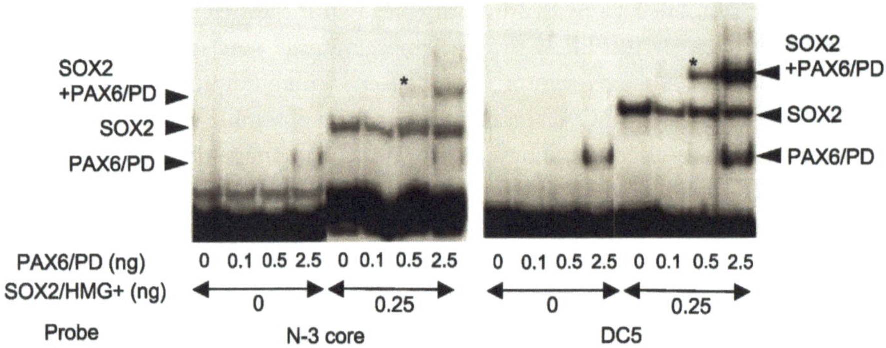

Figure 5 An electrophoretic mobility shift assay of SOX2 and PAX6 binding to the DC5 and N3 core enhancer sequences, showing SOX2-dependent cooperative binding of PAX6 to these sequences. The PAX6 paired domain (PD) alone barely bound to the N3 core (left) or DC5 (right) probe. However, the addition of the SOX2 HMG domain elicited cooperative binding of PAX6 PD to form the ternary complex with the probe even when PAX6 PD alone showed no detectable binding (asterisks). *Data were reproduced from Inoue et al. (2007).*

the Group B homology region (C-proximal polypeptide extension from the SOX2 HMG domain) (Uchikawa et al., 1999) interacts with the PAX6 N-proximal portion extending from the paired domain, which was confirmed by loss of SOX2-PAX6 co-binding after introducing a mutation in the predicted SOX2-PAX6 interface amino acid residues (Remenyi et al., 2003). Inserting short DNA sequences in DC5 between the SOX2- and PAX6-interacting DNA sequences disrupted the activation of the enhancer (Kamachi et al., 2001), which indicated that a highly organized interaction between the SOX2 and PAX6 amino acid residues is required to form the complex between SOX2 and PAX6 on DC5 and is essential to activate the enhancer.

Importantly, deviation in the PAX6-interacting sequence of the DC5 enhancer from the PAX6 standalone binding sequence is essential for eliciting enhancer activity. When the PAX6-interacting sequence was replaced by the PAX6 standalone consensus sequence (DC5con), the DC5con sequence bound both SOX2 and PAX6; however, the enhancer activity was negligible (Kamachi et al., 2001).

One study (Narasimhan et al., 2015) confirmed that deviations in the PAX6-interacting DNA sequences in DC5 and N3 from the PAX6 standalone consensus sequence are essential for cooperative binding of SOX2 and PAX6. The authors further conducted DC5 sequence mutational analysis and found that alterations in three DC5 sequence positions that better matched the PAX6 consensus sequence abrogated cooperative binding.

The same study investigated the overlapping genomic regions of the SOX2 and PAX6 ChIP-seq peaks in NPCs and enlisted potential SOX2-PAX6 heterodimer binding sites that share properties with DC5 (Narasimhan et al., 2015).

An electrophoretic mobility shift assay of the SOX2-PAX6–DC5 ternary complex indicated that when the ternary complex lacks enhancer activation, when using either the DC5con sequence or DC5 with base insertions, mobility of the ternary complexes decreases compared with that of the normal complex (Kamachi et al., 2001). This observation indicates that the functionally active SOX2-PAX6–DC5 ternary complex assumes a more compact form than that of the inactive complex.

A study evaluating DNA bending caused by SOX2 and PAX6 binding to DC5 DNA demonstrated a clear difference between normal DC5 and DC5con (Morimura et al., 2013). The degree of DNA bending was evaluated by placing nano-gold particles at both ends of DC5 DNA and measuring the red shift in the plasmon resonance wavelength, as determined by the peak wavelength of light scattered by the nano-gold pairs. The plasmon resonance wavelength increases when the distance between the nano-gold particles becomes shorter after DNA bending (Figure 6). SOX2 alone caused approximately 61° bending upon binding to either DC5 or DC5con, as expected from the HMG domain interaction with a minor groove of DNA (see Chapter 2). However, DC5 was further bent when SOX2 and PAX6 were bound together. Assuming that the second bend occurred at the center of the PAX6-interacting sequence, the bend angle was estimated to be 6°. In contrast, PAX6 did not induce further bending when SOX2 and PAX6 were bound to DC5con. This observation indicates that the tight association between SOX2 and PAX6 on DNA causes greater DNA bending than that caused by SOX2 alone.

SOX2-INTERACTING CHROMATIN MODIFICATION FACTORS AND COVALENT MODIFICATION OF THE SOX2 PROTEIN

The activity of the SOX2–partner factor complex may be modulated by interactions with cofactors or by covalent modifications. As discussed in Chapter 5, SOX2 interacts with CHD7 (Engelen et al., 2011), which is a major chromatin remodeling factor essential for normal ESC-specific gene expression (Schnetz et al., 2010). It is remarkable that CHD7 mutations phenocopy to a large extent the *Sox2* hypomorphism (Williamson et al., 2006; Engelen et al., 2011). This interaction between SOX2 with CHD7 may be part of the mechanism for SOX2 to act as a pioneer factor, as discussed in Chapter 9. In addition, Pax6 interacts with BAF, which is another class of chromatin remodeling complex (Ninkovic et al., 2013). The SOX2–partner factor complex may have various impacts on the local chromatin structure by recruiting chromatin remodeling factors.

SOX2 is covalently modified (phosphorylation, SUMOylation, and others), as summarized in Figure 7, under certain conditions that remain to be investigated (Kamachi and Kondoh, 2013). Most of these modifications affect the nuclear shuttling of SOX2 but possibly modulate interactions with partners and cofactors as well. It was shown that SUMOylation of SOX2 impairs interaction with POU5F1 in the regulation of *Nanog* (Wu et al., 2012).

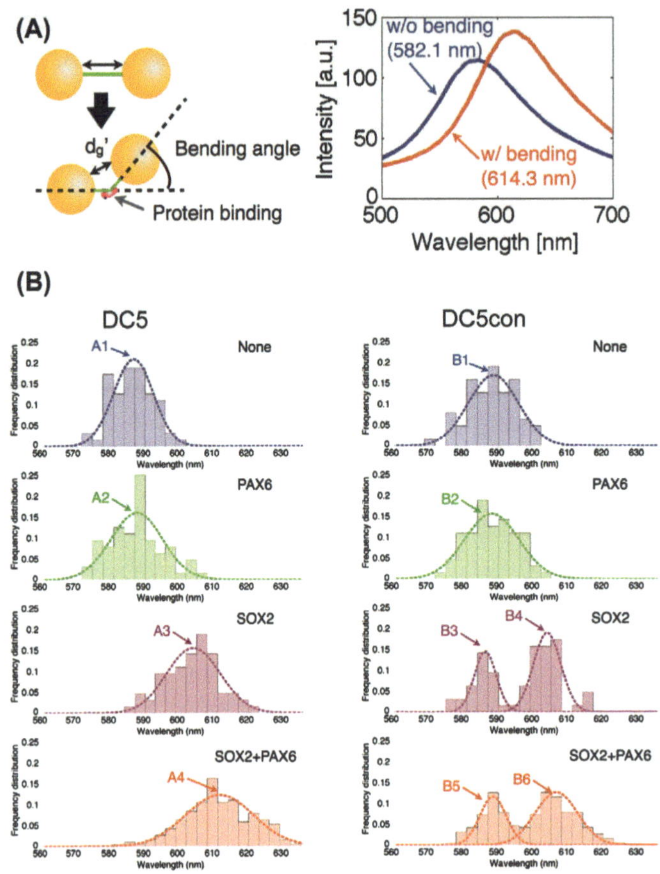

Figure 6 Detecting changes in DC5 DNA bending by the red shift in the plasmon resonance wavelength caused by nano-gold particles attached at both ends. (A) A diagram of the gold nanoparticles attached at both ends of DC5 (left) and the expected plasmon resonance wavelength shift upon DNA bending. (B) Changes in the plasmon resonance wavelength after binding of SOX2 and/or PAX6 to DC5 DNA (left) or DC5con DNA (right). The plasmon resonance wavelength without protein factor binding was similar for both DNAs, 586.8 ± 5.7 nm (peak A1) and 588.1 ± 7.1 nm (peak B1). The addition of Pax6 alone did not alter the peak wavelengths (peaks A2 and B2). The addition of SOX2 caused a similar increase in peak wavelengths to 604.1 ± 7.7 nm (peak A3) using DC5 and to 603.9 ± 3.9 nm (peak B4) using DC5con. The addition of SOX2 and PAX6 caused a further increase in the peak wavelength to 611.6 ± 9.8 nm (peak A4) using the DC5 sequence; however, using the DC5con sequence, the addition of SOX2 and PAX6 did not cause a further change compared with that caused by the addition of SOX2 alone (peak B6, 606.5 ± 6.1 nm). Peaks B3 and B5 represent factor-unbound DNA observed using DC5con. *The figure and data were reproduced from Morimura et al. (2013).*

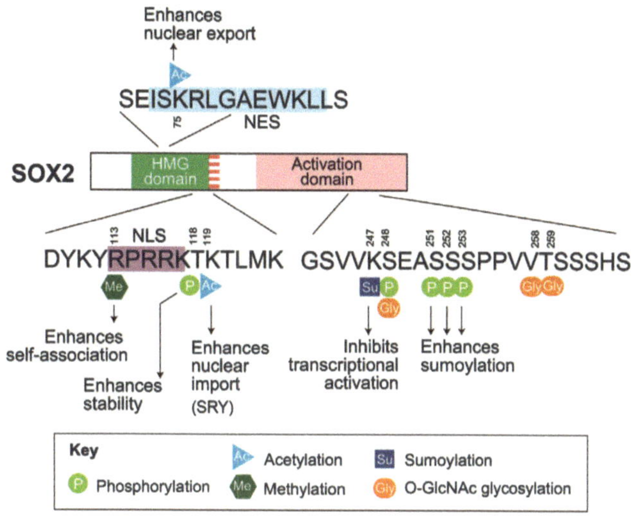

Figure 7 SOX2 protein covalent modifications. Types of posttranslational covalent modifications and their functional effects, based on studies using human or mouse SOX2, are indicated on the mouse SOX2 amino acid coordinate. K119 acetylation observed in SRY that presumably occurs in SOX2 is also indicated. The nuclear export signal (NES), nuclear localization signal (NLS), and Group B homology region are highlighted in blue, purple, and striped red, respectively. *The figure was adapted from Kamachi and Kondoh (2013).*

OTHER CASES OF GENE REGULATION THAT DEPEND ON SOX2–PARTNER FACTOR INTERACTION

In the previous sections, we focused on the partnership of SOX2 with POU factors or PAX6. A great variety of transcription factors have been registered as SOX2-interacting partners regulating target gene transcription. This issue is also discussed in Chapter 5.

MAF factors are SOX2 partners in the regulation of *γ-crystallin* genes via heterodimer binding to the promoters (Rajaram and Kerppola, 2004), and OTX2 serves as the SOX2 partner factor of SOX2 to activate *Rax* for retinal development (Danno et al., 2008), as discussed in Chapter 12. SALL4 and SOX2 bind next to each other in the core element of NOP1, a naso-otic specificity enhancer, as discussed in Chapter 7. SOX2 interacts with SIX1-EYA1 to regulate *Atoh1* during inner ear development (Ahmed et al., 2012) and with SOSTDC1 in the dermal papilla (Clavel et al., 2012).

SOX2-dependent regulation often intersects with sonic hedgehog (Shh) signaling, and the SOX2-GLI complex is suggested to be involved (Oosterveen et al., 2012; Peterson et al., 2012). The regulatory regions of transcription factor genes specifying the ventral spinal cord progenitors, such as *NK2 homeobox 2* (*Nkx2.2*), *oligodendrocyte transcription factor 2* (*Olig2*), *Nkx6.1*, and *Nkx6.2*, the GLI protein binding sequence for GLI proteins, the transcription factors downstream of shh signaling, are juxtaposed to the SOX2 binding sequence (Oosterveen et al., 2012; Peterson et al., 2012).

In addition to these gene-based investigations, genome-wide analysis of SOX2 and interacting transcription factors, for example, ChIP-seq analysis for overlapping binding peaks, has revealed novel SOX2 partners characteristic of cell types. In trophoblast stem cells derived from ESCs, TFAP2C rather than POU5F1 is the major partner of SOX2 (Adachi et al., 2013), whereas in squamous cell carcinoma P63 is the major SOX2 partner factor that jointly activates the oncogene *ETV4* (Watanabe et al., 2014). It has been reported (Hutchins et al., 2013) that SOX2-ESRRB complexes formed on a number of genomic sites differentiate ESCs from epiblast stem cells. In addition, it has been indicated that PRX1 functions with SOX2 to regulate neural stem cells (Shimozaki et al., 2013).

The variety of SOX2–partner factor interactions that occur during development and disease will increase with more investigation. We hope that the major principles of SOX2–POU5F1 and SOX2-PAX6 interactions with their target DNAs summarized in this chapter will help decipher interactions between transcription factors and their targets that regulate developmental and disease processes.

REFERENCES

Adachi, K., Nikaido, I., Ohta, H., Ohtsuka, S., Ura, H., Kadota, M., Wakayama, T., Ueda, H.R., Niwa, H., 2013. Context-dependent wiring of Sox2 regulatory networks for self-renewal of embryonic and trophoblast stem cells. Mol. Cell 52, 380–392.

Ahmed, M., Wong, E.Y., Sun, J., Xu, J., Wang, F., Xu, P.X., 2012. Eya1-Six1 interaction is sufficient to induce hair cell fate in the cochlea by activating Atoh1 expression in cooperation with Sox2. Dev. Cell 22, 377–390.

Aksoy, I., Jauch, R., Chen, J., Dyla, M., Divakar, U., Bogu, G.K., Teo, R., Ng, C.K.L., Herath, W., Lili, S., Hutchins, A.P., Robson, P., Kolatkar, P.R., Stanton, L.W., 2013. Oct4 switches partnering from Sox2 to Sox17 to reinterpret the enhancer code and specify endoderm. EMBO J. 32, 938–953.

Ambrosetti, D.C., Basilico, C., Dailey, L., 1997. Synergistic activation of the fibroblast growth factor 4 enhancer by Sox2 and Oct-3 depends on protein-protein interactions facilitated by a specific spatial arrangement of factor binding sites. Mol. Cell. Biol. 17, 6321–6329.

Aota, S., Nakajima, N., Sakamoto, R., Watanabe, S., Ibaraki, N., Okazaki, K., 2003. Pax6 autoregulation mediated by direct interaction of Pax6 protein with the head surface ectoderm-specific enhancer of the mouse Pax6 gene. Dev. Biol. 257, 1–13.

Boyer, L.A., Lee, T.I., Cole, M.F., Johnstone, S.E., Levine, S.S., Zucker, J.P., Guenther, M.G., Kumar, R.M., Murray, H.L., Jenner, R.G., Gifford, D.K., Melton, D.A., Jaenisch, R., Young, R.A., 2005. Core transcriptional regulatory circuitry in human embryonic stem cells. Cell 122, 947–956.

Chakravarthy, H., Boer, B., Desler, M., Mallanna, S.K., McKeithan, T.W., Rizzino, A., 2008. Identification of DPPA4 and other genes as putative Sox2:Oct-3/4 target genes using a combination of in silico analysis and transcription-based assays. J. Cell. Physiol. 216, 651–662.

Chen, J., Zhang, Z., Li, L., Chen, B.C., Revyakin, A., Hajj, B., Legant, W., Dahan, M., Lionnet, T., Betzig, E., Tjian, R., Liu, Z., 2014. Single-molecule dynamics of enhanceosome assembly in embryonic stem cells. Cell 156, 1274–1285.

Chen, X., Xu, H., Yuan, P., et al., 2008. Integration of external signaling pathways with the core transcriptional network in embryonic stem cells. Cell 133, 1106–1117.

Chew, J.L., Loh, Y.H., Zhang, W., Chen, X., Tam, W.L., Yeap, L.S., Li, P., Ang, Y.S., Lim, B., Robson, P., Ng, H.H., 2005. Reciprocal transcriptional regulation of Pou5f1 and Sox2 via the Oct4/Sox2 complex in embryonic stem cells. Mol. Cell. Biol. 25, 6031–6046.

Clavel, C., Grisanti, L., Zemla, R., Rezza, A., Barros, R., Sennett, R., Mazloom, A.R., Chung, C.Y., Cai, X., Cai, C.L., Pevny, L., Nicolis, S., Ma'ayan, A., Rendl, M., 2012. Sox2 in the dermal papilla niche controls hair growth by fine-tuning BMP signaling in differentiating hair shaft progenitors. Dev. Cell 23, 981–994.

Collignon, J., Sockanathan, S., Hacker, A., Cohen-Tannoudji, M., Norris, D., Rastan, S., Stevanovic, M., Goodfellow, P.N., Lovell-Badge, R., 1996. A comparison of the properties of Sox-3 with Sry and two related genes, Sox-1 and Sox-2. Development 122, 509–520.

Dailey, L., Yuan, H., Basilico, C., 1994. Interaction between a novel F9-specific factor and octamer-binding proteins is required for cell-type-restricted activity of the fibroblast growth factor 4 enhancer. Mol. Cell. Biol. 14, 7758–7769.

Danno, H., Michiue, T., Hitachi, K., Yukita, A., Ishiura, S., Asashima, M., 2008. Molecular links among the causative genes for ocular malformation: Otx2 and Sox2 coregulate Rax expression. Proc. Natl. Acad. Sci. U.S.A. 105, 5408–5413.

Di Rocco, G., Gavalas, A., Popperl, H., Krumlauf, R., Mavilio, F., Zappavigna, V., 2001. The recruitment of SOX/OCT complexes and the differential activity of HOXA1 and HOXB1 modulate the Hoxb1 auto-regulatory enhancer function. J. Biol. Chem. 276, 20506–20515.

Donner, A.L., Episkopou, V., Maas, R.L., 2007. Sox2 and Pou2f1 interact to control lens and olfactory placode development. Dev. Biol. 303, 784–799.

Engelen, E., Akinci, U., Bryne, J.C., et al., 2011. Sox2 cooperates with Chd7 to regulate genes that are mutated in human syndromes. Nat. Genet. 43, 607–611.

Epstein, J., Cai, J., Glaser, T., Jepeal, L., Maas, R., 1994. Identification of a Pax paired domain recognition sequence and evidence for DNA-dependent conformational changes. J. Biol. Chem. 269, 8355–8361.

Hutchins, A.P., Choo, S.H., Mistri, T.K., Rahmani, M., Woon, C.T., Ng, C.K., Jauch, R., Robson, P., 2013. Co-motif discovery identifies an Esrrb-Sox2-DNA ternary complex as a mediator of transcriptional differences between mouse embryonic and epiblast stem cells. Stem Cells 31, 269–281.

Inoue, M., Kamachi, Y., Matsunami, H., Imada, K., Uchikawa, M., Kondoh, H., 2007. PAX6 and SOX2-dependent regulation of the Sox2 enhancer N-3 involved in embryonic visual system development. Genes Cells 12, 1049–1061.

Iwafuchi-Doi, M., Matsuda, K., Murakami, K., Niwa, H., Tesar, P.J., Aruga, J., Matsuo, I., Kondoh, H., 2012. Transcriptional regulatory networks in epiblast cells and during anterior neural plate development as modeled in epiblast stem cells. Development 139, 3926–3937.

Iwafuchi-Doi, M., Yoshida, Y., Onichtchouk, D., Leichsenring, M., Driever, W., Takemoto, T., Uchikawa, M., Kamachi, Y., Kondoh, H., 2011. The Pou5f1/Pou3f-dependent but SoxB-independent regulation of conserved enhancer N2 initiates Sox2 expression during epiblast to neural plate stages in vertebrates. Dev. Biol. 352, 354–366.

Kamachi, Y., Kondoh, H., 1993. Overlapping positive and negative regulatory elements determine lens-specific activity of the delta 1-crystallin enhancer. Mol. Cell. Biol. 13, 5206–5215.

Kamachi, Y., Kondoh, H., 2013. Sox proteins: regulators of cell fate specification and differentiation. Development 140, 4129–4144.

Kamachi, Y., Sockanathan, S., Liu, Q., Breitman, M., Lovell-Badge, R., Kondoh, H., 1995. Involvement of SOX proteins in lens-specific activation of crystallin genes. EMBO J. 14, 3510–3519.

Kamachi, Y., Uchikawa, M., Collignon, J., Lovell-Badge, R., Kondoh, H., 1998. Involvement of Sox1, 2 and 3 in the early and subsequent molecular events of lens induction. Development 125, 2521–2532.

Kamachi, Y., Uchikawa, M., Kondoh, H., 2000. Pairing SOX off: with partners in the regulation of embryonic development. Trends Genet. 16, 182–187.

Kamachi, Y., Uchikawa, M., Tanouchi, A., Sekido, R., Kondoh, H., 2001. Pax6 and SOX2 form a co-DNA-binding partner complex that regulates initiation of lens development. Genes Dev. 15, 1272–1286.

Kondoh, H., Kamachi, Y., 2010. SOX-partner code for cell specification: regulatory target selection and underlying molecular mechanisms. Int. J. Biochem. Cell Biol. 42, 391–399.

Lodato, M.A., Ng, C.W., Wamstad, J.A., Cheng, A.W., Thai, K.K., Fraenkel, E., Jaenisch, R., Boyer, L.A., 2013. SOX2 co-occupies distal enhancer elements with distinct POU factors in ESCs and NPCs to specify cell state. PLoS Genet. 9, e1003288.

Morimura, H., Tanaka, S., Ishitobi, H., Mikami, T., Kamachi, Y., Kondoh, H., Inouye, Y., 2013. Nano-analysis of DNA conformation changes induced by transcription factor complex binding using plasmonic nanodimers. ACS Nano 7, 10733–10740.

Narasimhan, K., Pillay, S., Jayabal, S., Udayasuryan, B., Huang, Y., Veerapandian, V., Kolatkar, P., Cojocaru, K., Pervushin, K., & Jauch, R., 2015. DNA mediated cooperativity facilitates the co-selection of cryptic enhancer sequences by SOX2 and PAX6 transcription factors in neurogenesis. Nucleic Acids Res. 43, 1513–1528.

Ninkovic, J., Steiner-Mezzadri, A., Jawerka, M., et al., 2013. The BAF complex interacts with Pax6 in adult neural progenitors to establish a neurogenic cross-regulatory transcriptional network. Cell Stem Cell 13, 403–418.

Nishimoto, M., Fukushima, A., Okuda, A., Muramatsu, M., 1999. The gene for the embryonic stem cell coactivator UTF1 carries a regulatory element which selectively interacts with a complex composed of Oct-3/4 and Sox-2. Mol. Cell. Biol. 19, 5453–5465.

Oki, S., Maehara, K., Ohkawa, Y., Meno, C., 2014. SraTailor: graphical user interface software for processing and visualizing ChIP-seq data. Genes Cells 19, 919–926.

Oosterveen, T., Kurdija, S., Alekseenko, Z., Uhde, C.W., Bergsland, M., Sandberg, M., Andersson, E., Dias, J.M., Muhr, J., Ericson, J., 2012. Mechanistic differences in the transcriptional interpretation of local and long-range Shh morphogen signaling. Dev. Cell 23, 1006–1019.

Peterson, K.A., Nishi, Y., Ma, W., et al., 2012. Neural-specific Sox2 input and differential Gli-binding affinity provide context and positional information in Shh-directed neural patterning. Genes Dev. 26, 2802–2816.

Rajaram, N., Kerppola, T.K., 2004. Synergistic transcription activation by Maf and Sox and their subnuclear localization are disrupted by a mutation in Maf that causes cataract. Mol. Cell. Biol. 24, 5694–5709.

Remenyi, A., Lins, K., Nissen, L.J., Reinbold, R., Scholer, H.R., Wilmanns, M., 2003. Crystal structure of a POU/HMG/DNA ternary complex suggests differential assembly of Oct4 and Sox2 on two enhancers. Genes Dev. 17, 2048–2059.

Schnetz, M.P., Handoko, L., Akhtar-Zaidi, B., Bartels, C.F., Pereira, C.F., Fisher, A.G., Adams, D.J., Flicek, P., Crawford, G.E., Laframboise, T., Tesar, P., Wei, C.L., Scacheri, P.C., 2010. CHD7 targets active gene enhancer elements to modulate ES cell-specific gene expression. PLoS Genet. 6, e1001023.

Shimozaki, K., Clemenson Jr., G.D., Gage, F.H., 2013. Paired related homeobox protein 1 is a regulator of stemness in adult neural stem/progenitor cells. J. Neurosci. 33, 4066–4075.

Takahashi, K., Yamanaka, S., 2006. Induction of pluripotent stem cells from mouse embryonic and adult fibroblast cultures by defined factors. Cell 126, 663–676.

Takayama, Y., Clore, G.M., 2012. Impact of protein/protein interactions on global intermolecular translocation rates of the transcription factors Sox2 and Oct1 between DNA cognate sites analyzed by z-exchange NMR spectroscopy. J. Biol. Chem. 287, 26962–26970.

Tanaka, S., Kamachi, Y., Tanouchi, A., Hamada, H., Jing, N., Kondoh, H., 2004. Interplay of SOX and POU factors in regulation of the Nestin gene in neural primordial cells. Mol. Cell. Biol. 24, 8834–8846.

Tomioka, M., Nishimoto, M., Miyagi, S., Katayanagi, T., Fukui, N., Niwa, H., Muramatsu, M., Okuda, A., 2002. Identification of Sox-2 regulatory region which is under the control of Oct-3/4-Sox-2 complex. Nucleic Acids Res. 30, 3202–3213.

Uchikawa, M., Kamachi, Y., Kondoh, H., 1999. Two distinct subgroups of Group B Sox genes for transcriptional activators and repressors: their expression during embryonic organogenesis of the chicken. Mech. Dev. 84, 103–120.

Uwanogho, D., Rex, M., Cartwright, E.J., Pearl, G., Healy, C., Scotting, P.J., Sharpe, P.T., 1995. Embryonic expression of the chicken Sox2, Sox3 and Sox11 genes suggests an interactive role in neuronal development. Mech. Dev. 49, 23–36.

Watanabe, H., Ma, Q., Peng, S., et al., 2014. SOX2 and p63 colocalize at genetic loci in squamous cell carcinomas. J. Clin. Invest. 124, 1636–1645.

Williamson, K.A., Hever, A.M., Rainger, J., et al., 2006. Mutations in SOX2 cause anophthalmia-esophageal-genital (AEG) syndrome. Hum. Mol. Genet. 15, 1413–1422.

Woolfe, A., Goodson, M., Goode, D.K., et al., 2005. Highly conserved non-coding sequences are associated with vertebrate development. PLoS Biol. 3, e7.

Wu, Y., Guo, Z., Wu, H., Wang, X., Yang, L., Shi, X., Du, J., Tang, B., Li, W., Yang, L., Zhang, Y., 2012. SUMOylation represses Nanog expression via modulating transcription factors Oct4 and Sox2. PLoS One 7, e39606.

Xu, H.E., Rould, M.A., Xu, W., Epstein, J.A., Maas, R.L., Pabo, C.O., 1999. Crystal structure of the human Pax6 paired domain-DNA complex reveals specific roles for the linker region and carboxy-terminal subdomain in DNA binding. Genes Dev. 13, 1263–1275.

Yuan, H., Corbi, N., Basilico, C., Dailey, L., 1995. Developmental-specific activity of the FGF-4 enhancer requires the synergistic action of Sox2 and Oct-3. Genes Dev. 9, 2635–2645.

CHAPTER 9

Genomic Occupancy in Various Cellular Contexts and Potential Pioneer Factor Function of SOX2

Jonas Muhr
Ludwig Institute for Cancer Research, Department of Cell and Molecular Biology, Karolinska Institutet, Stockholm, Sweden

INTRODUCTION

Defining features of stem cells are their unlimited capacity to self-renew and their ability to generate rapidly dividing progenitor cells, which are destined to differentiate into one or more definitive cell types (Simons and Clevers, 2011). Stem cells are essential for the formation of organs and tissues at developmental stages, but are also retained in many adult organs to maintain tissue homeostasis and compensate for cell loss caused by damage. Although an extensive capacity for self-renewal is a common feature among all stem cells, their developmental potential can differ significantly. For instance, whereas the totipotent stem cells of the morula stage embryo can generate all cell types of both embryonic and extra-embryonic tissues, pluripotent stem cells in the inner cell mass (ICM) of the blastocyst embryo normally only generate cell types of the embryonic lineages and fail to generate extra-embryonic ectoderm. Tissue specific stem cells are further restricted in their developmental potential and have the capacity to generate progeny of only one particular organ or lineage. It is not fully understood how the unique features of stem cells are regulated, but emerging evidence indicates that the chromatin landscape, epigenetic constraints, and expression profile of gene regulatory transcription factors are the key determinants in the specification of stem cell behavior (Bonora et al., 2014). The transcription factor SOX2 has important regulatory roles in several stem cell populations and is involved in regulating processes such as stem cell maintenance and cell fate decisions, and is also necessary for the reprogramming of somatic cells into induced pluripotent stem (iPS) cells. Focusing on its DNA binding pattern, here I will review the regulatory roles of SOX2 in various types of stem and progenitor cells.

SOX2 FUNCTION AND BINDING PATTERN IN PLURIPOTENT STEM CELLS

In the early blastula-stage embryo, SOX2 is expressed in the ICM and in a subset of trophectoderm cells, which will later generate the embryo proper and placenta, respectively (Avilion et al., 2003; Keramari et al., 2010). SOX2 has an essential role at this stage of

Sox2
http://dx.doi.org/10.1016/B978-0-12-800352-7.00009-8

development, and embryos with homozygous deletion of *Sox2* fail to survive shortly after implantation (Avilion et al., 2003). One reason for the early lethality of *Sox2* mutant embryos is the failure of the ICM to generate pluripotent stem cells of the epiblast, from which the later embryonic lineages emerge. Without SOX2 expression the ICM is also unable to support the generation of pluripotent embryonic stem (ES) cells in vitro, and ES cells that have already been established cannot be preserved in a stem cell state upon SOX2 depletion (Masui et al., 2007). However, SOX2 is also necessary for a functional trophectoderm, which in its absence fails to generate extra-embryonic ectoderm (Keramari et al., 2010) or support the establishment of pluripotent trophoblast stem (TS) cells in vitro (Avilion et al., 2003). Hence, SOX2 has a key regulatory role in establishing and maintaining pluripotent stem cells both in vivo and in culture.

One question that arises from these findings is regarding what downstream genes SOX2 controls to achieve its regulatory functions in stem cells. The development of chromatin immunoprecipitation (ChIP) combined with deep-sequencing (ChIP-seq), or microarray (ChIP-on-Chip) analyses, has greatly enhanced our understanding of transcription factor function by allowing genome-wide identification of targeted regulatory regions (distal enhancers and promoters) and their associated genes. Moreover, by untangling how transcription factors co-occupy their targets with partner factors, co-repressors, and co-activators, or with histone or DNA modifications, these genome-wide techniques have helped to decipher their role in larger transcriptional regulatory networks. The binding pattern of SOX2 at a global level was first described in mouse and human ES cells. This pioneering work demonstrated that SOX2 binds promoters and more distal DNA-regulatory regions associated with thousands of different genes, but also that SOX2 co-binds many of its targets in ES cells together with other core pluripotency transcription factors such as OCT4 and NANOG (Marson et al., 2008; Boyer et al., 2005; Chen et al., 2008). Apart from binding to their own regulatory sequences, which indicates that SOX2, OCT4, and NANOG may collaborate in autoregulating their own expression (Boyer et al., 2005), these transcription factors co-occupy a large array of genes that are involved in controlling ES cell self-renewal and pluripotency. SOX2, OCT4, and NANOG bind the promoters of many genes encoding ES cell–expressed microribonucleic acids (RNAs) (Marson et al., 2008), transcription factors, and chromatin remodelers, as well as growth factors involved in ES cell regulation including WNT, transforming growth factor-β (TGFβ), and leukemia inhibitory factor or components of their downstream signaling cascades (Figure 1) (Orkin et al., 2008; Lim et al., 2007; Boyer et al., 2005; Marson et al., 2008; Chen et al., 2008). SOX2 has been proposed to promote the expression of its downstream genes in ES cells by recruiting the transcriptional co-activator and histone acetyl transferase P300 (Chen et al., 2008).

Moreover, the genome-wide binding of SOX2 has been characterized in pluripotent TS cells. Although the precursor cells for ES and TS cells, the ICM, and the trophectoderm,

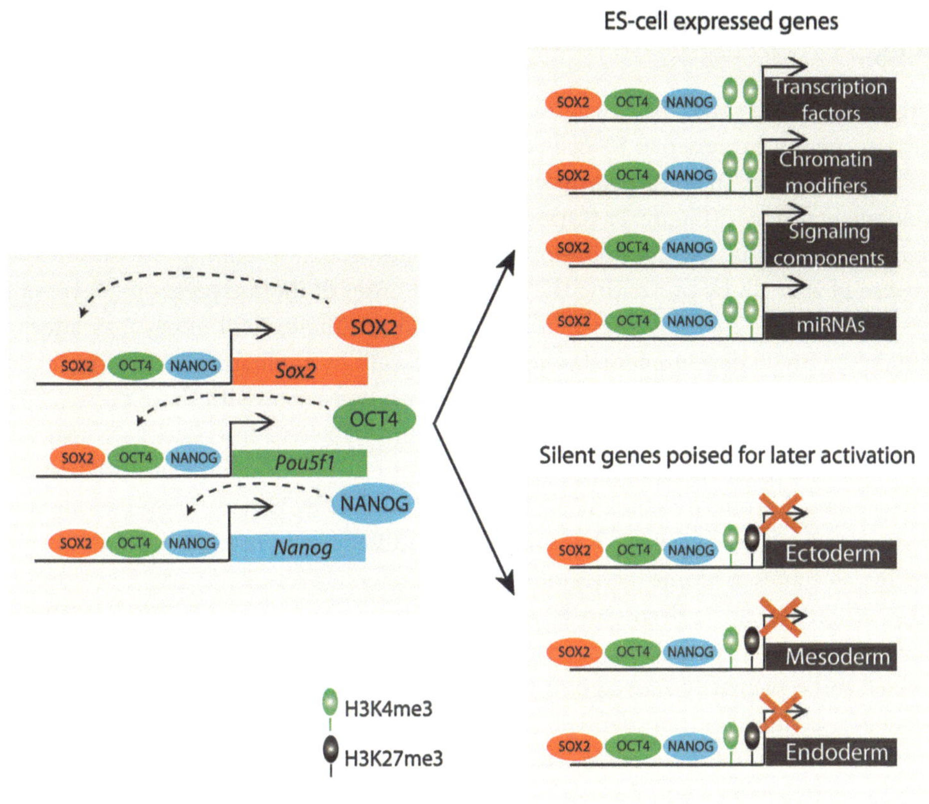

Figure 1 Model showing how SOX2, together with OCT4 and NANOG, are involved in regulating a key transcriptional network in ES cells. Many of the genes bound by SOX2 are co-occupied by OCT4 and NANOG. Apart from autoregulating their own expression, SOX2, OCT4, and NANOG also bind to a wide array of genes that are either actively expressed or silent in ES cells and first activated at later developmental stages. This transcriptional network appears to maintain pluripotency of ES cells by regulating genes promoting stem cells features, but also by prebinding developmental genes that may prevent their premature expression and facilitate their onset in later emerging lineages. Whereas expressed genes are associated with activating histone modifications (H3K4me3), several genes that are poised for later expression are associated with bivalent histone modifications (H3K4me3 and H327me3). *Figure adapted from Boyer et al. (2005).*

are developmentally related and are the first two lineages to be separated during embryogenesis, it is notable that the binding pattern of SOX2 is different in these pluripotent stem cell types (Adachi et al., 2013). For instance, although SOX2 binds to genes promoting stem cell maintenance in both ES cells and TS cells, SOX2 also enables a cell type–specific response to environmental stimuli by connecting growth factor signaling to a cell-specific set of transcription factors (Adachi et al., 2013).

PREBINDING OF LINEAGE-SPECIFIC GENES IN PLURIPOTENT STEM CELLS

Apart from binding genes that are actively transcribed in ES cells, SOX2 also targets many genes that are not expressed in ES cells, often together with OCT4 and NANOG, but that have important roles in regulating differentiation and cellular fate decisions at subsequent developmental stages (Figure 1) (Orkin et al., 2008; Boyer et al., 2005). One possibility is that the prebinding of silent developmental genes aids in sustaining the undifferentiated state of ES cells by preventing the premature onset of their expression (Boyer et al., 2005; Orkin et al., 2008). However, the prebinding of these silent genes may also help maintain the broad developmental potential of ES cells by facilitating their future activation in differentiating stem cell progeny. In any case, consistent with these possibilities is the observation that several genes that are prebound by SOX2 in ES cells are associated with bivalent histone domains consisting of a combination of transcriptional activating and repressing histone modifications (H3K4me3 and H3K27me3) (Boyer et al., 2005; Bernstein et al., 2006; Lee et al., 2006), which are subsequently resolved into monovalent domains as genes become activated or terminally repressed during development (Mikkelsen et al., 2007; Bernstein et al., 2006). Interestingly, genes in ES cells that are prebound by SOX2 and that are associated with bivalent histone domains have been demonstrated to be primarily activated in neural cells, and less often in cells of mesodermal or endodermal lineages (Bergsland et al., 2011). In contrast, no particular lineage enrichment could be found for genes that are marked by bivalent chromatin modifications and thus epigenetically prepared for activation, but that are lacking SOX2 binding. Thus, in addition to influencing the gene expression in ES cells and at later developmental stages, SOX2 prebinding may also predispose gene sets in ES cells to be preferentially activated during neural lineage development.

SOX2 FUNCTION IN TISSUE-SPECIFIC STEM CELLS

Besides regulating pluripotent stem cells of the blastocyst embryo or ES cells in vitro, SOX2 has important functions in several stem and progenitor populations at subsequent embryonic and adult stages. SOX2 is significantly expressed in most if not all stem and progenitor cells of the developing and adult central nervous system (CNS) (Collignon et al., 1996; Ferri, 2004; Favaro et al., 2009; Taranova et al., 2006) (see also Chapter 11), but also in stem and progenitor cells of the tongue epithelium (Okubo et al., 2006, 2009), dermal papilla of the hair follicle (Biernaskie et al., 2009; Driskell et al., 2009) (see also Chapter 16), the pituitary gland (Fauquier et al., 2008; Rizzoti et al., 2013) (see also Chapter 14), and gastrointestinal tract (Que et al., 2007; Arnold et al., 2011). In these different tissues SOX2 is involved in regulating processes important for organ development and homeostasis, such as precursor cell maintenance (Graham et al., 2003; Bylund et al., 2003; Ferri, 2004; Arnold et al., 2011) (see also Chapter 8), cell proliferation dynamics (Hagey and Muhr, 2014; Surzenko et al., 2013), and cell fate specification (Oosterveen et al., 2012; Peterson et al., 2012; Okubo et al., 2006; Driskell et al., 2009; Que et al., 2007). For instance, the loss of SOX2 in the mouse brain leads to premature differentiation of neural precursor cells (NPCs) and a reduction of the stem cell pools (Favaro et al., 2009; Ferri, 2004; Hagey and Muhr, 2014). In contrast, overexpression of SOX2, or SOX1 or SOX3, which are co-expressed with SOX2 in NPCs and together constitute the SOXB1 subgroup, preserve NPCs in a slowly dividing stem cell state and block the progression of neurogenesis (Bylund et al., 2003; Ferri, 2004; Oosterveen et al., 2012; Graham et al., 2003; Hagey and Muhr, 2014). SOX2 is expressed in NPCs of the healthy brain, but also in several types of brain-derived tumors (Holmberg et al., 2011). Consistent with its expression in stem and progenitor cells in the CNS, SOX2 expression has been identified in cancer cells that have stem cell properties and that can propagate tumors when transplanted into immunocompromised mice (Vanner et al., 2014). Moreover, genetic deletion of *Sox2*, or knockdown of *Sox2* expression using RNA interference, indicates that SOX2 is necessary for the maintenance of cancer stem cells that can support glioma and medulloblastoma formation and growth (Ikushima et al., 2009; Vanner et al., 2014; Favaro et al., 2009, 2014).

SOX2 BINDING PATTERN IN STEM AND PROGENITOR CELLS OF THE DEVELOPING CNS

Besides ES cells, the binding profile of SOX2 has been most extensively studied in NPCs, derived from either differentiated ES cells (Bergsland et al., 2011; Lodato et al., 2013; Peterson et al., 2012) or the embryonic mouse brain (Hagey and Muhr, 2014). In agreement with the redundant function of SOXB1 proteins in neural cells, these characterizations have revealed that the binding pattern of SOX2 and SOX3 are largely overlapping (>95%) (Bergsland et al., 2011; McAninch and Thomas, 2014) and that the similarity between SOX2 and SOX3 binding in NPCs is significantly higher compared with that of SOX2 in ES cells (McAninch and Thomas, 2014; Lodato et al., 2013; Bergsland et al., 2011; Hagey and Muhr, 2014). However, similar to the situation in ES cells, SOX2 and SOX3 bind tens of thousands of different sites in NPCs, which are predominantly located 10 kb or more from closest transcriptional start sites (TSS) (Lodato et al., 2013; Bergsland et al., 2011). Moreover, based on the co-occupancy of markers for distal enhancer elements, including the histone modifications H3K4me1, H3K27Ac or the transcriptional co-activator P300 (Bergsland et al., 2011; Lodato et al., 2013; Visel et al., 2009), a high proportion of the SOX2 targeted regions in NPCs function as active enhancers.

SOX2 and SOX3 binding in NPCs has been mapped to thousands of different genes. Even though approximately half of these genes also are bound by SOX2 in ES cells, these common genes are mostly targeted through distinct distal DNA regions, and only

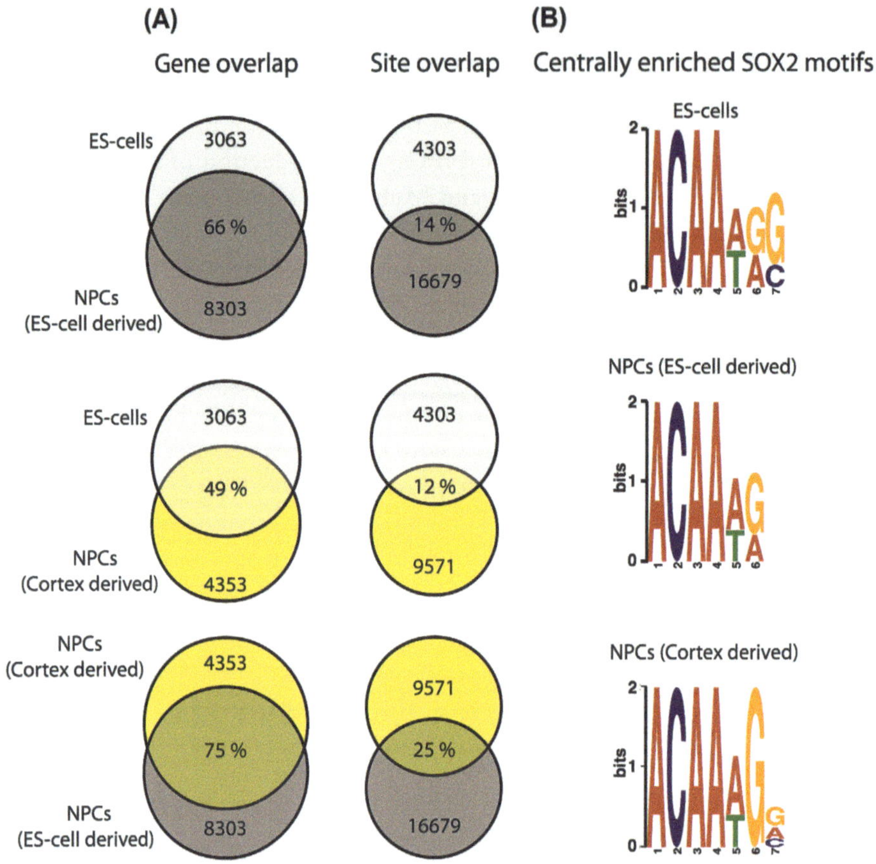

Figure 2 (A) Venn diagrams showing the overlap in genes and sites targeted by SOX2 in different cell types. Although the overlap of genes targeted by SOX2 in ES cells (Chen et al., 2008), ES cell–derived NPCs (Lodato et al., 2013), and cortex-derived NPCs (Hagey and Muhr, 2014) is substantial, the genes are mainly targeted through distinct regions, as determined by the low site overlap. (B) The centrally enriched *Sox2* motifs are similar in ChIP-Seq peaks in ES cells or NPCs derived from ES cells or embryonic cortices.

a minority (12%) of the regions bound by SOX2 in NPCs are targeted by SOX2 in ES cells (Figure 2) (Bergsland et al., 2011; Lodato et al., 2013; Hagey and Muhr, 2014). In fact, the selection of bound distal elements varies extensively depending on what cells are examined. Whereas nearly 75% of the SOX2 bound genes are common for the various types of NPCs examined (Bergsland et al., 2011; Lodato et al., 2013; Hagey and Muhr, 2014) less than one-quarter of the deoxyribonucleic acid (DNA) sites targeted by SOX2 in ES cell–derived NPCs are also bound by SOX2 in NPCs derived from embryonic cortices (Figure 2) (Hagey and Muhr, 2014; Bergsland et al., 2011). A comparison of SOX2 targets revealed that regions commonly bound in different types of NPCs tend to be located close to promoter regions, whereas uniquely bound regions are more often distally located (>50 kb from TSS) (Hagey and Muhr, 2014). Thus, these findings imply that enhancer elements, and not proximal promoter regions, primarily mediate cell type–specific gene regulatory features.

SPECIFICATION OF SOX2 BINDING

What, then, causes the distinct target selection of SOX2 in different types of cells? Genome-wide studies revealed that SOX2 binds to similar DNA sequences in various types of stem cells regardless of whether they are pluripotent or multipotent. The binding motif with the sequence composition ACAAWR (where W = A/T and R = A/G) is the most centrally enriched motif in SOX2-targeted regions in ES cells, as well as in regions bound by SOX2 and SOX3 in NPCs derived from ES cells or from embryonic cortices (Figure 2) (Lodato et al., 2013; Bergsland et al., 2011; Hagey and Muhr, 2014; Chen et al., 2008). Indeed, this target sequence is comparable to the most common binding motif of the SOXB proteins SOXNEURO and DICHAETE in *Drosophila* (Aleksic et al., 2013; Ferrero et al., 2014). Similar to the function of SOXB1 proteins in NPCs of the vertebrate CNS, both *SoxNeuro* and *Dichaete* are necessary for the equivalent neuroblast cells in the *Drosophila* embryo. SOXNEURO and DICHAETE have also been shown to interact with similar regulatory networks, as has been shown for SOX2 in mouse NPCs (Ferrero et al., 2014), and their functions can be rescued by the mammalian version of SOX2 (Soriano and Russell, 1998; Ferrero et al., 2014). Nevertheless, a determinant that is likely to be significant in specifying the target selection of SOX2 in different cell types is whether its binding motifs are within accessible or inaccessible chromatin (Soufi et al., 2012). Moreover, SOX proteins generally depend on collaboration with interacting partner transcription factors, which can increase the binding stability of SOX proteins to DNA and thereby help to specify their target selection and exhibit their gene regulatory functions (Kondoh and Kamachi, 2010; Chen et al., 2014; Jauch et al., 2011; Soufi et al., 2012). Thus, additional parameters that influence the binding pattern of SOX2 in different cell types are the expression of partner factors and the location of their target motifs. For instance, whereas OCT4 acts as an important partner factor to SOX2 in ES cells (Boyer et al., 2005; Ambrosetti et al., 1997; Yuan et al., 1995), OCT4 is not expressed in NPCs. Instead, bioinformatic, immunoprecipitation, and functional studies have identified a wide array of other proteins that appear to act as SOX2 partner factors in NPCs, including transcription factors of the TCF/LEF, POU, RFX, PAX, GLI, and homeodomain families (Lodato et al., 2013; Bergsland et al., 2011; Hagey and Muhr, 2014; Oosterveen et al., 2012; Peterson et al., 2012), together with chromatin-modifying and remodeling enzymes (Engelen et al., 2011). Hence, the specific binding pattern of SOX2 in different cell types is likely to be explained by several features including the chromatin accessibility and selection of expressed partner factors.

FUNCTIONS OF GENES BOUND BY SOXB1 PROTEINS IN NPCs

As mentioned previously, SOXB1 proteins have been implicated in regulating central processes in the developing and adult CNS. When genes targeted by SOX2 and SOX3 in NPCs are analyzed using gene ontology, terms such as stem cell development, negative regulation of neurogenesis, neural precursor proliferation, and neural patterning formation are highly enriched (Bergsland et al., 2011; Lodato et al., 2013; Hagey and Muhr, 2014). Consistent with this, many SOX2- and SOX3-targeted genes encode for proteins that are involved in regulating neurogenesis (including other SOX proteins, components of the NOTCH pathway, and proneural proteins) or encode for extracellular signaling components (such as components of the SONIC HEDGEHOG [SHH], TGFβ, fibroblast growth factor, and WNT signaling pathways), or transcription factors that are implicated in neural cell fate specification (including members of the homeodomain and PAX transcription factor families) (Bergsland et al., 2011; Hagey and Muhr, 2014).

Besides these functions, another role of SOXB1 proteins in NPCs is to confer neural specific response to environmental signaling. During embryogenesis, a limited number of secreted signaling proteins are involved in regulating a plethora of cellular processes, including cellular fate specification, growth, and survival. However, because a similar set of secreted factors is used in all three germ layers, one question is how these signaling events result in a tissue-specific response. In a study aiming to address this issue, a number of enhancers that are active in NPCs and regulated by SHH, retinoic acid (RA), or BONE MORPHOGENIC PROTEINS (BMPs) were characterized. Apart from binding sites for GLI, nuclear receptors, or SMAD proteins, which mediate the transcriptional response to SHH, RA, and BMPs, respectively, an overrepresentation of SOX binding motifs was found in their close vicinity (Oosterveen et al., 2012, 2013; Peterson et al., 2012). Because the binding of SOXB1 proteins to these motifs was a prerequisite for enhancer activity (Oosterveen et al., 2012; Peterson et al., 2012), it is interesting that these *Sox* motifs also restricted the enhancers to respond to SHH, BMP, and RA in a neural-specific manner. Consistent with this, when SOXB1 proteins were mis-expressed in limb buds of the chick embryo, co-transfected enhancers as well as neural-specific gene expression were ectopically activated in mesodermal cells in response to SHH, RA, or BMP signaling (Oosterveen et al., 2013). In mesodermal lineages, the specifically expressed transcription factors MYOD1 and PU.1 share the function of SOXB1 proteins in the developing CNS and specify the response to TGFβ signaling in myotubes and pro-B cells, respectively (Mullen et al., 2011). Thus, SOXB1 proteins are involved at several instances to orchestrate gene expression in the neural lineage, and apart from predisposing gene programs in pluripotent ES cells to be expressed in the emerging nervous system (see above), SOXB1 proteins target genes with key regulatory roles in NPCs and direct neural specific gene responses to inductive signaling events.

PREBINDING OF NEURONAL GENES BY SOXB1 PROTEINS IN NPC

Similar to the binding of SOX2 in ES cells, SOX2 and SOX3 only target genes that are actively transcribed in NPCs, they also prebind a large number of silent genes that are first activated as NPCs become postmitotic and commit to neuronal differentiation (Bergsland et al., 2011; Lodato et al., 2013). Upon neuronal differentiation, the expression of SOX2 is downregulated and most of its binding is replaced by the SOXC group protein SOX11, which results in transcriptional activation of the prebound neuronal genes (Figure 3) (Bergsland et al., 2011, 2006). Notably, sequentially acting SOXB1 and SOXC proteins have also been described during hematopoietic lineage development. In ES cells, SOX2 prebinds an intergenic enhancer of the silent *λ5-VpreB1* locus (Liber et al., 2010) which, analogous to the situation in neural lineage differentiation, is first activated after the expression of *Sox2* is downregulated and its binding has been replaced by the SOXC protein SOX4 in developing β-cells (Liber et al., 2010).

What, then, is the rationale for SOX2 to prebind genes in ES cells and NPCs that are first activated at subsequent stages of differentiation? SOX factors can function as architectural proteins that, upon binding to the minor groove of the DNA through their HMG domains, induce an approximately 80° bend of the double helix (Jauch and Kolatkar, 2013; Werner et al., 1995). One possibility is that this feature of SOX2, and the other SOXB1 members, promotes the formation of a permissive chromatin state around enhancers, which facilitates gene activation when the correct cellular context of activating

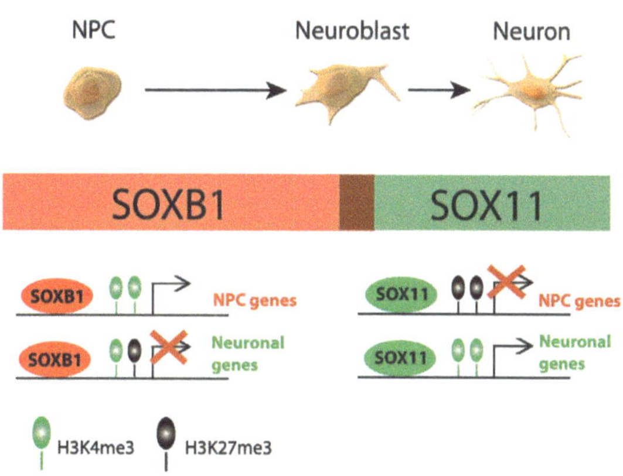

Figure 3 Model showing the sequential binding of SOX proteins to common downstream genes in stem cells differentiating along the neuronal lineage. In self-renewing stem cells, SOX2 and SOX3 have been demonstrated to bind to genes expressed in NPCs and to prebind silent genes that are first activated as the expression of SOX2 and SOX3 is downregulated, when their binding is replaced by SOX11 in differentiating neurons. Whereas actively expressed genes are associated with the activating histone modification H3K4me3, prebound genes tend to be associated with bivalent histone modifications consisting of both activating modifications and the repressive modification H3K27me3. *Figure adapted from Bergsland et al. (2011).*

transcription factors, such as alternative SOX proteins, has developed. In line with this, a large proportion of the enhancers that SOX2 prebinds in NPCs are associated with the active histone mark H3K4me1 (Lodato et al., 2013). Moreover, although genes prebound by SOX2 in ES cells and NPCs are not expressed, several of them nevertheless contain activating histone modifications (H3K4me3) at their proximal promoter regions (Bergsland et al., 2011; Lee et al., 2006; Bernstein et al., 2006; Boyer et al., 2005). Indeed, the establishment of activating H3K4 methylations at the *λ5-VpreB1* enhancer in ES cells is thought to depend on the prebinding of SOX2 (Liber et al., 2010), and ectopic expression of SOX3 in the mesodermal cell line C2C12 resulted in a significant increase in the activating H3K4me3 modification at SOX3-targeted regions (Bergsland et al., 2011). However, despite these changes in activating histone methylations upon SOX2 and SOX3 binding, no alterations could be detected in the expression of the associated genes (Bergsland et al., 2011; Liber et al., 2010).

Besides promoting the future expression of prebound genes, transactivation assays in vitro or in chick NPCs in vivo suggest that SOX2 and SOX3 may also prevent premature activation of prebound neuronal genes. In these experiments, both SOX2 and SOX3 or their DNA-binding HMG domains blocked the SOX11-mediated activation of enhancers (Bergsland et al., 2011). One explanation for this activity is that whereas SOX2 and SOX3 generally depend on the appropriate partner factors to activate gene expression (Kamachi and Kondoh, 2013), they are still able to repress the prebound genes either through competitive binding with other SOX proteins or through the recruitment of co-repressors. Thus, the prebinding of silent genes in ES cells and NPCs by SOXB1 proteins may have dual functions and may facilitate subsequent gene expression by promoting the formation of permissive chromatin as well as helping to maintain cells in an undifferentiated state by preventing the premature expression of differentiation genes.

PIONEERING ACTIVITY OF SOX2 DURING CELLULAR REPROGRAMMING

An important feature during development and during the reprogramming of somatic cells into iPS cells is that new DNase1-sensitive, open chromatin regions are established (Lara-Astiaso et al., 2014; Stergachis et al., 2013). As mentioned earlier, an example that highlights the role of SOX2 in the regulation of stem and progenitor features is the capacity of SOX2, together with OCT4, KLF4, and c-MYC, to reprogram somatic cells into iPS cells, a function it shares with SOX1 and SOX3 (Takahashi and Yamanaka, 2006; Nakagawa et al., 2007). To characterize how these factors may achieve this function, Zaret and colleagues (Soufi et al., 2012) set out to measure the binding pattern of SOX2, OCT4, KLF4 and c-MYC at early stages of the reprogramming process. Interestingly, in these experiments SOX2, OCT4 and KLF4 acted as pioneering transcription factors in that 48 h after mis-expression in human fibroblast cells they were found to target chromatin

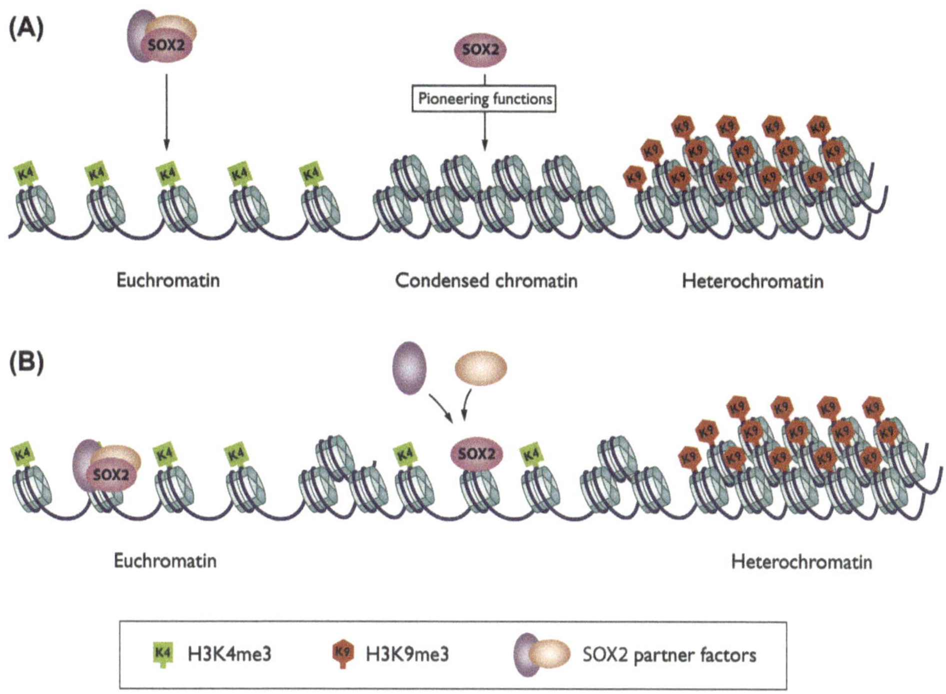

Figure 4 (A, B) Relaxed chromatin (euchromatin) is readily accessible for transcription factor binding. SOX2 possesses pioneering functions and can therefore also bind target sites with high nucleosome density (A), which enables other proteins that lack this function, such as partner factors, to access their target sites (B). Although SOX2 is able to bind chromatin that lacks DNase1 hypersensitivity, it fails to bind heterochromatin, as marked by the histone modification H3K9me3 (K9). *Figure adapted from Gaspar-Maia et al. (2011).*

regions that under normal conditions are closed, DNase1-insensitive, and lack markers for open chromatin such as H3K27Ac or H3K4me3 (Figure 4). Moreover, although c-MYC preferentially bound open chromatin regions associated with activating histone marks, it readily bound closed chromatin when mis-expressed with SOX2, OCT4, or KLF4 (Soufi et al., 2012; Iwafuchi-Doi and Zaret, 2014). Thus, during the reprogramming of fibroblasts into iPS cells, SOX2, OCT4, and KLF4 act as pioneering factors that on their own can bind closed, DNase1-resistant chromatin, which in turn facilitates the binding of additional transcription factors lacking this pioneering factor capacity, such as c-MYC (Figure 4) (Soufi et al., 2012). However, although in these assays SOX2, OCT4, and KLF4 acted as pioneering factors and could bind to closed chromatin, they failed to bind chromatin that possessed features of heterochromatin, including being more DNase1 resistant and containing a higher enrichment of the histone mark H3K9me3 (Figure 4).

To better determine how transcription factors find and bind their specific binding motifs, single molecule imaging experiments were conducted to quantitatively dissect the

DNA-binding dynamics of SOX2 and OCT4 in ES cells and in a cell-free model (Chen et al., 2014). The data indicated that the binding of SOX2 and OCT4 to chromatin DNA occurs in a hierarchically ordered fashion in which SOX2 serves as a lead factor that searches for and binds specific target genes, and subsequently assists the binding of OCT4 to DNA, which in turn stabilizes the ternary complex (Chen et al., 2014). Although the data support a role for SOX2 as a pioneering factor on the one hand, they also bring up the issue of whether the time SOX2 normally binds a specific target, which is reported to be around 12 s, is enough for SOX2 to fulfill a function as a pioneering factor on condensed chromatin. Furthermore, in contrast to the finding reported by Soufi et al. (2012), this study proposes a division in labor between SOX2 and OCT4, in which SOX2 acts as a lead and pioneering factor, whereas partner factors such as OCT4 may be better suited for target activation through its greater ability to interact with various co-activators. In any case, both the capacity of SOX2 to function as a lead factor, which searches the DNA for its appropriate target sequence, and its ability to bind to closed chromatin during the reprogramming of fibroblasts into iPS cells are features that need to be possessed by pioneering proteins.

CONCLUDING REMARKS

Methodological advancements in characterizing the binding pattern of transcription factors in relevant cell types have greatly improved our mechanistic understanding of how transcription factors regulate cellular processes and decisions. Rather than connecting one transcription factor to a particular enhancer, promoter, or downstream gene, these genome wide techniques have allowed transcription factors and their downstream genes to be categorized into larger regulatory networks. The composition of these regulatory networks has been further refined by parallel ChIP-Seq experiments that examined how transcription factors co-target many of their sites together with partner transcription factors, co-repressors, and co-activators, histone and chromatin modifications, or their modifying enzymes. This experimental approach has, for instance, revealed that SOX2 binds many of its targets together with OCT4 and NANOG in ES cells (Boyer et al., 2005), BRN2 in NPCs (Lodato et al., 2013), or the co-activator P300 in ES cells and NPCs (Chen et al., 2008; Bergsland et al., 2011). Apart from parallel ChIP-Seq experiments, screening of targeted DNA regions for the enrichment of associated transcription factor binding motifs has also identified partner factors of SOX2. In cortical NPCs, a high proportion of the SOX2-bound regions contains binding motifs for TCF/LEF transcription factors, with which SOX2 has also been shown to interact functionally (Hagey and Muhr, 2014). Furthermore, although the availability of partner factors also affects the binding pattern of SOX2, it is important that each ChIP-Seq experiment be based on defined cell populations. For instance, because the selection of partner factors in the developing CNS differs depending on what stage and region is being analyzed, the binding pattern of SOX2 is different in different types of NPCs.

In ES cells and NPCs, SOX2 and SOX3 target many silent genes that first are activated as the cells generate more differentiated progeny. As discussed, prebinding of these genes may have dual functions, and although it can be repressive and thus prevent their premature onset of expression (Bylund et al., 2003; Graham et al., 2003), it may also facilitate gene expression at subsequent developmental stages by maintaining the chromatin around enhancers in an open state. An interesting additional possibility is that prebinding, through the pioneering functions that have been ascribed to SOX2, establishes accessible chromatin around enhancers to which subsequent acting transcriptional regulators such as SOX11 can bind and activate gene expression. Thus, by enabling gene expression in differentiating progeny, prebinding by SOX2 and SOX3 may be important for the developmental potential of ES cells and NPCs. To better understand how prebinding and stem cell competence are connected, it would be informative to examine how the chromatin landscape and binding pattern of SOX2 correlate in different types of stem and progenitor cells.

REFERENCES

Adachi, K., et al., 2013. Context-dependent wiring of Sox2 regulatory networks for self-renewal of embryonic and trophoblast stem cells. Mol. Cell 52 (3), 380–392.

Aleksic, J., et al., 2013. The role of Dichaete in transcriptional regulation during *Drosophila* embryonic development. BMC Genomics 14 (1), 861.

Ambrosetti, D.C., Basilico, C., Dailey, L., 1997. Synergistic activation of the fibroblast growth factor 4 enhancer by Sox2 and Oct-3 depends on protein-protein interactions facilitated by a specific spatial arrangement of factor binding sites. Mol. Cell. Biol. 17 (11), 6321–6329.

Arnold, K., et al., 2011. $Sox2^+$ adult stem and progenitor cells are important for tissue regeneration and survival of mice. Cell Stem Cell 9 (4), 317–329.

Avilion, A.A., et al., 2003. Multipotent cell lineages in early mouse development depend on SOX2 function. Genes Dev. 17 (1), 126–140.

Bergsland, M., et al., 2011. Sequentially acting Sox transcription factors in neural lineage development. Genes Dev. 25 (23), 2453–2464.

Bergsland, M., et al., 2006. The establishment of neuronal properties is controlled by Sox4 and Sox11. Genes Dev. 20 (24), 3475–3486.

Bernstein, B.E., et al., 2006. A bivalent chromatin structure marks key developmental genes in embryonic stem cells. Cell 125 (2), 315–326.

Biernaskie, J., et al., 2009. SKPs derive from hair follicle precursors and exhibit properties of adult dermal stem cells. Cell Stem Cell 5 (6), 610–623.

Bonora, G., Plath, K., Denholtz, M., 2014. A mechanistic link between gene regulation and genome architecture in mammalian development. Curr. Opin. Genet. Dev. 27, 92–101.

Boyer, L.A., et al., 2005. Core transcriptional regulatory circuitry in human embryonic stem cells. Cell 122 (6), 947–956.

Bylund, M., et al., 2003. Vertebrate neurogenesis is counteracted by Sox1–3 activity. Nat. Neurosci. 6 (11), 1162–1168.

Chen, J., et al., 2014. Single-molecule dynamics of enhanceosome assembly in embryonic stem cells. Cell 156 (6), 1274–1285.

Chen, X., et al., 2008. Integration of external signaling pathways with the core transcriptional network in embryonic stem cells. Cell 133 (6), 1106–1117.

Collignon, J., et al., 1996. A comparison of the properties of Sox-3 with Sry and two related genes, Sox-1 and Sox-2. Development 122 (2), 509–520.

Driskell, R.R., et al., 2009. Sox2-positive dermal papilla cells specify hair follicle type in mammalian epidermis. Development 136 (16), 2815–2823.

Engelen, E., et al., 2011. Sox2 cooperates with Chd7 to regulate genes that are mutated in human syndromes. Nat. Genet. 43 (6), 607–611.

Fauquier, T., et al., 2008. SOX2-expressing progenitor cells generate all of the major cell types in the adult mouse pituitary gland. Proc. Natl. Acad. Sci. U.S.A. 105 (8), 2907–2912.

Favaro, R., et al., 2009. Hippocampal development and neural stem cell maintenance require Sox2-dependent regulation of Shh. Nat. Neurosci. 12 (10), 1248–1256.

Favaro, R., et al., 2014. Sox2 is required to maintain cancer stem cells in a mouse model of high-grade oligodendroglioma. Cancer Res. 74 (6), 1833–1844.

Ferrero, E., Fischer, B., Russell, S., 2014. SoxNeuro orchestrates central nervous system specification and differentiation in *Drosophila* and is only partially redundant with Dichaete. Genome Biol. 15 (5), R74.

Ferri, A.L.M., 2004. Sox2 deficiency causes neurodegeneration and impaired neurogenesis in the adult mouse brain. Development 131 (15), 3805–3819.

Gaspar-Maia, A., et al., 2011. Open chromatin in pluripotent and reprogramming. Nat. Rev. Mol. Cell Biol. 12 (4), 36–47.

Graham, V., et al., 2003. SOX2 functions to maintain neural progenitor identity. Neuron 39 (5), 749–765.

Hagey, D.W., Muhr, J., 2014. Sox2 acts in a dose-dependent fashion to regulate proliferation of cortical progenitors. Cell Rep. 9 (5), 1908–1920.

Holmberg, J., et al., 2011. In: Najbauer, J. (Ed.), Activation of Neural and Pluripotent Stem Cell Signatures Correlates with Increased Malignancy in Human Glioma. PloS One 6 (3), e18454.

Ikushima, H., et al., 2009. Autocrine TGF-β signaling maintains tumorigenicity of glioma-initiating cells through Sry-related HMG-box factors. Stem Cell. 5 (5), 504–514.

Iwafuchi-Doi, M., Zaret, K.S., 2014. Pioneer transcription factors in cell reprogramming. Genes Dev. 28 (24), 2679–2692.

Jauch, R., Kolatkar, P.R., 2013. What makes a pluripotency reprogramming factor? Curr. Mol. Med. 13 (5), 806–814.

Jauch, R., et al., 2011. Conversion of Sox17 into a pluripotency reprogramming factor by reengineering its association with Oct4 on DNA. Stem Cells 29 (6), 940–951.

Kamachi, Y., Kondoh, H., 2013. Sox proteins: regulators of cell fate specification and differentiation. Development 140 (20), 4129–4144.

Keramari, M., et al., 2010. In: Pera, M. (Ed.), Sox2 Is Essential for Formation of Trophectoderm in the Preimplantation Embryo. PloS One 5 (11), e13952.

Kondoh, H., Kamachi, Y., 2010. SOX-partner code for cell specification: regulatory target selection and underlying molecular mechanisms. Int. J. Biochem. Cell Biol. 42 (3), 391–399.

Lara-Astiaso, D., et al., 2014. Immunogenetics. Chromatin state dynamics during blood formation. Science (New York, N.Y.) 345 (6199), 943–949.

Lee, T.I., et al., 2006. Control of developmental regulators by Polycomb in human embryonic stem cells. Cell 125 (2), 301–313.

Liber, D., et al., 2010. Epigenetic priming of a pre-B cell-specific enhancer through binding of Sox2 and Foxd3 at the ESC stage. Cell Stem Cell 7 (1), 114–126.

Lim, L.S., et al., 2007. Zic3 is required for maintenance of pluripotency in embryonic stem cells. Mol. Biol. Cell 18 (4), 1348–1358.

Lodato, M.A., et al., 2013. In: Barsh, G.S. (Ed.), SOX2 Co-Occupies Distal Enhancer Elements with Distinct POU Factors in ESCs and NPCs to Specify Cell State. PLoS Genet. 9 (2), e1003288.

Marson, A., et al., 2008. Connecting microRNA genes to the core transcriptional regulatory circuitry of embryonic stem cells. Cell 134 (3), 521–533.

Masui, S., et al., 2007. Pluripotency governed by Sox2 via regulation of Oct3/4 expression in mouse embryonic stem cells. Nat. Cell Biol. 9 (6), 625–635.

McAninch, D., Thomas, P., 2014. In: Zheng, D. (Ed.), Identification of Highly Conserved Putative Developmental Enhancers Bound by SOX3 in Neural Progenitors Using ChIP-Seq. PloS One 9 (11), e113361.

Mikkelsen, T.S., et al., 2007. Genome-wide maps of chromatin state in pluripotent and lineage-committed cells. Nature 448 (7153), 553–560.

Mullen, A.C., et al., 2011. Master transcription factors determine cell-type-specific responses to TGF-β signaling. Cell 147 (3), 565–576.
Nakagawa, M., et al., 2007. Generation of induced pluripotent stem cells without Myc from mouse and human fibroblasts. Nat. Biotechnol. 26 (1), 101–106.
Okubo, T., Clark, C., Hogan, B.L.M., 2009. Cell lineage mapping of taste bud cells and keratinocytes in the mouse tongue and soft palate. Stem Cells 27 (2), 442–450.
Okubo, T., Pevny, L.H., Hogan, B.L.M., 2006. Sox2 is required for development of taste bud sensory cells. Genes Dev. 20 (19), 2654–2659.
Oosterveen, T., et al., 2012. Mechanistic differences in the transcriptional interpretation of local and long-range Shh morphogen signaling. Dev. Cell 23 (5), 1006–1019.
Oosterveen, T., et al., 2013. SoxB1-driven transcriptional network underlies neural-specific interpretation of morphogen signals. Proc. Natl. Acad. Sci. U.S.A. 110 (18), 7330–7335.
Orkin, S.H., et al., 2008. The transcriptional network controlling pluripotency in ES cells. Cold Spring Harbor Symp. Quan. Biol. 73, 195–202.
Peterson, K.A., et al., 2012. Neural-specific Sox2 input and differential Gli-binding affinity provide context and positional information in Shh-directed neural patterning. Genes Dev. 26 (24), 2802–2816.
Que, J., et al., 2007. Multiple dose-dependent roles for Sox2 in the patterning and differentiation of anterior foregut endoderm. Development 134 (13), 2521–2531.
Rizzoti, K., Akiyama, H., Lovell-Badge, R., 2013. Mobilized adult pituitary stem cells contribute to endocrine regeneration in response to physiological demand. Cell Stem Cell 13 (4), 419–432.
Simons, B.D., Clevers, H., 2011. Strategies for homeostatic stem cell self-renewal in adult tissues. Cell 145 (6), 851–862.
Soriano, N.S., Russell, S., 1998. The *Drosophila* SOX-domain protein Dichaete is required for the development of the central nervous system midline. Development 125 (20), 3989–3996.
Soufi, A., Donahue, G., Zaret, K.S., 2012. Facilitators and impediments of the pluripotency reprogramming factors' initial engagement with the genome. Cell 151 (5), 994–1004.
Stergachis, A.B., et al., 2013. Developmental fate and cellular maturity encoded in human regulatory DNA landscapes. Cell 154 (4), 888–903.
Surzenko, N., et al., 2013. SOX2 maintains the quiescent progenitor cell state of postnatal retinal Muller glia. Development 140 (7), 1445–1456.
Takahashi, K., Yamanaka, S., 2006. Induction of pluripotent stem cells from mouse embryonic and adult fibroblast cultures by defined factors. Cell 126 (4), 663–676.
Taranova, O.V., et al., 2006. SOX2 is a dose-dependent regulator of retinal neural progenitor competence. Genes Dev. 20 (9), 1187–1202.
Vanner, R., et al., 2014. Quiescent $Sox2^+$ cells drive hierarchical growth and relapse in sonic hedgehog subgroup medulloblastoma. Cancer Cell 26 (1), 33–47.
Visel, A., et al., 2009. ChIP-seq accurately predicts tissue-specific activity of enhancers. Nature 457 (7231), 854–858.
Werner, M.H., et al., 1995. Molecular basis of human 46X,Y sex reversal revealed from the three-dimensional solution structure of the human SRY-DNA complex. Cell 81 (5), 705–714.
Yuan, H., et al., 1995. Developmental-specific activity of the FGF-4 enhancer requires the synergistic action of Sox2 and Oct-3. Genes Dev. 9 (21), 2635–2645.

PART 3

Sox2 Regulatory Functions in Specific Cells and Tissues

CHAPTER 10

SOX2-Dependent Regulation of Pluripotent Stem Cells

Frederick C.K. Wong, Ian Chambers, Nicholas P. Mullin*
MRC Centre for Regenerative Medicine, Institute for Stem Cell Research, School of Biological Sciences, University of Edinburgh, Edinburgh, Scotland, UK
*Corresponding author: Telephone: (44) 131 651 9561; Fax: (44) 131 651 9501; nick.mullin@ed.ac.uk

INTRODUCTION

The transcriptional circuitry regulating the self-renewal of embryonic stem (ES) and induced pluripotent stem (iPS) cells is a complex web of protein–deoxyribonucleic acid (DNA) and protein–protein interactions involving a large number of sequence-specific and general transcription factors, ribonucleic acid (RNA) polymerase, and associated transcriptional regulatory proteins that work together with epigenetic regulators and chromatin modifiers. Of the sequence-specific transcription factors, SOX2 is considered to operate at the core of the pluripotency gene regulatory network alongside OCT4 and NANOG. Indeed, a complex interplay of positive and negative regulatory loops links the three factors (Navarro et al., 2012).

In the mouse, the Sry box (Sox) family has 20 members. Mouse SOX2 protein is 319 amino acids in length and contains a high mobility group (HMG) DNA binding domain of 80 amino acids, with the remainder of the sequence subdivided into regions based on function or amino acid prevalence (Figure 1). The non-HMG sequence of SOX2 can be divided into an N-terminal region and a C-terminal transactivation domain that contains a serine-rich region (Ambrosetti et al., 2000; Nowling et al., 2000). Comparison of different SOX proteins demonstrates that the HMG domains share a sequence identity of at least 50% with Sry (Pevny and Lovell-Badge, 1997), and in some cases it is possible to switch the HMG domains between different Sox family members without interfering with protein function (Bergstrom et al., 2000).

SOX2 is one of only a few Sox genes expressed in pluripotent cells. In vivo, SOX2 messenger RNA (mRNA) and protein can be detected in the oocyte and in the two-cell embryo (Avilion et al., 2003; Keramari et al., 2010). Subsequently, *Sox2* mRNA is enriched in the inner cells of 16-cell morulae before becoming restricted to cells of the inner cell mass in early blastocysts (Guo et al., 2010). By day 7.5, SOX2 expression becomes restricted to the neuroectoderm (Wood and Episkopou, 1999). Although expressed early in development, investigation of the contribution of SOX2 to early embryonic development was initially hampered by the persistence of detectable levels of maternal protein up

Sox2
http://dx.doi.org/10.1016/B978-0-12-800352-7.00010-4

Figure 1 ***Diagram representing main features of Sox2 protein.*** N: N-terminal domain, HMG: high mobility group domain, TAD: transactivation domain, Ser rich: serine-rich region.

to and including the blastocyst stage (Avilion et al., 2003). Consequently, knockout studies initially showed that SOX2 is essential to establish the postimplantation epiblast (Avilion et al., 2003). More recently, small interfering RNA (siRNA) studies that deplete both maternal and embryonic mRNA indicate that a lack of SOX2 results in arrest at the morula stage and a failure to cavitate and form trophectoderm (Keramari et al., 2010).

SOX2 is not restricted to a role in early development; it is also important later during neurogenesis and lens placode formation (Miyagi et al., 2009) and is expressed in many stem and progenitor cells throughout adult life (Arnold et al., 2011). However, in this chapter we restrict our discussion principally to consideration of pluripotent cells. Moreover, although a number of reviews exist covering the regulatory mechanisms involved in the maintenance of pluripotency (Ng and Surani, 2011; Chambers and Tomlinson, 2009; Jaenisch and Young, 2008), our focus is on SOX2.

SOX2–PROTEIN PARTNER INTERACTIONS

Interaction of SOX2 with OCT4

It has become clear in recent years that transcription factors rarely act alone, but rather are bound to DNA together with other factors. Indeed, the identification of super-enhancers that have large numbers of factors concentrated in a small region of DNA demonstrates the importance of co-localization between different proteins for control of gene expression (Whyte et al., 2013; Loven et al., 2013). Nevertheless, at most regulatory regions where multiple factors are found, the biochemical basis of the molecular interactions between factors remains unclear. However, the interactions between SOX2 and partner proteins have been widely characterized and can be thought of as a paradigm for the combinatorial interaction of transcription factors allowing for differential gene expression in different cell types. In chick embryos, SOX2 regulates gene expression during lens differentiation (Kamachi et al., 1998). Sequences adjacent to the SOX2 binding sites are essential for correct regulation of certain target genes, including δ-crystallin (Kamachi et al., 1995). PAX6 is the transcription factor that acts in conjunction with SOX2 to regulate the DC5 enhancer of δ-crystallin (Kamachi et al., 2001). Electrophoretic mobility shift assays (EMSAs) using the DC5 element show that both SOX2 and PAX6 bind DNA. However, when both proteins are included in the EMSA reaction, a cooperative interaction is observed and a ternary SOX2–PAX6–DNA complex is formed. Luciferase assays using the DC5 element demonstrate the synergistic nature of the cooperative interaction between SOX2 and PAX6 with upregulation of transcription observed only in the presence of both SOX2 and PAX6 (Kamachi et al., 2001). In mouse neural development, SOX2-dependent gene regulation requires interaction with transcription factors of the POU (Pit, Oct, Unc) family, most notably BRN2 (*Pou3f2*) (Tanaka et al., 2004), OCT6 (Nishimoto et al., 1999), and OCT1 (*Pou2f1*) (Di Rocco et al., 2001). It is in ES cells, however, that the interactions of SOX2 have been most thoroughly studied, particularly the interaction of SOX2 with OCT4.

The initial observations that SOX2 could act cooperatively with OCT4 were made at the *Fgf4* enhancer in EC cells (Yuan et al., 1995). Cooperative DNA binding between SOX2 and OCT4 was then described for a number of other genes in ES cells including *Utf1* (Nishimoto et al., 1999) and *Fbxo15* (Tokuzawa et al., 2003). Subsequently, the OCT4–SOX2 complex was shown to regulate the transcription of both *Oct4* and *Sox2* (Chew et al., 2005; Tomioka et al., 2002) and was implicated in the regulation of *Nanog* (Rodda et al., 2005; Kuroda et al., 2005). The molecular details of the OCT4 and SOX2 interaction were initially probed by studying the interaction of different truncated versions of the protein in vitro (Ambrosetti et al., 1997). This demonstrated that the HMG DNA binding domain of SOX2 and the POU-specific (POU_S) DNA binding domain of OCT4 mediate the interaction between the proteins. Although the structure of the OCT4–SOX2–DNA complex has not been determined, the interaction of OCT1/SOX2/DNA has been studied at atomic resolution and is highly informative with respect to the likely interaction of OCT4 and SOX2 (Remenyi et al., 2003; Williams et al., 2004). These studies demonstrate that side chain interactions between Oct and Sox proteins are responsible for stabilizing the interaction between the POU_S domain and the DNA. In the absence of SOX2, two conformations of the OCT4–DNA complex are present: one in which both the POU-homeodomain (POU_{HD}) and the POU_S domains are in contact with the DNA and one in which the POU_{HD} alone is bound to DNA. However, in the presence of SOX2, the only complex found is that in which both DNA binding domains of OCT4 are bound to the DNA. Thus, the interaction between SOX2 and OCT4 is responsible for holding the POU_S domain of OCT4 onto the DNA. The details of this interaction are described in more detail in Chapter 2 of this book.

These studies demonstrated that the OCT4–SOX2 complex has a role in transcriptional regulation of ES cells but were restricted to a small number of genes. However, the advent of genome-wide methods for examining chromatin binding of proteins has allowed a wider investigation of the roles of various transcription factors. One such ChIP-seq study examined the binding of OCT4 and NANOG in mouse ES cells and found 965 OCT4 and 2544 NANOG binding sites, respectively. Although SOX2 was not included in this study, de novo motif discovery algorithms found a composite OCT4–SOX2 motif (Loh et al., 2006) under more than 70% of the high-confidence OCT4 peaks. Subsequent sequential ChIP of OCT4 and SOX2 at six of these composite sites showed that both proteins were indeed bound at these positions. This work was complemented by a study coupling ChIP with a microarray-based analysis of human ES cells (Boyer et al., 2005). This dataset included ChIP of SOX2, OCT4, and NANOG and showed that the proteins were bound at a large number of locations (2260 sites), and that there was a large amount of overlap between sites bound by each factor. Thus, OCT4 and SOX2 were co-bound at 404 (18%) sites; interestingly 87% of these OCT4–SOX2 sites are also bound by NANOG. The presence of NANOG at the vast majority of the OCT4–SOX2 sites is intriguing, and together with the results of the OCT4–NANOG ChIP (Loh et al., 2006) suggests potential synergy between DNA binding of OCT4, SOX2, and NANOG.

Also, although the literature is highly weighted to the study and discussion of the contribution of OCT4–SOX2, the human ES cell study suggests that only 2% of the total number of binding sites are co-bound by OCT4 and SOX2 without NANOG binding (Boyer et al., 2005). In contrast, the most commonly co-bound factors are SOX2 and NANOG (without OCT4), which form 22% of the total number of sites. A number of these observations were corroborated by a subsequent ChIP-seq study in mouse ES cells that examined the distribution of 13 transcription factors (including OCT4, SOX2, and NANOG) throughout the genome and a ChIP-promoter microarray study of nine transcription factors (Chen et al., 2008; Kim et al., 2008a). In the Chen study, 1124 SOX2 binding sites were detected, of which 606 (54%) were co-bound by OCT4 and NANOG. A much smaller fraction of the SOX2 bound sites (72 sites; 6%) were co-bound by OCT4 in the absence of NANOG, whereas the number of SOX2 bound sites co-bound by NANOG without OCT4 was considerably higher (364 sites; 33%). Indeed, the frequency of NANOG co-binding with SOX2 is so marked that motif discovery algorithms applied to the NANOG binding sites identify the sequence CATTGTNATG, a composite of a complete SOX2 (CATTGTN) with the first three bases of the octamer motif (ATGCAAAT).

The number of SOX2 binding sites identified by these various global studies suggests that a large number of genes that form a significant portion of the genome are regulated by SOX2 and other ES cell–specific transcription factors. However, the complexities of assigning a ChIP-seq peak to the transcriptional regulation of a particular gene, including the arbitrary placement of a threshold to distinguish signal from noise in bioinformatics analysis and the fact that the overlap of binding sites between studies is weak, mean that it is unlikely that all binding sites identified to date are of functional importance (Chambers and Tomlinson, 2009). An example of the disparity between binding sites and regulation is offered by NANOG, which in independent ChIP-seq studies (Kim et al., 2008a; Chen et al., 2008) identified a common set of more than 5000 genes (Festuccia et al., 2012). However, in a separate study only 64 genes showed an alteration in expression of 1.5-fold after NANOG induction in ES cells (Festuccia et al., 2012). These results reinforce the difficulty in assigning transcriptional responsiveness of target genes on the basis of ChIP data alone.

The chromatin localization studies described in this section reinforce the concept that individual transcription factors are rarely found bound to chromatin in isolation. However, ChIP studies alone do not distinguish between factors that colocalize independently and those in which chromatin binding is cooperative. Further biochemical characterization of colocalized factors is required to establish whether an interaction is cooperative.

Interaction of SOX2 and NANOG

The high proportion of OCT4–SOX2 sites that are co-bound by NANOG and the prevalence of NANOG–SOX2 sites over OCT4/SOX2 sites suggest that to consider the OCT4–SOX2 interaction as the defining feature of SOX2-mediated regulation in ES cells is an oversimplification. In particular, the frequency of SOX2 and NANOG co-binding implies an important function for this interaction. The role of NANOG in SOX2 function was investigated by a proteomic study that identified SOX2 as a binding partner of NANOG and that further characterized the SOX2–NANOG interaction (Gagliardi et al., 2013). This demonstrated that NANOG and SOX2 could form a complex in the absence of either OCT4 or DNA. Furthermore, SELEX identified a unique DNA binding motif for the complex that was close to a number of ES cell–specific genes that were transcriptionally perturbed by expression of a mutant form of SOX2 unable to interact with NANOG (Gagliardi et al., 2013).

The ability of the SOX2–NANOG complex to form in the absence of DNA is an interesting point given the seemingly contradictory results concerning the ability of OCT4 and SOX2 to interact in the absence of DNA. Some of the first work that characterized OCT4–SOX2 synergy on the FGF4 enhancer showed that in vitro translated OCT4 and SOX2 could interact with one another, which suggested that DNA binding is not essential for the two proteins to interact (Ambrosetti et al., 1997). Subsequently, co-immunoprecipitation of OCT4 and SOX2 was demonstrated in the presence of benzonase, a highly active nuclease (van den Berg et al., 2010), which again suggests that the presence of DNA is not necessary for the interaction. However, more recent work that employed fusions of OCT4 and SOX2 with fluorescent proteins and biophysical techniques to probe interactions showed that in the absence of a cognate DNA sequence, the interaction between OCT4 and SOX2 was not detectable (Lam et al., 2012). The contradictions between the different results may reflect a weak interaction between the proteins in the absence of DNA that is not detectable in the system employed by Lam et al. Such an explanation is supported by the demonstration that the HMG domains of other Sox proteins also mediate weak interactions with protein partners that are easily perturbed by relatively low salt concentrations and that do not survive the relatively benign conditions used in EMSAs (Wissmuller et al., 2006). Alternatively, the fusion of fluorescent proteins to either OCT4 or SOX2, or both, may be sufficiently destabilizing to result in a loss of interaction in the absence of DNA. Whatever the explanation for the contradictory results with respect to OCT4 and SOX2, the NANOG–SOX2 interaction is fundamentally different in strength from the OCT4–SOX2 interaction as the NANOG–SOX2 complex remains intact during purification by ion-exchange and elution at relatively high sodium chloride concentrations. Perhaps the high affinity of the NANOG–SOX2 interaction relative to other SOX2–protein interactions may have some functional implication yet to be discovered. For example, it is possible that the formation of the NANOG–SOX2 complex is a means of regulating the amount of one or both of the molecules available for other protein partners. Because NANOG levels fluctuate within undifferentiated ES cells (Chambers et al., 2007), it is possible that at high NANOG concentrations, a large fraction of SOX2 forms the NANOG–SOX2 complex and is unavailable for other interactions. Further work will be required to determine whether regulating the extent of NANOG–SOX2 complex formation can tune the transcriptional output of ES cells.

Other SOX2–Protein Interactions

The importance of protein–protein interactions in the function of transcription factors has been underscored by the discovery that transcription factor interactomes contain many proteins (Chuang et al., 2012; Rudra et al., 2012; Steunou et al., 2013). In ES cells, interactomes have been determined for OCT4, NANOG, ESRRB, NR0B1, and SALL4 (Gagliardi et al., 2013; van den Berg et al., 2010; Pardo et al., 2010; Wang et al., 2006). In each case, an individual transcription factor interacts with multiple partners, and in some cases, hundreds of these (van den Berg et al., 2010; Gagliardi et al., 2013; Wang et al., 2006; Costa et al., 2013; Pardo et al., 2010). The published ES cell–related interactomes for SOX2 are either from cells in the early stages of differentiation (Mallanna et al., 2010) or from self-renewing cells overexpressing SOX2, OCT4, KLF4, and c-myc (Gao et al., 2012). This latter study, identifying an interactome of 71 proteins, is the most relevant to a discussion of the role of SOX2 in ES cells. NANOG and OCT4 are present in this interactome but so, too, are other transcription factors including ESRRB, KLF4, SALL1, and SALL4, all of which have been demonstrated to function in the maintenance of ES cell self-renewal (Festuccia et al., 2012; Sakaki-Yumoto et al., 2006; Zhang et al., 2006; Karantzali et al., 2011; Zhang et al., 2010b). The degree of overlap between the SOX2 interactome and those of other factors is marked; 40% of proteins in the SOX2 interactome have previously been shown to interact with another ES cell transcription factor. For example, ESRRB is present in the interactomes of SOX2, NANOG, OCT4, and SALL4. Likewise, SALL4 is common to the interactomes of NANOG, SOX2, OCT4, ESRRB, and NR0B1. The degree of overlap between the different interactomes suggests that a consideration of the binary interactions between any two factors in isolation may miss a myriad of other interactions that contribute to the function of the pluripotency gene regulatory network. One route to a fuller understanding may be to integrate the proteomic data with ChIP-based datasets. This was done on a limited basis by Orkin and colleagues, who showed that the promoters of many proteins in the protein interaction network were also co-bound by multiple factors (Kim et al., 2008a). A similar but large-scale integration collated more than 200,000 ChIP-seq interactions and 1000 protein–protein interactions in the Embryonic Stem Cell Atlas of Pluripotency Evidence database (Xu et al., 2013). Systematic interrogation of these data may identify novel interaction hubs that can be further perturbed experimentally to deepen understanding of control of pluripotent identity by protein networks.

Another noteworthy feature of the SOX2 interactome is that transcription factors are not the dominant functional group of partners. The other proteins can be categorized into various functional groups including chromatin remodeling (e.g., NuRD, Swi/Snf), DNA repair, and DNA replication machinery. A similar situation is observed with other ES cell transcription factor interactomes in which the dominant feature is not the presence of other transcription factors but rather the large number of proteins associated with a wide variety of additional functions. Although it is intuitively obvious how in general some families of proteins found in the interactomes, such as chromatin remodeling proteins, may have a role in transcription factor function, a detailed understanding of how these functions are linked is currently missing. In addition, the functions of other families of proteins are not obvious. Indeed, the annotated functions for particular proteins may be misleading because they may have different roles in different contexts. For example, the SOX2 interactome contains the XPC–nucleotide repair complex. Although the potential function of this complex in the context of a transcriptional regulator is not immediately clear, detailed examination has demonstrated that the complex acts a transcriptional co-activator in ES cells, a function for which the nucleotide repair activity of the complex is not required (Fong et al., 2011). Other proteins in the interactome may also have such a dual function and act as co-activators or co-repressors in an ES cell context, although the extent to which this example is generally pervasive is unknown. Perhaps it is more likely that many protein partners are involved in pathways that have yet to be studied in pluripotent cells.

One of the most striking aspects of the SOX2 interactomes from ES cells is the large difference between the spectrum of protein partners in self-renewing cells and cells exposed to differentiation medium for 24 h (Mallanna et al., 2010; Gao et al., 2012). These interactomes were obtained from cells expressing the same levels of SOX2, in the same laboratory using the same purification methods and the same proteomic platform. As such, differences between the interactomes are more likely to reflect true biological function rather than being consequences of differences in methodologies and levels of protein expression. A comparison of the interactomes shows that although the size of the interactomes is similar, the identities of the interacting proteins differ, with less than one-third of the proteins common to the two interactomes. In the future, it may be informative to consider how these changes in interactomes relate to cellular commitment points during differentiation. Although the change in interactome is partially the results of the change in expression levels of some of the partners during the onset of differentiation, the fact that the expression levels of most of interactors do not change suggests that other mechanisms regulate protein–protein interactions. Such mechanisms may include covalent modifications of both SOX2 and partners and are discussed in the following section.

REGULATION OF SOX2

The need for SOX2 levels to be maintained and controlled means that SOX2 regulation must sustain the correct SOX2 concentration while remaining responsive to external cues that initiate commitment into particular lineages. The means by which SOX2 is regulated can be considered in three ways.

Transcriptional Regulation

The defining feature of the transcriptional regulation of the three core regulators of ES cell pluripotency is that the regulatory regions of each gene are bound by each of the three as determined by genome-wide ChIP analysis (Boyer et al., 2005; Chen et al., 2008; Kim et al., 2008a; Marson et al., 2008; Loh et al., 2006). This initially suggested that a positive autoregulatory circuit might exist, with each member of the core positively regulating the genes encoding all three core members. Indeed, using a combination of luciferase and EMSAs, OCT4 and SOX2 have been demonstrated to contribute to positive regulation of NANOG and OCT4 (Rodda et al., 2005; Kuroda et al., 2005; Okumura-Nakanishi et al., 2005). Moreover, mutation of the OCT4 binding site in the Oct–Sox motif at the endogenous *Nanog* locus results in decreased *Nanog* expression (Karwacki-Neisius et al., 2013). However, the assumption that all three factors upregulate transcription of all three genes is incorrect because NANOG acts autorepressively and the contribution of NANOG to regulation of OCT4 and SOX2 is minimal (Navarro et al., 2012; Fidalgo et al., 2012). However, OCT4 and SOX2 binding can be demonstrated in the enhancers of both *Oct4* and *Sox2* and knocking down either OCT4 or SOX2 results in decreased expression of both molecules (Chew et al., 2005). The current view of the core regulatory circuit is shown in Figure 2. The presence of feed-forward loops maintains a robust level of gene expression but allows the system to be switched on and off by environmental signals (Mangan and Alon, 2003; Chickarmane

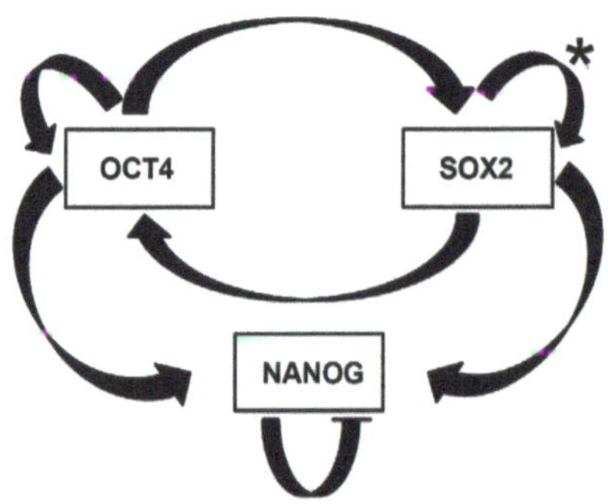

Figure 2 ***Model of architecture of core transcriptional network in ES cells.*** *SOX2 may act both positively and negatively on the *Sox2* gene. Overexpression of SOX2 decreases endogenous *Sox2* mRNA expression (Kopp et al., 2008). On the other hand, mutations within the Sox component of a Sox–Oct motif at the SRR2 enhancer of *Sox2* decrease activity in luciferase assays (Tomioka et al., 2002). *Adapted from Navarro et al. (2012).*

et al., 2006; Boyer et al., 2005; Doncic and Skotheim, 2013). The architecture of the circuitry may also indicate how the exogenous expression of OCT4 and SOX2 as part of a reprogramming cocktail can ignite the expression of endogenous factors to set up a robust expression of ES cell factors. However, another study has suggested that elevating the SOX2 level represses the enhancer activity of target genes containing Oct–Sox motifs such as NANOG and OCT4 (Boer et al., 2007). Further work will be required to determine whether the regulatory activity of SOX2 on target genes is context-dependent, allowing the circuitry to collapse in response to differentiation cues.

Posttranscriptional Regulation

MicroRNAs (miRNAs) are noncoding RNAs that bind to complementary sites in messenger RNA and cause their degradation or repression of translation. MicroRNA regulatory systems can be complex, with one miRNA regulating more than one gene and one gene being regulated by multiple miRNAs. By examining miRNA levels in undifferentiated and retinoic acid treated cells, miRNAs that changed upon initiation of differentiation were identified and proposed to be involved in modulating differentiation (Tay et al., 2008a). One of these miRNAs (miR-134) was predicted to bind to five different regions within the coding sequence of SOX2 mRNA and was shown to decrease the expression level of SOX2 protein markedly when expressed in ES cells. The potential complexity of the miRNA system is illustrated by the fact that miR-134 also binds to the UTR of NANOG and attenuates NANOG expression (Tay et al., 2008b).

Another example of miRNA-mediated modulation of expression is seen in human ES cells in which miR-145 is highly upregulated during differentiation and contributes to repression of SOX2 by targeting the UTR of SOX2. As is the case with the example above, the system is not restricted to a single protein and the same miRNA also downregulates expression of OCT4 and KLF4 (Xu et al., 2009). Further complexity is added by the fact that the miR-145 promoter contains an OCT4 binding site, with miR-145 expression repressed upon OCT4 binding (Xu et al., 2009). More discussions on these points are found in Chapter 4.

Posttranslational Regulation

A number of posttranslational covalent modifications of SOX2 also have a part in controlling SOX2 function. These modifications are described in detail in Chapter 8 and are known to include phosphorylation, O-GlcNAcylation, ubiquitination, SUMOylation, acetylation, methylation, and ADP-ribosylation (Jeong et al., 2010; Van Hoof et al., 2009; Swaney et al., 2009; Adams et al., 2007; Tsuruzoe et al., 2006; Jang et al., 2012; Baltus et al., 2009; Zhao et al., 2011; Gao et al., 2009). Posttranslational modifications of SOX2 are concentrated in two main areas: one within the DNA binding domain and one within the transactivation domain. Such a localization of modifications suggests that the main thrust of posttranslational regulation is to modulate both the binding of SOX2 to DNA and the ability of SOX2 to activate downstream effects after DNA binding.

Rather than each modification system existing in isolation, cross-talk between posttranslational modification systems can result in interdependency between posttranslational modifications. This can be illustrated by the phosphorylation and SUMOylation systems, which intersect around Lys247, a substrate for SUMOylation (Tsuruzoe et al., 2006). Three nearby serine residues (Ser251–253) are phosphorylation substrates, and mutation of these three serines to aspartates, a change thought to mimic phosphorylation, increases the level of SUMOylation (Van Hoof et al., 2009). Therefore, phosphorylation may increase SUMOylation. Because SUMOylation can in turn inhibit SOX2 binding to the FGF4 enhancer in vitro (Tsuruzoe et al., 2006), these posttranslational modifications may be linked to function. However, it will be important to determine whether this effect of SUMOylation on SOX2 binding is general or specific to the FGF4 locus. Furthermore, because Ser248 is a substrate for O-GlcNAcylation, other posttranslational modifications might also affect function.

CONSEQUENCES OF ALTERING SOX2 LEVELS

SOX2 Deletion

The presence of Oct–Sox motifs in the regulatory regions of so many ES cell genes demonstrates the pivotal role of SOX2 in ES cell function and embryonic development. An initial study that addressed this generated a targeted mutation of one of the two *Sox2* alleles in vivo (Avilion et al., 2003). This showed that embryos homozygous for the SOX2 deletion were not viable and failed to develop an epiblast postimplantation. Although it is likely that the ability of the embryos to advance as far as the late blastocyst stage resulted from the presence of the maternal SOX2 protein, the fact that development cannot continue postimplantation is a clear indication of the importance of SOX2 for epiblast function. The role of SOX2 in ES cells was directly investigated by an siRNA screen for factors affecting ES cell self-renewal (Ivanova et al., 2006). This study showed that SOX2 knockdown caused ES cells to change morphology, lose alkaline phosphatase activity, and differentiate toward trophectoderm and epiblast-derived lineages. However, the interpretation of both of these studies is complicated by either the presence of maternal protein (Avilion et al., 2003) or potential incomplete knockdown and/or off-target effects of siRNA methods (Ivanova et al., 2006).

Any question about the role of SOX2 in maintaining of pluripotency was answered by a knockout study showing that SOX2 was indispensable for maintaining ES cells with knockout cells acquiring trophectoderm-like identity (Masui et al., 2007). However, this study also demonstrated that, contrary to expectations, the transcriptional output from genes containing composite Oct–Sox motifs in their regulatory regions such as FGF4, LEFTY1, OCT4, and UTF1 were not adversely affected in the first 24–48h after SOX2 deletion. The expression levels of these genes changed only at the point where the cells showed high levels of trophectoderm markers such as CSH1 and TPBPA. FGF4, LEFTY1, OCT4, and UTF1 all had previously been shown to react quickly to the deletion of OCT4, which suggested that

SOX2 may be dispensable for the regulation of Oct–Sox motifs (Niwa et al., 2000, 2005; Matoba et al., 2006). The importance of the interaction between OCT4 and SOX2 for stabilization of the OCT4–DNA interaction described above therefore implies that other factors may be able to substitute for SOX2 upon deletion. To investigate this possible redundancy, the expression of other Sox proteins in ES cells was assessed (Masui et al., 2007). The Sox protein family can be divided into 10 groups with SOX2 in Group B1. No other SoxB1 group proteins could be detected in ES cells. However, three Sox proteins were detectable: namely, SOX4 and 11 (which are Group C proteins) and SOX15 (Group G). The level of expression of these proteins in ES cells is not known, although SOX15 and SOX2 mRNAs are expressed at comparable levels (Maruyama et al., 2005). All three proteins have been demonstrated to act in concert with OCT4 to activate Oct–Sox motifs (Maruyama et al., 2005; Wiebe et al., 2003; Masui et al., 2007). However, deletion of SOX15 has no effect on pluripotency (Maruyama et al., 2005). Furthermore, each of these proteins can be detected bound to the Oct–Sox motifs of *Fgf4*, *Oct4*, *Nanog*, and *Lefty1* in cells that also express SOX2 (Masui et al., 2007). These results suggest that redundancy between these Sox proteins has a role in regulation of Oct–Sox motifs. Further understanding of the role of these additional Sox proteins awaits more detailed biochemical analysis of SOX4, 11, and 15 binding to DNA. Together with experiments to assess whether the overexpression of these Sox proteins can prevent differentiation of SOX2 null cells, this should reveal the full extent of functional redundancy between these Sox proteins.

Alignment between SOX2, 4, 11, and 15 is shown in Figure 3. The DNA binding HMG domains show the largest degree of similarity; the rest of the proteins are highly divergent.

Figure 3 ***Alignment of amino acid sequences of Sox2, Sox4, Sox11, and Sox15.*** *Residues conserved in all four proteins. Conservative and semiconservative alterations are marked with colons and periods, respectively. The HMG domains are highlighted in red.

```
B1|SOX2   EKRPFIDEAKRLRALHMKEHPDYKYRP 114
B1|SOX1   EKRPFIDEAKRLRALHMKEHPDYKYRP 122
B1|SOX3   EKRPFIDEAKRLRAVHMKEYPDYKYRP 140
C |SOX4   DKIPFIQEAERLRLKHMADYPDYKYRP 130
C |SOX11  EKIPFIREAERLRLKHMADYPDYKYRP 120
F |SOX17  EKRPFVEEAERLRVQHMQDHPNYKYRP 139
G |SOX15  EKRPFVEEAKRLRARHLRDYPDYKYRP 118
          :* **: **:***  *  ::*:***:*
```

Figure 4 ***Alignment of a portion of the HMG domain of Sox family members.*** The five residues identified by Williams et al. as critical for the interaction of SOX2 with OCT1 are highlighted in green on SOX2 and in the other Sox family members when conserved. The Lys (K) to Glu (E) substitution seen in a number of Sox family members is highlighted in black. Residues conserved in all proteins are marked by an asterisk. Conservative and semiconservative alterations are marked with colons and periods, respectively.

Even within the HMG domains there are subtle differences in the sequences. Most notably there are differences in the five residues identified in structural studies by Williams and colleagues as being critical for interaction with OCT1 (Williams et al., 2004) (Figure 4). The fact that SOX4, 11, and 15 all can bind to Oct–Sox motifs and activate reporters driven by Oct–Sox motifs demonstrates that an association with OCT4 can still occur. However, in the absence of structure–function data, the subtleties of the effects of substitution of the residues involved in the Oct–Sox interface remain uncertain. The importance of performing these experiments is illustrated by an elegant study that showed the rational conversion of SOX17 into a reprogramming factor (Jauch et al., 2011). SOX17 is essential for endoderm formation, and overexpression in ES cells drives the cells to differentiate into an endoderm-like fate (Niakan et al., 2010). SOX17 can dimerize with OCT4, but its target is a compressed Oct–Sox motif to which OCT4 and SOX2 do not bind cooperatively (Jauch et al., 2011). Comparison of the sequences of SOX2 and SOX17 HMG domains shows that a lysine residue critical for contact of SOX2 with Oct proteins (Williams et al., 2004) is changed to a glutamic acid in SOX17 (Glu122) (Figure 4) (Jauch et al., 2011). In fact, this lysine-to-glutamic acid substitution is observed in a number of Sox family proteins. Mutagenesis of Glu122 of SOX17 to a lysine converts SOX17 into a protein that can interact with OCT4 on the classical Oct–Sox motif both in vitro and in vivo. Remarkably, when the mutated version of SOX17 is used in place of SOX2 in the classical four-factor reprogramming of mouse embryonic fibroblasts (MEFs), reprogramming is observed and is in fact more efficient than reprogramming with SOX2 (Jauch et al., 2011).

The ability of SOX17 to substitute for SOX2 upon mutation of a single residue and the potential for other Sox proteins to compensate functionally for loss of SOX2 upon genetic deletion is particularly intriguing given the disparate nature of the proteins sequences. SOX2, SOX4, SOX11, SOX15, and SOX17 share only notable sequence homology over the HMG domains. Such large differences between protein sequences suggest that the spectrum of protein partners bound by each protein would be different. It is therefore something of a conundrum how such disparate proteins can functionally compensate for the deletion of SOX2. One possibility is that proteins that interact with the HMG domain are sufficient to activate expression of target genes from Oct–Sox motifs. Alternatively, proteins interacting with OCT4 alone may suffice to activate Oct–Sox-directed expression without a contribution from other Sox-interacting proteins. Another aspect of this could be that simply binding of a Sox protein to the Sox consensus is sufficient to bend the DNA and enable a productive OCT4 interaction. Irrespective of how the downstream effects of binding to Oct–Sox motifs are mediated, it is clear from the SOX17 study that the interaction of OCT4 with a Sox protein in the correct orientation is critical to overcome other sequence differences even when such differences are significant.

Evidence from genetic deletion that SOX2 is not essential for the expression of reporters driven by Oct–Sox motifs raises the question of what the precise role is of SOX2 because SOX2 deletion has such deleterious effects on ES cell self-renewal. To address this question, microarrays were used to identify the genes affected by SOX2 deletion (Masui et al., 2007). This showed that the expression of a large number of genes was affected by SOX2 loss, with 1271 genes changing expression more than 1.5-fold within 24 h of deletion (almost equal numbers were upregulated and downregulated). Within this list, inducers of differentiation such as Eomes and Esx1 were increased by 260% and 190%, respectively, upon SOX2 deletion. Also present in the list are two known regulators of OCT4: NR5A2 and NR2F2. Both are members of the nuclear receptor family of transcription factors and act as positive and negative regulators of OCT4 expression, respectively. Upon SOX2 removal, NR5A2 transcript levels were decreased sixfold by 24 h, whereas NR2F2 was upregulated 1.7-fold. This suggested the hypothesis that the role of SOX2 is to promote the expression of positive regulators of OCT4 such as NR5A2 and simultaneously repress negative regulators of OCT4 such as NR2F2 and mediators of differentiation (e.g., Eomes). In this situation, the removal of SOX2 would cause decreased expression of OCT4 and upregulate the expression of inducers of differentiation. In such a scenario, it may therefore be expected that ectopic expression of OCT4 would rescue the SOX2 null phenotype; this is exactly what is observed when an OCT4 transgene is introduced after the induction of SOX2 excision. The scenario described by these results is depicted in Figure 5. However, there are scant data pertaining to the role of NR5A2 and NR2F2 in ES cells or their functions in SOX2-deleted cells. Moreover, many other genes were identified as SOX2 targets by the SOX2 null microarray. These putative SOX2 target genes are implicated in a wide variety of processes including DNA binding, chromatin remodeling, and metabolism. The roles of most of these genes in maintaining ES cell self-renewal remains to be determined.

SOX2 Overexpression

Overexpression of the factors controlling ES cell self-renewal does not have a single phenotype. It may be intuitive to consider that overexpression would lead to a highly robust self-renewing state; this is indeed what is observed in the case of NANOG (Chambers et al., 2003). However, OCT4 does not have the same characteristics. One

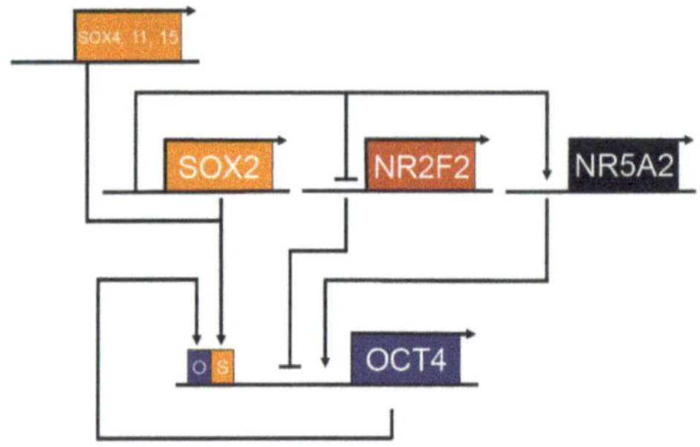

Figure 5 ***Hypothetical regulatory interactions between Sox proteins and Oct4.*** *Sox2* and *Oct4* genes produce proteins that bind to and positively regulate Oct–Sox motifs in target genes (illustrated here at the OCT4 locus as OS). These Oct–Sox motifs can also bind SOX4, 11, and 15 and associated genes are positively regulated by these Sox proteins. Also indicated is the negative regulation of the OCT4 repressor NR2F2 by SOX2 and the positive regulation of the OCT4 activator NR5A2 by SOX2.

study used immunoblotting and suggested that the concentration of OCT4 needed to be restricted between 50% and 150% of the wild-type level (Niwa et al., 2000). However, studies indicated that a level of OCT4 expression below 50% of the wild-type level is compatible with ES cell self-renewal (Radzisheuskaya et al., 2013). Another study demonstrated that OCT4 heterozygote ($Oct4^{+/-}$) ES cells exhibit a stabilized pluripotent state demonstrated by homogeneous expression of pluripotency transcription factors including NANOG, ESRRB, and KLF4 (Karwacki-Neisius et al., 2013). Moreover, reduction in the OCT4 level results in increased genome-wide binding of OCT4, particularly at enhancers of pluripotency-associated genes. Importantly, $Oct4^{+/-}$ ES cells exhibit enhanced responsiveness to leukemia inhibitory factor signaling, independence from bone morphogenetic protein 4–serum, and delayed differentiation kinetics, which indicates that these cells with reduced OCT4 expression have a robust self-renewal capacity (Karwacki-Neisius et al., 2013). In fact, the differentiation observed by OCT4 overexpression (Niwa et al., 2000) is obvious even at wild-type OCT4 levels (Karwacki-Neisius et al., 2013). These studies indicate that wild-type ES cells encode departure from the pluripotent state in their OCT4 expression level.

Initial reports of the effect of constitutive SOX2 overexpression observed no deleterious effects on ES cells self-renewal even in cell lines expressing many fold higher levels of SOX2 than wild-type. Instead, it was demonstrated that the overexpressing cells were biased toward the neural lineage upon culturing in conditions that favor differentiation (Zhao et al., 2004). However, a subsequent study using inducible SOX2 overexpression reported contradictory results with a small induction in the level of SOX2 causing loss of self-renewal and differentiation of cells into neuroectodermal, mesodermal, and trophectodermal lineages (Kopp et al., 2008). Interestingly, introduction of exogenous SOX2 has been demonstrated to reduce endogenous *Sox2* transcription (Kopp et al., 2008; Boer et al., 2007). Whether such negative regulation contributes to the discrepancy in these studies or whether a difference in constitutive versus induced overexpression underlies these differences will require further studies.

SOX2 IN REPROGRAMMING

Initial groundbreaking reports describing the reprogramming of somatic cells to an ES-like state showed that only four transcription factors were required to achieve this remarkable feat (Takahashi and Yamanaka, 2006). These four proteins or Yamanaka factors are SOX2, OCT4, KLF4, and c-myc. Subsequent work examined the contribution of the individual factors, improved efficiencies, and examined different combinations of factors. It has since become clear that SOX2 and OCT4 form the core of the reprogramming cocktail with a number of other accessory factors that can be included to improve efficiency but that are not absolutely necessary for reprogramming. For example, NANOG and SALL4 will both increase efficiencies of reprogramming MEFs (Silva et al., 2009; Hanna et al., 2009; Tsubooka et al., 2009). Furthermore, the related SoxB1 group proteins SOX1 and SOX3 can substitute for the need for ectopic SOX2 in reprogramming, but this capacity is not shared by more distantly related family members such as SOX7, SOX15, SOX17, and SOX18 (Nakagawa et al., 2008). This suggests that perhaps the reprogramming process is less tolerant of subtle differences in DNA sequence recognition by Sox proteins and the complexes they form with OCT4 in contrast to self-renewing ES cells discussed previously. Although SOX2 is one of the core reprogramming factors, reprogramming can be achieved in the absence of SOX2. For example, replacing SOX2 with a transforming growth factor-β inhibitor is reported to reprogram fibroblasts via upregulation of *Nanog* expression (Ichida et al., 2009). A study also demonstrated that fibroblasts lacking the tumor suppressor p27 can be reprogrammed with only OCT4 and KLF4 (Li, 2012). Nonetheless, endogenous SOX2 expression is activated in the cells that successfully achieved reprogramming. The significance of endogenous SOX2 expression was demonstrated in several studies showing that cells expressing endogenous SOX2 such as neural stem cells (NSCs) and melanocytes can be reprogrammed in the absence of ectopic SOX2 (Kim et al., 2008b; Silva et al., 2008; Eminli et al., 2008; Utikal et al., 2009). Retroviral integration of the SOX2 transgene is also observed to be frequently absent in NSC-derived iPS cells using the four Yamanaka factors. Furthermore, the efficiency of reprogramming was enhanced when three Yamanaka factors were used without SOX2, which suggests that high levels of SOX2 have detrimental effects during reprogramming (Silva et al., 2008).

The role of SOX2, OCT4, KLF4, and c-myc in the reprogramming process is to activate the pluripotency gene regulatory network. When global ChIP studies of cells undergoing reprogramming are analyzed, a situation similar to that observed in ES cells is observed with all the factors strongly overlapping in their binding sites. Following single cells during reprogramming with the four Yamanaka factors revealed that the transcriptional profile changes occur in two phases (Buganim et al., 2012). Cells early in the process are transcriptionally heterogeneous and many cells do not achieve programming. Activation of endogenous SOX2 expression occurs late in this process in cells likely to achieve reprogramming and is associated with a deterministic sequence of gene expression to activate the pluripotency circuitry.

SOX2 IN EPIBLAST STEM CELLS

Self-renewing pluripotent stem cells can be derived from the epiblast of postimplantation embryonic day 5.5 embryos (Tesar et al., 2007; Brons et al., 2007) and are termed epiblast stem cells (EpiSCs). These cultured pluripotent cells retain teratocarcinoma forming capacity when ectopically grafted into kidney capsules, but unlike ES cells, they do not have the ability to contribute to the developing blastocyst (Tesar et al., 2007; Brons et al., 2007). One reason for this loss of ability to integrate into development when EpiSCs are introduced into a preimplantation embryo is a reduction in E-cadherin expression in EpiSCs (Ohtsuka et al., 2012), which results in altered adherence properties. Consistent with this notion, when grafted to the postimplantation epiblast, EpiSCs can readily contribute to the developing embryo, differentiate into all three primary germ layers, and form primordial germ cells (Huang et al., 2012; Kojima et al., 2013).

Epiblast stem cells are different from ES cells in several aspects, although pluripotency is common to both. They exhibit distinct growth factor dependence for self-renewal and are maintained in the presence of Activin/Nodal and FGF signaling. One crucial difference compared with mouse ES cells is that female EpiSCs have repressive histone trimethylation (H3K27me3) characteristic of an inactive X chromosome (Guo et al., 2009). This strongly underlines their resemblance to the developmentally advanced postimplantation epiblast (Tan et al., 1993).

Although the circuitry maintaining EpiSC pluripotency is largely different from that maintaining ES cells, the expression of the core pluripotency transcription factors OCT4, SOX2, and NANOG are common to both cell types. However, SOX2 and NANOG transcript levels are reduced in EpiSCs compared with ES cells (Guo et al., 2009; Guo and Smith, 2010; Zhang et al., 2010a; Tesar et al., 2007; Brons et al., 2007; Greber et al., 2010). Indeed, the identification of a SOX2–ESRRB DNA binding motif in the regulatory elements of a number of genes important for the ES program suggests that the abrogation of the complex between SOX2, ESRRB, and DNA is important for the ES-to-EpiSC transition (Hutchins et al., 2013). Several other different markers are enriched specifically in ES cells, such as ESRRB, REX1 (ZFP42), DAX1 (NR0B1), FGF4, NR5A2, Stella (Dppa3), KLF2, and KLF4, whereas EpiSCs are enriched in the postimplantation epiblast-specific marker FGF5, as well as several markers of differentiating lineages, such as T/brachyury, FOXA2, SOX1, SOX17, and GATA6. In EpiSCs, OCT4 still binds at the *Sox2*, *Nanog*, and *Oct4* loci as in ES cells, although the overlap between the global OCT4-bound targets in EpiSCs and ES cells is small (Tesar et al., 2007). Indeed, there is a relocalization of OCT4 binding to a subset of target gene enhancers when ES cells differentiate under EpiSC conditions (Buecker et al., 2014). Enhancers that are occupied by ESRRB in ES cells lose both ESRRB and OCT4 during this transition, with other enhancers that bind OTX2 gaining OCT4 binding (Buecker et al., 2014). Furthermore, genes common to ES cells and EpiSCs can have distinct enhancers that are differentially active in ES cells and EpiSCs (Factor et al., 2014). These studies indicate that the pluripotency circuitry regulated by OCT4 is substantially different in EpiSCs. Likewise, SOX2 supports the epiblast state by regulating a spectrum of genes different from those that it appears to control in ES cells (Iwafuchi-Doi et al., 2012).

Iwafuchi-doi et al. attempted to address the importance of SOX2 in the postimplantation epiblast by targeted deletion of the epiblast-specific N2 enhancer of SOX2 (Iwafuchi-Doi et al., 2011). Embryonic day 7.5 embryos lacking the N2 enhancer have a 10-fold reduction in SOX2 expression specifically in the epiblast but without obvious morphological abnormalities. There are several potential explanations for this result. Perhaps the residual SOX2 level is sufficient for function; perhaps other Sox proteins in the epiblast can compensate for reduced SOX2, or possibly the role of SOX2 in the epiblast may be to specify cells toward neuroectoderm development rather than to sustain pluripotency. Because EpiSCs share defining features with the postimplantation epiblast (Tesar et al., 2007; Brons et al., 2007; Kojima et al., 2013), it will be important in future experiments to further determine whether EpiSCs share the absolute dependence on SOX2 seen in ES cells. In particular, experiments to capture postimplantation pluripotent cells in culture from different SOX2-expressing epiblast regions may illuminate how EpiSC pluripotency is regulated and how it is ultimately dismantled.

PERSPECTIVES

The requirement for SOX2 in maintaining ES cells is clear. However, the detailed mechanisms by which SOX2 functions in ES cells are less transparent. Although a large number of SOX2 chromatin targets have been identified, the functional implications of SOX2 binding at the vast majority of these sites remain to be elucidated. Likewise, whereas proteomic studies have identified many SOX2 protein partners, the biological role of most of these interactions remains to be elucidated. In addition, there is likely to be much still to uncover about posttranscriptional regulation of SOX2. Finally, further extension of the study of SOX2 in EpiSCs will allow a comparison of preimplantation and postimplantation pluripotency that can be expected to narrow down the essential contribution of SOX2 to pluripotency. Answers to these questions will enrich our knowledge of the potentially diverse roles that SOX2 has in different stem cell types and will deepen our understanding of what it means to be pluripotent.

ACKNOWLEDGMENTS

We thank Tapan Mistri and Simon Tomlinson for comments on the manuscript. Research in our lab is funded by the Medical Research Council of the United Kingdom.

REFERENCES

Adams, V.H., Mcbryant, S.J., Wade, P.A., Woodcock, C.L., Hansen, J.C., 2007. Intrinsic disorder and autonomous domain function in the multifunctional nuclear protein, MeCP2. J. Biol. Chem. 282, 15057–15064.

Ambrosetti, D.C., Basilico, C., Dailey, L., 1997. Synergistic activation of the fibroblast growth factor 4 enhancer by Sox2 and Oct-3 depends on protein-protein interactions facilitated by a specific spatial arrangement of factor binding sites. Mol. Cell. Biol. 17, 6321–6329.

Ambrosetti, D.C., Scholer, H.R., Dailey, L., Basilico, C., 2000. Modulation of the activity of multiple transcriptional activation domains by the DNA binding domains mediates the synergistic action of Sox2 and Oct-3 on the fibroblast growth factor-4 enhancer. J. Biol. Chem. 275, 23387–23397.

Arnold, K., Sarkar, A., Yram, M.A., Polo, J.M., Bronson, R., Sengupta, S., Seandel, M., Geijsen, N., Hochedlinger, K., 2011. Sox2^{+} adult stem and progenitor cells are important for tissue regeneration and survival of mice. Cell Stem Cell 9, 317–329.

Avilion, A.A., Nicolis, S.K., Pevny, L.H., Perez, L., Vivian, N., Lovell-Badge, R., 2003. Multipotent cell lineages in early mouse development depend on SOX2 function. Genes Dev. 17, 126–140.

Baltus, G.A., Kowalski, M.P., Tutter, A.V., Kadam, S., 2009. A positive regulatory role for the mSIN3a/HDAC complex in pluripotency through Nanog and Sox2. J. Biol. Chem. 284, 6998–7006.

van den Berg, D.L., Snoek, T., Mullin, N.P., Yates, A., Bezstarosti, K., Demmers, J., Chambers, I., Poot, R.A., 2010. An Oct4-centered protein interaction network in embryonic stem cells. Cell Stem Cell 6, 369–381.

Bergstrom, D.E., Young, M., Albrecht, K.H., Eicher, E.M., 2000. Related function of mouse SOX3, SOX9, and SRY HMG domains assayed by male sex determination. Genesis 28, 111–124.

Boer, B., Kopp, J., Mallanna, S., Desler, M., Chakravarthy, H., Wilder, P.J., Bernadt, C., Rizzino, A., 2007. Elevating the levels of Sox2 in embryonal carcinoma cells and embryonic stem cells inhibits the expression of Sox2:Oct-3/4 target genes. Nucleic Acids Res. 35, 1773–1786.

Boyer, L.A., Lee, T.I., Cole, M.F., Johnstone, S.E., Levine, S.S., Zucker, J.P., Guenther, M.G., Kumar, R.M., Murray, H.L., Jenner, R.G., Gifford, D.K., Melton, D.A., Jaenisch, R., Young, R.A., 2005. Core transcriptional regulatory circuitry in human embryonic stem cells. Cell 122, 947–956.

Brons, I.G., Smithers, L.E., Trotter, M.W., Rugg-Gunn, P., Sun, B., Chuva De Sousa Lopes, S.M., Howlett, S.K., Clarkson, A., Ahrlund-Richter, L., Pedersen, R.A., Vallier, L., 2007. Derivation of pluripotent epiblast stem cells from mammalian embryos. Nature 448, 191–195.

Buecker, C., Srinivasan, R., Wu, Z., Calo, E., Acampora, D., Faial, T., Simeone, A., Tan, M., Swigut, T., Wysocka, J., 2014. Reorganization of enhancer patterns in transition from naive to primed pluripotency. Cell Stem Cell 14, 838–853.

Buganim, Y., Faddah, D.A., Cheng, A.W., Itskovich, E., Markoulaki, S., Ganz, K., Klemm, S.L., Oudenaarden, A.V., Jaenisch, R., 2012. Single-cell expression analyses during cellular reprogramming reveal an early stochastic and a late hierarchic phase. Cell 150, 1209–1222.

Chambers, I., Tomlinson, S.R., 2009. The transcriptional foundation of pluripotency. Development 136, 2311–2322.

Chambers, I., Colby, D., Robertson, M., Nichols, J., Lee, S., Tweedie, S., Smith, A., 2003. Functional expression cloning of Nanog, a pluripotency sustaining factor in embryonic stem cells. Cell 113, 643–655.

Chambers, I., Silva, J., Colby, D., Nichols, J., Nijmeijer, B., Robertson, M., Vrana, J., Jones, K., Grotewold, L., Smith, A., 2007. Nanog safeguards pluripotency and mediates germline development. Nature 450, 1230–1234.

Chen, X., Xu, H., Yuan, P., Fang, F., Huss, M., Vega, V.B., Wong, E., Orlov, Y.L., Zhang, W., Jiang, J., Loh, Y.H., Yeo, H.C., Yeo, Z.X., Narang, V., Govindarajan, K.R., Leong, B., Shahab, A., Ruan, Y., Bourque, G., Sung, W.K., Clarke, N.D., Wei, C.L., Ng, H.H., 2008. Integration of external signaling pathways with the core transcriptional network in embryonic stem cells. Cell 133, 1106–1117.

Chew, J.L., Loh, Y.H., Zhang, W., Chen, X., Tam, W.L., Yeap, L.S., Li, P., Ang, Y.S., Lim, B., Robson, P., Ng, H.H., 2005. Reciprocal transcriptional regulation of Pou5f1 and Sox2 via the Oct4/Sox2 complex in embryonic stem cells. Mol. Cell. Biol. 25, 6031–6046.

Chickarmane, V., Troein, C., Nuber, U.A., Sauro, H.M., Peterson, C., 2006. Transcriptional dynamics of the embryonic stem cell switch. PLoS Comput. Biol. 2, e123.

Chuang, L.S., Lai, S.K., Murata-Hori, M., Yamada, A., Li, H.Y., Gunaratne, J., Ito, Y., 2012. RUNX3 interactome reveals novel centrosomal targeting of RUNX family of transcription factors. Cell Cycle 11, 1938–1947.

Costa, Y., Ding, J., Theunissen, T.W., Faiola, F., Hore, T.A., Shliaha, P.V., Fidalgo, M., Saunders, A., Lawrence, M., Dietmann, S., Das, S., Levasseur, D.N., Li, Z., Xu, M., Reik, W., Silva, J.C., Wang, J., 2013. Nanog-dependent function of TET1 and TET2 in establishment of pluripotency. Nature 495, 370–374.

Di Rocco, G., Gavalas, A., Popperl, H., Krumlauf, R., Mavilio, F., Zappavigna, V., 2001. The recruitment of SOX/OCT complexes and the differential activity of HOXA1 and HOXB1 modulate the Hoxb1 auto-regulatory enhancer function. J. Biol. Chem. 276, 20506–20515.

Doncic, A., Skotheim, J.M., 2013. Feedforward regulation ensures stability and rapid reversibility of a cellular state. Mol. Cell 50, 856–868.

Eminli, S., Utikal, J., Arnold, K., Jaenisch, R., Hochedlinger, K., 2008. Reprogramming of neural progenitor cells into induced pluripotent stem cells in the absence of exogenous Sox2 expression. Stem Cells 26, 2467–2474.

Factor, D.C., Corradin, O., Zentner, G.E., Saiakhova, A., Song, L., Chenoweth, J.G., Mckay, R.D., Crawford, G.E., Scacheri, P.C., Tesar, P.J., 2014. Epigenomic comparison reveals activation of "Seed" enhancers during transition from naive to primed pluripotency. Cell Stem Cell 14, 854–863.

Festuccia, N., Osorno, R., Halbritter, F., Karwacki-Neisius, V., Navarro, P., Colby, D., Wong, F., Yates, A., Tomlinson, S.R., Chambers, I., 2012. Esrrb is a direct Nanog target gene that can substitute for Nanog function in pluripotent cells. Cell Stem Cell 11, 477–490.

Fidalgo, M., Faiola, F., Pereira, C.F., Ding, J., Saunders, A., Gingold, J., Schaniel, C., Lemischka, I.R., Silva, J.C., Wang, J., 2012. Zfp281 mediates Nanog autorepression through recruitment of the NuRD complex and inhibits somatic cell reprogramming. Proc. Natl. Acad. Sci. U.S.A. 109, 16202–16207.

Fong, Y.W., Inouye, C., Yamaguchi, T., Cattoglio, C., Grubisic, I., Tjian, R., 2011. A DNA repair complex functions as an Oct4/Sox2 coactivator in embryonic stem cells. Cell 147, 120–131.

Gagliardi, A., Mullin, N.P., Ying Tan, Z., Colby, D., Kousa, A.I., Halbritter, F., Weiss, J.T., Felker, A., Bezstarosti, K., Favaro, R., Demmers, J., Nicolis, S.K., Tomlinson, S.R., Poot, R.A., Chambers, I., 2013. A direct physical interaction between Nanog and Sox2 regulates embryonic stem cell self-renewal. EMBO J. 32, 2231–2247.

Gao, F., Kwon, S.W., Zhao, Y., Jin, Y., 2009. PARP1 poly(ADP-ribosyl)ates Sox2 to control Sox2 protein levels and FGF4 expression during embryonic stem cell differentiation. J. Biol. Chem. 284, 22263–22273.

Gao, Z., Cox, J.L., Gilmore, J.M., Ormsbee, B.D., Mallanna, S.K., Washburn, M.P., Rizzino, A., 2012. Determination of protein interactome of transcription factor Sox2 in embryonic stem cells engineered for inducible expression of four reprogramming factors. J. Biol. Chem. 287, 11384–11397.

Greber, B., Wu, G., Bernemann, C., Joo, J.Y., Han, D.W., Ko, K., Tapia, N., Sabour, D., Sterneckert, J., Tesar, P., Schöler, H.R., 2010. Conserved and divergent roles of FGF signaling in mouse epiblast stem cells and human embryonic stem cells. Cell Stem Cell 6, 215–226.

Guo, G., Smith, A., 2010. A genome-wide screen in EpiSCs identifies Nr5a nuclear receptors as potent inducers of ground state pluripotency. Development 137, 3185–3192.

Guo, G., Yang, J., Nichols, J., Hall, J.S., Eyres, I., Mansfield, W., Smith, A., 2009. Klf4 reverts developmentally programmed restriction of ground state pluripotency. Development 136, 1063–1069.

Guo, G., Huss, M., Tong, G.Q., Wang, C., Sun, L.L., Clarke, N.D., Robson, P., 2010. Resolution of cell fate decisions revealed by single-cell gene expression analysis from zygote to blastocyst. Dev. Cell 18, 675–685.

Hanna, J., Saha, K., Pando, B., Van Zon, J., Lengner, C.J., Creyghton, M.P., Van Oudenaarden, A., Jaenisch, R., 2009. Direct cell reprogramming is a stochastic process amenable to acceleration. Nature 462, 595–601.

Huang, Y., Osorno, R., Tsakiridis, A., Wilson, V., 2012. In vivo differentiation potential of epiblast stem cells revealed by chimeric embryo formation. Cell Rep. 2, 1571–1578.

Hutchins, A.P., Choo, S.H., Mistri, T.K., Rahmani, M., Woon, C.T., Ng, C.K., Jauch, R., Robson, P., 2013. Co-motif discovery identifies an Esrrb-Sox2-DNA ternary complex as a mediator of transcriptional differences between mouse embryonic and epiblast stem cells. Stem Cells 31, 269–281.

Ichida, J.K., Blanchard, J., Lam, K., Son, E.Y., Chung, J.E., Egli, D., Loh, K.M., Carter, A.C., Di Giorgio, F.P., Koszka, K., Huangfu, D., Akutsu, H., Liu, D.R., Rubin, L.L., Eggan, K., 2009. A small-molecule inhibitor of tgf-Beta signaling replaces Sox2 in reprogramming by inducing Nanog. Cell Stem Cell 5, 491–503.

Ivanova, N., Dobrin, R., Lu, R., Kotenko, I., Levorse, J., Decoste, C., Schafer, X., Lun, Y., Lemischka, I.R., 2006. Dissecting self-renewal in stem cells with RNA interference. Nature 442, 533–538.

Iwafuchi-Doi, M., Yoshida, Y., Onichtchouk, D., Leichsenring, M., Driever, W., Takemoto, T., Uchikawa, M., Kamachi, Y., Kondoh, H., 2011. The Pou5f1/Pou3f-dependent but SoxB-independent regulation of conserved enhancer N2 initiates Sox2 expression during epiblast to neural plate stages in vertebrates. Dev. Biol. 352, 354–366.

Iwafuchi-Doi, M., Matsuda, K., Murakami, K., Niwa, H., Tesar, P.J., Aruga, J., Matsuo, I., Kondoh, H., 2012. Transcriptional regulatory networks in epiblast cells and during anterior neural plate development as modeled in epiblast stem cells. Development 139, 3926–3937.

Jaenisch, R., Young, R., 2008. Stem cells, the molecular circuitry of pluripotency and nuclear reprogramming. Cell 132, 567–582.

Jang, H., Kim, T.W., Yoon, S., Choi, S.Y., Kang, T.W., Kim, S.Y., Kwon, Y.W., Cho, E.J., Youn, H.D., 2012. O-GlcNAc regulates pluripotency and reprogramming by directly acting on core components of the pluripotency network. Cell Stem Cell 11, 62–74.

Jauch, R., Aksoy, I., Hutchins, A.P., Ng, C.K., Tian, X.F., Chen, J., Palasingam, P., Robson, P., Stanton, L.W., Kolatkar, P.R., 2011. Conversion of Sox17 into a pluripotency reprogramming factor by reengineering its association with Oct4 on DNA. Stem Cells 29, 940–951.

Jeong, C.H., Cho, Y.Y., Kim, M.O., Kim, S.H., Cho, E.J., Lee, S.Y., Jeon, Y.J., Lee, K.Y., Yao, K., Keum, Y.S., Bode, A.M., Dong, Z., 2010. Phosphorylation of Sox2 cooperates in reprogramming to pluripotent stem cells. Stem Cells 28, 2141–2150.

Kamachi, Y., Sockanathan, S., Liu, Q., Breitman, M., Lovell-Badge, R., Kondoh, H., 1995. Involvement of SOX proteins in lens-specific activation of crystallin genes. EMBO J. 14, 3510–3519.

Kamachi, Y., Uchikawa, M., Collignon, J., Lovell-Badge, R., Kondoh, H., 1998. Involvement of Sox1, 2 and 3 in the early and subsequent molecular events of lens induction. Development 125, 2521–2532.

Kamachi, Y., Uchikawa, M., Tanouchi, A., Sekido, R., Kondoh, H., 2001. Pax6 and SOX2 form a co-DNA-binding partner complex that regulates initiation of lens development. Genes Dev. 15, 1272–1286.

Karantzali, E., Lekakis, V., Ioannou, M., Hadjimichael, C., Papamatheakis, J., Kretsovali, A., 2011. Sall1 regulates embryonic stem cell differentiation in association with Nanog. J. Biol. Chem. 286, 1037–1045.

Karwacki-Neisius, V., GöKe, J., Osorno, R., Halbritter, F., Ng, J.H., WEIßE, A.Y., Wong, F.C.K., Gagliardi, A., Mullin, N.P., Festuccia, N., Colby, D., Tomlinson, S.R., Ng, H.-H., Chambers, I., 2013. Reduced Oct4 expression directs a robust pluripotent state with distinct signaling activity and increased enhancer occupancy by Oct4 and Nanog. Cell Stem Cell 12, 531–545.

Keramari, M., Razavi, J., Ingman, K.A., Patsch, C., Edenhofer, F., Ward, C.M., Kimber, S.J., 2010. Sox2 is essential for formation of trophectoderm in the preimplantation embryo. PLoS One 5, e13952.

Kim, J., Chu, J., Shen, X., Wang, J., Orkin, S.H., 2008a. An extended transcriptional network for pluripotency of embryonic stem cells. Cell 132, 1049–1061.

Kim, J.B., Zaehres, H., Wu, G., Gentile, L., Ko, K., Sebastiano, V., Arauzo-Bravo, M.J., Ruau, D., Han, D.W., Zenke, M., Scholer, H.R., 2008b. Pluripotent stem cells induced from adult neural stem cells by reprogramming with two factors. Nature 454, 646–650.

Kojima, Y., Kaufman-Francis, K., Studdert, J.B., Steiner, K.A., Power, M.D., Loebel, D.A.F., Jones, V., Hor, A., Alencastro, G.D., Logan, G.J., Teber, E.T., Tam, O.H., Stutz, M.D., Alexander, I.E., Pickett, H.A., Tam, P.P.L., 2013. The transcriptional and functional properties of mouse epiblast stem cells resemble the anterior primitive streak. Cell Stem Cell 14, 107–120.

Kopp, J.L., Ormsbee, B.D., Desler, M., Rizzino, A., 2008. Small increases in the level of Sox2 trigger the differentiation of mouse embryonic stem cells. Stem Cells 26, 903–911.

Kuroda, T., Tada, M., Kubota, H., Kimura, H., Hatano, S.Y., Suemori, H., Nakatsuji, N., Tada, T., 2005. Octamer and Sox elements are required for transcriptional cis regulation of Nanog gene expression. Mol. Cell. Biol. 25, 2475–2485.

Lam, C.S., Mistri, T.K., Foo, Y.H., Sudhaharan, T., Gan, H.T., Rodda, D., Lim, L.H., Chou, C., Robson, P., Wohland, T., Ahmed, S., 2012. DNA-dependent Oct4-Sox2 interaction and diffusion properties characteristic of the pluripotent cell state revealed by fluorescence spectroscopy. Biochem. J. 448, 21–33.

Li, H., Collado, M., Villasante, A., Matheu, A., Lynch, C.J., Canamero, M., Rizzoti, K., Carneiro, C., Martinez, G., Vidal, A., Lovell-Badge, R., Serrano, M., 2012. p27(Kip1) directly represses Sox2 during embryonic stem cell differentiation. Cell Stem Cell 11, 845–852.

Loh, Y.H., Wu, Q., Chew, J.L., Vega, V.B., Zhang, W., Chen, X., Bourque, G., George, J., Leong, B., Liu, J., Wong, K.Y., Sung, K.W., Lee, C.W., Zhao, X.D., Chiu, K.P., Lipovich, L., Kuznetsov, V.A., Robson, P., Stanton, L.W., Wei, C.L., Ruan, Y., Lim, B., Ng, H.H., 2006. The Oct4 and Nanog transcription network regulates pluripotency in mouse embryonic stem cells. Nat. Genet. 38, 431–440.

Loven, J., Hoke, H.A., Lin, C.Y., Lau, A., Orlando, D.A., Vakoc, C.R., Bradner, J.E., Lee, T.I., Young, R.A., 2013. Selective inhibition of tumor oncogenes by disruption of super-enhancers. Cell 153, 320–334.

Mallanna, S.K., Ormsbee, B.D., Iacovino, M., Gilmore, J.M., Cox, J.L., Kyba, M., Washburn, M.P., Rizzino, A., 2010. Proteomic analysis of Sox2-associated proteins during early stages of mouse embryonic stem cell differentiation identifies Sox21 as a novel regulator of stem cell fate. Stem Cells 28, 1715–1727.

Mangan, S., Alon, U., 2003. Structure and function of the feed-forward loop network motif. Proc. Natl. Acad. Sci. U.S.A. 100, 11980–11985.

Marson, A., Levine, S.S., Cole, M.F., Frampton, G.M., Brambrink, T., Johnstone, S., Guenther, M.G., Johnston, W.K., Wernig, M., Newman, J., Calabrese, J.M., Dennis, L.M., Volkert, T.L., Gupta, S., Love, J., Hannett, N., Sharp, P.A., Bartel, D.P., Jaenisch, R., Young, R.A., 2008. Connecting microRNA genes to the core transcriptional regulatory circuitry of embryonic stem cells. Cell 134, 521–533.

Maruyama, M., Ichisaka, T., Nakagawa, M., Yamanaka, S., 2005. Differential roles for Sox15 and Sox2 in transcriptional control in mouse embryonic stem cells. J. Biol. Chem. 280, 24371–24379.

Masui, S., Nakatake, Y., Toyooka, Y., Shimosato, D., Yagi, R., Takahashi, K., Okochi, H., Okuda, A., Matoba, R., Sharov, A.A., Ko, M.S., Niwa, H., 2007. Pluripotency governed by Sox2 via regulation of Oct3/4 expression in mouse embryonic stem cells. Nat. Cell Biol. 9, 625–635.

Matoba, R., Niwa, H., Masui, S., Ohtsuka, S., Carter, M.G., Sharov, A.A., Ko, M.S., 2006. Dissecting Oct3/4-regulated gene networks in embryonic stem cells by expression profiling. PLoS One 1, e26.

Miyagi, S., Kato, H., Okuda, A., 2009. Role of SoxB1 transcription factors in development. Cell. Mol. Life Sci. 66, 3675–3684.

Nakagawa, M., Koyanagi, M., Tanabe, K., Takahashi, K., Ichisaka, T., Aoi, T., Okita, K., Mochiduki, Y., Takizawa, N., Yamanaka, S., 2008. Generation of induced pluripotent stem cells without Myc from mouse and human fibroblasts. Nat. Biotechnol. 26, 101–106.

Navarro, P., Festuccia, N., Colby, D., Gagliardi, A., Mullin, N.P., Zhang, W., Karwacki-Neisius, V., Osorno, R., Kelly, D., Robertson, M., Chambers, I., 2012. OCT4/SOX2-independent Nanog autorepression modulates heterogeneous Nanog gene expression in mouse ES cells. EMBO J. 31, 4547–4562.

Ng, H.H., Surani, M.A., 2011. The transcriptional and signalling networks of pluripotency. Nat. Cell Biol. 13, 490–496.

Niakan, K.K., Ji, H., Maehr, R., Vokes, S.A., Rodolfa, K.T., Sherwood, R.I., Yamaki, M., Dimos, J.T., Chen, A.E., Melton, D.A., Mcmahon, A.P., Eggan, K., 2010. Sox17 promotes differentiation in mouse embryonic stem cells by directly regulating extraembryonic gene expression and indirectly antagonizing self-renewal. Genes Dev. 24, 312–326.

Nishimoto, M., Fukushima, A., Okuda, A., Muramatsu, M., 1999. The gene for the embryonic stem cell coactivator UTF1 carries a regulatory element which selectively interacts with a complex composed of Oct-3/4 and Sox-2. Mol. Cell. Biol. 19, 5453–5465.

Niwa, H., Miyazaki, J., Smith, A.G., 2000. Quantitative expression of Oct-3/4 defines differentiation, dedifferentiation or self-renewal of ES cells. Nat. Genet. 24, 372–376.

Niwa, H., Toyooka, Y., Shimosato, D., Strumpf, D., Takahashi, K., Yagi, R., Rossant, J., 2005. Interaction between Oct3/4 and Cdx2 determines trophectoderm differentiation. Cell 123, 917–929.

Nowling, T.K., Johnson, L.R., Wiebe, M.S., Rizzino, A., 2000. Identification of the transactivation domain of the transcription factor Sox-2 and an associated co-activator. J. Biol. Chem. 275, 3810–3818.

Ohtsuka, S., Nishikawa-Torikai, S., Niwa, H., 2012. E-cadherin promotes incorporation of mouse epiblast stem cells into normal development. PLoS One 7, e45220.

Okumura-Nakanishi, S., Saito, M., Niwa, H., Ishikawa, F., 2005. Oct-3/4 and Sox2 regulate Oct-3/4 gene in embryonic stem cells. J. Biol. Chem. 280, 5307–5317.

Pardo, M., Lang, B., Yu, L., Prosser, H., Bradley, A., Babu, M.M., Choudhary, J., 2010. An expanded Oct4 interaction network: implications for stem cell biology, development, and disease. Cell Stem Cell 6, 382–395.

Pevny, L.H., Lovell-Badge, R., 1997. Sox genes find their feet. Curr. Opin. Genet. Dev. 7, 338–344.

Radzisheuskaya, A., Chia, G.L.B., Santos, R.L.D., Theunissen, T.W., Castro, L.F.C., Nichols, J., Silva, J.C.R., 2013. A defined Oct4 level governs cell state transitions of pluripotency entry and differentiation into all embryonic lineages. Nat. Cell Biol. 15, 579–590.

Remenyi, A., Lins, K., Nissen, L.J., Reinbold, R., Scholer, H.R., Wilmanns, M., 2003. Crystal structure of a POU/HMG/DNA ternary complex suggests differential assembly of Oct4 and Sox2 on two enhancers. Genes Dev. 17, 2048–2059.

Rodda, D.J., Chew, J.L., Lim, L.H., Loh, Y.H., Wang, B., Ng, H.H., Robson, P., 2005. Transcriptional regulation of Nanog by OCT4 and SOX2. J. Biol. Chem. 280, 24731–24737.

Rudra, D., Deroos, P., Chaudhry, A., Niec, R.E., Arvey, A., Samstein, R.M., Leslie, C., Shaffer, S.A., Goodlett, D.R., Rudensky, A.Y., 2012. Transcription factor Foxp3 and its protein partners form a complex regulatory network. Nat. Immunol. 13, 1010–1019.

Sakaki-Yumoto, M., Kobayashi, C., Sato, A., Fujimura, S., Matsumoto, Y., Takasato, M., Kodama, T., Aburatani, H., Asashima, M., Yoshida, N., Nishinakamura, R., 2006. The murine homolog of SALL4, a causative gene in Okihiro syndrome, is essential for embryonic stem cell proliferation, and cooperates with Sall1 in anorectal, heart, brain and kidney development. Development 133, 3005–3013.

Silva, J., Barrandon, O., Nichols, J., Kawaguchi, J., Theunissen, T.W., Smith, A., 2008. Promotion of reprogramming to ground state pluripotency by signal inhibition. PLoS Biol. 6, e253.

Silva, J., Nichols, J., Theunissen, T.W., Guo, G., Van Oosten, A.L., Barrandon, O., Wray, J., Yamanaka, S., Chambers, I., Smith, A., 2009. Nanog is the gateway to the pluripotent ground state. Cell 138, 722–737.

Steunou, A.L., Ducoux-Petit, M., Lazar, I., Monsarrat, B., Erard, M., Muller, C., Clottes, E., Burlet-Schiltz, O., Nieto, L., 2013. Identification of the hypoxia-inducible factor 2alpha nuclear interactome in melanoma cells reveals master proteins involved in melanoma development. Mol. Cell. Proteomics 12, 736–748.

Swaney, D.L., Wenger, C.D., Thomson, J.A., Coon, J.J., 2009. Human embryonic stem cell phosphoproteome revealed by electron transfer dissociation tandem mass spectrometry. Proc. Natl. Acad. Sci. U.S.A. 106, 995–1000.

Takahashi, K., Yamanaka, S., 2006. Induction of pluripotent stem cells from mouse embryonic and adult fibroblast cultures by defined factors. Cell 126, 663–676.

Tan, S.-S., Williams, E.A., Tam, P.P.L., 1993. X-chromosome inactivation occurs at different times in different tissues of the post-implantation mouse embryo. Nat. Genet. 3, 170–174.

Tanaka, S., Kamachi, Y., Tanouchi, A., Hamada, H., Jing, N., Kondoh, H., 2004. Interplay of SOX and POU factors in regulation of the Nestin gene in neural primordial cells. Mol. Cell. Biol. 24, 8834–8846.

Tay, Y., Zhang, J., Thomson, A.M., Lim, B., Rigoutsos, I., 2008a. MicroRNAs to Nanog, Oct4 and Sox2 coding regions modulate embryonic stem cell differentiation. Nature 455, 1124–1128.

Tay, Y.M., Tam, W.L., Ang, Y.S., Gaughwin, P.M., Yang, H., Wang, W., Liu, R., George, J., Ng, H.H., Perera, R.J., Lufkin, T., Rigoutsos, I., Thomson, A.M., Lim, B., 2008b. MicroRNA-134 modulates the differentiation of mouse embryonic stem cells, where it causes post-transcriptional attenuation of Nanog and LRH1. Stem Cells 26, 17–29.

Tesar, P.J., Chenoweth, J.G., Brook, F.A., Davies, T.J., Evans, E.P., Mack, D.L., Gardner, R.L., Mckay, R.D., 2007. New cell lines from mouse epiblast share defining features with human embryonic stem cells. Nature 448, 196–199.

Tokuzawa, Y., Kaiho, E., Maruyama, M., Takahashi, K., Mitsui, K., Maeda, M., Niwa, H., Yamanaka, S., 2003. Fbx15 is a novel target of Oct3/4 but is dispensable for embryonic stem cell self-renewal and mouse development. Mol. Cell. Biol. 23, 2699–2708.

Tomioka, M., Nishimoto, M., Miyagi, S., Katayanagi, T., Fukui, N., Niwa, H., Muramasu, M., Okuda, A., 2002. Identification of Sox-2 regulatory region which is under the control of Oct-3/4-Sox-2 complex. Nucleic Acids Res. 30, 3202–3213.

Tsubooka, N., Ichisaka, T., Okita, K., Takahashi, K., Nakagawa, M., Yamanaka, S., 2009. Roles of Sall4 in the generation of pluripotent stem cells from blastocysts and fibroblasts. Genes Cells 14, 683–694.

Tsuruzoe, S., Ishihara, K., Uchimura, Y., Watanabe, S., Sekita, Y., Aoto, T., Saitoh, H., Yuasa, Y., Niwa, H., Kawasuji, M., Baba, H., Nakao, M., 2006. Inhibition of DNA binding of Sox2 by the SUMO conjugation. Biochem. Biophys. Res. Commun. 351, 920–926.

Utikal, J., Maherali, N., Kulalert, W., Hochedlinger, K., 2009. Sox2 is dispensable for the reprogramming of melanocytes and melanoma cells into induced pluripotent stem cells. J. Cell Sci. 122, 3502–3510.

Van Hoof, D., Munoz, J., Braam, S.R., Pinkse, M.W., Linding, R., Heck, A.J., Mummery, C.L., Krijgsveld, J., 2009. Phosphorylation dynamics during early differentiation of human embryonic stem cells. Cell Stem Cell 5, 214–226.

Wang, J., Rao, S., Chu, J., Shen, X., Levasseur, D.N., Theunissen, T.W., Orkin, S.H., 2006. A protein interaction network for pluripotency of embryonic stem cells. Nature 444, 364–368.

Whyte, W.A., Orlando, D.A., Hnisz, D., Abraham, B.J., Lin, C.Y., Kagey, M.H., Rahl, P.B., Lee, T.I., Young, R.A., 2013. Master transcription factors and mediator establish super-enhancers at key cell identity genes. Cell 153, 307–319.

Wiebe, M.S., Nowling, T.K., Rizzino, A., 2003. Identification of novel domains within Sox-2 and Sox-11 involved in autoinhibition of DNA binding and partnership specificity. J. Biol. Chem. 278, 17901–17911.

Williams Jr., D,C., Cai, M., Clore, G.M., 2004. Molecular basis for synergistic transcriptional activation by Oct1 and Sox2 revealed from the solution structure of the 42-kDa Oct1 Sox2 Hoxb1-DNA ternary transcription factor complex. J. Biol. Chem. 279, 1449–1457.

Wissmuller, S., Kosian, T., Wolf, M., Finzsch, M., Wegner, M., 2006. The high-mobility-group domain of Sox proteins interacts with DNA-binding domains of many transcription factors. Nucleic Acids Res. 34, 1735–1744.

Wood, H.B., Episkopou, W., 1999. Comparative expression of the mouse Sox1, Sox2 and Sox3 genes from pre-gastrulation to early somite stages. Mech. Dev. 86, 197–201.

Xu, N., Papagiannakopoulos, T., Pan, G., Thomson, J.A., Kosik, K.S., 2009. MicroRNA-145 regulates OCT4, SOX2, and KLF4 and represses pluripotency in human embryonic stem cells. Cell 137, 647–658.

Xu, H., Baroukh, C., Dannenfelser, R., Chen, E.Y., Tan, C.M., Kou, Y., Kim, Y.E., Lemischka, I.R., Ma'ayan, A., 2013. ESCAPE: database for integrating high-content published data collected from human and mouse embryonic stem cells. Database (Oxford) 2013, bat045.

Yuan, H., Corbi, N., Basilico, C., Dailey, L., 1995. Developmental-specific activity of the FGF-4 enhancer requires the synergistic action of Sox2 and Oct-3. Genes Dev. 9, 2635–2645.

Zhang, J., Tam, W.L., Tong, G.Q., Wu, Q., Chan, H.Y., Soh, B.S., Lou, Y., Yang, J., Ma, Y., Chai, L., Ng, H.H., Lufkin, T., Robson, P., Lim, B., 2006. Sall4 modulates embryonic stem cell pluripotency and early embryonic development by the transcriptional regulation of Pou5f1. Nat. Cell Biol. 8, 1114–1123.

Zhang, K., Li, L., Huang, C., Shen, C., Tan, F., Xia, C., Liu, P., Rossant, J., Jing, N., 2010a. Distinct functions of BMP4 during different stages of mouse ES cell neural commitment. Development 137, 2095–2105.

Zhang, P., Andrianakos, R., Yang, Y., Liu, C., Lu, W., 2010b. Kruppel-like factor 4 (Klf4) prevents embryonic stem (ES) cell differentiation by regulating Nanog gene expression. J. Biol. Chem. 285, 9180–9189.

Zhao, S., Nichols, J., Smith, A.G., Li, M., 2004. SoxB transcription factors specify neuroectodermal lineage choice in ES cells. Mol. Cell. Neurosci. 27, 332–342.

Zhao, H.Y., Zhang, Y.J., Dai, H., Zhang, Y., Shen, Y.F., 2011. CARM1 mediates modulation of Sox2. PLoS One 6, e27026.

CHAPTER 11

Sox2-Dependent Regulation of Neural Stem Cells and CNS Development

Jessica Bertolini*, Sara Mercurio*, Rebecca Favaro*, Jessica Mariani, Sergio Ottolenghi, Silvia K. Nicolis
Department of Biotechnology and Biosciences, University of Milano-Bicocca, Milano, Italy
*joint first authors

INTRODUCTION

Sox2 has become widely known for its role in maintenance of the pluripotent stem cells of the early embryo and for its ability to reprogram differentiated cells to pluripotent (iPS) cells (Avilion et al., 2003; Takahashi and Yamanaka, 2006). However, shortly after its initial discovery (Gubbay et al., 1990), *Sox2* was noticed to be a panneural gene whose expression marks the developing central nervous system (CNS) from the earliest embryogenesis stages (Collignon, 1992; Collignon et al., 1996). The complete *Sox2* knockout in mouse, causing loss of the pluripotent cells of the epiblast, leads to early embryonic lethality before the onset of neural development (Avilion et al., 2003) (see also Chapter 10). However, through time, generation of hypomorphic and conditional knockout mutations in mouse and loss- and gain-of-function experiments performed in mouse and other animal models and in neural stem cells (NSC) have uncovered unique functions for *Sox2* in the development of the CNS and in NSC.

In this chapter, we will first examine how the transcriptional regulation of *Sox2* in NSC integrates the activity of important transcription factors, cell cycle regulators, micro ribonucleic acids (RNAs) (miRNAs), and long noncoding RNAs (lnRNAs). We will then address the study of *Sox2* functional roles in animal genetic systems (mainly the mouse). Finally, we will connect *Sox2* function to its direct regulation of target genes, which are beginning to be identified in neural cells by candidate gene, as well as by genome-wide approaches.

SOX2 EXPRESSION MARKS THE DEVELOPING CNS

Sox2 Panneural Expression Is Present from the Earliest Stages of CNS Development

In situ hybridization detects *Sox2* in the mouse CNS starting from the earliest stages of its embryogenesis, widely overlapping with that of *Sox1* and *Sox3*, its closely related homologs (Collignon, 1992; Wood and Episkopou, 1999). In mouse, expression of *Sox2*

Sox2
http://dx.doi.org/10.1016/B978-0-12-800352-7.00011-6

(although not *Sox1* and *3*) is already present in the blastocyst inner cell mass and epiblast, and is thus maintained in the developing neuroectoderm, while being downregulated in the developing mesoderm and endoderm (Avilion et al., 2003). The *Sox2* early panneural expression is conserved throughout the evolution of vertebrates and is detected in chordates (amphioxus) (Cattell et al., 2012; Meulemans and Bronner-Fraser, 2007; Pevny and Nicolis, 2010).

Sox2 Is Expressed in NSC

NSC are defined by their ability to self-renew and give rise, by terminal differentiation, to the three cell types that constitute the mature CNS: neurons, astrocytes, and oligodendrocytes.

NSC were initially demonstrated in in vitro cultures from primary neural tissue (Reynolds and Weiss, 1992), by clonogenic assays of self-renewal and multipotency (ability to produce neurons and glia), building on concepts and assays originally developed in the study of hematopoiesis (Anderson, 1989). *Sox2* expression within cells having functional NSC properties in vitro was first demonstrated by the use of a *beta-geo* reporter transgene, driven by *Sox2* promoter/5′ enhancer flanking sequences (Zappone et al., 2000) or knocked-in at the *Sox2* locus (*Sox2beta-geo*) (Ferri et al., 2004); *beta-geo* expressing cells, defined by their resistance to G418, could be cultured from embryonic or adult brain in conditions favoring NSC self-renewal and expanded in long-term cultures, retaining their ability to differentiate terminally into neurons, astrocytes, and oligodendrocytes (Ferri et al., 2004; Zappone et al., 2000). Similarly, cells expressing a *Sox2* green fluorescent protein (GFP) knock-in allele demonstrated in vitro NSC properties, with GFP-high cells being the most NSC-enriched fraction (Ellis et al., 2004). Viral transduction of a GFP transgene guided by the *Sox2* promoter/5′ enhancer within primary human fetal neural tissue also allowed selection for NSC capable of expansion and self-renewal in culture (Wang et al., 2010).

NSC Cultured from Different CNS Regions Show Differential Activity of Transgenes Guided by Region-Specific *Sox2* Enhancers

Sox2 expression within the developing CNS is guided by several distinct enhancers acting in specific CNS regions (see Chapter 7). The first of such enhancers, identified in mouse, drives expression of a *beta-geo* transgene specifically to the developing telencephalon (Zappone et al., 2000). When NSC are cultured from the telencephalon of this transgenic mouse, they retain transgene expression over long-term in vitro culture (40 passages) in the absence of G418 selective pressure. By contrast, NSC cultured from more posterior regions (spinal cord, not expressing the transgene in vivo) never express the transgene although they expand equally well in culture. Consistently, genes expressed in vivo in a telencephalic-specific way are expressed in telencephalic but not spinal cord-derived NSCs, which express genes of more posterior fates (Zappone et al., 2000). These findings indicate a region-specific identity retained by NSC in long-term culture and the evolution, within the *Sox2* locus, of regulatory elements active in specific subsets of NSC displaying region-specific (telencephalic) character. Of note, ex vivo telencephalic-derived NSC retain, together with the expression of genes encoding important telencephalic regulators, a genomic pattern of long-range interactions in chromatin involving these genes, as detected by chromatin interaction analysis by paired-end tagsequencing (ChIA-PET) (Zhang et al., 2013).

Sox2 Expression in NSC In Vivo

After their initial demonstration in vitro, NSC have been demonstrated in vivo and proposed to have important roles in CNS development (Temple, 2001). In the mouse developing cortex and postnatal hippocampus, radial glia cells labeled with a GFP-expressing retrovirus proved able to self-renew and generate neurons; these cells expressed *Sox2* (Suh et al., 2007). More recently, an NSC type particularly abundant in human developing cortex (and proposed to have a role in its selective enlargement in evolution), termed outer subventricular zone NSC, also showed *Sox2* expression (LaMonica et al., 2012). NSC persist within a limited number of neurogenic niches found in the adult brain, the subventricular zone (SVZ), and the hippocampus dentate gyrus (DG); in vivo, stem cells at these locations, characterized by glial fibrillary acidic protein (GFAP) and nestin expression, also express *Sox2* (Ferri et al., 2004). *Sox2* expression, although present in NSC, is not exclusive to them, because *Sox2* is detected also within proliferating progenitors downstream to NSCs (Andreu et al., 2015; Pevny and Nicolis, 2010).

TRANSCRIPTIONAL REGULATION OF SOX2 EXPRESSION IN NEURAL CELLS

Sox2 Regulation in NSC of the Developing Brain

Positive Direct Regulators of Sox2 Expression

Sox2 is expressed in NSC and early neural progenitors throughout the neural system; when neural cells undergo differentiation to finally exit the cell cycle, *Sox2* is repressed, disappearing from most but not all neurons. Its regulation can thus be considered from different points of view: initiation, maintenance, and repression. Abundant evidence about factors regulating *Sox2* expression has emerged mainly from studies of identified *Sox2* enhancers and from the analysis of *Sox2* involvement in cell cycle regulation. Recent work suggests that initiation of *Sox2* expression in the neural system substantially differs mechanistically from later regulation.

Telencephalic Regulation of *Sox2*

Different *Sox2* enhancers are active in different regions of the developing CNS and their relative importance varies through developmental time (see Chapter 7). In addition to regulating *Sox2* expression via both region-restricted and more ubiquitous neural transcription factors, these enhancers also integrate cell cycle regulation in a more general way.

In the developing telencephalon and in NSC therein, two enhancers are active from embryogenesis to adult stages: one (5′ enhancer, also termed SRR1, or N2) lies 5′; another (termed SRR2) lies 3′ to the *Sox2* gene (Mariani et al., 2012; Miyagi et al., 2004; Pevny and Nicolis, 2010; Uchikawa et al., 2003; Zappone et al., 2000) (Figure 1). POU transcription factors (TF) are important for the activity of both, as seen with transgenic and cell transfection studies (Catena et al., 2004; Josephson et al., 1998; Mariani et al., 2012; Miyagi et al., 2004; Zappone et al., 2000). The 5′ enhancer is bound and activated by BRN1/2 (POU3F3, POU3F2) in NSC, in vitro and in vivo; integrity of the sites is required for enhancer activity in the developing telencephalon in transgenic mice (Catena et al., 2004; Miyagi et al., 2004; Takahashi and Yamanaka, 2006). BRN factors are also critical for regulation of the 3′ enhancer, where they bind together with *Sox2* itself, leading to increased activity (Catena et al., 2004;

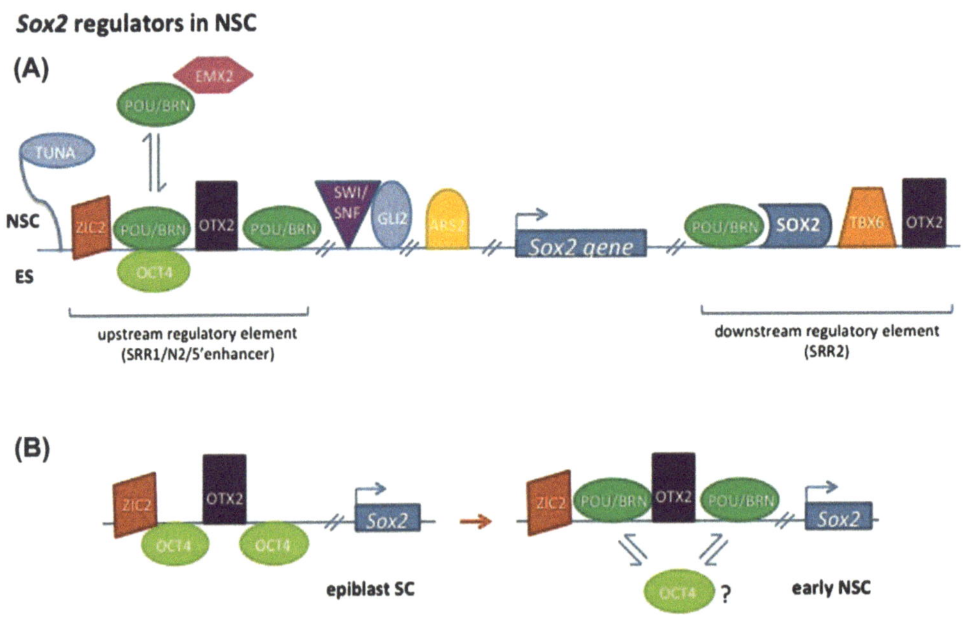

Figure 1 ***Sox2 regulators in neural stem cells.*** Transcriptional regulators that directly bind to the *Sox2* upstream and downstream transcriptional regulatory regions (SRR1/N2 and SRR2) are shown. (A) Positive regulators binding to *Sox2* in NSC/neural cells are shown above the line. EMX2, a negative *Sox2* regulator, antagonizes binding of BRN2 activator to DNA. TUNA is a long noncoding RNA. Under the line, the binding position of OCT4/POU5F1 in ES cells is shown. (B) Transcription factors binding to the *Sox2* 5′ regulatory region at the transition between pluripotent ES cells and NSCs, as represented by EpiSC and early neural plate/NSC. OCT4, still present in the early neural plate, might pave the way for binding of BRN2, which has just been activated (see text).

Mariani et al., 2012; Miyagi et al., 2004; Takahashi and Yamanaka, 2006). The 3′ enhancer is thus an example of a SOX2 target sequence, acting in concert with BRN2 in a positive feedback mechanism. This is consistent with studies on the nestin gene neural enhancer, demonstrating that mutagenesis of POU (or SOX2) binding sites led to decreased enhancer activity in transgenic and transfection experiments (Josephson et al., 1998; Mariani et al., 2012; Miyagi et al., 2004; Tanaka et al., 2004).

Interestingly, EMX2, a transcription factor important for telencephalic development, acts as a *repressor* of both *Sox2* enhancers (SRR1, 2) active in telencephalic NSC. EMX2 acts by antagonizing the binding of activator POU TF BRN2 to both enhancers (Mariani et al., 2012) (Figure 1). This regulation appears to contribute to *Sox2* function in brain development, because hemizygosity for a null *Emx2* mutation partially rescues the NSC defect seen in the hippocampus of *Sox2* hypomorphic mutants (see below) (Mariani et al., 2012). Because *Emx2* expression is maintained in more differentiated neural cells, this repression may be part of the mechanisms that switch off *Sox2* in differentiation (Mariani et al., 2012). Zinc-finger protein arsenite-resistance protein 2 (ARS2; also known as SRRT) was identified as an important direct positive regulator of *Sox2* (Andreu-Agullo and Maurin, 2012). ARS2 binds to the *Sox2* promoter/5′ enhancer region (Figure 1); this region is required for ARS2-mediated activation of *Sox2*, indicating a direct regulation (Andreu-Agullo and Maurin, 2012). *Ars2* is expressed in neurogenic zones of the adult brain, and its knockdown within GFAP-positive cells of the subventricular zone (which includes adult NSCs) results in loss of NSC self-renewal in vivo, which can be rescued by ectopic SOX2 expression. In vitro cultured *Ars2*-knockout NSCs (neurospheres) are deficient in self-renewal and differentiation, which are rescued by *Sox2* readdition.

Finally, a new kind of regulator acting on the *Sox2* promoter in stem cells is lnRNA TUNA (*Tcl1* Upstream Neuron-Associated lnRNA, or megamind), expressed in pluripotent cells and later CNS-restricted. Interestingly, TUNA knockdown inhibited neural differentiation of mouse ES cells and caused impaired locomotor function in zebrafish (Ng et al., 2013).

Transition from *Sox2* Epiblastic Expression to Early Neural Plate Expression

In the early mouse embryo, *Sox2* is expressed in the early blastocyst and in the epiblast. In embryonic stem (ES) cells (in vitro early epiblast derivatives), *Sox2* is highly expressed and maintained via a robust network of transcription factors interactions, in which *Oct4* (*Pou5f1*) is a pivotal element (see also Chapter 10). How does the transition to neural plate expression occur? In principle, it is possible that *Sox2* expression is simply maintained during neural specification, and is instead repressed in other derivatives. Intriguingly, OCT4 is able to bind to sites in the 5′/SRR1/N2 enhancer that are later bound by neural POU factors in the neuroepithelium (Figure 1) (Catena et al., 2004; Iwafuchi-Doi et al., 2012, 2011; Mariani et al., 2012). OCT4 binding might carry on the open state on these enhancers, progressively delivering their maintenance to other POU factors active at later stages (Catena et al., 2004; Dillon, 2012). *Oct4* expression is indeed present in the early neural tube and within leukemia inhibitory factor–dependent early NSC isolated from the embryonic day (E)5.5 mouse embryo (Akamatsu et al., 2009); *Oct4* conditional knockout at early neural stages (E7.5) leads to important defects of the neural tube (Akamatsu et al., 2009). Some POU factors, in particular POU3F factors (OCT6, BRN1, and BRN2), have been implicated at the anterior neural plate stage as critical regulators of the 5′/SRR1/N2 *Sox2* enhancer (using epiblast stem sells cultured in vitro, and their early neural derivatives) (Iwafuchi-Doi et al., 2012, 2011). These factors seem to act in conjunction with OTX2 (an anterior neural plate/forebrain and midbrain restricted factor) and ZIC transcription factors, which bind to the N2 enhancer in close proximity to the POU sites. Indeed, in $Otx2^{-/-}$ embryos, 5′/SRR1/N2 enhancer activity is lost, together with Sox2 expression, in the neural plate at E7.75, which points to *otx2* as a critical early regulator of *Sox2* (Iwafuchi-Doi et al., 2012, 2011). These results point to factors positively regulating *Sox2* as mediators of its expression in the early nervous system.

However, at this stage, *Sox2* is also subject to a type of negative regulation acting within axial stem cells. These cells are neuromesodermal progenitors that show self-renewal properties and express *Sox2* together with the mesoderm determinant T; they are bipotent progenitors to mesoderm or to posterior neural tube (Gouti et al., 2014; Wilson et al., 2009). SOX2 has an important role as a mediator of the function of TF TBX6 in the commitment of axial stem cells to the neural, as opposed to mesodermal, fate (Kondoh and Takemoto, 2012; Takemoto et al., 2011). Mice mutant in the *Tbx6* gene generate ectopic neural tubes, at the expense of paraxial mesoderm; in these mutants, *Sox2* fails to be repressed in mesoderm-fated axial stem cell progeny. This repression depends on enhancer N1, regulated by TBX6 in paraxial mesoderm (Figure 1). Indeed, homozygous deletion of this enhancer, in embryos also carrying the *Tbx6* homozygous deletion, prevented ectopic neural tube development. Conversely, misexpression of *Sox2* in the paraxial mesoderm of wild-type embryos caused development of ectopic neural tubes. These findings imply that in axial stem cells, factors capable of *Sox2* activation exist but are kept in check by *Tbx6*.

Finally, positive and negative regulators of *Sox2* early expression in the neural plate have been investigated in the chick embryo, acting on the chick N2 enhancer homolog (repressors ERNI and HP1γ, preventing premature *Sox2* expression, and activators Geminin and BERT, an ERNI antagonist) (Papanayotou et al., 2008). For a comprehensive report on the function of *Sox2* enhancers in chick, see Chapter 7.

Mitotic Stimuli and Cell Cycle Regulators Acting on *Sox2* Transcription

Transcriptional regulators involved in control of the cell cycle have been shown to regulate *Sox2* directly; furthermore, *Sox2* appears to mediate some of their biological functions (Figure 2).

Transcription factors E2f3a and E2f3b are isoforms of E2f3 and belong to a transcription factor family (E2f) that interacts with the retinoblastoma (Rb) tumor suppressor proteins to control cell cycle progression. Both isoforms are expressed in the neurogenic regions of the developing and postnatal brain, their isoform-specific knockouts have an opposite effect on NPC: E2f3a loss causes increased NPC expansion and reduced neural differentiation both in vivo and in vitro (including defects in adult hippocampal neurogenesis and memory formation), whereas E2f3b loss leads to reduced NPC expansion and increased differentiation (Julian et al., 2013). These phenotypes are accompanied by variation in *Sox2* expression, of opposing sign, in the two mutants (*E2f3a* knockout leads to *Sox2* activation and *E2f3b* knockout to *Sox2* repression). Both E2f3a and E2f3b directly bind to *Sox2* upstream regions (within 800 base pairs from the transcription start site) but regulate *Sox2* transcription in opposing manners: E2f3a (acting together with

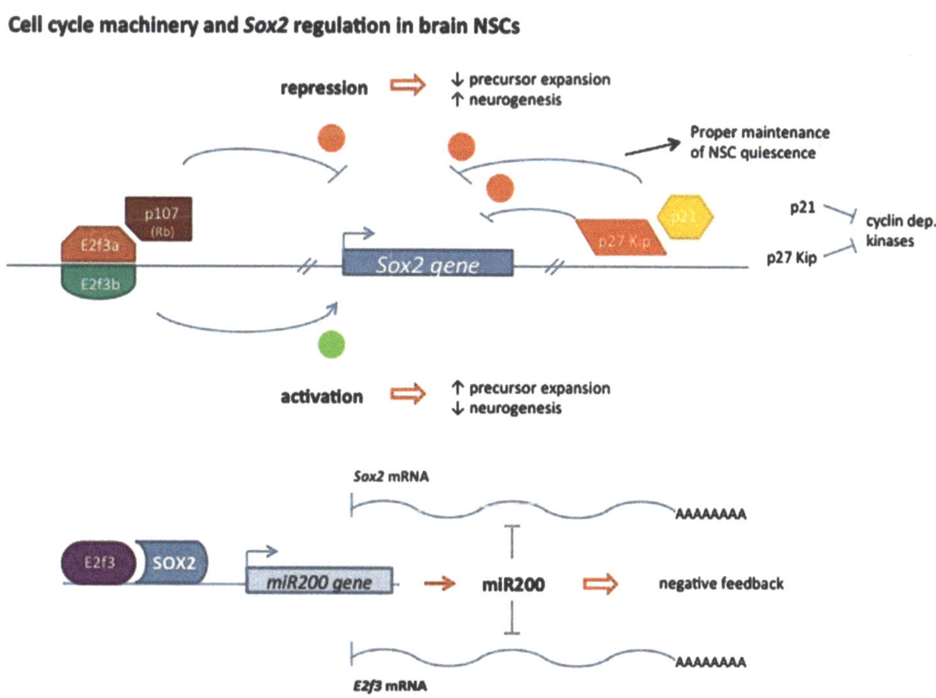

Figure 2 ***Cell cycle machinery and Sox2 regulation in neural stem cells.*** Top: Cell cycle regulators directly binding to *Sox2* are shown. Red dots highlight Sox2 repressors; green dots highlight *Sox2* activators. Bottom: Posttranscriptional regulation of *Sox2* and *E2f3* mRNAs by miR200 is shown; the miR200 gene is in turn regulated by SOX2 and E2F3 (see text).

Rb-type protein p107) functions as a repressor and E2f3b as an activator (in in vitro transfection experiments). In vitro, *Sox2* knockdown in *E2f3*a$^{-/-}$ cells abolished the excess expansion of NPC detected in the *E2f3a*$^{-/-}$ mutant, indicating that it acts as a mediator of *E2f3a* function (Julian et al., 2013).

Both the E2f3 and *Sox2* mRNAs are targets of miR-200, an miRNA that is highly expressed during CNS development (Figure 2) (see also Chapter 4). In addition, both SOX2 and E2F directly bind to the *miR-200* gene promoter, acting as positive regulators. Both in vivo and in vitro, loss of miR-200 causes reduced cell cycle exit and neuronal differentiation (Peng et al., 2012), whereas miR200 upregulation leads to a (very moderate) increase in neuronal differentiation. Micro R200 may thus contribute to the balance between NSC self-renewal and differentiation, acting on *Sox2* posttranscriptional regulation and as a *Sox2* target (Peng et al., 2012).

Cyclin-dependent kinase inhibitor p21 represents a further link between cell cycle regulation in NSC and *Sox2*. p21 associates to the 3′ SRR2 *Sox2* enhancer, as shown by chromatin immunoprecipitation (ChIP) (although this may be mediated by a cofactor directly binding to DNA, such as E2F, C-MYC, or STAT3) and acts as a *Sox2* repressor in adult NSC (Marques-Torrejon et al., 2013). p21$^{-/-}$ NSC fail to maintain the proper quiescent state characteristic of the NSC pool and display an increase in proliferation in culture, which is accompanied by increased *Sox2* expression; *Sox2* knockdown diminishes the excess proliferation observed in the mutant. In the long term, p21 knockout leads to NSC exhaustion; this is suggested to be related to excess *Sox2*-dependent replicative stress, with DNA damage, and p53 activation. In this model, high p21 levels promote NSC quiescence by repression of *Sox2*; S-phase entry is caused by transiently diminished p21 levels and increased *Sox2* expression.

Finally, p27, another cell cycle inhibitor, is also a direct *Sox2* repressor by associating to the SRR2 enhancer, together with a repressive p130–E2F4–SIN3a complex. Although this activity was directly demonstrated in ES cells, studies of p27 null mice show elevated levels of *Sox2* in the retina and the brain (Marques-Torrejon et al., 2013).

Overall, the regulatory network discussed above points to two antagonistic functions related to *Sox2* expression: *Sox2* stimulatory factors, leading to active *Sox2* expression and cell proliferation are antagonized by downstream negative feedbacks, decreasing *Sox2* activity. It is conceivable the proper balance of these opposing activities controls the maintenance of stem cells; in contrast, prevailing negative feedback may be important during the transition to cell cycle exit and neuronal differentiation.

As discussed below, *Sox2* regulation varies in different brain parts. This raises the question of which diffusible factors regulate *Sox2* transcription during development and postnatally. One important factor may be Sonic Hedgehog (*Shh*), which is both upstream and downstream, genetically, to *Sox2* (Ferri et al., 2013) and has effects on both stem cell proliferation and differentiation (Ferri et al., 2013 and see below). A link to *Shh* function is provided by GLI2, a known transcription factor in the Shh signaling pathway.

GLI2 binds to the *Sox2* 5′ enhancer (Takanaga et al., 2009), activating it. This regulation appears to be meaningful for *Sox2* function, because overexpression of a truncated, dominant-negative GLI2 or of *Gli2* shRNAs reduces NPC proliferation and *Sox2* expression. These findings suggest that Shh signaling may be an upstream regulator of *Sox2* function; this provides an interesting complement to *Sox2* conditional knockout experiments showing that the *Shh* pathway is an important downstream effector of *Sox2* function in NSC maintenance.

SOX2 FUNCTIONS AND MOLECULAR TARGETS IN CNS DEVELOPMENT

Studies in *Xenopus* and Chick

Experimental manipulation of *Sox2* activity to address its function in neurogenesis was first performed in frog (*Xenopus laevis*) and chick (*Gallus gallus*) (Kishi et al., 2000); reviewed in Pevny and Nicolis (2010). Overexpression of *Sox2* in the developing chick neuroepithelium antagonized cell cycle exit and neural differentiation, whereas dominant-negative *SoxB1* mutants (affecting *Sox1* and *Sox3* in addition to *Sox2* activity) led to forced cell cycle exit with increased differentiation. These experiments first indicated *Sox2* as being functionally involved in maintenance of the NS/progenitor cell state (Bylund et al., 2003; Graham et al., 2003). More recently, an informatic study led to define cis-regulatory modules within promoters of genes whose activity was associated with Shh signaling within the developing neuroepithelium; these modules were enriched in *SoxB1* binding sites, and neural interpretation of morphogen signaling was linked to binding of *SoxB1* TFs, together with morphogen-mediatory factors (GLI TFs). Indeed, ectopic expression of SOXB1 proteins in the limb bud gave mesodermal cells the potential to activate neural-specific target genes upon SHH, retinoid, or bone morphogenetic protein (BMP) signaling (Oosterveen et al., 2013) These findings have a correlate in genome-wide studies of SOX2 targets in neural cells that detect its frequent co-binding with GLI1 (Peterson et al., 2012) (see below).

In Vivo Studies in Mouse

Complete *Sox2* knockout leads to early embryonic lethality owing to *Sox2* requirement by pluripotent cells of the epiblast and of the extraembryonic ectoderm (Avilion et al., 2003). Investigation of *Sox2* function in the CNS by gene targeting thus requires the generation of hypomorphic and conditional mouse mutants.

Sox2 Hypomorphic Mouse Mutants

The discovery of a telencephalic-specific enhancer in the *Sox2* locus (Zappone et al., 2000) represented the basis for a first strategy to address *Sox2* function in the brain: the targeted deletion of this enhancer in the mouse germline (Ferri et al., 2004; Zappone et al., 2000). This deletion (*delta-enh* mutation) leads to a reduction of expression of the linked *Sox2* gene; in compound *null/delta-enh* heterozygotes, Sox2 expression is about 30% of that seen in the wild-type developing brain (Ferri et al., 2004).

A different hypomorphic mutation in *Sox2* (*Sox2*LP) was obtained after the insertion of a neo-cassette in the *Sox2* locus (Taranova et al., 2006). Finally, targeted deletion of the early neural *Sox2* enhancer N2 was obtained, decreasing *Sox2* levels in early neurogenesis stages (Iwafuchi-Doi et al., 2011).

Sox2 Has Dosage-Sensitive Functions in Brain that Mirror Defects Seen in *SOX2*-Deficient Patients

Sox2 null/delta-enh mutants showed brain defects (Ferri et al., 2004). Radial glia in the postnatal hippocampus DG was substantially reduced and neurogenesis was diminished in both the hippocampus DG and SVZ. Intriguingly, neurodegeneration with intracellular aggregates was observed in the ventral forebrain, particularly in the thalamus and septum, where important tissue loss was seen. Epilepsy was also detected in these mutants (Ferri et al., 2004). Of note, hippocampal defects and epilepsy are part of the CNS abnormalities detected in patients heterozygous for loss-of-function mutations in human *Sox2* (see Chapter 13).

Abnormalities of the hypothalamus, together with eye coloboma, were detected in *Sox2* hypomorphic mutants obtained by neo-cassette insertion (*Sox2* HYP); these defects also recapitulate abnormalities seen in *SOX2*-deficient patients (Langer et al., 2012).

Sox2 Functions in GABAergic Neuron Differentiation

Sox2 hypomorphic mutants (*Sox2 null/delta enh* genotype) also uncovered an unexpected function of *Sox2* downstream to the NSC, in differentiating GABAergic neurons. Whereas in vitro NSC self-renewal was essentially normal, terminal differentiation of neurons was defective, with reduced branching and failure to activate terminal differentiation markers (Cavallaro et al., 2008).

In vivo, GABAergic neurons were also reduced, failed to complete migration from the developing ventral telencephalon to their final locations in the cortex, and had reduced branching. In vitro, the defect could be rescued by *Sox2* lentiviral transduction at early but not late stages of differentiation, which pointed to a *Sox2*-dependent transcriptional program within early differentiating neurons. Expression of *Sox2* persisted in young GABAergic neurons (Cavallaro et al., 2008), a finding confirmed by gene expression analysis of GABAergic neurons sorted from the mouse brain by cell type–specific GFP transgenes (Sugino et al., 2006). These findings are in agreement with the observation by Karow et al. (2012) (see below, *Sox2* reactivation) that pericytes can be reprogrammed by *Sox2* together with MASH1, specifically to GABAergic neurons; in this process, GABAergic specificity results from *Sox2*, because MASH1 acts as a general neuronal inducer able to give rise to glutamatergic neurons with *Brn2* and *Myt1l* (Vierbuchen, 2010, 2011) and DOPA-minergic neurons with *Nurr1/Nr4a2* and *Lmx1a* (Caiazzo et al., 2011; Amamoto and Arlotta, 2014).

Conditional *Sox2* Mutants

The study of the effect of *complete Sox2* deletion on CNS development required the generation of conditional models of *Sox2* deletion by the Cre-lox system.

Sox2 Functions in the Developing Eye, a Model for Human Sox2-dependent Eye Disease

The first reported conditional deletion of *Sox2*, by means of a floxed *Sox2* allele, was performed with a Cre recombinase gene driven to the developing eye (Taranova et al., 2006). This is a CNS region in which *Sox2*, but not *Sox1* or *Sox3*, is expressed. Eye defects (microphtalmy or anophtalmy) are the pathological hallmark of patients carrying heterozygous *SOX2* loss-of-function mutations. This pathology was faithfully reproduced in these mutant animals. Interestingly, important *Sox2* functions were documented within retina stem/precursor cells, but also within differentiated cells (Mueller glia), similar to what had been found in the brain (see above). The failure of *Sox2* heterozygous null mutations to produce an eye phenotype became an interesting term of comparison for the human situation; it has been attributed to a difference in the degree of sensitivity to *Sox2* levels between mouse and human, or possibly to the presence of additional, unidentified genetic factors in human patients, predisposing them to develop microphtalmia in *Sox2* heterozygotes (Taranova et al., 2006); for a detailed discussion, see Pevny and Nicolis (2010).

Sox2 function in eye development and in human genetic disease is addressed in more detail in Chapters 12 and 14, respectively.

Sox2 Functions and Targets in the Developing Brain and Neural Tube

Effects of Sox2 Panneural Deletion Are Very Different in Different Regions of the Developing CNS

Deletion of *Sox2* throughout the developing neural tube was first achieved in mice homozygous for a *Sox2flox* conditional mutation, carrying a *nestin-Cre* transgene, causing complete *Sox2* loss at midgestation by E11.5 (Favaro et al., 2009). These mutants showed a severe deficiency in postnatal DG development mirroring the hippocampal hypoplasia seen in *Sox2*-deficient human patients (Favaro et al., 2009). However, the remainder of the brain and CNS was comparatively unaffected despite complete (panneural) *Sox2* deletion by *nestin-Cre*. This observation showed that *Sox2* is uniquely required by some brain regions although not by others.

A quantitative decrease in cell proliferation was also noted at E14.5 in a different study, using a *Sox2flox* mutation that retained the neo-resistance cassette within the targeted allele. However, these mice did not survive past birth possibly because of a knockdown effect caused by the neo-cassette before *Sox2* excision, which prevented the study of postnatal abnormalities (Miyagi et al., 2008).

Sox2 Neural Deletion at Midgestation Leads to NSC Loss in the Hippocampus

At the cellular level, the hippocampus of *nestin-Cre Sox2*-deleted mutants was essentially normal at birth but failed to maintain the radial glia-like NSC that develop postnatally in the DG, fueling the neurogenesis that continues in the DG throughout life (Favaro et al., 2009). These NSC, which are immunopositive for GFAP and nestin, express *Sox2* (Ferri et al., 2004; Suh et al., 2007). They also express *Sox3* (Rogers et al., 2013), which suggests that the unique *Sox2* requirement by these cells does not depend on their failure to express other SoxB factors.

Sox2 Early Deletion Leads to Loss of the Ventral Telencephalon and to Severe Hippocampal Defects

In subsequent experiments, *Sox2* was deleted in the developing telencephalon by a BF1-Cre knockin (Ferri et al., 2013). This deletion was earlier than that obtained with *nestin-Cre*, and was complete in the telencephalon by E10. In these mutants, the development of the ventral telencephalon, in particular the medial ganglionic eminences (primordia of the caudate/putamen basal ganglia regions), was severely impaired by early stages (E12), with spreading of the expression of dorsal markers (e.g., *Pax6*) toward the ventral aspect.

These observations point to a stage-specific, early requirement for *Sox2* in the generation of this brain region. The fact that no such abnormalities are seen with *nestin-Cre* deletion (complete in the telencephalon at E11.5) points to an early stage-specific *Sox2* function within the gene regulatory program that initiates the development of these brain structures and suggests that *Sox2* later becomes dispensable for these regions (Figure 3). No alteration in the expression of *Sox1* and *3*, coexpressed with *Sox2* in the developing ventral telencephalon, was detected in these mutants, which suggests that *Sox2* has nonredundant roles in the early ventral telencephalon (Ferri et al., 2013). Because the *BF1-Cre* knockin destroys one copy of the endogenous *BF1* gene, this might contribute, in principle, to the phenotype; however, we found no significant decrease in BF1 mRNA by in situ hybrization. Moreover, compound *Sox2/BF1* heterozygous mutants had no phenotype, which indicated that complete Sox2 loss is necessary to develop the phenotype. Interestingly, in *BF1-Cre Sox2* conditional mutants, a severe loss of DG was detectable at E18, in contrast to the normal appearance in the *Nestin-Cre* late mutants (Ferri et al., 2013 and work in preparation). This suggests that the early deletion may affect targets and/or cell populations other than the later deletion.

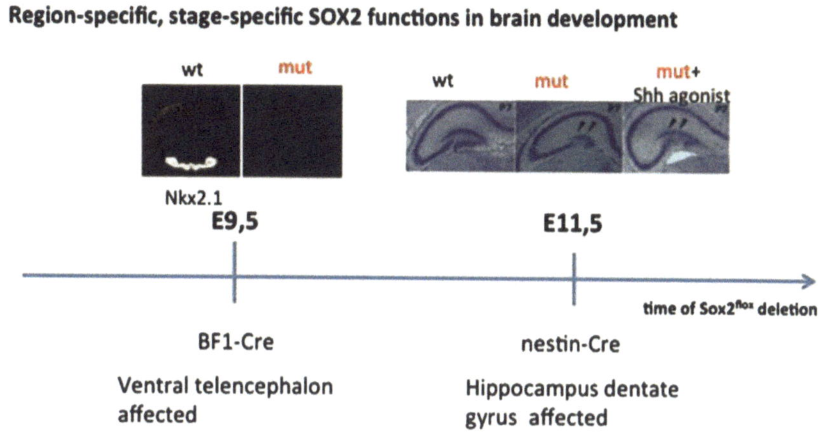

Figure 3 ***Region- and stage-specific Sox2 functions in brain development.*** *Sox2* deletion via *nestin-Cre* at E11.5 leads to defective postnatal development of the hippocampus dentate gyrus, but *Sox2* deletion 2 days earlier (E9.5) via *BF1-Cre* leads to a severely defective development of the ventral telencephalon. Images show: left, in situ hybridization of *BF1-Cre Sox2* mutant and control for *Nkx2.1*, a master regulator of ventral telencephalic development, and a direct *Sox2* target, at E10.5; right, histological section of the hippocampus in nestin-Cre *Sox2* mutants: control (wt), mutant (mut), and mutant rescued with SHH agonist, are shown. Arrowheads point to the hippocampus dentate gyrus. *Modified from Favaro et al. (2009), Ferri et al. (2013).*

Sox2 Reactivation and Neural Reprogramming

Sox2 expression is generally lost with terminal differentiation of stem cell progeny. However, it can be reactivated in the process of reprogramming of differentiated cells to iPS cells (Takahashi and Yamanaka, 2006). A related, earlier observation is the reactivation of *Sox2* transcription after reprogramming of oligodendrocyte progenitor cells (OPC), cultured from the rat optic nerve, to NSC. This reprogramming process involves initial OPC exposure to *platelet-derived growth factor* (PDGF) and BMP, converting them into type 2 astrocytes (2As); 2As, when further cultured in basic fibroblast growth factor (bFGF), are transformed into multipotent neural stem-like cells (NSLCs) able to self-renew and differentiate into neurons, astrocytes, and oligodendrocytes (Kondo and Raff, 2004). Chromatin remodeling factor BRAHMA and tumor suppressor BRCA1 are recruited to the *Sox2* 5′ telencephalic stem cells enhancer as OPC are converted to 2As, and contribute to *Sox2* reactivation; furthermore, *Sox2* knockdown by RNAi in OPCs, followed by induction of their conversion into NSLCs, resulted in loss of proliferation of NSLCs and in activation of neuronal differentiation markers (Kondo and Raff, 2004). Intriguingly, the role of PDGF in *Sox2* activation and neural conversion parallels the reprogramming of neural cells to cancer stem cells. This may occur in PDGF-dependent oligodendroglioma; in a mouse oligodendroglioma model induced by PDGF, *Sox2* was shown to be required for cancer stem cell maintenance (Favaro et al., 2014).

More recently, *Sox2* ectopic activation in vivo within specific differentiated, non-neuronal cell types of the adult brain demonstrated the ability of *Sox2* to reprogram them to neuroblasts (Niu et al., 2013) or neurons (Karow et al., 2012). After in vivo *Sox2* delivery and ectopic expression in astrocytes via a viral vector injected into the striatum (Niu et al., 2013), resident astrocytes of the adult mouse brain were converted into proliferative neuroblasts. These neuroblasts could then differentiate to neurons in vivo, integrating into the local electrophysiologically active neuronal network, after administration of BDNF (a neurotrophic factor not expressed in the striatum) and noggin (a BMP antagonist promoting adult neurogenesis) (Niu et al., 2013).

Karow et al. (2012) observed that cells within the human cerebral cortex, termed pericytes (involved in the establishment of the blood–brain barrier and the regulation of blood flow), could be reprogrammed specifically to neuronal GABAergic cells (iN cells) by the retroviral-mediated coexpression of *Sox2* and *Mash1/Ascl1*.

Finally, studies on the Axolotl, a salamander that retains in adulthood the ability to fully regenerate a tail, including the lesioned spinal cord, demonstrated an essential role of *Sox2* in neural regeneration. Here, CRISPR-mediated knockout of the resident *Sox2* gene inhibited the NSC proliferation normally seen after tail damage and led to spinal cord–specific regeneration failure (Fei et al., 2014). Normal CNS development was not notably impaired; interestingly, *Sox3* expression was detected in CNS development but not in regeneration, which suggests that it might act redundantly with *Sox2* in embryogenesis but not in the transcriptional program of regeneration.

Overall, reprogramming and regeneration studies confirm the conclusions obtained by loss-of-function experiments, pointing to critical roles of *Sox2* in neural cells. Moreover, the finding that both astrocytes and pericytes, which are abundant cell types in the brain, can be converted by *Sox2* to neural cells, adds hope regarding the possible use of Sox2-mediated reprogramming to neurons to repair neuronal damage.

Search for *SOX2* Targets Mediating Its Function in Brain Development and NSC

Candidate Gene Approaches Identify Shh and Nkx2.1 as Sox2 Targets

In *Sox2* mutants obtained by *nestin-Cre*–mediated *Sox2* deletion, NSC loss is accompanied by a sharp decrease in cell proliferation and an increase in cell death postnatally in the mutant hippocampus DG (Favaro et al., 2009). This phenotype was reminiscent of that seen in conditional mouse mutants in the gene encoding the SHH cytokine, or its co-receptor SMO (Machold et al., 2003; Hébert and Fishell, 2008).

This candidate gene approach led an analysis of *Shh* expression, which was found to be importantly decreased in *nestin-Cre* mutants in the (late) developing telencephalon and postnatal hippocampus (Favaro et al., 2009).

The most prominent defect seen in *BF1-Cre Sox2*-deleted mutants, impaired development of the ventral telencephalon, also recapitulated defects seen in *BF1-Cre*–mediated deletion of *Smo*, encoding the SHH receptor. Indeed, in situ hybridization demonstrated a sharp decrease in *Shh* expression by E10.5 in these mutants. The mutant phenotype also bore similarity to that of null mutants in the gene encoding TF *Nkx2.1*, which is essential for ventral telencephalic development and a direct regulator of the *Shh* gene. In situ hybridization showed that *Nkx2.1* fails to be initially activated in the early mutant telencephalon, showing extremely reduced expression at E9.5 and E10.5 in mutants (Ferri et al., 2013).

Shh Signaling Is an Important *Sox2* Functional Effector in the Developing and Postnatal Brain

Because downregulation of *Shh* and its regulator *Nkx2.1* is observed in late (*nestin-Cre*) and early (*BF1-Cre*) *Sox2*-deleted mutants, rescuing experiments were performed with an *Shh* pharmacological agonist (SHH-Ag) mimicking *Shh* function (Favaro et al., 2009; Ferri et al., 2013). In *nestin-Cre Sox2* mutants, *Shh*-Ag partly prevented the postnatal loss of hippocampal NSC, antagonizing the decrease in cell proliferation and increase in cell death and leading to partial morphological recovery of the DG (Favaro et al., 2009). In *BF1-Cre* mutants, to some extent *Shh*-Ag prevented the loss of *medial ganglionic eminence* and ventral telencephalic markers, allowing partial morphological recovery (Ferri et al., 2013). Thus, at least in this case, a drug mimicking the effect of a single gene deregulated in mutants had a biologically significant effect. Another signaling molecule, Wnt3a, was also reduced in the nestin-cre mutants, although functional tests were not carried out. On the other hand, it is highly likely that the function of *Sox2* is exerted through many additional target genes, with effects varying from important to subtle (see below) (Engelen et al., 2011).

Sox2, Together With *Sox3*, Regulates Shh Signaling in the Developing Hypothalamus

Involvement of *Sox2* in the development of a different brain territory, the hypothalamus, was found in mouse through the study of a mutant in which the Shh gene was deleted within the developing hypothalamus by a *Cre* transgene driven by an Shh enhancer (Zhao et al., 2012). This mouse has characteristics reminiscent of septo-optic dysplasia (SOD), a disease caused in rare instances by the mutation of *SOX2*, *SOX3*, or *HESX-1* (see Chapter 14). Important features of SOD (pituitary hypoplasia; absence of optic disc) are recapitulated in these mutants. Male mice mutant in *Sox3*, an X-linked gene (*Sox3*Y/−), or heterozygous mutant in *Sox2*, showed little change in *Shh* expression in the hypothalamus, but *Sox3*Y/−; *Sox2*± compound mutants showed important reduction of *Shh* (and *Six6*) expression in the developing anterior hypothalamus. This indicates a function for *Sox2* in the developing hypothalamus and a likely involvement of *Sox2*-dependent regulation of *Shh* in the pathogenesis of SOD in patients carrying *SOX2* or *SOX3* mutations (Zhao et al., 2012).

SOX2 Molecular Targets: SOX2 Directly Regulates Shh and Nkx2.1 in Early and Later Brain Development (Forebrain; Hypothalamus)

The deregulation of *Shh* and of *Nkx2.1* (a transcription factor that activates *Shh* and is also activated by *Shh* signaling) (Ferri et al., 2013; Hébert and Fishell, 2008) in *Sox2* mutant mice suggested the possibility of direct regulation of the corresponding genes by *Sox2*.

The Shh gene is regulated by a set of regionally acting long-range enhancers distributed over 2 MB of genome surrounding the *Shh* coding region in both mice and humans. In transgenic mice, these enhancers drive activity of a lacZ transgene to specific subregions of the *Shh* expression domain (Jeong et al., 2006). *Electrophoretic mobility shift assay* (EMSA) and ChIP experiments detected SOX2 binding on *Shh* enhancers active in the telencephalon and the diencephalon, the two brain regions where abnormalities had been found in *Sox2* mutants, which could be functionally related to *Shh* deficiency (Favaro et al., 2009; Ferri et al., 2013; Zhao et al., 2012) (Figure 4).

Mutation of consensus SOX2 binding sites within the *Shh* telencephalic and hypothalamic enhancers impaired SOX2 binding in EMSA (Favaro et al., 2009; Zhao et al., 2012) and reduced activity of the SBE2 hypothalamic enhancer (Zhao et al., 2012) and SBE3 telencephalic enhancer (our unpublished data) in transfection assays; mutant

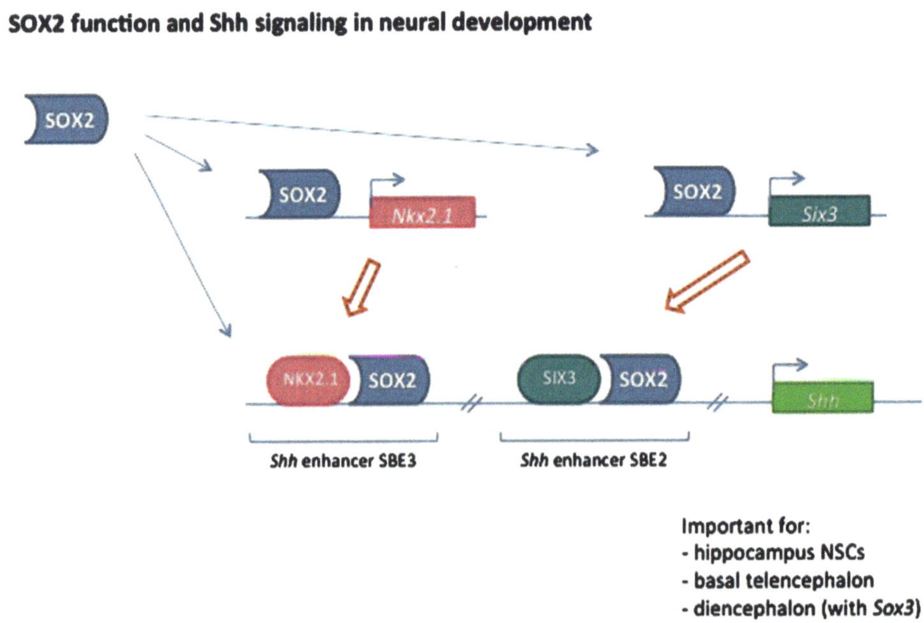

Figure 4 ***Sox2 function and Shh signaling in neural development.*** Sox2 directly binds and regulates *Nkx2.1* and *Six3* (*six3.2* in medakafish; see text), two *Shh* activators, and *Shh* itself. *Sox2* regulation of the *Shh* signaling is an important mediator of *Sox2* function in the maintenance of the hippocampus dentate gyrus, the ventral telencephalon, and (with *Sox3*) the diencephalon (see text).

SBE2 activity was also reduced in transgenic mice at E10.5 with compared to the wild-type sequence (Zhao et al., 2012).

These findings point to a role for *Sox2* in regulating *Shh* neural enhancers. The hypothalamus study suggests a contribution of *Sox3* at least in the regulation of the hypothalamic enhancer; the relative importance of *Sox2* and *Sox3*, widely coexpressed within SHH-expressing regions, may vary between different enhancers and its understanding could benefit from a comparison of *Sox3* and *Sox2* null mutants in the different regions of activity of the *Shh* enhancers.

Nkx2.1, a Master Regulator of Ventral Telencephalic Development, Is an Early Direct SOX2 Target in Mouse and Fish

Early deletion of *Sox2* by *BF1-Cre* identified another *Sox2* target, *Nkx2.1*, a master regulator of ventral telencephalic development and an activator of the *Shh* gene (Ferri et al., 2013) (Figure 4). Expression of *Nkx2.1* in *Sox2-BF1Cre* mutants is severely downregulated, from its initial expression in the ventral telencephalon (E9) (Ferri et al., 2013). *Nkx2.1* was known to act both upstream and downstream to SHH signaling; however, mosaic deletion of *Sox2* within the early telencephalon (using a *Sox2CreERT2* inducible Cre), still allowing normal brain development and maintenance of SHH, showed that *Nkx2.1* expression was retained only in cells that also retained SOX2, pointing to a cell-autonomous action of SOX2 on *Nkx2.1* expression. SOX2 was found to bind on Sox consensus sites within one of the *Nkx2.1* promoters in ChIP experiments, and this region drove expression of a GFP transgene to the developing forebrain of transgenic medakafish, and of in utero electroporated mouse embryos. The *Nkx2.1* promoter was activated by SOX2 in cotransfection assays, and mutation of the three *Sox2* consensus sites impaired its activity in cotransfection and in transgenic fish. *Nkx2.1* transgenes were stimulated by co-injected *Sox2* mRNA and negatively regulated by *Sox2* depletion; the endogenous *Nkx2.1* responded in the same way to variation of *Sox2* levels, pointing to *Sox2* as an important regulator of medakafish *Nkx2.1* in vivo, similar to mouse (Ferri et al., 2013).

Sox2 reduction in the mouse hypothalamus (coupled to *Sox3* loss) does not lead to an impairment of *Nkx2.1* expression as dramatic as that seen in *Bf1-Cre Sox2* mutants (Zhao et al., 2012). This may depend on differences in the regulatory networks driving *Nkx2.1* in telencephalon and hypothalamus, but also on the different timing of *Sox2* deletion in the two mutants (earlier in the *BF1-Cre* mutants). *Sox2* deletion with an *Nkx2.1-Cre* mouse, acting in both regions, may clarify this issue.

From Mouse Phenotype Back to Disease: *Sox2* Mutant Mice Have Some Features of Holoprosencephaly, a Disease Caused by *Shh* (or *Six3*) Mutation

Interestingly, a point mutation in the SBE2 *Shh* enhancer (see previous discussion) had been described in a patient with holoprosencephaly, a multifactorial disease that can be caused by mutation in *SHH*, or genes in its pathway (Jeong et al., 2008; Roessler and Muenke, 2010). This mutation impaired binding of SIX3 to the SBE2 enhancer, and reduced SBE2 activity in transgenic embryos (Jeong et al., 2008), similar to the effect observed after mutation of the *SoxB1* site (see above) (Zhao et al., 2012). Mutations in *SIX3* can also cause holoprosencephaly (Roessler and Muenke, 2010); importantly, SOX2 was also shown to be a direct regulator of *Six3.2* (one of the two *Six3* loci) in medakafish (Beccari et al., 2012), which lends further support to its possible involvement in *Shh* regulation through SIX3.

Involvement of Sox2 in *Shh* regulation (in part through its enhancers and partly through regulation of *Shh* activators NKX2.1 and possibly SIX3) led us to ask whether *Sox2* mutation could also lead to holoprosencephaly. Indeed, in *BF1-Cre Sox2*-deleted mice, facial abnormalities that are characteristic of holoprosencephaly (e.g., fused anterior nasal plate) were observed (Ferri et al., 2013). This raises the interesting question as to whether polymorphic variation in *Sox2* activity may cooperate with mutations in other genes to cause holoprosencephaly in humans.

SOX2 FUNCTIONS IN NSC IN VITRO

Sox2 Neural Deletion Prevents NSC Maintenance in Long-Term In Vitro Cultures

Sox2 conditional deletion allowed a function for *Sox2* to be observed also in the maintenance of NSC self-renewal in long-term in vitro NSC cultures, which paralleled its function in in vivo NSC (Favaro et al., 2009). NSC cultured as neurospheres from the telencephalon of *nestin-Cre Sox2*-deleted mice were devoid of *Sox2*; they self-renewed for several passages in culture but then underwent a decrease in growth, with progressive culture exhaustion. *Sox2* loss led to activation of differentiation marker beta-tubulin, and to reduction in cell proliferation and increased cell death, paralleling the defects observed in in vivo (hippocampal) NSC. Sphere formation could be rescued by lentiviral *Sox2*. Exponential growth in culture was partially rescued by medium that had been previously conditioned by wild-type cells, which pointed to (so far unidentified) *Sox2*-dependent diffusible factors as important contributors to in vitro NSC maintenance. Shh mRNA is present, at low levels, in wild-type NSC; recombinant *Shh* (important for in vivo NSC) (see above), only moderately stimulated the growth of wild-type or *Sox2*-mutant NSC, which indicated the importance of other factors for in vitro growth of mutant NSC (Favaro et al., 2009).

A subsequent study also detected a requirement for *Sox2* for the maintenance of in vitro NSC using embryo-derived radial glia-like NS cells, growing in adherence, as a model system. Heterozygous *Sox2* loss caused moderate proliferative defects, and complete loss of *Sox2* by acute in vitro deletion caused exit from the cell cycle despite the expression of *Sox3* (Gómez-López et al., 2011). This gives important confirmation of a unique requirement for *Sox2* to maintain in vitro NSC self-renewal within a different experimental system (NS cells).

Telencephalic-derived wild type NSC (Favaro et al., 2009) were used in a study that determined the genome-wide pattern of long-range interactions in NSC chromatin mediated by RNA polymerase II by ChIA-PET (Zhang et al., 2013). Long-range interactions connected genes to regions with enhancer potential in vivo, as demonstrated in transgenic zebrafish. Significantly, regions with enhancer potential were often (36% of the time in ES cells) connected not to the most proximal gene, but to genes lying far away, past the most proximal genes (Zhang et al., 2013). This indicates that an important proportion of NSC enhancers may lie far from the regulated genes at positions that are not predictable by proximity criteria, but need to be addressed experimentally by connectivity studies. Telencephalic-derived NSC retain their expected region-specific gene expression (in contrast to ES-derived NS cells, which do not show such specificity); importantly, neural genes expressed in the telencephalon in telencephalic NSC usually showed chromatin connections, whereas neural genes expressed in regions outside the telencephalon (hindbrain and spinal cord) did not show connections (Zhang et al., 2013). This further suggests that the observed connections may be of functional significance. It will be interesting to determine the effects of *Sox2* loss on the pattern of long-range interactions in NSC chromatin.

SOX2 TARGETS IN NSC

SOX2 Directly Regulates Important Target Genes in In Vitro NSC

Target genes regulated by SOX2 that mediate its functions in neural development and NSC are starting to be identified by studies at the single-gene level as well as by genomic approaches (Figures 4 and 5).

We already discussed the SOX2-mediated regulation of *Shh* and *Nkx2.1* (see above and Figure 4). A study identified RNA-binding protein LIN 28, a repressor of the biogenesis (maturation) of specific miRNA families (including let7), as a direct SOX2 target at the chromatin level (Cimadamore et al., 2013). In vivo in mouse, *Lin28* was expressed in the developing neural tube and postnatal hippocampus DG and *Sox2* ablation by Cre led to partial loss of *Lin28* expression. In vitro, in human ES cell (hES)-derived NSCs, *LIN28* was also expressed, and *SOX2* downregulation led to a reduction of *LIN28* and of cell proliferation (about 50–60% decrease in a short-term assay); reexpression of *LIN28* rescued the proliferative defect of Sox2-downregulated cells. As expected, *LIN28* targets *LET7I* and *LET7B* were upregulated in *SOX2* knockdown (Cimadamore et al., 2013). *LET 7B* overexpression by itself led to a reduction in cell proliferation and *LET7I* overexpression led to reduced activation of neurogenic genes *NGN1* and *MASH1* and reduced neuronal differentiation (Map2-neurons). These results suggest a model in which by activating *LIN28*, the *LET7* biogenesis repressor, *SOX2* prevents the inhibitory effects of these microRNAs on proliferation and (at late stages) differentiation. This model is in accordance with findings regarding defective cell proliferation and differentiation in *Sox2*-downregulated cells

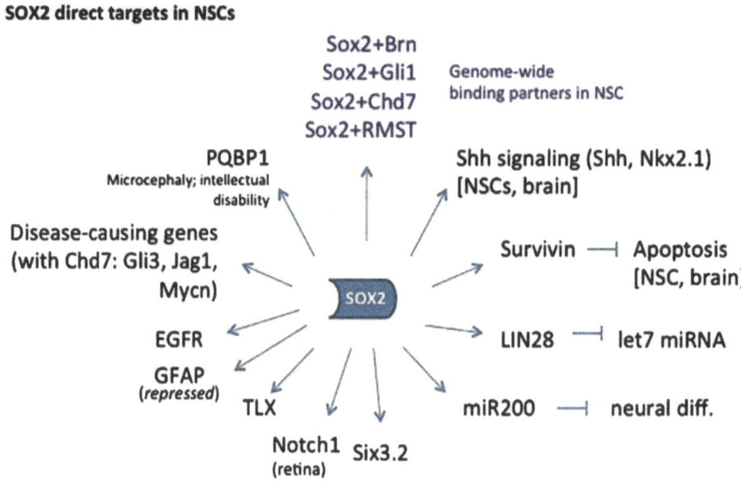

Figure 5 ***Sox2 direct targets in neural stem cells.*** In black, genes directly bound and regulated by Sox2 in NSC, possible mediators of *Sox2* functions. In blue, Sox2 binding partners in NSC, identified by genomic (ChIP-seq) approaches.

(Cavallaro et al., 2008; Favaro et al., 2009). However, it is unclear whether rescue by *LIN28* overexpression of NPC proliferation after acute *SOX2* deletion observed in short-term experiments (Cimadamore et al., 2013) would be able to antagonize the slow and progressive decline of NPC function that is observed in vivo and in long-term culture of telencephalon-derived P0 and E14.5 *Sox2*-ablated cells (Favaro et al., 2009). Importantly, the latter cells express little or no *lin28* RNA (our unpublished data); this discrepancy may reflect differences between human and mouse cells or in vivo-derived versus ES-derived NSC cells.

A recently identified target of SOX2 is survivin (*Birc5* gene), a molecule that inhibits mitochondria-dependent apoptosis in NSC (Feng et al., 2013). Loss of survivin leads to activation of caspases 9 and 3, resulting in NSC apoptosis in mouse cortex. SOX2 directly binds to the Survivin gene promoter (−902 to −530) and transactivates it in cotransfection experiments. The survivin promoter binds SOX2 also in NSC chromatin (ChIP-seq data) (Engelen et al., 2011; Lodato et al., 2013) (our unpublished data). Survivin ectopic expression protected NSCs from apoptosis induced by *Sox2* downregulation in vitro and also in vivo (embryonic telencephalon at E14.5) (Feng et al., 2013). In *Sox2* conditional mutants (see above), increased apoptosis represents an early consequence of *Sox2* loss in vivo in both the hippocampus (Favaro et al., 2009) and ventral telencephalon (Ferri et al., 2013) and in NSC cultured in vitro (Favaro et al., 2009).

Another interesting SOX2 direct target is nuclear receptor tailless (*Tlx*, also known as *Nr2e1*), an important regulator of NSC maintenance and self-renewal in embryonic and adult mouse brain (Shimozaki et al., 2012). SOX2 was found to bind the *Tlx* promoter and to transactivate it in transfection assays. TLX, in turn, binds to its own promoter, repressing it, and SOX2 antagonizes this binding (Shimozaki et al., 2012). Because *Sox2* hypomorphic (Ferri et al., 2004) and conditional knockout mutants (Favaro et al., 2009) show impaired adult neurogenesis, similar to the phenotype of *Tlx* mouse mutants, these findings raise the hypothesis that TLX mediates part of SOX2 function in adult NSC maintenance.

Encoding polyglutamine binding protein 1 (PQBP1), a nuclear-cytoplasmic shuttling protein involved in RNA metabolism and transcription, is a microcephaly and intellectual disability-causing gene in humans. The PQBP1 promoter is bound by SOX2, and SOX2 transactivates it in transfection assays (Li et al., 2013). Encoding polyglutamine binding protein 1 is expressed in the developing cerebral cortex, predominantly in the ventricular zone; postnatally, its expression is retained in the subventricular zone and in the cerebellum external and internal granule layers (Qi et al., 2005). The mechanism of microcephaly in patients with defective PQPB1 is not known. Intriguingly, mouse *Pqbp1* expression was reduced in the brain of *Sox2* heterozygous knockout animals in vivo (Li et al., 2013). *Sox2* conditional mutants show a reduction in telencephalic size (Ferri et al., 2013), which suggests that a decrease in *Pqbp1* expression contributes to this aspect of the Sox2-mutant phenotype.

It is interesting that PQBP1 was initially identified as an interacting factor and inhibitor of transcription factor BRN2, and its overexpression suppresses cell growth and enhances cell susceptibility to stress (Feng et al., 2013). Because BRN2 is both a positive regulator of *Sox2* (see above) and a SOX2 cofactor on target genes (Lodato et al., 2013) (see below) in NSC, it is possible that PQBP1 activated by SOX2 controls *Sox2* itself via a negative feedback to prevent its overexpression, known to lead to stress and growth arrest (Marques-Torrejon et al., 2013).

Epidermal growth factor receptor (EGFR) is another direct target of SOX2 in neural progenitor cells; SOX2 binds to the Egfr gene promoter and transactivates it. In turn, *Sox2* is positively stimulated by epidermal growth factor (EGF), thus giving rise to a positive feedback loop. Because EGF provides a mitogenic stimulus for NSC self-renewal (Hu et al., 2010), it is possible that SOX2 function in NSC maintenance is mediated in part by its action on EGFR. It will be interesting to determine whether EGFR expression in vivo depends on *Sox2* function; no differences in EGFR expression were detected in NSC cultured in vitro from conditional *Sox2*-deleted brains (Favaro et al., 2009) (our unpublished results), which suggests that *Egfr* loss is not the primary cause of the failure of *Sox2*-deleted NSC to undergo long-term self-renewal in culture (Favaro et al., 2009).

Genomic Perspectives on SOX2 Targets in NSC

Chromatin immunoprecipitation coupled with high-throughput sequencing of immunoprecipitated DNA (ChIP-seq) has made the genome-wide, unbiased determination of SOX2 targets possible. This has given the unprecedented opportunity to determine SOX2 molecular targets in NSC in an unbiased way (in experimental NSC systems that can provide sufficiently high cell numbers).

Genomic targets of SOX2 are treated in detail in Chapter 9. Here, we will address some aspects of genome-wide SOX2 targets determination in NSC that are directly related to the issues of *Sox2* function in the CNS and NSC discussed above.

Combinatorial Logic of *Sox2* Function on Its NSC Genomic Targets

SOX2 Genomic Targets Differ Between ES and NSC in Relation to Different SOX2 Partners

Studies at the level of individual genes indicated that SOX2 binds to DNA and regulates transcription in combination with POU factors (see above; e.g., the nestin enhancer, *Sox2* 3′ enhancer) (Figure 4), including OCT4 in ES cells and BRN1/2 in NSC, as discussed in Chapters 8, 9, and 10.

Chromatin immunoprecipitation sequencing experiments done in NS cells for SOX2 and BRN2 confirmed that SOX2 binding was widely associated with that of BRN2 (Lodato et al., 2013) (Figure 5). Co-binding of SOX2 with POU factor OCT4 had also been found by ChIP-seq experiments in ES cells (Chen et al., 2008). However, the genomic targets of SOX2 differed widely between ES cells and NSC (Lodato et al., 2013). By misexpressing BRN2 in ES cells, it was shown that it could relocalize SOX2 binding in a genome-wide way, taking it closer to the binding pattern seen in NS cells. The neural SOX/POU binding site differed from the ES cell site, as could be determined by comparison of SOX2–OCT4 and SOX2–BRN2 sites at the genomic level (Lodato et al., 2013). This proves that a specific neural POU factor, BRN2, already known to be a SOX2 cofactor, can function in determining the genomic binding sites for SOX2 at the genome-wide level.

Another ChIP-seq study for SOX2, BRN2, or BRN1 in NS-5 cells (Bery et al., 2014) retrieved a list of putative enhancers, which were reduced, through a pipeline involving gene ontology and expression patterns, to 20 sequences. These sequences were linked to a minimal promoter driving GFP expression and were tested by ex vivo electroporation of GFP reporter constructs in the mouse telencephalon. A total of 35% of the sequences showed enhancer activity and mutagenesis of SOX/POU sites showed that these contributed to transgene activity.

A different ChIP-seq study mapped SOX2 binding sites in ES-derived neural progenitors, treated with a SHH agonist to activate the Shh pathway, together with binding sites for transcription factor GLI1 (ChIP-seq for FLAG-Gli1 transgene-FLAG transgene), an important downstream effector of Shh signaling (Peterson et al., 2012). The sequences co-bound by SOX2 and GLI1 had enhancer activity in the nervous system of transgenic mice. SOX2 binding by itself was not associated with chromatin marks of active enhancers, yet co-binding with GLI3 was. These findings indicate that *Sox2* may specify a neural context for the function of Shh signaling (Oosterveen et al., 2013; Peterson et al., 2012).

In a different study, SOX2 was also found to co-bind to target genes with its interactor CHD7 (a protein involved in chromatin opening) in ES cell–derived NS cell chromatin at the genome-wide level (Engelen et al., 2011) (see also Chapter 5). Genes belonging to the

Shh, Notch, and Myc pathways were bound by both SOX2 and CHD7, and were downregulated after Sox2 or CHD7 downmodulation with shRNAs. One interesting finding was the co-binding of Sox2 and Chd7 to genes, whose mutation causes genetic diseases such as CHARGE, Alagille, and other syndromes, displaying phenotypes that partially overlap with those seen with mutations in SOX2 or CHD7 (Engelen et al., 2011). These findings suggest a new perspective on the involvement of *Sox2* in neural genetic disease: *Sox2* may act as a modulator of disease-causing genes. This suggests that mutations affecting SOX2 levels, and SOX2-regulated enhancers and promoters, could contribute to multigenic genetic disease.

Finally, a study defined a functional interaction between SOX2 and lnRNA RMST. SOX2 and RMST physically interact and co-bind to target genes genome-wide to co-regulate NSC neurogenesis (Ng et al., 2013). Interestingly, RMST is required to allow transcription of various SOX2-bound regulatory genes, which indicates that Sox2-lnRNA complexes represent a novel mechanism to modulate SOX2 target regulation (Ng et al., 2013).

SOX2 Function on Gene Expression: Positive Control of NSC Maintenance, or Repression of Differentiation?

Genomic studies of gene expression, coupled to *Sox2* experimental downregulation in NSC by shRNAs gave evidence of both an activator and a repressor role for SOX2 (Bergsland et al., 2011; Engelen et al., 2011; Lodato et al., 2013). This reflects previous observations at the single gene level, in which SOX2 was found to act as activator (*Sox2* 3′ enhancer; *Nkx2.1*; *Shh*) (see above), but also as repressor of differentiation genes (e.g., GFAP) (Cavallaro et al., 2008; Liu et al., 2014).

Sox2 Functions in Oligodendrocytes and Astrocytes

Sox2 hypomorphic mutants had shed light on a function for *Sox2* within a differentiated cell type, GABAergic neurons (Ferri et al., 2004; see above). Recent work also proved that *Sox2* controls the production of another important differentiated cell type, oligodendrocytes, as shown in mice in which complete loss of *Sox2* and *Sox3* was induced within developing oligodendrocytes (Hoffmann et al., 2014). *Sox2* and *Sox3* are expressed until early phases of oligodendrocyte terminal differentiation; their combined deletion impairs their terminal differentiation in the spinal cord. In transfection studies, *Sox2* and *Sox3* regulate the promoter of myelin protein P0, an important oligodendrocytic terminal differentiation gene; interestingly, they also negatively regulate the expression of *miR145*, an miRNA that inhibits cell differentiation. These findings indicate a new function for Sox2 in oligodendrocyte differentiation and point to *miR145* as a novel target potentially important in mediating *Sox2* functions in differentiation (Hoffmann et al., 2014). In addition, in NSC, together with HDAC2, Sox2 binds to a repressive site in a major enhancer of *Sox10*, a key regulator of oligodendrocyte differentiation, possibly collaborating with HDAC2 to maintain *Sox10* downregulation in undifferentiated cells; in turn, after differentiation to oligodendrocytes, SOX10 downregulates *Sox2* to maintain the differentiated state (Castelo-Branco et al., 2014); this observation is in accordance with the known role of *Sox2* in reversing oligodendrocytic differentiation in certain experimental conditions (Kondo and Raff, 2004). Analysis of the differentiation of neural cells from *Sox2* hypomorphic mutants also showed that mutant NSC failed to repress glial markers completely within differentiating neurons, and SOX2 negatively regulated the promoter of the astroglial gene GFAP (Cavallaro et al., 2008), a confirmed finding pointing to the ability of SOX2 to repress some of its target genes (Liu et al., 2014).

OPEN QUESTIONS AND PERSPECTIVES

Sox2 Functions in the Developing Brain

The experimental manipulation of *Sox2* clearly demonstrated an essential role of *Sox2* in maintaining NSC (in vivo and in vitro) in several species and important defects in the development of the brain in *Sox2* mutant mice. Yet, although *Sox2* deletion was obtained at early stages (E9.5–12.5), the overall development of the brain and nervous system was not affected; important morphological changes were confined to the ventral telencephalon and hippocampus (in the *BF1-Cre* mutant) and to the postnatal DG in the *Nestin-Cre* mutant. This implies that mutant NSC were able to support an essentially normal growth of the neural system for many cell cycles. Why are the defects confined to restricted brain regions, in contrast to the panneural expression of *Sox2* in NSC? In the case of the ventral telencephalon defects, it is possible that interference with the *Nkx2.1–Shh* axis is sufficient to explain the phenotype, which largely overlaps with that of the *Nkx2.1* and *Shh* mutants (Ferri et al., 2013); however, the important prenatal defects of the hippocampus (Ferri et al., 2013 and unpublished data) that follow the early *Sox2* deletion are not explained and are difficult to reconcile with the late appearance of DG defects in the *Nestin-Cre* mutants (Favaro et al., 2009). It will be important to precisely define the early cell types affected in the *BF1-Cre* mutant, as well as the molecular targets of SOX2 in these cells, whose alterations may influence the complex cell division/migration/differentiation events involved in hippocampal genesis. In this mutant, the brain volume and cortex are also reduced, yielding a (moderate) microcephalic phenotype (Ferri et al., 2013); this points to additional molecular targets and possible candidate genes.

Sox2 Functions in NSC

Sox2 mutant NSC are progressively exhausted during in vitro and in vivo growth (Cavallaro et al., 2008; Favaro et al., 2009; Ferri et al., 2013). The mechanisms of their failure to self-renew and their increased apoptotic death have not yet been clearly linked to the *Sox2* defect. Intriguingly, we have shown (Favaro et al., 2009) that wild-type cell–derived conditioned medium can support *Sox2* mutant cells via diffusible factors; we envisage such factors as being able to sustain in mutant cells the defective signaling pathways normally dependent on *Sox2*. Identification of these factors will be important for understanding of *Sox2*-dependent NSC maintenance.

Interestingly, it is also possible that in vivo, NSC maintenance depends on cell-intrinsic pathways and/or diffusible factors that are not strictly dependent on *Sox2*, which may explain the relatively long persistence in vivo of *Sox2* mutant NSC.

Sox2 and Cell Cycle Regulation

As shown in Figures 2 and 5, *Sox2* is at the center of a complex network involved in the control of mitosis, in which *Sox2* both regulates and is controlled by transcription factors acting on cell proliferation. In particular, *Sox2* is stimulated by Shh and EGF-dependent signals and controls the expression of molecules related to Shh, Notch, and EGF signaling. This network may control NSC quiescence, proliferation, and exit from proliferation, concurrent with commitment to terminal differentiation. The stimuli regulating this interaction network are still incompletely known: for example, which factors affect NSC and *Sox2*-dependent processes postnatally, and particularly during aging, when NSC are known to decline (Encinas et al., 2011). What happens after acute brain lesions, and would it be possible to induce some regeneration via manipulations of *Sox2*-dependent pathways?

When *Sox2* is decreased (knockdown mutant, Cavallaro et al., 2008), GFAP, a glial marker, is activated and partially differentiated cells with mixed neuronal/glia characteristics arise. This finding suggests that in NSC, *Sox2* may have inhibitory functions on genes related to differentiation (as also shown in ESC) (Avilion et al., 2003; Lee et al., 2006). It will be important to investigate whether *Sox2* decline in differentiation activates any negative regulators of the NSC state while causing the loss of positive effects on cell cycle. Genome-wide studies of RNA expression in wild-type and *Sox2* mutant cells will provide answers to this question.

Sox2 and Genome Architecture

Recent work allowed the genome-wide detection of long-range interactions in the chromatin of NSC; such interactions connect genes to DNA regions that have properties of enhancers in vivo in zebrafish transgenic assays (Zhang et al., 2013). Some loci, including *Sox2*, were multiconnected, showing long-range interactions with a high number of different genomic loci. This may reflect a multiplicity of connected regulatory elements involved in the regulation of *Sox2*; alternatively, *Sox2* (and similarly multiconnected regions, including *Olig1* and *2*) may act as regulatory hubs, contributing to the coordinated regulation of connected genes (Phillips-Cremins et al., 2013; Zhang et al., 2013). Three-dimensional genome architecture in the nucleus represents an important layer in gene regulation; *Sox2* may have important roles at this level as well. *Sox2*-deleted NSC cultured in vitro and their wild-type counterpart provide a tool to investigate at the genomic level the consequences of *Sox2* loss on long-range chromatin interactions and, in parallel, on epigenetic enhancer marks and gene expression. It is possible that some of the nonredundant functions of *Sox2* in CNS development and NSC will find novel explanations in an architectural role in chromatin.

REFERENCES

Akamatsu, W., Deveale, B., Okano, H., Cooney, A.J., van der Kooy, D., 2009. Suppression of Oct4 by germ cell nuclear factor restricts pluripotency and promotes neural stem cell development in the early neural lineage. J. Neurosci. 29, 2113–2124.

Amamoto, R., Arlotta, P., 2014. Development-inspired reprogramming of the mammalian central nervous system. Science 343, 1239882.

Anderson, D.J., 1989. The neural crest cell lineage problem: neuropoiesis? Neuron 3, 1–12.

Andreu, Z., Khan, M.A., González-Gómez, P., Negueruela, S., Hortigüela, R., San Emeterio, J., Ferrón, S.R., Martínez, G., Vidal, A., Fariñas, I., Lie, D.C., Mira, H., 2015. The cyclin-dependent kinase inhibitor p27 kip1 regulates radial stem cell quiescence and neurogenesis in the adult hippocampus. Stem Cells 33, 219–229. http://dx.doi.org/10.1002/stem.1832.

Andreu-Agullo, C., Maurin, T., 2012. Ars2, an essential player in neural stem cell identity. Med. Sci. (Paris) 28, 459–462.

Avilion, A.A., Nicolis, S.K., Pevny, L.H., Perez, L., Vivian, N., Lovell-Badge, R., 2003. Multipotent cell lineages in early mouse development depend on SOX2 function. Genes Dev. 17, 126–140.

Beccari, L., Conte, I., Cisneros, E., Bovolenta, P., 2012. Sox2-mediated differential activation of Six3.2 contributes to forebrain patterning. Development 139, 151–164.

Bergsland, M., Ramsköld, D., Zaouter, C., Klum, S., Sandberg, R., Muhr, J., 2011. Sequentially acting Sox transcription factors in neural lineage development. Genes Dev. 25, 2453–2464.

Bery, A., Martynoga, B., Guillemot, F., Joly, J.S., Retaux, S., 2014. Characterization of enhancers active in the mouse embryonic cerebral cortex suggests Sox/Pou cis-regulatory logics and heterogeneity of cortical progenitors. Cereb. Cortex 24, 2822–2834.

Bylund, M., Andersson, E., Novitch, B.G., Muhr, J., 2003. Vertebrate neurogenesis is counteracted by Sox1-3 activity. Nat. Neurosci. 6, 1162–1168.

Caiazzo, M., Dell'anno, M.T., Dvoretskova, E., Lazarevic, D., Taverna, S., Leo, D., Sotnikova, T.D., Menegon, A., Roncaglia, P., Colciago, G., Russo, G., Carninci, P., Pezzoli, G., Gainetdinov, R.R., Gustincich, S., Dityatev, A., Broccoli, V., 2011. Direct generation of functional dopaminergic neurons from mouse and human fibroblasts. Nature 476, 224–227.

Castelo-Branco, G., Lilja, T., Wallenborg, K., Falcao, A.M., Marques, S.C., Gracias, A., Solum, D., Paap, R., Walfridsson, J., Teixeira, A.I., Rosenfeld, M.G., Jepsen, K., Hermanson, O., 2014. Neural stem cell differentiation is dictated by distinct actions of nuclear receptor corepressors and histone deacetylases. Stem Cell Rep. 3, 502–515.

Catena, R., Tiveron, C., Ronchi, A., Porta, S., Ferri, A., Tatangelo, L., Cavallaro, M., Favaro, R., Ottolenghi, S., Reinbold, R., Scholer, H., Nicolis, S.K., 2004. Conserved POU binding DNA sites in the Sox2 upstream enhancer regulate gene expression in embryonic and neural stem cells. J. Biol. Chem. 279, 41846–41857.

Cattell, M.V., Garnett, A.T., Klymkowsky, M.W., Medeiros, D.M., 2012. A maternally established SoxB1/SoxF axis is a conserved feature of chordate germ layer patterning. Evol. Dev. 14, 104–115.

Cavallaro, M., Mariani, J., Lancini, C., Latorre, E., Caccia, R., Gullo, F., Valotta, M., Debiasi, S., Spinardi, L., Ronchi, A., Wanke, E., Brunelli, S., Favaro, R., Ottolenghi, S., Nicolis, S.K., 2008. Impaired generation of mature neurons by neural stem cells from hypomorphic Sox2 mutants. Development 135, 541–557.

Chen, X., Xu, H., Yuan, P., Fang, F., Huss, M., Vega, V.B., Wong, E., Orlov, Y.L., Zhang, W., Jiang, J., Loh, Y.H., Yeo, H.C., Yeo, Z.X., Narang, V., Govindarajan, K.R., Leong, B., Shahab, A., Ruan, Y., Bourque, G., Sung, W.K., Clarke, N.D., Wei, C.L., Ng, H.H., 2008. Integration of external signaling pathways with the core transcriptional network in embryonic stem cells. Cell 133, 1106–1117.

Cimadamore, F., Amador-Arjona, A., Chen, C., Huang, C.T., Terskikh, A.V., 2013. Sox2-Lin28/let-7 pathway regulates proliferation and neurogenesis in neural precursors. Proc. Natl. Acad. Sci. U.S.A. 110, E3017–E3026.

Collignon, J., Sockanathan, S., Hacker, A., Cohen-Tannoudji, M., Norris, D., Rastan, S., Stevanovic, M., Goodfellow, P.N., Lovell-Badge, R., 1996. A comparison of the properties of Sox-3 with Sry and two related genes, Sox-1 and Sox-2. Development 122, 509–520.

Collignon, J., 1992. (Ph.D. thesis).

Dillon, N., 2012. Factor mediated gene priming in pluripotent stem cells sets the stage for lineage specification. Bioessays 34, 194–204.

Ellis, P., Fagan, B.M., Magness, S.T., Hutton, S., Taranova, O., Hayashi, S., Mcmahon, A., Rao, M., Pevny, L., 2004. Sox2, a persistent marker for multipotential neural stem cells derived from embryonic stem cells, the embryo or the adult. Dev. Neurosci. 26, 148–165.

Encinas, J.M., Michurina, T.V., Peunova, N., Park, J.H., Tordo, J., Peterson, D.A., Fishell, G., Koulakov, A., Enikolopov, G., 2011. Division-coupled astrocytic differentiation and age-related depletion of neural stem cells in the adult hippocampus. Cell Stem Cell 8, 566–579.

Engelen, E., Akinci, U., Bryne, J.C., Hou, J., Gontan, C., Moen, M., Szumska, D., Kockx, C., Van Ijcken, W., Dekkers, D.H., Demmers, J., Rijkers, E.J., Bhattacharya, S., Philipsen, S., Pevny, L.H., Grosveld, F.G., rottier, R.J., Lenhard, B., Poot, R.A., 2011. Sox2 cooperates with Chd7 to regulate genes that are mutated in human syndromes. Nat. Genet. 43, 607–611.

Favaro, R., Valotta, M., Ferri, A.L., Latorre, E., Mariani, J., Giachino, C., Lancini, C., Tosetti, V., Ottolenghi, S., Taylor, V., Nicolis, S.K., 2009. Hippocampal development and neural stem cell maintenance require Sox2-dependent regulation of Shh. Nat. Neurosci. 12, 1248–1256.

Favaro, R., Appolloni, I., Pellegatta, S., Sanga, A.B., Pagella, P., Gambini, E., Pisati, F., Ottolenghi, S., Foti, M., Finocchiaro, G., Malatesta, P., Nicolis, S.K., 2014. Sox2 is required to maintain cancer stem cells in a mouse model of high-grade oligodendroglioma. Cancer Res. 74, 1833–1844.

Fei, J.F., Schuez, M., Tazaki, A., Taniguchi, Y., Roensch, K., Tanaka, E.M., 2014. CRISPR-mediated genomic deletion of Sox2 in the axolotl shows a requirement in spinal cord neural stem cell amplification during tail regeneration. Stem Cell Rep. 3, 444–459.

Feng, R., Zhou, S., Liu, Y., Song, D., Luan, Z., Dai, X., Li, Y., Tang, N., Wen, J., Li, L., 2013. Sox2 protects neural stem cells from apoptosis via up-regulating survivin expression. Biochem. J. 450, 459–468.

Ferri, A.L., Cavallaro, M., Braida, D., Di Cristofano, A., Canta, A., Vezzani, A., Ottolenghi, S., Pandolfi, P.P., Sala, M., Debiasi, S., Nicolis, S.K., 2004. Sox2 deficiency causes neurodegeneration and impaired neurogenesis in the adult mouse brain. Development 131, 3805–3819.

Ferri, A., Favaro, R., Beccari, L., Bertolini, J., Mercurio, S., Nieto-Lopez, F., Verzeroli, C., La Regina, F., De Pietri Tonelli, D., Ottolenghi, S., Bovolenta, P., Nicolis, S.K., 2013. Sox2 is required for embryonic development of the ventral telencephalon through the activation of the ventral determinants Nkx2.1 and Shh. Development 140, 1250–1261.

Gómez-López, S., Wiskow, O., Favaro, R., Nicolis, S.K., Price, D.J., Pollard, S.M., Smith, A., 2011. Sox2 and Pax6 maintain the proliferative and developmental potential of gliogenic neural stem cells in vitro. Glia 59, 1588–1599.

Gouti, M., Tsakiridis, A., Wymeersch, F.J., Huang, Y., Kleinjung, J., Wilson, V., Briscoe, J., 2014. In vitro generation of neuromesodermal progenitors reveals distinct roles for wnt signalling in the specification of spinal cord and paraxial mesoderm identity. PLoS Biol. 12, e1001937.

Graham, V., Khudyakov, J., Ellis, P., Pevny, L., 2003. SOX2 functions to maintain neural progenitor identity. Neuron 39, 749–765.

Gubbay, J., Collignon, J., Koopman, P., Capel, B., Economou, A., Munsterberg, A., Vivian, N., Goodfellow, P., Lovell-Badge, R., 1990. A gene mapping to the sex-determining region of the mouse Y chromosome is a member of a novel family of embryonically expressed genes. Nature 346, 245–250.

Hébert, J.M., Fishell, G., 2008. The genetics of early telencephalon patterning: some assembly required. Nat. Rev. Neurosci. 9, 678–685.

Hoffmann, S.A., Hos, D., Kuspert, M., Lang, R.A., Lovell-Badge, R., Wegner, M., Reiprich, S., 2014. Stem cell factor Sox2 and its close relative Sox3 have differentiation functions in oligodendrocytes. Development 141, 39–50.

Hu, Q., Zhang, L., Wen, J., Wang, S., Li, M., Feng, R., Yang, X., Li, L., 2010. The EGF receptor-sox2-EGF receptor feedback loop positively regulates the self-renewal of neural precursor cells. Stem Cells 28, 279–286.

Iwafuchi-Doi, M., Yoshida, Y., Onichtchouk, D., Leichsenring, M., Driever, W., Takemoto, T., Uchikawa, M., Kamachi, Y., Kondoh, H., 2011. The Pou5f1/Pou3f-dependent but SoxB-independent regulation of conserved enhancer N2 initiates Sox2 expression during epiblast to neural plate stages in vertebrates. Dev. Biol. 352, 354–366.

Iwafuchi-Doi, M., Matsuda, K., Murakami, K., Niwa, H., Tesar, P.J., Aruga, J., Matsuo, I., Kondoh, H., 2012. Transcriptional regulatory networks in epiblast cells and during anterior neural plate development as modeled in epiblast stem cells. Development 139, 3926–3937.

Jeong, Y., El-Jaick, K., Roessler, E., Muenke, M., Epstein, D.J., 2006. A functional screen for sonic hedgehog regulatory elements across a 1 Mb interval identifies long-range ventral forebrain enhancers. Development 133, 761–772.

Jeong, Y., Leskow, F.C., El-Jaick, K., Roessler, E., Muenke, M., Yocum, A., Dubourg, C., Li, X., Geng, X., Oliver, G., Epstein, D.J., 2008. Regulation of a remote Shh forebrain enhancer by the Six3 homeoprotein. Nat. Genet. 40, 1348–1353.

Josephson, R., Muller, T., Pickel, J., Okabe, S., Reynolds, K., Turner, P.A., Zimmer, A., Mckay, R.D., 1998. POU transcription factors control expression of CNS stem cell-specific genes. Development 125, 3087–3100.

Julian, L.M., Vandenbosch, R., Pakenham, C.A., Andrusiak, M.G., Nguyen, A.P., Mcclellan, K.A., Svoboda, D.S., Lagace, D.C., Park, D.S., Leone, G., Blais, A., Slack, R.S., 2013. Opposing regulation of Sox2 by cell-cycle effectors E2f3a and E2f3b in neural stem cells. Cell Stem Cell 12, 440–452.

Karow, M., Sanchez, R., Schichor, C., Masserdotti, G., Ortega, F., Heinrich, C., Gascon, S., Khan, M.A., Lie, D.C., Dellavalle, A., Cossu, G., Goldbrunner, R., Gotz, M., Berninger, B., 2012. Reprogramming of pericyte-derived cells of the adult human brain into induced neuronal cells. Cell Stem Cell 11, 471–476.

Kishi, M., Mizuseki, K., Sasai, N., Yamazaki, H., Shiota, K., Nakanishi, S., Sasai, Y., 2000. Requirement of Sox2-mediated signaling for differentiation of early *Xenopus* neuroectoderm. Development 127, 791–800.

Kondo, T., Raff, M., 2004. Chromatin remodeling and histone modification in the conversion of oligodendrocyte precursors to neural stem cells. Genes. Dev. 18, 2963–2972.

Kondoh, H., Takemoto, T., 2012. Axial stem cells deriving both posterior neural and mesodermal tissues during gastrulation. Curr. Opin. Genet. Dev. 22, 374–380.

LaMonica, B.E., Lui, J.H., Wang, X., Kriegstein, A.R., 2012. OSVZ progenitors in the human cortex: an updated perspective on neurodevelopmental disease. Curr. Opin. Neurobiol. 22, 747–753.

Langer, L., Taranova, O., Sulik, K., Pevny, L., 2012. SOX2 hypomorphism disrupts development of the prechordal floor and optic cup. Mech. Dev. 129, 1–12.

Lee, T.I., Jenner, R.G., Boyer, L.A., Guenther, M.G., Levine, S.S., Kumar, R.M., Chevalier, B., Johnstone, S.E., Cole, M.F., Isono, K., Koseki, H., Fuchikami, T., Abe, K., Murray, H.L., Zucker, J.P., Yuan, B., Bell, G.W., Herbolsheimer, E., Hannett, N.M., Sun, K., Odom, D.T., Otte, A.P., Volkert, T.L., Bartel, D.P., Melton, D.A., Gifford, D.K., Jaenisch, R., Young, R.A., 2006. Control of developmental regulators by polycomb in human embryonic stem cells. Cell 125, 301–313.

Li, C., Ito, H., Fujita, K., Shiwaku, H., Qi, Y., Tagawa, K., Tamura, T., Okazawa, H., 2013. Sox2 transcriptionally regulates PQBP1, an intellectual disability-microcephaly causative gene, in neural stem progenitor cells. PLoS One 8, e68627.

Liu, Y.R., Laghari, Z.A., Novoa, C.A., Hughes, J., Webster, J.R., Goodwin, P.E., Wheatley, S.P., Scotting, P.J., August 8, 2014. Sox2 acts as a transcriptional repressor in neural stem cells. BMC Neurosci. 15, 95. http://dx.doi.org/10.1186/1471-2202-15-95 (Medline in process).

Lodato, M.A., Ng, C.W., Wamstad, J.A., Cheng, A.W., Thai, K.K., Fraenkel, E., Jaenisch, R., Boyer, L.A., 2013. SOX2 co-occupies distal enhancer elements with distinct POU factors in ESCs and NPCs to specify cell state. PLoS Genet. 9, e1003288.

Machold, R., Hayashi, S., Rutlin, M., Muzumdar, M.D., Nery, S., Corbin, J.G., Gritli-Linde, A., Dellovade, T., Porter, J.A., Rubin, L.L., Dudek, H., Mcmahon, A.P., Fishell, G., 2003. Sonic hedgehog is required for progenitor cell maintenance in telencephalic stem cell niches. Neuron 39, 937–950.

Mariani, J., Favaro, R., Lancini, C., Vaccari, G., Ferri, A.L., Bertolini, J., Tonoli, D., Latorre, E., Caccia, R., Ronchi, A., Ottolenghi, S., Miyagi, S., Okuda, A., Zappavigna, V., Nicolis, S.K., 2012. Emx2 is a dose-dependent negative regulator of Sox2 telencephalic enhancers. Nucleic Acids Res. 40, 6461–6476.

Marques-Torrejon, M.A., Porlan, E., Banito, A., Gomez-Ibarlucea, E., Lopez-Contreras, A.J., Fernandez-Capetillo, O., Vidal, A., Gil, J., Torres, J., Farinas, I., 2013. Cyclin-dependent kinase inhibitor p21 controls adult neural stem cell expansion by regulating Sox2 gene expression. Cell Stem Cell 12, 88–100.

Meulemans, D., Bronner-Fraser, M., 2007. The amphioxus SoxB family: implications for the evolution of vertebrate placodes. Int. J. Biol. Sci. 3, 356–364.

Miyagi, S., Saito, T., Mizutani, K., Masuyama, N., Gotoh, Y., Iwama, A., Nakauchi, H., Masui, S., Niwa, H., Nishimoto, M., Muramatsu, M., Okuda, A., 2004. The Sox-2 regulatory regions display their activities in two distinct types of multipotent stem cells. Mol. Cell. Biol. 24, 4207–4220.

Miyagi, S., Masui, S., Niwa, H., Saito, T., Shimazaki, T., Okano, H., Nishimoto, M., Muramatsu, M., Iwama, A., Okuda, A., 2008. Consequence of the loss of Sox2 in the developing brain of the mouse. FEBS Lett. 582, 2811–2815.

Ng, S.Y., Bogu, G.K., Soh, B.S., Stanton, L.W., 2013. The long noncoding RNA RMST interacts with SOX2 to regulate neurogenesis. Mol. Cell 51, 349–359.

Niu, W., Zang, T., Zou, Y., Fang, S., Smith, D.K., Bachoo, R., Zhang, C.L., 2013. In vivo reprogramming of astrocytes to neuroblasts in the adult brain. Nat. Cell Biol. 15, 1164–1175.

Oosterveen, T., Kurdija, S., Enstero, M., Uhde, C.W., Bergsland, M., Sandberg, M., Sandberg, R., Muhr, J., Ericson, J., 2013. SoxB1-driven transcriptional network underlies neural-specific interpretation of morphogen signals. Proc. Natl. Acad. Sci. U.S.A. 110, 7330–7335.

Papanayotou, C., Mey, A., Birot, A.M., Saka, Y., Boast, S., Smith, J.C., Samarut, J., Stern, C.D., 2008. A mechanism regulating the onset of Sox2 expression in the embryonic neural plate. PLoS Biol. 6, e2.

Peng, C., Li, N., Ng, Y.K., Zhang, J., Meier, F., Theis, F.J., Merkenschlager, M., Chen, W., Wurst, W., Prakash, N., 2012. A unilateral negative feedback loop between miR-200 microRNAs and Sox2/E2F3 controls neural progenitor cell-cycle exit and differentiation. J. Neurosci. 32, 13292–13308.

Peterson, K.A., Nishi, Y., Ma, W., Vedenko, A., Shokri, L., Zhang, X., Mcfarlane, M., Baizabal, J.M., Junker, J.P., Van Oudenaarden, A., Mikkelsen, T., Bernstein, B.E., Bailey, T.L., Bulyk, M.L., Wong, W.H., Mcmahon, A.P., 2012. Neural-specific Sox2 input and differential Gli-binding affinity provide context and positional information in Shh-directed neural patterning. Genes Dev. 26, 2802–2816.

Pevny, L.H., Nicolis, S.K., 2010. Sox2 roles in neural stem cells. Int. J. Biochem. Cell Biol. 42, 421–424.

Phillips-Cremins, J.E., Sauria, M.E., Sanyal, A., Gerasimova, T.I., Lajoie, B.R., Bell, J.S., Ong, C.T., Hookway, T.A., Guo, C., Sun, Y., Bland, M.J., Wagstaff, W., Dalton, S., Mcdevitt, T.C., Sen, R., Dekker, J., Taylor, J., Corces, V.G., 2013. Architectural protein subclasses shape 3D organization of genomes during lineage commitment. Cell 153, 1281–1295.

Qi, Y., Hoshino, M., Wada, Y., Marubuchi, S., Yoshimura, N., Kanazawa, I., Shinomiya, K., Okazawa, H., 2005. PQBP-1 is expressed predominantly in the central nervous system during development. Eur. J. Neurosci. 22, 1277–1286.

Reynolds, B.A., Weiss, S., 1992. Generation of neurons and astrocytes from isolated cells of the adult mammalian central nervous system. Science 255, 1707–1710.

Roessler, E., Muenke, M., 2010. The molecular genetics of holoprosencephaly. Am. J. Med. Genet. C Semin. Med. Genet. 154C, 52–61.

Rogers, N., Cheah, P.S., Szarek, E., Banerjee, K., Schwartz, J., Thomas, P., 2013. Expression of the murine transcription factor SOX3 during embryonic and adult neurogenesis. Gene Expression Patterns 13, 240–248.

Shimozaki, K., Zhang, C.L., Suh, H., Denli, A.M., Evans, R.M., Gage, F.H., 2012. SRY-box-containing gene 2 regulation of nuclear receptor tailless (Tlx) transcription in adult neural stem cells. J. Biol. Chem. 287, 5969–5978.

Sugino, K., Hempel, C.M., Miller, M.N., Hattox, A.M., Shapiro, P., Wu, C., Huang, Z.J., Nelson, S.B., 2006. Molecular taxonomy of major neuronal classes in the adult mouse forebrain. Nat. Neurosci. 9, 99–107.

Suh, H., Consiglio, A., Ray, J., Sawai, T., D'amour, K.A., Gage, F.H., 2007. In vivo fate analysis reveals the multipotent and self-renewal capacities of $Sox2^{+}$ neural stem cells in the adult hippocampus. Cell stem Cell 1, 515–528.

Takahashi, K., Yamanaka, S., 2006. Induction of pluripotent stem cells from mouse embryonic and adult fibroblast cultures by defined factors. Cell 126, 663–676.

Takanaga, H., Tsuchida-Straeten, N., Nishide, K., Watanabe, A., Aburatani, H., Kondo, T., 2009. Gli2 is a novel regulator of sox2 expression in telencephalic neuroepithelial cells. Stem Cells 27, 165–174.

Takemoto, T., Uchikawa, M., Yoshida, M., Bell, D.M., Lovell-Badge, R., Papaioannou, V.E., Kondoh, H., 2011. Tbx6-dependent Sox2 regulation determines neural or mesodermal fate in axial stem cells. Nature 470, 394–398.

Tanaka, S., Kamachi, Y., Tanouchi, A., Hamada, H., Jing, N., Kondoh, H., 2004. Interplay of SOX and POU factors in regulation of the Nestin gene in neural primordial cells. Mol. Cell. Biol. 24, 8834–8846.

Taranova, O.V., Magness, S.T., Fagan, B.M., Wu, Y., Surzenko, N., Hutton, S.R., Pevny, L.H., 2006. SOX2 is a dose-dependent regulator of retinal neural progenitor competence. Genes Dev. 20, 1187–1202.

Temple, S., 2001. The development of neural stem cells. Nature 414, 112–117.

Uchikawa, M., Ishida, Y., Takemoto, T., Kamachi, Y., Kondoh, H., 2003. Functional analysis of chicken Sox2 enhancers highlights an array of diverse regulatory elements that are conserved in mammals. Dev. Cell 4, 509–519.

Vierbuchen, T., Ostermeier, A., Pang, Z.P., Kokubu, Y., Südhof, T.C., Wernig, M., 2010. Direct conversion of fibroblasts to functional neurons by defined factors. Nature 463, 1035–1041. http://dx.doi.org/10.1038/nature08797.

Vierbuchen, T., Wernig, M., 2011. Direct lineage conversions: unnatural but useful? Nat. Biotechnol. 29, 892–907. http://dx.doi.org/10.1038/nbt.1946. Review.

Wang, S., Chandler-Militello, D., Lu, G., Roy, N.S., Zielke, A., Auvergne, R., Stanwood, N., Geschwind, D., Coppola, G., Nicolis, S.K., Sim, F.J., Goldman, S.A., 2010. Prospective identification, isolation, and profiling of a telomerase-expressing subpopulation of human neural stem cells, using sox2 enhancer-directed fluorescence-activated cell sorting. J. Neurosci. 30, 14635–14648.

Wilson, V., Olivera-Martinez, I., Storey, K.G., 2009. Stem cells, signals and vertebrate body axis extension. Development 136, 1591–1604.

Wood, H.B., Episkopou, V., 1999. Comparative expression of the mouse Sox1, Sox2 and Sox3 genes from pre-gastrulation to early somite stages. Mech. Dev. 86, 197–201.

Zappone, M.V., Galli, R., Catena, R., Meani, N., De Biasi, S., Mattei, E., Tiveron, C., Vescovi, A.L., Lovell-Badge, R., Ottolenghi, S., Nicolis, S.K., 2000. Sox2 regulatory sequences direct expression of a (beta)-geo transgene to telencephalic neural stem cells and precursors of the mouse embryo, revealing regionalization of gene expression in CNS stem cells. Development 127, 2367–2382.

Zhang, Y., Wong, C.H., Birnbaum, R.Y., Li, G., Favaro, R., Ngan, C.Y., Lim, J., Tai, E., Poh, H.M., Wong, E., Mulawadi, F.H., Sung, W.K., Nicolis, S., Ahituv, N., Ruan, Y., Wei, C.L., 2013. Chromatin connectivity maps reveal dynamic promoter-enhancer long-range associations. Nature 504, 306–310.

Zhao, L., Zevallos, S.E., Rizzoti, K., Jeong, Y., Lovell-Badge, R., Epstein, D.J., 2012. Disruption of SoxB1-dependent Sonic hedgehog expression in the hypothalamus causes septo-optic dysplasia. Dev. Cell 22, 585–596.

CHAPTER 12

Multiple Roles for SOX2 in Eye Development

Hisato Kondoh[1], Masanori Uchikawa[2], Yasuo Ishii[1]
[1]Faculty of Life Sciences, Kyoto Sangyo University, Kyoto, Japan; [2]Graduate School of Frontier Biosciences, Osaka University, Suita, Osaka, Japan

INTRODUCTION

The eye tissues develop from two distinct primordia: that is, the first, a portion of the embryonic central nervous system (CNS) that gives rise to the neural retina, pigmented epithelium, ciliary body, iris, and optic stalk; and the second, the median head ectoderm that gives rise to the lens and corneal epithelium (Kondoh, 2002; Heavner and Pevny, 2012). Because a variety of tissues develop from the eye primordia, many steps of tissue differentiation are involved in their development; in almost all of these steps, SOX2 function is involved. These variable contributions of SOX2 are ascribed to the nature of this transcription factor, which specify its functions by association with a wide range of partner transcription factors. Congenital malformations of eye tissues including anophthalmia occur as a consequence of *Sox2* mutation and/or reduced SOX2 expression levels (see Chapter 14). The etiology of these SOX2-dependent abnormalities should be understood as reflecting many steps of tissue regulation involving SOX2 functions.

STEPS INVOLVED IN EMBRYONIC EYE DEVELOPMENT

Initial Contact of Retinal Primordium and Surface Ectoderm

Eyes are composed of two major tissue components, retina and lens, which interact during the early morphogenetic process of eye development. The optic vesicle, the retinal primordium, initially develops as a protrusion of the wall of the embryonic CNS and contacts the head surface ectoderm (Figure 1(A)). Lens tissues develop from the region of surface ectoderm where the optic vesicle makes contact. The requirement of this contact for lens development was first indicated by the classic experiment of optic vesicle ablation of salamander embryos performed by Spemann (1901). This mechanism that spatially coordinates the development of two eye tissues has evolved, presumably to guarantee that the lens and retinal tissues share an optical axis. The process, called "lens induction" (Kondoh, 2010), can occur widely in the dorsoventrally median region of the surface ectoderm once it is contacted by the optic vesicle. An extreme case is the cyclopic eye with a normal lens that develops when an optic cup develops along the midline.

Sox2
http://dx.doi.org/10.1016/B978-0-12-800352-7.00012-8

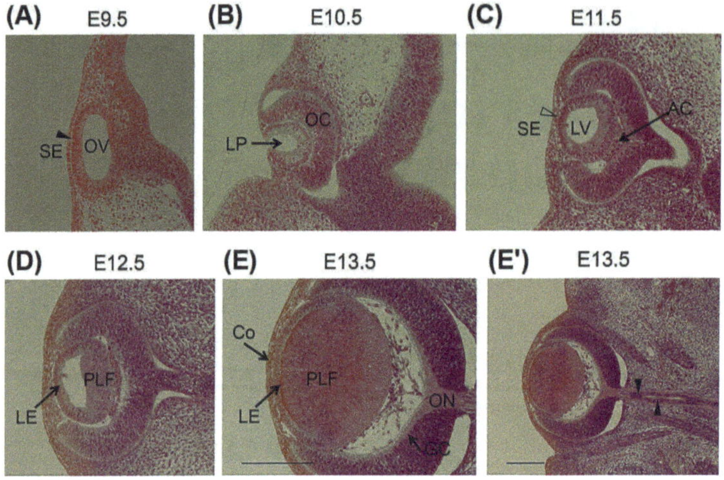

Figure 1 Development of eye tissues in mouse embryos from embryonic day (E)9.5 to E13.5. Hematoxylin–eosin-stained cross-sections through an embryonic eye at various stages, fixed with Bouin's fixative. (A) At E9.5, the head surface ectoderm (SE) is apposed by the optic vesicle (OV). Note that no cells exist between the tightly apposed tissues (arrowhead). The lens placode develops immediately. (B) At E10.5, OV has invaginated to form the optic cup (OC). The lens placode has already invaginated and is about to form a lens vesicle while leaving the lens pit (LP) still unclosed. (C) At E11.5, the lens vesicle (LV) is completed and separated from SE, which develops into the corneal epithelium. Mutant embryos deficient in transcription factors essential for lens fiber development (e.g., Sox1 and cMaf; Nishiguchi et al., 1998; Kawauchi et al., 1999; Kim et al., 1999) arrest lens development at this stage. Note that the cells that later constitute the corneal stroma (open arrowhead) have migrated into the space between SE and LV, and that the population of angiogenic cells (AC) has migrated into the posterior eye chamber through the unclosed ventrally located optic fissure. (D) At E12, the primary lens fibers (PLF) have started to develop in the posterior half of the lens vesicle, while the anterior half remain as lens epithelium (LE). (E and E') At E13.5, the development of PLF has completed, filling the vesicular space. After this stage, the secondary lens fibers develop to wrap around PLF from the peripheral LE located in the equatorial bow region. In the neural retina, GC have fully developed and send massive axons to form optic nerve (ON) along the optic stalk now developing into the glial cells of ON. The nuclei along ON (E', arrowheads) are those of glia. At this stage, the corneal tissue (Co) consisting of the outer epithelium and the inner stroma is established and separated from the lens. Bars indicate 200 µm.

Eye Tissue Invagination

After contact of the optic vesicle with the median head ectoderm, these tissues interact with each other. First, the contacted area of head ectoderm becomes thickened to form a lens placode, which expresses placode-characteristic cell adhesion molecules (CAM) N-cadherin and N-CAM (Matsumata et al., 2005). The size of the lens placode area is determined by the size of the contacted area, as demonstrated by surgical operations on optic vesicles in chicken embryos (Li et al., 1994; Kamachi et al., 1998). Subsequently, both the lens placode and the contacted area of the optic vesicle start to invaginate and develop into the lens pit and the optic cup, respectively (Figure 1(B)). Initiation of this invagination process appears to depend on transient signaling from the head ectoderm, because its surgical removal before lens placode development impairs optic cup invagination, whereas removal at later stages does not (Coulombre and Coulombre, 1964; Hyer et al., 2003). Invagination of both tissues occurs concordantly, partly assisted by the contractile filopodia that tether the invaginating lens vesicle and inner optic cup (Chauhan et al., 2009).

Formation of the Lens Vesicle and Its Separation from the Surface Ectoderm

The lens vesicle eventually closes to form an epithelial lens vesicle and separates from the remaining surface ectoderm covering the eye that later develops into the corneal epithelium (Figure 1(C)). The latter process involves the action of many transcription factors, and genetic impairment in one of the factors can cause persistent connection of the lens and corneal epithelial tissues, a structure called a "lens stalk," which is a form of Peters' anomaly (see Chapter 13).

Distinct Developmental Fates of Three Different Compartments of the Optic Cup

Whereas the optic vesicle invaginates, cell growth is more active in the ventral half, resulting in displacement of the initially dorsal-most region to cover the entire outer layer of the optic cup that differentiates into the pigmented epithelium (Kondoh, 2002). As invagination proceeds to appose the inner and outer layers of the optic cup, the inner layer differentiates into the neural retina, and the remaining proximal region differentiates into the optic stalk (Figure 1(E)), the future source of glial cells of the optic nerve (Figure 1(E′)). At this stage, the cup and stalk have a ventral opening called the choroid fissure, from where precursors for the eye vascular systems migrate (Figure 1(C)). The fissure is then closed, but failure of this process results in a coloboma. The distal rim of the optic cup later develops into the iris and the ciliary body.

Lens Development

After lens vesicle formation, the cells in the anterior (closer to the cornea) and posterior compartments develop differently. The anterior compartment remains epithelial and continues to proliferate, whereas the cells in the posterior compartment elongate to form the primary lens fibers that fill the vesicular space (Figure 1(D and E)). Proliferation of lens epithelial cells results in the accumulation of cells in the equatorial region, where the cells start to elongate and form secondary lens fibers that wrap around the primary fibers. This phase of lens development continues even after birth.

Lens development is accompanied by the synthesis of crystallins that maintain the lens as a transparent and highly refractive tissue throughout life. Lens epithelium expresses α-crystallins (αA and αB), whereas lens fibers express β- and γ-crystallins. However, in avians, γ-crystallins are completely absent, and instead the taxon-unique δ-crystallin is expressed (Kondoh et al., 1983). δ-Crystallin expression is activated in the lens placode and continues in the lens vesicle, lens epithelium, and lens fibers, where δ-crystallin expression is further augmented (Kamachi et al., 1998).

EXPRESSION OF *Sox2* IN EYE TISSUES AND ITS REGULATION

Sox2 is expressed in both lens and retinal tissues from the earliest stages of eye development (Kamachi et al., 1998). All cells in the optic vesicle express *Sox2*, as in other parts of the neural tube.

The head ectoderm initially expresses low levels of *Sox2* with a ventral to dorsal gradient (Figure 2(A-a) and (B-a)), and contact with the optic vesicle elicits local augmentation of *Sox2* expression in the contacted area; this continues to the lens placode (Figure 2(A-b,c) and (B-b)). In mammalian lenses *Sox2* expression continues to the stages of invagination of the lens pit, lens vesicle, and lens epithelium (Kamachi et al., 1998). However, at the stage of lens fiber development, *Sox2* is replaced by *Sox1* (Nishiguchi et al., 1998). This situation is unique to mammals; in the chicken, *Sox2* continues to be expressed in the lens fiber as well (Figure 2(B-c)) (Table 1).

Invagination of the optic vesicle results in generation of the inner and outer layer of the optic cup. Before invagination, *Sox2* expression is downregulated in the dorsal region of the optic vesicle (Figure 2(B-a)). This low-*Sox2* region expands to form the outer layer of the optic cup (Figure 2(A-d) and (B-b,c)) and later differentiates into the pigmented epithelium (Figure 2(A-e)). The inner layer of the optic cup expresses *Sox2* strongly and develops into the neural retina (Kamachi et al., 1998; Ishii et al., 2009).

Once the retinal neuronal cells begin to differentiate in the inner layer, the neuroblastic retinal progenitor cells (RPCs) become the major SOX2 expression domain. Among the differentiated cells, ganglion cells (cholinergic), a cholinergic subset of amacrine cells, and Müller cells also express SOX2 (Lin et al., 2009).

ENHANCERS THAT REGULATE *Sox2* EXPRESSION DURING EYE DEVELOPMENT

The *Sox2* gene expressed at these different tissue domains is regulated by an array of enhancers that have distinct spatiotemporal specificities (Uchikawa et al., 2003), as detailed in Chapter 7. *Sox2* expression in the anterior CNS, including a part of the optic vesicle protrusion, is initially activated by the N2 enhancer, which is regulated by OTX2, ZIC2/3, and POU3F factors (Iwafuchi-Doi et al., 2011, 2012). This is followed by the activation of an ocular system-dedicated N3 enhancer, which shows activity in the diencephalon, optic vesicle, and lens placode. The N3 enhancer is activated by the combined action of SOX2 and PAX6 (Inoue et al., 2007) as discussed in Chapter 8, consistent with its strong participation in ocular development.

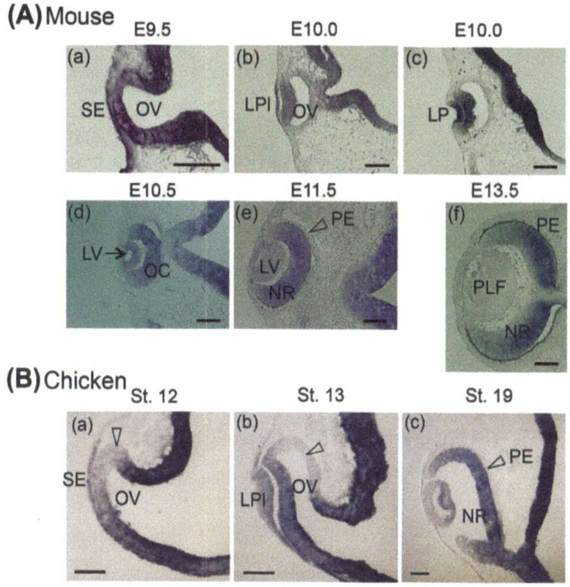

Figure 2 Expression of *Sox2* during the early stages of eye development, as indicated by in situ hybridization. (A) Mouse embryonic eyes. (a) From before their apposition, optic vesicle (OV) and surface ectoderm (SE) have expressed *Sox2*. This section is overstained to show relatively low *Sox2* expression in SE. (b and c) At E10.0 the lens placode (LPl) before invagination (b) and after the start of invagination to form the lens pit (LP) (c), as well as the apposed area of the optic vesicle (OV), express *Sox2* strongly. (d) At E10.5, when the lens vesicle (LV) is closed, LV and the inner layer of the optic cup (OC) express *Sox2* strongly, but SE and the outer layer of OC are low in *Sox2* expression. (e) At E11.5, the tightly associated layers of the neural retina (NR) and the pigmented epithelium (PE) are established, where *Sox2* expression is undetectable in the latter. (f) At E13.5, the primary lens fiber (PLF) lacking *Sox2* expression fills the vesicular space of the lens, NR starts to differentiate, and PE has already developed melanin pigmentation. Note that *Sox2* is expressed in the NR layer with a proximo-distal gradient. (This section is understained to show the gradient.) Bars indicate 100 μm. (B) Chicken embryonic eyes. (a) At Hamburger–Hamilton stage (St.) 12 when SE and OV are apposed, *Sox2* expression in the apposed area of SE is increased, whereas that in the dorsal part of OV (open arrowhead) is decreased. (b) At St. 13, both LPl and OV start to invaginate. At this stage, the *Sox2*-low region of OV is displaced to the proximal side of OV that will form the outer layer of OC. (c) At St. 19, the tight association between NR and PE is established, whereas *Sox2* expression is undetectable in the latter. Note that the developing PLF cells express *Sox2*, in contrast to its absence in the mouse lens (A (f)). Bars indicate 200 μm.

Sox2 expression in the inner layer of the optic cup to promote retinal development depends on a fibroblast growth factor (FGF) signal (Vogel-Hopker et al., 2000; Ishii et al., 2009; Nishihara et al., 2012). Neither enhancers that mediate FGF-dependent *Sox2* activation nor those responsible for *Sox2* activation in specific retinal neurons (Lin et al., 2009) have been identified. Because a variety of enhancer sequences have been identified in genomic locations remote from the *Sox2* gene (Chapter 7), it is possible that some of them participate in neural retina-specific *Sox2* regulation.

Table 1 Expression of *Sox2*, *Sox1*, and *Sox3* in embryonic eye tissues of mouse and chicken embryos

		Sox2	***Sox1***	***Sox3***
Mouse				
	Optic vesicle	**Yes**	**Yes**	**Yes**
	Optic cup	**Yes**	No	No
	Lens epithelium	**Yes**	No	No
	Lens fiber	No	**Yes**	No
Chicken				
	Optic vesicle	**Yes**	Yes but low	**Yes**
	Optic cup	**Yes**	Yes but low	Yes but low
	Lens epithelium	**Yes**	**Yes**	**Yes**
	Lens fiber	**Yes**	**Yes**	**Yes**

In lens development, the initially low-level *Sox2* expression in the head ectoderm is presumably regulated by the N4 enhancer that bears many regulatory modules (Saigou et al., 2010). Once the head ectoderm is contacted by the optic vesicle, the N3 enhancer is activated (Inoue et al., 2007), which accounts for the local surge of *Sox2* expression in the forming lens placode. The N3 activity is present until the lens epithelial cell stage and is absent in the developing lens fibers. *Sox2* expression in the chicken lens fibers is activated by a specific enhancer, L, which is missing in mammalian genomes (Uchikawa et al., 2003).

TAXON-DEPENDENT EXPRESSION OF SOXB1 FACTORS AND UNIQUE CONTRIBUTION OF SOX2 IN MAMMALIAN EYE DEVELOPMENT

As in many other organs, two other SOXB1 factors, SOX1 and SOX3, have redundant functions with SOX2 when they are expressed (Wood and Episkopou, 1999; Uchikawa et al., 2011). Whereas *Sox2* is consistently expressed in the eye tissue primordia, the expression of *Sox1* and *Sox3* in the eye tissues is taxon dependent, as summarized in Table 1. Chicken is unique in that three SOXB1 factors appear to act redundantly to promote the neural retinal fate in the optic cup (Ishii et al., 2009). However, in mammals (Wood and Episkopou, 1999; Uchikawa et al., 2011) as well as in zebrafish and Xenopus (Nitta et al., 2006; Okuda et al., 2006), *Sox2* is the major *SoxB1* expressed in the neural retina. In mammalian lens development as well, *Sox2* is uniquely expressed from the lens placode stage to the lens epithelium stage, after which *Sox1* expression is activated in the lens fiber without accompanying *Sox2* expression. In other species also, the selection of *SoxB1* genes to be expressed in eye tissues is taxon dependent. (Okuda et al., 2006). This is one reason for genetic conditions that impose low-SOX2 activities in developing eye tissues to cause congenital eye abnormalities in humans, as discussed in Chapter 13. This situation allows for an evaluation of the minimal requirement of *Sox2* expression levels in the absence of other *SoxB1* expression for proper tissue development using combinations of various hypomorphic *Sox2* alleles (Taranova et al., 2006). Retinal tissues were found to withstand *Sox2* expression levels at as low as 40% of the wild-type level for their normal development, but at lower *Sox2* expression levels, retinal tissues displayed progressively more severe abnormalities.

ROLES FOR SOX2 AT DISTINCT STAGES OF RETINAL DEVELOPMENT

As a transcriptional activator, SOX2 and other SOX transcription factors do not function alone, but only combined with partner transcription factors binding to nearby genomic sites. This mechanism specifies the regulatory target genes and controls the cell state, a mechanism called the SOX–partner code (Kamachi et al., 2000; Kondoh and Kamachi, 2010; Kamachi and Kondoh, 2013), as discussed in Chapter 8. Not all partner factors required for SOX functions in eye tissues have been identified, but SOX2 functions should be understood on the basis of SOX2–partner interactions.

Cooperation of SOX2 and OTX2 to Activate *Rax* (*Rx*) in the Early Optic Vesicle

The optic vesicle originates from a specific region of anterior neural plate called the eye field, which expresses a unique combination of transcription factors that are also involved in eye development (Zuber et al., 2003). Among these transcription factors, RAX (RX), a paired-type homeodomain protein, appears to have a major role in the earliest process of retinal development (Andreazzoli et al., 1999; Ohuchi et al., 1999; Zhang et al., 2000). Indeed, *Rax*-deficient mutant mouse embryos, Xenopus embryos injected with *Rax*-specific morpholino oligonucleotides, and other related animal models show defects in optic vesicle development (Mathers et al., 1997; Andreazzoli et al., 1999; Bailey et al., 2004). It has been demonstrated that the *Rax* gene is regulated by a combination of SOX2 and OTX2, which binds to a phylogenetically conserved enhancer element containing juxtaposed binding sequences for respective factors (Danno et al., 2008). Indeed, from the stage of the prevesicle eye field to the stage immediately before contact with the surface ectoderm, the entire optic vesicle expresses *Sox2* and *Otx2* uniformly. Considering the entire loss of optic vesicle development in *Rax*-deficient embryos, SOX2–OTX2 interaction is essential for the initiation of eye development.

Cross-Repression between SOX2 and OTX2 to Differentiate Neural Retina and Pigmented Epithelium in the Optic Cup

After contact of the optic vesicle with the head ectoderm, *Sox2* expression is downregulated in the dorsal aspect of the vesicle, which corresponds to the future outer layer of the optic cup (Figure 2(B-a,b)), whereas *Otx2* expression is downregulated in other domains of the invaginating optic vesicle (Bovolenta et al., 1997; Martinez-Morales et al., 2001; Ishii et al., 2009; Nishihara et al., 2012). These complementary processes eventually generate distinct types of cells in the inner and outer layers of the optic cup, that is, the *Sox2*-expressing thickened neural retina and the *Sox2*-negative and *Otx2*-expressing pigmented epithelium (Figure 2(A-d,e) and (B-c)). Experimental evidence indicates that *Sox2* and *Otx2* cross-repress each other, producing distinct neural and pigmented layers (Ishii et al., 2009; Nishihara et al., 2012). More specifically, inhibition of the OTX2 function in the pigmented epithelial layer transforms the cells into the neural retina, concomitant with activation of *Sox2* and *Fgf8*, which promotes neural retina development (Nishihara et al., 2012). Consistently, *Otx2*-hypomorphic mouse embryos develop neural retinal tissues in the outer layer of the optic cup (Martinez-Morales et al., 2001). Conversely, overexpression of SOX2 in the pigmented epithelial layer inhibits expression of OTX2 and MITF1, a downstream transcription factor of OTX2 (Martinez-Morales et al., 2001), and elicits neural retina development in the outer layer (Ishii et al., 2009) (Figure 3).

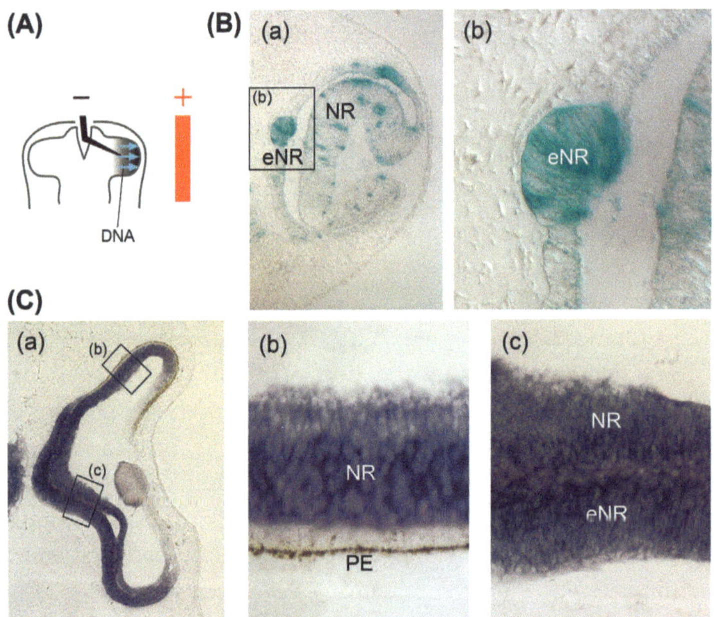

Figure 3 Ectopic expression of *Sox2* or *Sox2*-activating FGF in the pigmented epithelium (PE) transforms PE into neural retina. (A) Electroporation into the optic vesicle of St. 11 chicken embryo to express *Sox2* or *Fgf4* locally. (B) Effect of *Sox2* expression examined at St. 19. Electroporated cells were stained blue using X-gal. (a) Expression of *Sox2* causes development of ectopic neural retina (eNR) in the outer layer of the optic cup, which otherwise develops into PE. (b) Enlargement of the rectangle area in (a). (C) Effect of *Fgf4* expression in the posterior side of the optic vesicle examined at St. 22 by *Sox2* in situ hybridization using a horizontal section. (a) Anterior region of the PE that did not receive the *Fgf4* vector developed normally, as shown by the rectangle (b), whereas the posterior domain with FGF4 expression was transformed into eNR, as exemplified by the area in the rectangle (c), that expressed *Sox2* strongly and gained the tissue thickness of a normal neural retina (NR). (b) and (c) Enlargement of rectangle areas in (a). *Data reproduced from Ishii et al. (2009).*

Counteraction of SOX2 and PAX6 in the Proximo-Distal Differentiation of the Retinal Inner Layer

PAX6 is known as an essential transcription factor for overall eye development. *Pax6*-deficient mouse mutant embryos develop optic vesicles that may be enlarged, but the vesicles do not develop further (Hill et al., 1991; Walther and Gruss, 1991; Grindley et al., 1995). Apart from this early function, PAX6 and SOX2 counteract each other to determine the area of ciliary epithelium located at the distal end of the neural retina (Matsushima et al., 2011). In the optic cup, *Sox2* expression has a shallow gradient, with its lowest level in the distal margin (Figure 2(A-e)), whereas *Pax6* expression exhibits a complementary gradient (Matsushima et al., 2011). When the *Sox2* gene was ablated in the distal half of the neural retina (with Cre using the specificity of the *Pax6* P0 enhancer), strong *Pax6* expression was expanded to the *Sox2*-ablated region and the region developed into the ciliary epithelium, rather than the neural retina, in the postnatal period (Matsushima et al., 2011). These observations indicate that *Pax6* expression is suppressed under a high level of SOX2, and that the high PAX6/low SOX2 condition that occurs at the distal margin of the optic cup inner layer directs the cells to develop into the ciliary epithelium.

ROLES FOR SOX2 IN REGULATING RPC MAINTENANCE INVOLVING NOTCH1 ACTIVATION AND IN THE CELL IDENTITY OF GANGLION CELLS, A SUBSET OF AMACRINE CELLS, AND MÜLLER CELLS

Retinal neuronal cells differentiate from RPCs that occur in sequence (with temporal overlap), and from ganglion cells, horizontal cells, cones, amacrine cells, and rods, to the bipolar cells; this process is followed by differentiation of the retina-unique glia, Müller cells (Cepko et al., 1996; Heavner and Pevny, 2012), resulting in the development of beautifully arranged cell layers (Figure 4). During this period, SOX2 is expressed in RPCs, ganglion cells, a subset of amacrine cells, and Müller cells (Le Rouedec et al., 2002; Taranova et al., 2006; Lin et al., 2009).

Conditional inactivation of *Sox2* in RPCs results in total loss of cell proliferation and cell differentiation that normally occur in a cell-autonomous manner (Taranova et al., 2006), indicating an essential function of SOX2 in the regulation of RPCs. An important mechanism in RPC regulation is the direct activation of *Notch1* by SOX2 (Taranova et al., 2006). Many studies have established that Notch signaling promotes RPC maintenance and Müller cell differentiation, whereas Notch signal downregulation promotes differentiation of neuronal cells (Austin et al., 1995; Ahmad et al., 1997; Henrique et al., 1997; Jadhav et al., 2006; Nelson et al., 2007).

Müller cells maintain their RPC potential, and upon injury of retinal tissue, they resume cell proliferation and produce multiple neuronal cell types (Fischer and Reh, 2001; Goldman, 2014). Therefore, Müller cells can be regarded as RPC storage in the form of differentiated glial cells. Expression of *Sox2* in the Müller cells is consistent with this notion; indeed, ablation of the *Sox2* gene in Müller cells resulted in the loss of RPC characteristics (Surzenko et al., 2013).

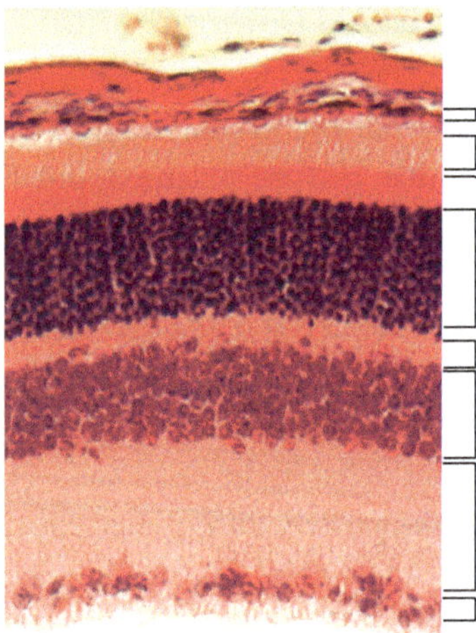

Figure 4 Anatomy of mature retina and arrangement of seven cell types of neural retina. Outer nuclear layers contain nuclei of photoreceptors, rods, and cones, and inner nuclear layers contain nuclei of horizontal cells, bipolar cells, amacrine cells, and Müller cells. Although Müller cells have their nuclei in the inner nuclear layer, the cell body spans all layers of neurons to the external limiting membrane. Plexiform layers consist primarily of neuritis of the neurons. *Photograph reproduced from Kondoh (2002).*

After Notch signal downregulation to produce retinal neurons, the second phase of the SOX2-dependent process occurs, promoting the development of ganglion cells and a cholinergic subset of amacrine cells (Lin et al., 2009). In the *Sox2*-hypomorphic mouse retina (<30% of the normal *Sox2* expression level), the development of the ganglion cells was severely impaired (Taranova et al., 2006). On the other hand, overexpression of SOX2 in the E17 retina explant caused overproduction of cholinergic amacrine cells at ectopic positions (Lin et al., 2009).

COOPERATION OF SOX2 AND PAX6 IN LENS DEVELOPMENT

Involvement of SOX in lens development was first indicated by its participation in *δ-crystallin* gene activation. *δ-Crystallin* in the chicken is activated at the lens placode stage and provides a good indication of lens states of cell differentiation (Kamachi et al., 1998). The *δ-crystallin* gene was correctly regulated after introduction into mouse cells in culture and into developing mouse embryos (Kondoh et al., 1983, 1987; Takahashi et al., 1988) although the mouse lacks the endogenous *δ-crystallin* gene. This gene is regulated by a lens-specific enhancer located in the third intron (Hayashi et al., 1987), and trimming of nonessential regions of the enhancer sequence (Goto et al., 1990; Kamachi and Kondoh, 1993) left a minimal lens-specific enhancer element of 35 bp, called DC5, that contained two juxtaposed deoxyribonucleic acid (DNA) elements bound by different transcription factors. One of these factors was identified as SOX2 (and related SOX1 and SOX3 in chicken embryos) (Kamachi et al., 1995), and the second as PAX6 (Kamachi et al., 2001). Binding of only one of these factors did

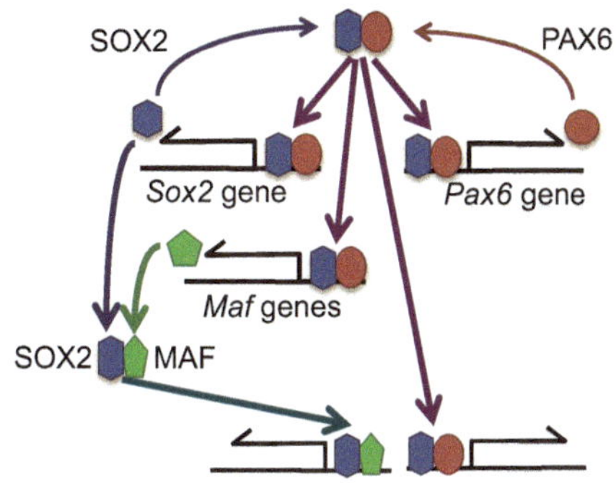

Figure 5 Co-autoregulatory loop between *Sox2* and *Pax6* and downstream regulatory cascades that determine the lens cell state. Heterodimers of SOX2 and PAX6 bind to the enhancer elements of both respective genes, maintain activation of *Sox2* and *Pax6*, and stabilize the lens state. The SOX2-PAX6 heterodimers also activate *Maf* transcription factor genes. Both SOX2-PAX6 and SOX2-MAF heterodimers participate in the activation of *crystallin* genes, the *N-cadherin* gene, and other lens characteristic genes.

not activate the DC5 enhancer, and their simultaneous binding to the DC5 DNA was required to activate the enhancer (Kamachi et al., 1995, 2001). In fact, SOXB1 transcription factors and PAX6 are always expressed together throughout the stages of lens development (Kamachi et al., 1998).

Also, this observation, together with analysis of the *Fgf4* enhancer activated by SOX2 and POU5F1 (OCT3/4) (Yuan et al., 1995), provided the first experimental evidence for the model that Sox proteins do not function alone, but in combination with partner transcription factors (Kamachi et al., 2000; Kondoh and Kamachi, 2010; Kamachi and Kondoh, 2013), as discussed in greater detail in Chapter 8.

Another class of crystallin genes expressed in the lens fiber stage, *γ-crystallins*, was also found to be activated by a combination of SOXB1 and the second factor (Kamachi et al., 1995), which was identified as c-MAF (Rajaram and Kerppola, 2004). c-MAF in the lens is activated by PAX6 (Sakai et al., 2001) and L-Maf in the chicken lens is expressed downstream of the SOX2-PAX6 combination (Ogino et al., 2012). It was also found that the N3 enhancer of the *Sox2* gene that activates *Sox2* in the lens placode and lens vesicle (Uchikawa et al., 2003) is activated by the combination of SOX2 and PAX6 (Inoue et al., 2007). Furthermore, *Pax6* expression in the lens is regulated by the upstream lens epithelium (LE) enhancer, which is activated by the combined action of SOX2 and PAX6 (Aota et al., 2003). These observations indicated the model in which SOX2 and PAX6, in combination, initiate and maintain the lens state, and also activate MAF transcription factors. SOX2-PAX6 and SOX2-MAF pairs then function together to activate N-cadherin and other lens-characteristic genes (Figure 5).

The contribution of SOX2-PAX6 interaction to initiating lens development is supported by the following observations. When the lateral aspect of the head ectoderm is already expressing low levels of *Sox2* and *Pax6*, both *Sox2* and *Pax6* are strongly

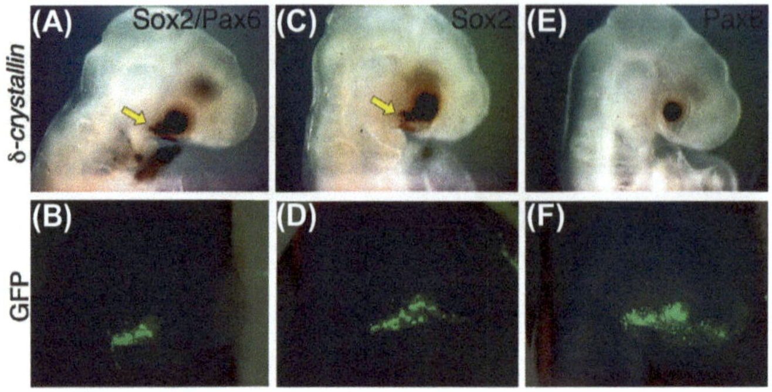

Figure 6 Ectopic expression of *Sox2* and *Pax6* outside the eye of the head ectoderm elicits development of numerous miniature lens placodes at the ectopic sites. (A and B) When expression vectors for *Sox2* and *Pax6* were electroporated into the surface ectoderm at St. 11, with labeling of electroporated cells by green fluorescent protein (GFP) (B), many islets of lens placodal cells that are characterized by thickened placodal morphology (Kamachi et al., 2001) and *δ-crystallin* expression were detected by in situ hybridization (A) (yellow arrow). (C and D) The result was similar when the *Sox2* expression vector was electroporated alone (yellow arrow), because at St. 11, *Pax6* expression is still ongoing in the surface ectoderm, and via the mechanism schematized in Figure 5, exogenous SOX2 activates the endogenous *Pax6* gene, causing the same effects as *Sox2–Pax6* co-electroporation (Kamachi et al., 2001). (E and F) By contrast, with electroporation of *Pax6* alone, no ectopic lens placodes developed. Note that the cells electroporated with *Sox2* plus *Pax6* (B) or with *Sox2* (D) (marked by GFP fluorescence) tend to form tight clusters, owing to *N-cadherin* activation, in contrast to the scattered distribution of GFP-expressing cells in (F), electroporated with *Pax6*. *Adapted from Kamachi et al. (2001).*

activated in the contacted area of the head ectoderm. *Sox2* activation is ascribed to N3 enhancer activation (Inoue et al., 2007), whereas *Pax6* activation is possibly mediated by the SOX2-dependent LE enhancer of *Pax6* that responds to SOX2 activation. Indeed, forced expression of SOX2 in the *Pax6*-expression domain of the head ectoderm, outside the eye region, elicited the development of the lens placode tissues that expressed δ-crystallin (Kamachi et al., 2001) (Figure 6). Stage-focused inactivation of *Sox2* or *Pax6* using conditional alleles confirmed that expression of *Sox2* and *Pax6* becomes interdependent only after the lens placode stage (Smith et al., 2009).

Involvement of SOXB1-MAF interactions in the later stages of lens development are in fact indicated by the phenotypes of respective mutant mouse embryos: Mutations inactivating *c-Maf* or *Sox1* in the mouse genome resulted in the arrest of lens development at the lens vesicle stage (Nishiguchi et al., 1998; Kawauchi et al., 1999; Kim et al., 1999). All of these observations support the model shown in Figure 5.

MECHANISMS UNDERLYING LENS TRANSDIFFERENTIATION AND LENS REGENERATION

The major pathway of transcriptional regulation for lens development as summarized in Figure 5 also applies to many cases of lens development that occur in contexts other than normal eye development. A typical example is lens regeneration after lentectomy from the dorsal part of the iris as observed in newts (Hayashi et al., 2008). Lens regeneration proceeds in two discrete steps: (1) FGF2-dependent alteration of the entire periphery of the iris into the lens vesicle-like state of normal eye development (Hayashi et al., 2004), and (2) Wnt signal-dependent lens fiber development occurring only in the dorsal part of the iris, where the Wnt signal is available (Hayashi et al., 2006). In the first step, SOX2, PAX6, and other transcription factors expressed in the embryonic lens vesicle are activated.

The adenohypophysis precursor (Rathke's pouch) transdifferentiates into the lens when the hedgehog signal fails to act on the precursor. Known examples are *yot* and *iguana* mutants of zebrafish (Kondoh et al., 2000) (Figure 7(A) and (B)), coding for dominant-negative GLI2 (Karlstrom et al., 1999) and a zinc finger protein required for hedgehog signaling (Sekimizu et al., 2004; Wolff et al., 2004), respectively, and also the *Talpid3* mutant of the chicken (Ede and Kelly, 1964; Davey et al., 2006). Adenohypophysis precursors strongly express *Sox2* and *Pax6* (Kamachi et al., 1998), together with additional transcription factor genes such as *Six3* and *Pitx3* (Kondoh et al., 2000) and even express a low level of δ-crystallin (Barabanov 1977; Kamachi et al., 1998); however, they do not develop into the lens in normal embryogenesis. Under the normal process of adenohypophysis development, SOX2 and PAX6 presumably interact with many other transcription factors that are involved in adenohypophyseal development, resulting in the limited occasions of SOX2-PAX6 pairing that may be sufficient for low-level activation of the *δ-crystallin* gene, but insufficient for overt lens development. However, when hedgehog signaling is inhibited, those transcription factors that are dedicated to adenohypophysis development are lost, resulting in a strong interaction of SOX2 and PAX6 that initiates lens development after the regulatory steps of lens development, as summarized in Figure 7(C).

As discussed above, the head ectoderm expresses a low level of *Sox2* and *Pax6* before contact by the optic vesicle. Isolated pieces of the head surface ectoderm can give rise to lens tissue when isolated and placed under appropriate tissue culture conditions (Sullivan et al., 2004; Bailey et al., 2006). In culture, the expression levels of *Sox2* and *Pax6* gradually increase, presumably by operation of the process summarized in Figure 5. However, this is a slow process requiring several days before lens development, compared with the normal lens development that proceeds in half a day in the case of the chicken embryo (Kondoh, 2010). In addition, lens development from isolated head ectoderm pieces is completely suppressed when combined with cephalic neural crest cells (Bailey et al., 2006; Grocott et al., 2011) that migrate underneath the head ectoderm, leaving the area contacted by the optic vesicle (Sullivan et al., 2004). Therefore, it is likely that the inhibitory effect of the cephalic neural crest on the lens development, mediated by transforming growth factor-β and Wnt signaling (Grocott et al., 2011), secures development of lenses only in association with retinal tissue. In other words, contact of the optic vesicle with the head ectoderm has two different roles: quick activation of high-level *Sox2* expression confined to the contacted area of the head ectoderm, and exclusion of the cephalic neural crest cells, which may otherwise interfere with normal lens development, from the contacted area.

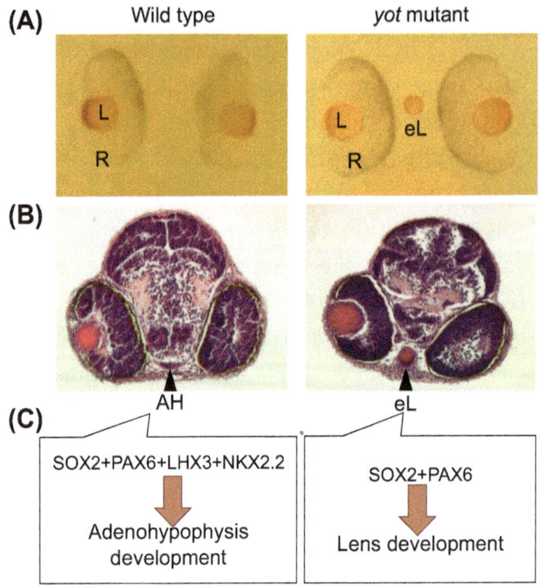

Figure 7 Ectopic lens development from adenohypophysis precursors in zebrafish mutants deficient in hedgehog signaling. At 72 h after fertilization, *yot* mutant embryos develop ectopic lenses (eL) on the midline at the expense of the adenohypophysis precursor. (A) Whole-mount staining of embryos using anti-β-crystallins. In contrast to the wild-type embryo (left) with a pair of β-crystallin–expressing lenses (L) surrounded by the retina (R), the *yot* mutant embryo (right) developed an eL on the midline. (B) Hematoxylin–eosin-stained cross-sections across the eyes. The eL of the *yot* mutant developed, replacing AH found in the wild-type embryo. (C) Model of interactions of transcription factors that promote adenohypophysis development in wild-type and lens development in *yot* mutants. *Adapted from Kondoh et al. (2000).*

REFERENCES

Ahmad, I., Dooley, C.M., Polk, D.L., 1997. Delta-1 is a regulator of neurogenesis in the vertebrate retina. Dev. Biol. 185, 92–103.

Andreazzoli, M., Gestri, G., Angeloni, D., Menna, E., Barsacchi, G., 1999. Role of Xrx1 in Xenopus eye and anterior brain development. Development 126, 2451–2460.

Aota, S., Nakajima, N., Sakamoto, R., Watanabe, S., Ibaraki, N., Okazaki, K., 2003. Pax6 autoregulation mediated by direct interaction of Pax6 protein with the head surface ectoderm-specific enhancer of the mouse Pax6 gene. Dev. Biol. 257, 1–13.

Austin, C.P., Feldman, D.E., Ida Jr., J.A., Cepko, C.L., 1995. Vertebrate retinal ganglion cells are selected from competent progenitors by the action of Notch. Development 121, 3637–3650.

Bailey, A.P., Bhattacharyya, S., Bronner-Fraser, M., Streit, A., 2006. Lens specification is the ground state of all sensory placodes, from which FGF promotes olfactory identity. Dev. Cell 11, 505–517.

Bailey, T.J., El-Hodiri, H., Zhang, L., Shah, R., Mathers, P.H., Jamrich, M., 2004. Regulation of vertebrate eye development by Rx genes. Int. J. Dev. Biol. 48, 761–770.

Barabanov, V.M., 1977. Detection of delta-crystallins in the adenohypophysis of chick embryos. Dokl. Akad. Nauk SSSR 234, 195–198.

Bovolenta, P., Mallamaci, A., Briata, P., Corte, G., Boncinelli, E., 1997. Implication of OTX2 in pigment epithelium determination and neural retina differentiation. J. Neurosci. 17, 4243–4252.

Cepko, C.L., Austin, C.P., Yang, X., Alexiades, M., Ezzeddine, D., 1996. Cell fate determination in the vertebrate retina. Proc. Natl. Acad. Sci. U.S.A. 93, 589–595.

Chauhan, B.K., Disanza, A., Choi, S.Y., Faber, S.C., Lou, M., Beggs, H.E., Scita, G., Zheng, Y., Lang, R.A., 2009. Cdc42- and IRSp53-dependent contractile filopodia tether presumptive lens and retina to coordinate epithelial invagination. Development 136, 3657–3667.

Coulombre, A.J., Coulombre, J.L., 1964. Lens development. I. Role of the lens in eye growth. J. Exp. Zool. 156, 39–47.

Danno, H., Michiue, T., Hitachi, K., Yukita, A., Ishiura, S., Asashima, M., 2008. Molecular links among the causative genes for ocular malformation: Otx2 and Sox2 coregulate Rax expression. Proc. Natl. Acad. Sci. U.S.A. 105, 5408–5413.

Davey, M.G., Paton, I.R., Yin, Y., Schmidt, M., Bangs, F.K., Morrice, D.R., Smith, T.G., Buxton, P., Stamataki, D., Tanaka, M., Munsterberg, A.E., Briscoe, J., Tickle, C., Burt, D.W., 2006. The chicken talpid3 gene encodes a novel protein essential for Hedgehog signaling. Genes Dev. 20, 1365–1377.

Ede, D.A., Kelly, W.A., 1964. Developmental abnormalities in the head region of the talpid3 mutant of the fowl. J. Embryol. Exp. Morphol. 12, 161–182.

Fischer, A.J., Reh, T.A., 2001. Muller glia are a potential source of neural regeneration in the postnatal chicken retina. Nat. Neurosci. 4, 247–252.

Goldman, D., 2014. Muller glial cell reprogramming and retina regeneration. Nat. Rev. Neurosci. 15, 431–442.

Goto, K., Okada, T.S., Kondoh, H., 1990. Functional cooperation of lens-specific and nonspecific elements in the delta 1-crystallin enhancer. Mol. Cell. Biol. 10, 958–964.

Grindley, J.C., Davidson, D.R., Hill, R.E., 1995. The role of Pax-6 in eye and nasal development. Development 121, 1433–1442.

Grocott, T., Johnson, S., Bailey, A.P., Streit, A., 2011. Neural crest cells organize the eye via TGF-beta and canonical Wnt signalling. Nat. Commun. 2, 265.

Hayashi, S., Goto, K., Okada, T.S., Kondoh, H., 1987. Lens-specific enhancer in the third intron regulates expression of the chicken delta 1-crystallin gene. Genes Dev. 1, 818–828.

Hayashi, T., Mizuno, N., Kondoh, H., 2008. Determinative roles of FGF and Wnt signals in iris-derived lens regeneration in newt eye. Dev. Growth Differ. 50, 279–287.

Hayashi, T., Mizuno, N., Takada, R., Takada, S., Kondoh, H., 2006. Determinative role of Wnt signals in dorsal iris-derived lens regeneration in newt eye. Mech. Dev. 123, 793–800.

Hayashi, T., Mizuno, N., Ueda, Y., Okamoto, M., Kondoh, H., 2004. FGF2 triggers iris-derived lens regeneration in newt eye. Mech. Dev. 121, 519–526.

Heavner, W., Pevny, L., 2012. Eye development and retinogenesis. Cold Spring Harbor Perspect. Biol. 4, a008391.

Henrique, D., Hirsinger, E., Adam, J., Le Roux, I., Pourquie, O., Ish-Horowicz, D., Lewis, J., 1997. Maintenance of neuroepithelial progenitor cells by Delta-Notch signalling in the embryonic chick retina. Curr. Biol. 7, 661–670.

Hill, R.E., Favor, J., Hogan, B.L., Ton, C.C., Saunders, G.F., Hanson, I.M., Prosser, J., Jordan, T., Hastie, N.D., van Heyningen, V., 1991. Mouse small eye results from mutations in a paired-like homeobox-containing gene. Nature 354, 522–525.

Hyer, J., Kuhlman, J., Afif, E., Mikawa, T., 2003. Optic cup morphogenesis requires pre-lens ectoderm but not lens differentiation. Dev. Biol. 259, 351–363.

Inoue, M., Kamachi, Y., Matsunami, H., Imada, K., Uchikawa, M., Kondoh, H., 2007. PAX6 and SOX2-dependent regulation of the Sox2 enhancer N-3 involved in embryonic visual system development. Genes Cells 12, 1049–1061.

Ishii, Y., Weinberg, K., Oda-Ishii, I., Coughlin, L., Mikawa, T., 2009. Morphogenesis and cytodifferentiation of the avian retinal pigmented epithelium require downregulation of Group B1 Sox genes. Development 136, 2579–2589.

Iwafuchi-Doi, M., Matsuda, K., Murakami, K., Niwa, H., Tesar, P.J., Aruga, J., Matsuo, I., Kondoh, H., 2012. Transcriptional regulatory networks in epiblast cells and during anterior neural plate development as modeled in epiblast stem cells. Development 139, 3926–3937.

Iwafuchi-Doi, M., Yoshida, Y., Onichtchouk, D., Leichsenring, M., Driever, W., Takemoto, T., Uchikawa, M., Kamachi, Y., Kondoh, H., 2011. The Pou5f1/Pou3f-dependent but SoxB-independent regulation of conserved enhancer N2 initiates Sox2 expression during epiblast to neural plate stages in vertebrates. Dev. Biol. 352, 354–366.

Jadhav, A.P., Cho, S.H., Cepko, C.L., 2006. Notch activity permits retinal cells to progress through multiple progenitor states and acquire a stem cell property. Proc. Natl. Acad. Sci. U.S.A. 103, 18998–19003.

Kamachi, Y., Kondoh, H., 1993. Overlapping positive and negative regulatory elements determine lens-specific activity of the delta 1-crystallin enhancer. Mol. Cell. Biol. 13, 5206–5215.

Kamachi, Y., Kondoh, H., 2013. Sox proteins: regulators of cell fate specification and differentiation. Development 140, 4129–4144.

Kamachi, Y., Sockanathan, S., Liu, Q., Breitman, M., Lovell-Badge, R., Kondoh, H., 1995. Involvement of SOX proteins in lens-specific activation of crystallin genes. EMBO J. 14, 3510–3519.

Kamachi, Y., Uchikawa, M., Collignon, J., Lovell-Badge, R., Kondoh, H., 1998. Involvement of Sox1, 2 and 3 in the early and subsequent molecular events of lens induction. Development 125, 2521–2532.

Kamachi, Y., Uchikawa, M., Kondoh, H., 2000. Pairing SOX off: with partners in the regulation of embryonic development. Trends Genet. 16, 182–187.

Kamachi, Y., Uchikawa, M., Tanouchi, A., Sekido, R., Kondoh, H., 2001. Pax6 and SOX2 form a co-DNA-binding partner complex that regulates initiation of lens development. Genes Dev. 15, 1272–1286.

Karlstrom, R.O., Talbot, W.S., Schier, A.F., 1999. Comparative synteny cloning of zebrafish you-too: mutations in the Hedgehog target gli2 affect ventral forebrain patterning. Genes Dev. 13, 388–393.

Kawauchi, S., Takahashi, S., Nakajima, O., Ogino, H., Morita, M., Nishizawa, M., Yasuda, K., Yamamoto, M., 1999. Regulation of lens fiber cell differentiation by transcription factor c-Maf. J. Biol. Chem. 274, 19254–19260.

Kim, J.I., Li, T., Ho, I.C., Grusby, M.J., Glimcher, L.H., 1999. Requirement for the c-Maf transcription factor in crystallin gene regulation and lens development. Proc. Natl. Acad. Sci. U.S.A. 96, 3781–3785.

Kondoh, H., 2002. Development of the eye. In: Rossant, J., Tam, P.P.L. (Eds.), Mouse Development. Academic Press, Orland, pp. 519–538.

Kondoh, H., 2010. Lens induction. In: Encyclopedia of Life Sciences (ELS). John Wiley & Sons, Ltd., Chichester, p. a0001142.

Kondoh, H., Kamachi, Y., 2010. SOX-partner code for cell specification: regulatory target selection and underlying molecular mechanisms. Int. J. Biochem. Cell Biol. 42, 391–399.

Kondoh, H., Katoh, K., Takahashi, Y., Fujisawa, H., Yokoyama, M., Kimura, S., Katsuki, M., Saito, M., Nomura, T., Hiramoto, Y., et al., 1987. Specific expression of the chicken delta-crystallin gene in the lens and the pyramidal neurons of the piriform cortex in transgenic mice. Dev. Biol. 120, 177–185.

Kondoh, H., Uchikawa, M., Yoda, H., Takeda, H., Furutani-Seiki, M., Karlstrom, R.O., 2000. Zebrafish mutations in Gli-mediated hedgehog signaling lead to lens transdifferentiation from the adenohypophysis anlage. Mech. Dev. 96, 165–174.

Kondoh, H., Yasuda, K., Okada, T.S., 1983. Tissue-specific expression of a cloned chick delta-crystallin gene in mouse cells. Nature 301, 440–442.

Le Rouedec, D., Rayner, K., Rex, M., Wigmore, P.M., Scotting, P.J., 2002. The transcription factor cSox2 and Neuropeptide Y define a novel subgroup of amacrine cells in the retina. J. Anat. 200, 51–56.

Li, H.S., Yang, J.M., Jacobson, R.D., Pasko, D., Sundin, O., 1994. Pax-6 is first expressed in a region of ectoderm anterior to the early neural plate: implications for stepwise determination of the lens. Dev. Biol. 162, 181–194.

Lin, Y.P., Ouchi, Y., Satoh, S., Watanabe, S., 2009. Sox2 plays a role in the induction of amacrine and Muller glial cells in mouse retinal progenitor cells. Invest. Ophthalmol. Vis. Sci. 50, 68–74.

Martinez-Morales, J.R., Signore, M., Acampora, D., Simeone, A., Bovolenta, P., 2001. Otx genes are required for tissue specification in the developing eye. Development 128, 2019–2030.

Mathers, P.H., Grinberg, A., Mahon, K.A., Jamrich, M., 1997. The Rx homeobox gene is essential for vertebrate eye development. Nature 387, 603–607.

Matsumata, M., Uchikawa, M., Kamachi, Y., Kondoh, H., 2005. Multiple N-cadherin enhancers identified by systematic functional screening indicate its Group B1 SOX-dependent regulation in neural and placodal development. Dev. Biol. 286, 601–617.

Matsushima, D., Heavner, W., Pevny, L.H., 2011. Combinatorial regulation of optic cup progenitor cell fate by SOX2 and PAX6. Development 138, 443–454.

Nelson, B.R., Hartman, B.H., Georgi, S.A., Lan, M.S., Reh, T.A., 2007. Transient inactivation of Notch signaling synchronizes differentiation of neural progenitor cells. Dev. Biol. 304, 479–498.

Nishiguchi, S., Wood, H., Kondoh, H., Lovell-Badge, R., Episkopou, V., 1998. Sox1 directly regulates the gamma-crystallin genes and is essential for lens development in mice. Genes Dev. 12, 776–781.

Nishihara, D., Yajima, I., Tabata, H., Nakai, M., Tsukiji, N., Katahira, T., Takeda, K., Shibahara, S., Nakamura, H., Yamamoto, H., 2012. Otx2 is involved in the regional specification of the developing retinal pigment epithelium by preventing the expression of sox2 and fgf8, factors that induce neural retina differentiation. PLoS One 7, e48879.

Nitta, K.R., Takahashi, S., Haramoto, Y., Fukuda, M., Onuma, Y., Asashima, M., 2006. Expression of Sox1 during Xenopus early embryogenesis. Biochem. Biophys. Res. Commun. 351, 287–293.

Ogino, H., Ochi, H., Reza, H.M., Yasuda, K., 2012. Transcription factors involved in lens development from the preplacodal ectoderm. Dev. Biol. 363, 333–347.

Ohuchi, H., Tomonari, S., Itoh, H., Mikawa, T., Noji, S., 1999. Identification of chick rax/rx genes with overlapping patterns of expression during early eye and brain development. Mech. Dev. 85, 193–195.

Okuda, Y., Yoda, H., Uchikawa, M., Furutani-Seiki, M., Takeda, H., Kondoh, H., Kamachi, Y., 2006. Comparative genomic and expression analysis of group B1 sox genes in zebrafish indicates their diversification during vertebrate evolution. Dev. Dyn. 235, 811–825.

Rajaram, N., Kerppola, T.K., 2004. Synergistic transcription activation by Maf and Sox and their subnuclear localization are disrupted by a mutation in Maf that causes cataract. Mol. Cell. Biol. 24, 5694–5709.

Saigou, Y., Kamimura, Y., Inoue, M., Kondoh, H., Uchikawa, M., 2010. Regulation of Sox2 in the preplacodal cephalic ectoderm and central nervous system by enhancer N-4. Dev. Growth Differ. 52, 397–408.

Sakai, M., Serria, M.S., Ikeda, H., Yoshida, K., Imaki, J., Nishi, S., 2001. Regulation of c-maf gene expression by Pax6 in cultured cells. Nucleic Acids Res. 29, 1228–1237.

Sekimizu, K., Nishioka, N., Sasaki, H., Takeda, H., Karlstrom, R.O., Kawakami, A., 2004. The zebrafish iguana locus encodes Dzip1, a novel zinc-finger protein required for proper regulation of Hedgehog signaling. Development 131, 2521–2532.

Smith, A.N., Miller, L.A., Radice, G., Ashery-Padan, R., Lang, R.A., 2009. Stage-dependent modes of Pax6-Sox2 epistasis regulate lens development and eye morphogenesis. Development 136, 2977–2985.

Spemann, H., 1901. Ueber Correlationen in der Entwickelung des Auges. Verh. Anat. Ges. 61–79.

Sullivan, C.H., Braunstein, L., Hazard-Leonards, R.M., Holen, A.L., Samaha, F., Stephens, L., Grainger, R.M., 2004. A re-examination of lens induction in chicken embryos: in vitro studies of early tissue interactions. Int. J. Dev. Biol. 48, 771–782.

Surzenko, N., Crowl, T., Bachleda, A., Langer, L., Pevny, L., 2013. SOX2 maintains the quiescent progenitor cell state of postnatal retinal Muller glia. Development 140, 1445–1456.

Takahashi, Y., Hanaoka, K., Hayasaka, M., Katoh, K., Kato, Y., Okada, T.S., Kondoh, H., 1988. Embryonic stem cell-mediated transfer and correct regulation of the chicken delta-crystallin gene in developing mouse embryos. Development 102, 259–269.

Taranova, O.V., Magness, S.T., Fagan, B.M., Wu, Y., Surzenko, N., Hutton, S.R., Pevny, L.H., 2006. SOX2 is a dose-dependent regulator of retinal neural progenitor competence. Genes Dev. 20, 1187–1202.

Uchikawa, M., Ishida, Y., Takemoto, T., Kamachi, Y., Kondoh, H., 2003. Functional analysis of chicken Sox2 enhancers highlights an array of diverse regulatory elements that are conserved in mammals. Dev. Cell 4, 509–519.

Uchikawa, M., Yoshida, M., Iwafuchi-Doi, M., Matsuda, K., Ishida, Y., Takemoto, T., Kondoh, H., 2011. B1 and B2 Sox gene expression during neural plate development in chicken and mouse embryos: universal versus species-dependent features. Dev. Growth Differ. 53, 761–771.

Vogel-Hopker, A., Momose, T., Rohrer, H., Yasuda, K., Ishihara, L., Rapaport, D.H., 2000. Multiple functions of fibroblast growth factor-8 (FGF-8) in chick eye development. Mech. Dev. 94, 25–36.

Walther, C., Gruss, P., 1991. Pax-6, a murine paired box gene, is expressed in the developing CNS. Development 113, 1435–1449.

Wolff, C., Roy, S., Lewis, K.E., Schauerte, H., Joerg-Rauch, G., Kirn, A., Weiler, C., Geisler, R., Haffter, P., Ingham, P.W., 2004. iguana encodes a novel zinc-finger protein with coiled-coil domains essential for Hedgehog signal transduction in the zebrafish embryo. Genes Dev. 18, 1565–1576.

Wood, H.B., Episkopou, V., 1999. Comparative expression of the mouse Sox1, Sox2 and Sox3 genes from pre-gastrulation to early somite stages. Mech. Dev. 86, 197–201.

Yuan, H., Corbi, N., Basilico, C., Dailey, L., 1995. Developmental-specific activity of the FGF-4 enhancer requires the synergistic action of Sox2 and Oct-3. Genes Dev. 9, 2635–2645.

Zhang, L., Mathers, P.H., Jamrich, M., 2000. Function of Rx, but not Pax6, is essential for the formation of retinal progenitor cells in mice. Genesis 28, 135–142.

Zuber, M.E., Gestri, G., Viczian, A.S., Barsacchi, G., Harris, W.A., 2003. Specification of the vertebrate eye by a network of eye field transcription factors. Development 130, 5155–5167.

CHAPTER 13

Congenital Abnormalities and *SOX2* Mutations

Veronica van Heyningen
MRC Human Genetics Unit, Institute of Genetics and Molecular Medicine, University of Edinburgh, Edinburgh, GBR

IDENTIFICATION OF *SOX2* AS A KEY GENE MUTATED IN BILATERAL ANOPHTHALMIA AND SEVERE MICROPHTHALMIA CASES

Eye development is a complex process (see Chapter 12). Charles Darwin had noted that the eye is an organ of "extreme perfection and complication," meaning that it needs to be finely tuned to fulfill its function correctly (Darwin, 1859). It is therefore not surprising that abnormalities of the eye are cumulatively a relatively common phenomenon.

Anophthalmia (absence of the ocular globe) is one of the most severe ocular developmental malformations. Microphthalmia can be nearly equally deleterious functionally in terms of vision loss (Morrison et al., 2002; Verma and Fitzpatrick, 2007). Relatedness of the two phenotypes is clearly demonstrated by the fact that in a single individual there may be unilateral absence of the globe while the other eye presents with microphthalmia. Identification of the single-exon *SOX2* gene (Fantes et al., 2003) as a major mutational target implicated in these phenotypes was an exciting early finding in the field. Previously, the only microphthalmia-associated genes identified had been the homeobox transcription factor (TF) *VSX2* (previously *CHX10*), a rare cause of recessive or biallelic microphthalmia (Ferda Percin et al., 2000). Other genes identified around the same time or subsequently include the TF *RAX* (biallelic) (Voronina, 2003) and *OTX2* (heterozygous, haploinsufficient) (Ragge et al., 2005a). With the advent of improved gene finding, a number of additional anophthalmia and microphthalmia coloboma genes have been identified (Williamson and FitzPatrick, 2014). In addition to developmental regulatory TFs, several genes belonging to the retinoic acid synthesis pathway are mutated in some severe developmental eye abnormalities (Williamson and FitzPatrick, 2014). A comprehensive list of microphthalmia/anophthalmia genes can also be found at [http://ghr.nlm.nih.gov/condition/microphthalmia and http://ghr.nlm.nih.gov/search?query=anophthalmia&Search=], with links to the relevant literature. As increasing numbers of anophthalmia, microphthalmia, and coloboma cases have been subjected to the most recent types of mutation analysis, encompassing sequencing, including whole exome sequencing, and copy number variation analysis by array-based comparative genomic hybridization, *SOX2* has emerged as the most frequently mutated gene in bilateral anophthalmia and severe microphthalmia. Forty percent of these severe malformation cases

Sox2
http://dx.doi.org/10.1016/B978-0-12-800352-7.00013-X

have heterozygous *SOX2* mutations leading to a severe loss of function (Gerth-Kahlert et al., 2013; Chassaing et al., 2013; Williamson and FitzPatrick, 2014).

DECIPHERING THE ROLE OF *SOX2* IN DEVELOPMENT AND DISEASE

The first case implicating *SOX2* was found to carry a deletional translocation in which one copy of the single-exon *SOX2* gene was lost, which suggested a dosage-sensitive role for this gene in eye development (Fantes et al., 2003). Surprisingly, the targeted knock-out of *Sox2* in mice, produced around the time when the first anophthalmia-related *SOX2* mutation was identified in humans, did not reveal a severe eye phenotype in the heterozygotes (Avilion et al., 2003) whereas homozygotes showed early postimplantation lethality. Subsequently, however, *Sox2* dosage was shown to be critical in the mouse, too, when anophthalmia-like phenotypes were observed upon further regional reduction of *Sox2* dosage in a conditional mutant mouse (Taranova et al., 2006).

Defining the gene expression pattern for SOX2 helps to understand the complex phenotypes observed, with multiple extraocular abnormalities associated with some cases. The gene is expressed from early placode stages in eye development (Chapter 12) (Furuta and Hogan, 1998). A partial overlap of expression with Pax6 reflects the synergistic role these genes have in eye development (Figure 1). Expression in the developing surface ectoderm is followed by lens expression, whereas retinal progenitor cells also require the presence of fully functional *Sox2*. *Sox2* is significantly but not completely coexpressed with *Pax6* in the developing eye (Chapter 12).

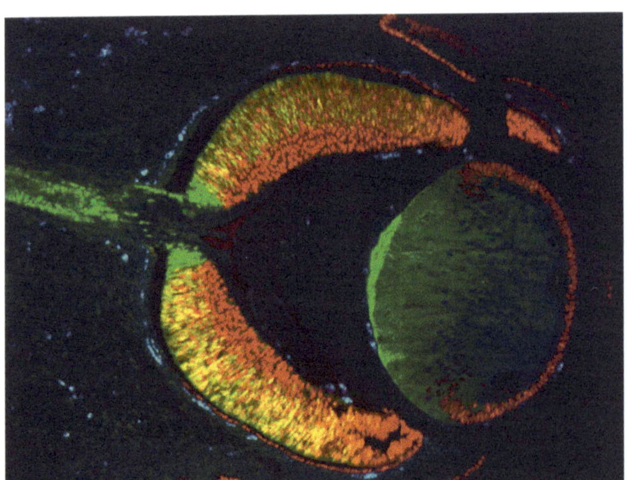

Figure 1 Expression of SOX2 (green) and PAX6 (red) in an E13.5 mouse embryonic eye, shown by immunofluorescence. The specimen was also stained with DAPI for nuclear staining. *Reproduced from Hever et al. (2006).*

Lens placode and optic cup development are mutually interdependent; if one fails, all eye development fails (Ashery-Padan et al., 2000; Stemmer et al., 2015). *Pax6* and *Sox2* autoregulate themselves and cross-regulate each other. Disruption of their multilevel interaction results in the failure of eye development (Smith et al., 2009). At least in part, they fulfill their regulatory role as a dimer, binding to closely adjacent deoxyribonucleic acid (DNA) sites (see Chapter 8) such that the PAX6 "paired" domain interacts with the SOX2 HMG domain, whereas these domains also bind the DNA binding sites. SOX2 interacts with many other homeodomain TFs, and together these TFs fulfill a myriad of finely tuned developmental tasks. Precise dosage and stoichiometry of SOX2 with the other TFs are critical for the roles they fulfill together, binding to many different target sequences to trigger enhancer (and perhaps repressor) activity. Distinct co-crystal structures of SOX2 and POU5F1 co-binding to two different target sites have helped us to define functionality of the HMG domain (Reményi et al., 2003). Many other genes are targets of SOX2 and partner genes (Sarkar and Hochedlinger, 2013): the partners often possess a homeodomain. SOX2 and PAX6, for example, regulate the expression δ-crystallin in chick, whereas SOX2 and MAF regulate γ-crystallin in mouse required for lens formation at a later stage. They also regulate the eye gene *MAF*, whose mutation also causes microphthalmia and coloboma (Jamieson et al., 2002). The predicted SOX2 structures and associated protein interaction features are summarized in Figure 2(A) (FitzPatrick, 2008).

Sox2 is best known as a gene required for totipotency in embryonic stem cells. It is also expressed in primordial germ cells, and this HMG box TF is a key regulator of neural progenitor cells. A number of other tissues require *Sox2*, particularly in proliferative phases. Multiple areas of the later developing brain, such as the hippocampus (Sisodiya et al., 2006), express *Sox2* as well foregut endodermal tissues, including the trachea and esophagus (Chapter 17) (Sarkar and Hochedlinger, 2013). The non-ocular anomalies frequently associated with *SOX2* mutation in humans reflect this diverse expression pattern. Thus, learning disability and seizures are frequently present in *SOX2* mutation patients (Fantes et al., 2003; Ragge et al., 2005b; Sisodiya et al., 2006; Bardakjian and Schneider, 2011); a smaller proportion of cases present with tracheoesophageal fistula requiring urgent surgery neonatally (Williamson et al., 2006; Bardakjian and Schneider, 2005; Slavotinek, 2011). Males are frequently reported to have cryptorchidism and hypospadias (Williamson and FitzPatrick, 2014; FitzPatrick, 2008), which may be responsible for the observed low frequency of vertical transmission for heterozygous *SOX2* mutations. Female fertility may also be compromised because SOX2 is expressed in primordial germ cells of both (Chapter 14) ovary and testis (Sarkar and Hochedlinger, 2013). Postnatal growth retardation is another *SOX2* mutation-associated feature with pituitary endocrine deficiency, reflecting the key role of *Sox2* in the development of the hypothalamo–pituitary–gonadal axis (Kelberman et al., 2006; Jayakody et al., 2012). The complexity of *Sox2* expression in so many different tissues is thought to be regulated by a large number

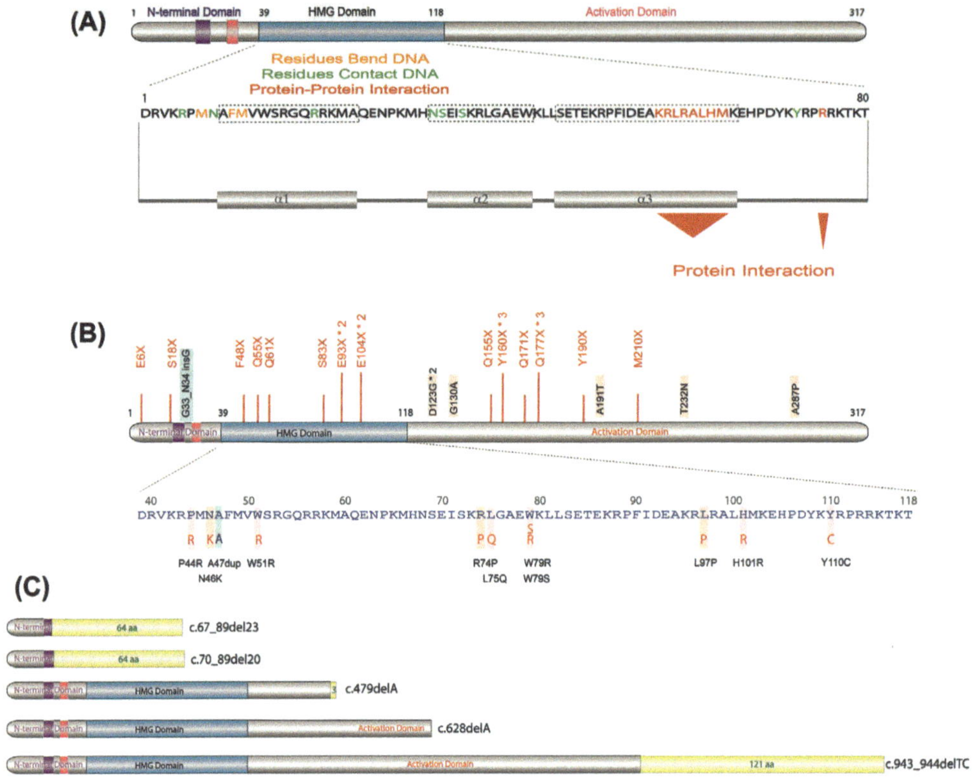

Figure 2 *SOX2*-encoded protein domains and human mutations causing congenital abnormalities. (A) SOX2 protein domain organizations. (B) *SOX2* point mutations causing ocular abnormalities. Direct stop codon mutations are shown in red above the drawing of the protein. Most missense changes are found in the HMG domain sequence depicted in blue below the protein. (C) Examples of frame-shift mutants of *SOX2* with the predicted encoded proteins. *Revised version, after FitzPatrick (2008).*

of different enhancers present in the highly conserved gene-flanking noncoding region around the single-exon gene (Chapter 7). The *Sox2* gene itself lies within an intron of a spliced, polyadenylated noncoding ribonucleic acid (RNA) gene, *Sox2-OT* (Fantes et al., 2003; Kamachi et al., 2009). The functions of this RNA gene are not known.

THE SPECTRUM OF *SOX2* MUTATIONS IN OCULAR MALFORMATION AND RELATED ANOMALIES

Many groups around the world have reported *SOX2* mutations in cases of eye disease. These mutations are collated in a standardized database (Fokkema et al., 2011) (http://lsdb.hgu.mrc.ac.uk/home.php?select_db=SOX2). Currently, 121 independent mutations have been recorded with phenotypic information where available and linked to the relevant publications. A total of 75 variants are unique, which suggests a high rate of de novo mutations. Deletions of the gene clearly lead to a loss of gene expression from one allele. We know that this is a frequent mechanism of mutation (Williamson and FitzPatrick, 2014). Other types of intragenic changes are documented in Figure 2(B) and (C). Most mutations give rise to predicted protein truncation. This can arise by direct stop codon mutations, of which 20 examples are shown in Figure 2(B). Some have arisen repeatedly; others are unique changes. One of the most frequent categories of mutation is frame-shift, in which deletion or insertion of a number of nucleotides occurs that is not divisible by 3, so a predicted out-of-frame protein is produced in some cases, with an extension composed of alien amino acids. Occasionally a frame-shift mutation can produce a direct stop. In multi-exonic proteins, premature termination codons can trigger nonsense-mediated decay (Miller and Pearce, 2014) and so can give rise to pure haploinsufficiency. However, single-exon genes such as *SOX2* are not subject to this mechanism of aberrant RNA destruction. Thus, we believe that truncated proteins are made. No clear evidence has been provided for this assumption because SOX2-producing cells are not readily accessible.

The final category of mutations is missense changes in which, usually through a single nucleotide change, an abnormal amino acid is inserted. These are merely the predicted outcomes of mutation; in most cases, we do not know for sure what happens to the predicted protein. The proteins with missense changes are expected to be produced and may exert some effect on function: for example, by misfolding, which in turn could lead to accelerated protein clearance and haploinsufficiency. Alternatively, the altered protein may have some deleterious gain of function or simply altered stability and turnover, which can give rise to phenotypic effects. Protein truncation can also result in abnormal loss of a functional subdomain: for example, when the activation domain is lost (Figure 2(C)); such fragments can exert adverse dominant negative effects.

LIMITED GENOTYPE–PHENOTYPE CORRELATIONS

Generally, there is no clear correlation between the type of mutation observed and the phenotype in patients. As discussed, most of the bilateral anophthalmia and severe microphthalmia cases carry severe loss of function changes in SOX2 (Gerth-Kahlert et al., 2013; Williamson and FitzPatrick, 2014; Chassaing et al., 2013; Schilter et al., 2013; Mauri et al., 2015). When families have been studied, do novo mutations causing variant phenotypes between generations have been observed. One mild missense change (D123G in Figure 2(B)) has been reported twice in families with vertical transmission and some relatively mild phenotypes, including iris coloboma, more reminiscent of *PAX6* mutations (Mihelec et al., 2009; Gerth-Kahlert et al., 2013). The mutation lies in the region just C-terminal to the HMG domain. In one case in which a child with anophthalmia/microphthalmia inherited the mutation from her mother with a milder phenotype (Figure 3, middle row), the mutation arose de novo in the mother (Gerth-Kahlert et al., 2013), which supports the deduction that this is a disease-causing mild missense

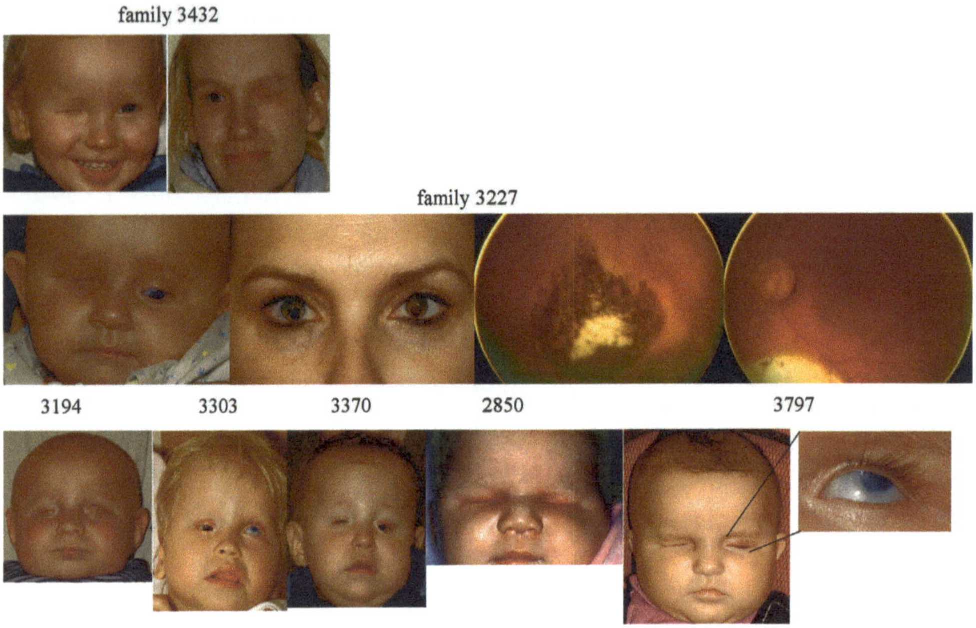

Figure 3 Spectrum of ocular abnormalities caused by *SOX2* mutations. *Reproduced from Gerth-Kahlert et al. (2013).*

mutation. Interestingly, in the Gerth–Kahlert cohort of cases, there is a second incidence of vertical transmission; in this case, the most commonly recurring *SOX2* mutation was identified in a mother and son. This is a frame-shift with a deletion starting in the repetitive glycine-rich region of SOX2 (c.70_89del20), leading to premature termination (Figure 2(C)). In the earlier reported family, both mild and severe phenotypes and even nonpenetrance are shown to segregate within the family (Mihelec et al., 2009).

REFERENCES

Ashery-Padan, R., et al., 2000. Pax6 activity in the lens primordium is required for lens formation and for correct placement of a single retina in the eye. Genes Dev. 14 (21), 2701–2711. Available at: http://www.ncbi.nlm.nih.gov/pubmed/11069887.

Avilion, A.A., et al., 2003. Multipotent cell lineages in early mouse development on SOX2 function. Genes Dev. 17, 126–140. Available at: http://www.ncbi.nlm.nih.gov/pubmed/12514105.

Bardakjian, T.M., Schneider, A., 2005. Association of anophthalmia and esophageal atresia: four new cases identified by the anophthalmia/microphthalmia clinical registry. Am. J. Med. Genet. 132A (1), 54–56.

Bardakjian, T.M., Schneider, A., 2011. The genetics of anophthalmia and microphthalmia. Curr. Opin. Ophthalmol. 22 (5), 309–313.

Chassaing, N., et al., 2013. Molecular findings and clinical data in a cohort of 150 patients with anophthalmia/microphthalmia. Clin. Genet. 86 (4), 326–334.

Darwin, C.R., 1859. On the Origin of Species. John Murray, London.

Fantes, J., et al., 2003. Mutations in SOX2 cause anophthalmia. Nat. Genet. 33 (4), 461–463. Available at: http://www.ncbi.nlm.nih.gov/pubmed/12612584 (accessed 01.10.13).

Ferda Percin, E., et al., 2000. Human microphthalmia associated with mutations in the retinal homeobox gene CHX10. Nat. Genet. 25 (4), 397–401. Available at: http://www.ncbi.nlm.nih.gov/pubmed/10932181 (accessed 12.04.15).

FitzPatrick, D.R., 2008. SOX2 and malformations of the eye, brain, foregut and male genital tract. In: Epstein, C., Erickson, R., Wynshaw-Boris, A. (Eds.), Inborn Errors of Development. Oxford University Press, New York, pp. 894–900.

Fokkema, I.F., et al., 2011. LOVD v.2.0: the next generation in gene variant databases. Hum. Mutat. 32 (5), 557–563.

Furuta, Y., Hogan, B.L.M., 1998. BMP4 is essential for lens induction in the mouse embryo. Genes Dev. 12 (23), 3764–3775. Available at: http://www.ncbi.nlm.nih.gov/pubmed/985198.

Gerth-Kahlert, C., et al., 2013. Clinical and mutation analysis of 51 probands with anophthalmia and/or severe microphthalmia from a single center. Mol. Genet. Genomic Med. 1 (1), 15–31. Available at: http://doi.wiley.com/10.1002/mgg3.2 (accessed 06.10.13).

Hever, A., et al., 2006. Developmental malformations of the eye: the role of *PAX6, SOX2* and *OTX2*. Clin. Genet. 69 (6), 459–470.

Jamieson, R.V., et al., 2002. Domain disruption and mutation of the bZIP transcription factor, MAF, associated with cataract, ocular anterior segment dysgenesis and coloboma. Hum. Mol. Genet. 11 (1), 33–42. Available at: http://www.ncbi.nlm.nih.gov/pubmed/11772997.

Jayakody, S.A., et al., 2012. SOX2 regulates the hypothalamic-pituitary axis at multiple levels. J. Clin. Invest. 122 (10), 3635–3646.

Kamachi, Y., et al., 2009. Evolution of non-coding regulatory sequences involved in the developmental process: reflection of differential employment of paralogous genes as highlighted by Sox2 and group B1 Sox genes. Proc. Jpn. Acad. Ser. B Phys. Biol. Sci. 85 (2), 55–68.

Kelberman, D., et al., 2006. Mutations within Sox2/SOX2 are associated with abnormalities in the hypothalamo-pituitary-gonadal axis in mice and humans. J. Clin. Invest. 116 (9), 2442–2455.

Mauri, L., et al., 2015. SOX2, OTX2 and PAX6 analysis in subjects with anophthalmia and microphthalmia. Eur. J. Med. Genet. 58 (2), 66–70. Available at: http://linkinghub.elsevier.com/retrieve/pii/S176972121400216X.

Mihelec, M., et al., 2009. Novel SOX2 partner-factor domain mutation in a four-generation family. Eur. J. Hum. Genet. : EJHG 17 (11), 1417–1422. Available at: http://dx.doi.org/10.1038/ejhg.2009.79.

Miller, J.N., Pearce, D.A., 2014. Nonsense-mediated decay in genetic disease: friend or foe? Mutat. Res. Rev. Mutat. Res. 762, 52–64. Available at: http://linkinghub.elsevier.com/retrieve/pii/S1383574214000350.

Morrison, D., et al., 2002. National study of microphthalmia, anophthalmia, and coloboma (MAC) in Scotland: investigation of genetic aetiology. J. Med. Genet. 39 (1), 16–22. Available at: http://www.pubmedcentral.nih.gov/articlerender.fcgi?artid=1734963&tool=pmcentrez&rendertype=abstract.

Ragge, N.K., Brown, A.G., et al., 2005a. Heterozygous mutations of OTX2 cause severe ocular malformations. Am. J. Hum. Genet. 76 (6), 1008–1022. Available at: http://www.pubmedcentral.nih.gov/articlerender.fcgi?artid=1196439&tool=pmcentrez&rendertype=abstract (accessed 06.10.13).

Ragge, N.K., Lorenz, B., et al., 2005b. SOX2 anophthalmia syndrome. Am. J. Med. Genet. Part A 135 (1), 1–7. discussion 8. Available at: http://www.ncbi.nlm.nih.gov/pubmed/15812812 (accessed 06.10.13).

Reményi, A., et al., 2003. Crystal structure of a POU/HMG/DNA ternary complex suggests differential assembly of Oct4 and Sox2 on two enhancers. Genes Dev. 17 (16), 2048–2059. Available at: http://www.ncbi.nlm.nih.gov/pubmed/12923055.

Sarkar, A., Hochedlinger, K., 2013. The Sox family of transcription factors: Versatile regulators of stem and progenitor cell fate. Cell Stem Cell 12 (1), 15–30. Available at: http://dx.doi.org/10.1016/j.stem.2012.12.007.

Schilter, K., et al., 2013. Whole-genome copy number variation analysis in anophthalmia and microphthalmia. Clin. Genet. 84 (5), 473–481.

Sisodiya, S.M., et al., 2006. Role of SOX2 mutations in human hippocampal malformations and epilepsy. Epilepsia 47 (3), 534–542. Available at: http://www.ncbi.nlm.nih.gov/pubmed/16529618 (accessed 06.10.13).

Slavotinek, A.M., 2011. Eye development genes and known syndromes. Mol. Genet. Metab. 104 (4), 448–456. Available at: http://dx.doi.org/10.1016/j.ymgme.2011.09.029.

Smith, A.N., et al., 2009. Stage-dependent modes of Pax6-Sox2 epistasis regulate lens development and eye morphogenesis. Development 136 (17), 2977–2985. Available at: http://www.ncbi.nlm.nih.gov/pubmed/19666824.

Stemmer, M., et al., 2015. Cavefish eye loss in response to an early block in retinal differentiation progression. Development 142 (4), 743–752. Available at: http://dev.biologists.org/cgi/doi/10.1242/dev.114629.

Taranova, O.V., et al., 2006. SOX2 is a dose-dependent regulator of retinal neural progenitor competence. Genes Dev. 20 (9), 1187–1202. Available at: http://www.ncbi.nlm.nih.gov/pubmed/16651659.

Verma, A.S., Fitzpatrick, D.R., 2007. Anophthalmia and microphthalmia. Orphanet J. Rare Dis. 2, 47. Available at: http://www.ncbi.nlm.nih.gov/pubmed/18039390.

Voronina, V.A., 2003. Mutations in the human RAX homeobox gene in a patient with anophthalmia and sclerocornea. Hum. Mol. Genet. 13 (3), 315–322. Available at: http://www.ncbi.nlm.nih.gov/pubmed/14662654 (accessed 12.04.15).

Williamson, K.A., FitzPatrick, D.R., 2014. The genetic architecture of microphthalmia, anophthalmia and coloboma. Eur. J. Med. Genet. 57 (8), 369–380. Available at: http://www.sciencedirect.com/science/article/pii/S1769721214001141.

Williamson, K.A., et al., 2006. Mutations in SOX2 cause anophthalmia-esophageal-genital (AEG) syndrome. Hum. Mol. Genet. 15 (9), 1413–1422. Available at: http://www.ncbi.nlm.nih.gov/pubmed/16543359.

CHAPTER 14

Role of SOX2 in the Hypothalamo–Pituitary Axis

Karine Rizzoti, Robin Lovell-Badge
The Crick Institute, London, UK

INTRODUCTION

The hypothalamo–pituitary axis is an important regulator of homeostasis and of essential physiological functions in vertebrates (growth in the young and in puberty, pregnancy, and lactation, to name a few). The hypothalamus, which can be considered to be on top of the axis, is localized within the ventral diencephalon and serves to centralize information from other regions of the brain (mainly the brain stem), as well as from the periphery, and to regulate pituitary hormonal secretions. (It also regulates aspects of the autonomic nervous system, but this will not be discussed here.) In turn, pituitary hormones control the function of endocrine glands within the periphery, whereas several also act directly on a range of other tissues and organs (Figure 1).

The pituitary is located at the base of the skull, just under the hypothalamus. It is composed of three different lobes: anterior, intermediate, and posterior. The anterior lobe is the largest and contains five different populations of endocrine cells. The intermediate lobe, located between the anterior and the posterior lobes, contains only one endocrine population; in humans it is hypoplastic. The posterior lobe contains no endocrine cells but is composed of glial cells and hypothalamic axonal termini. The pituitary is connected to the hypothalamus by the pituitary stalk, which contains both neural and vascular components. The stalk is the structural mediator of the hypothalamic control on the pituitary. The different nuclei or group of neurons that constitute the hypothalamus secrete neurohormones directly into the pituitary from axonal termini that reach the gland through the pituitary stalk, or into a bed of capillaries located at the base of the third ventricle, in the median eminence, where the hypophyseal portal system, also located within the stalk, will deliver them to the gland (Figure 1).

Pituitary hormone imbalances, either deficits or excesses, have pleiotropic consequences. They are associated with significant morbidity and even mortality. Deficiencies can be congenital or acquired, and they can affect the pituitary, the hypothalamus, or both. When affecting more than one hormone, the conditions are referred as combined pituitary hormone deficiencies. Most congenital cases are of unknown etiology and their prevalence is 1 in 4000–10,000 births (Kelberman et al., 2009; Alatzoglou et al., 2014).

Sox2
http://dx.doi.org/10.1016/B978-0-12-800352-7.00014-1

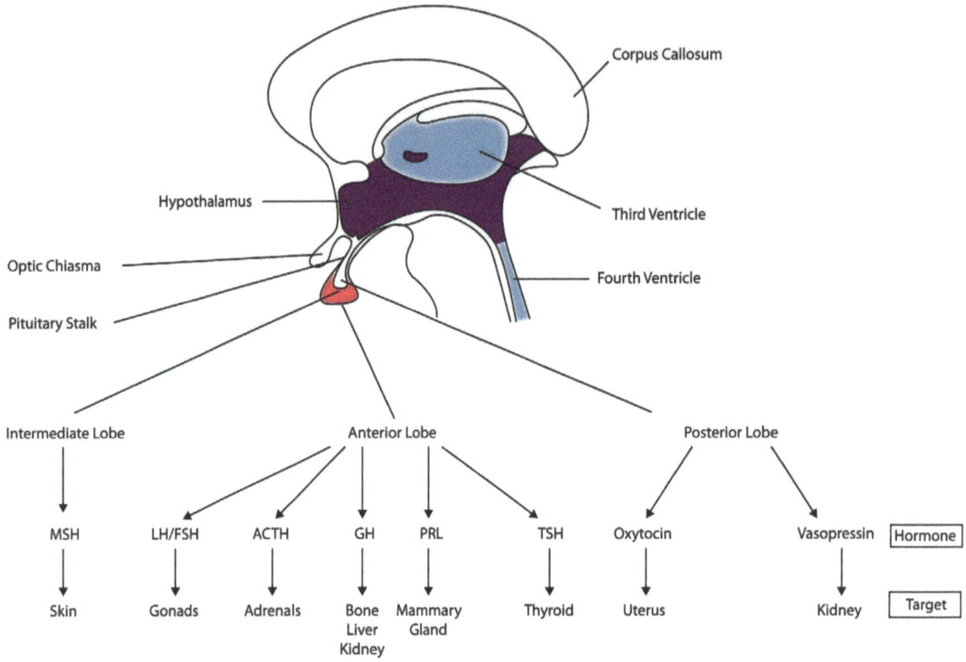

Figure 1 ***Schematic structure and function of the hypothalamo–pituitary axis.*** Within the ventral diencephalon, the hypothalamus centralizes peripheral information and controls pituitary endocrine secretions by releasing hypophysiotropic hormones. Some hypothalamic peptide hormones reach the gland directly (oxytocin and vasopressin secreted directly in the posterior lobe), and others via the hypophyseal portal system, such as peptides (GnRH, gonadotrophin releasing hormone, GHRH, GH releasing hormone, TRH, TSH releasing hormone, CRH, corticotropin releasing hormone and the inhibitory SST, somatostatin), are secreted at the median eminence and collected by capillaries. Anterior pituitary endocrine cells are composed of lactotrophs (producing prolactin, Prl), gonadotrophs (producing luteinizing hormone, LH, and follicle-stimulating hormone, FSH), thyrotrophs (producing thyroid-stimulating hormone, TSH), corticotrophs (producing adrenocorticotropic hormone, ACTH, proteolytically cleaved from proopiomelanocortin, POMC), and somatotrophs (producing GH). The intermediate lobe, hypoplastic in humans, contains only one endocrine cell type, melanotrophs (producing melanocyte-stimulating hormone, MSH, also proteolytically cleaved from POMC but with a different convertase).

Later in life, pituitary endocrine imbalances are mainly associated with the development of pituitary tumors. These are not metastatic but they can be invasive, impair the overall function of the pituitary and/or the hypothalamus, and affect vision as a result of optic nerve compression. These adenomas can be surgically removed but some recur.

The regulative role of the axis implies a constant adaptation of both hypothalamic and pituitary secretions to changing physiological contexts; therefore, plasticity is an important feature of the system. One level at which plasticity is achieved is through modulating cell numbers, both neurons in the hypothalamus and endocrine cells in the pituitary. Stem cell populations have been characterized in each compartment; these express SOX2 from the earliest stages of embryonic development. The protein, as we will describe here, is required for proper development of both compartments of the axis in the embryo. It is also probably required postnatally; however, much less is known about the role of SOX2 in this context. Consequently, mutations affecting the gene in both humans and mice are associated with conditions in which different aspects of morphogenesis and/or function of the axis are altered (Kelberman et al., 2009; Jayakody et al., 2012).

EXPRESSION OF SOX2 IN THE HYPOTHALAMO–PITUITARY AXIS

From the Hypophyseal Placode to the Pituitary

Pituitary development is initiated at about 8 days postcoitum (dpc) in the mouse embryo, as the neural plate forms (Figure 2). It first appears as an ectodermal thickening in the rostral most position, in the anterior neural ridge just in front of the future ventral diencephalon. This domain, the hypophyseal placode, forms along with the olfactory, optic, and otic placodes, where SOX2 is expressed in all (Schlosser et al., 2014; Soukup et al., 2013). In the olfactory placode, among other cell types, SOX2-positive progenitors will give rise to gonadotrophin-releasing hormone (GnRH) neurons (Rawson et al., 2010). These will migrate through the developing brain to reach their destination in the hypothalamus, where they control reproductive functions (Stevenson et al., 2013). They are the only hypothalamic neurons with an origin outside the central nervous system (CNS).

At 9 dpc in the mouse, the presumptive telencephalon has formed by fusion of the anterior midline. Cell proliferation is more intense in this domain than in the presumptive diencephalon, so the latter is now curved and appears shifted posteriorly. Because the hypophyseal placode is physically in contact with the ventral diencephalon, it is still present just underneath, but it has become the pituitary anlagen or Rathke's pouch (RP), an epithelial invagination that forms in continuity with the oral ectoderm and extends dorsally toward the ventral diencephalon. SOX2 is ubiquitously expressed in RP progenitors (Fauquier et al., 2008), in the oral ectoderm, and in the ventral diencephalon. Moreover, RP is in contact with the anterior border of foregut endoderm (Khonsari et al., 2013), where SOX2 is also present (Rizzoti and Lovell-Badge, 2007; Que et al., 2007, Chapter 17). Secreted signals from these different territories are important for proper RP morphogenesis; therefore, SOX2 is likely to have both direct and indirect roles during morphogenesis of the axis.

Rathke's pouch grows rapidly, and by 11.5 dpc it also separates from the oral ectoderm. Dorsally, the region of the ventral diencephalon directly in contact with the pouch starts to evaginate toward the pouch. This territory is the infundibulum, which gives rise to the median eminence, pituitary stalk, and posterior lobe (Pearson and Placzek, 2013), whereas the ventral diencephalon gives rise to the hypothalamus, and RP will form the anterior and intermediate lobes of the gland. Therefore, the three pituitary

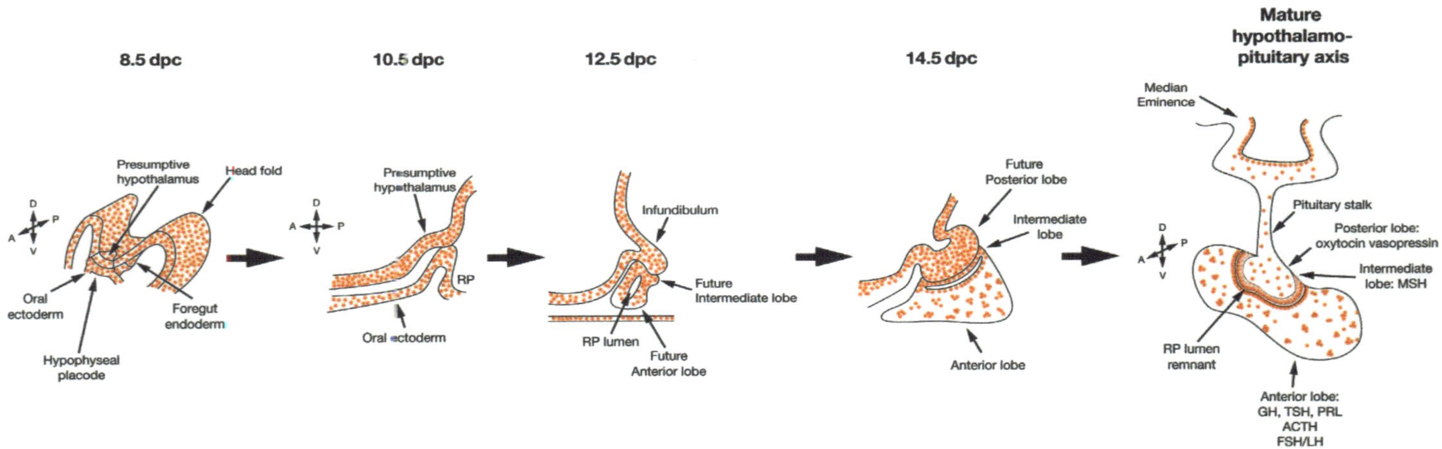

Figure 2 ***Expression of SOX2 during morphogenesis of the hypothalamo–pituitary axis.*** SOX2 (expression pattern represented in red dots) is present from the earliest stage of pituitary development, in the hypophyseal placode. It is also present at this stage in the neural plate and foregut endoderm. The placode later invaginates to become the Rathke's pouch (RP), still in contact in its dorsal-most part with the infundibulum, within the ventral diencephalon; SOX2 is present in both tissues. The infundibulum then evaginates toward RP. It will give rise to the median eminence, the pituitary stalk, and the posterior lobe while the anterior and intermediate lobes originate from RP. SOX2 expression starts to decrease in the ventral region of the pouch as the first endocrine cells differentiate. At 14.5 dpc, SOX2 expression is maintained in pituitary progenitors, mainly dorsally. In the CNS, it is present in neural progenitors and pituicytes. Postnatally, SOX2 is present in pituitary stem cells, mainly in the epithelium lining the cleft, or RP lumen remnant, and scattered in nonendocrine cells in the anterior lobe. It is also maintained at low levels in melanotrophs in the intermediate lobe. In the hypothalamus it is present in stem cells lining the third ventricle, in neurons and glial cells scattered in the parenchyma, and in the median eminence. *Adapted from Rizzoti and Lovell-Badge (2011).*

lobes do not have the same embryonic origin: anterior and intermediate lobes are ectodermal derivatives whereas the posterior one has a neuroectodermal origin. SOX2 along with SOX3 is present in the infundibulum and ventral diencephalon (Kelberman et al., 2006; Rizzoti et al., 2004), but SOX2 remains the only SOXB1 protein to be expressed in the pouch.

A large proportion of RP progenitors exit the cell cycle between 11.5 and 13.5 dpc (Davis et al., 2011); differentiation, characterized by expression of secreted hormones, follows (Japon et al., 1994). Lineage tracing experiments using a *Sox2*^{creERT2} allele have shown that SOX2-positive progenitors give rise to all endocrine cell types in the pituitary (Andoniadou et al., 2013; Rizzoti et al., 2013). SOX2 remains present in proliferative progenitors, around the RP lumen, dorsally, and scattered in the presumptive anterior lobe while differentiating cells rapidly downregulate its expression (Fauquier et al., 2008). Toward the end of gestation, by 14.5 dpc, SOX9, a SOXE protein, is upregulated in SOX2-positive cells. Its expression correlates with a decreased proliferative potential (KR and Carlotta Pires, unpublished data). Moreover, lineage tracing experiments using a *Sox9*$^{ires\text{-}creERT2}$ allele or the *Sox2*^{creERT2} allele but inducing it late show that the SOX9;SOX2-positive progenitors also tend to differentiate less often (Rizzoti et al., 2013).

SOX2 and SOX9 are maintained in the postnatal pituitary in a heterogeneous population comprising stem cells (Fauquier et al., 2008; Rizzoti et al., 2013). These are located, as in the embryo, in the epithelium surrounding the remnant of RP, and also in the anterior lobe, sometimes organized as rosettes. These encompass the population of S-100-positive folliculostellate cells (Fauquier et al., 2008; Andoniadou et al., 2013). These cells appear postnatally and have been proposed to have supporting roles for endocrine cells and progenitor properties (Denef, 2008). It is important to better characterize the different subpopulations contained within adult SOX2-positive pituitary cells. There are fewer SOX2-positive cells in the adult compared with early postnatal stages, and their proliferative and differentiative potentials decrease as the animal ages (Gremeaux et al., 2012; Rizzoti et al., 2013; Andoniadou et al., 2013). However, they have been shown to give rise to all endocrine cell types of the pituitary (Rizzoti et al., 2013; Andoniadou et al., 2013).

From the Embryonic Ventral Diencephalon to the Hypothalamus

All three SOXB1 proteins are present throughout the developing CNS from the neural plate stage (Wood and Episkopou, 1999). However, as the CNS begins to regionalize and certainly once progenitors begin to differentiate, their expression is no longer uniform and both quantitative and qualitative differences become apparent. Of relevance here, the expression of SOX2 and SOX3 is maintained at high levels in the developing ventral forebrain, comprising the prospective hypothalamus and infundibulum, whereas that of SOX1 is downregulated in the floor of the diencephalon (Wood and Episkopou, 1999).

As progenitors differentiate into neurons and glia, expression of the SOXB1 genes tends to go off, but there are exceptions (Ekonomou et al., 2005); nevertheless, they remain at high levels in neural stem and progenitor cells (Hutton and Pevny, 2011).

The hypothalamus develops from the ventral diencephalon, with the exception of GnRH neurons. It is organized around the third ventricle as groups of neurons or nuclei. Each nucleus regulates different physiological functions such as the circadian rhythms for the suprachiasmatic nucleus and feeding behavior for the arcuate. Its morphogenesis is complex because of this organization and we know relatively little compared with other regions of the brain where the anatomy is somehow simpler (Bedont et al., 2015). SOX2 expression is maintained in cells lining the ventricle during neurogenesis, but its precise pattern of expression in the developing parenchyma has not been reported.

In the hypothalamus, SOX2 and SOX3 are mostly present in cells lining the third ventricle, comprising hypothalamic stem cells (Lee et al., 2012b; Li et al., 2012b; Robins et al., 2013a,b; Haan et al., 2013); they are also observed scattered in the parenchyma, in some neurons and glial cells (Galichet et al., in preparation). In contrast with other hypothalamic nuclei, SOX2 is particularly highly expressed in the suprachiasmatic nucleus, which regulates circadian rhythms, where it is mostly expressed in neurons, some of which are light activated (Hoefflin and Carter, 2014).

ROLE OF SOX2 IN HYPOTHALAMO–PITUITARY AXIS MORPHOGENESIS

SOX2 Regulates Placode Development

SOX2 is important for placodal development and its role has been extensively studied in the context of the lens placode, where it has been shown to act cooperatively with the paired domain transcription factor PAX6 to induce expression of the γ1-crystallin gene (Kondoh et al., 2004 chapter 12). In the olfactory, lens, and hypophyseal placodes, it has also been shown to interact with the POU domain transcription factor OCT-1 (also known as POU2F1). The SOX2–OCT-1 complex is likely to activate expression of the transcription factor PAX6 (Donner et al., 2007). PAX6 is in turn required for maintenance of SOX2 expression. Interactions among these three factors are crucial for development of the placodes because compound *Oct-1*$^{-/-}$; *Sox2*$^{+/-}$ mutants display severe defects in induction and development of the three placodes (Donner et al., 2007). Later, in the olfactory epithelium, SOX2 is not homogeneously expressed, and according to its concentration it can either repress or activate expression of the gene encoding for the homeobox protein MEIS. As a consequence, SOX2 dose is important for genesis of the three neuronal types emerging from the olfactory placode: olfactory, vomeronasal, and GnRH (Tucker et al., 2010). Conditional deletion of *Sox2* using a targeted *Hesx1*^{Cre} allele, expressed in the anterior neural ridge (Dattani et al., 1998), results in loss of GnRH neurons from the earliest stages of their migration in the brain. This suggests that loss of the protein in the placode prevents generation of the neurons (Jayakody et al., 2012).

SOX2 and Patterning of the Ventral Diencephalon

Secreted signals, mostly from the infundibulum, are crucial for induction and maintenance of RP. Infundibulum formation is complex, but progress has been made unraveling the molecular events allowing its emergence. SOX2 and SOX3 have both been shown to be important players in this context.

The secreted ligand bone morphogenetic protein-4 (BMP4), made by the diencephalon from 8.5 dpc, is the earliest signaling molecule known to be required for RP development (Takuma et al., 1998; Ericson et al., 1998). Indeed, in the few murine *Bmp4*$^{-/-}$ mutants surviving at 10 dpc, no pouch or even ectodermal thickening is observed; therefore, BMP4 is proposed to be responsible for RP induction (Takuma et al., 1998). Its activity must be precisely regulated for proper pituitary patterning, as suggested by the simultaneous expression of several BMP inhibitors (Treier et al., 1998; Davis and Camper, 2007). The FGFs are the second family of secreted molecules important in mediating infundibular signaling toward RP. *Fgf 8*, *10*, and *18* are expressed in a domain overlapping that of *Bmp4*, but starting later, at 9.5 dpc (Ericson et al., 1998; Treier et al., 1998, 2001). In chicks, it has been shown that FGFs are required for expansion of infundibular cell numbers (Pearson et al., 2011). In the mouse, FGFs have been shown to induce dorsal RP progenitor proliferation and maintenance. As for BMPs, precise regulation of their activities is required for proper RP development. In the absence of *Sox3*, diencephalic cells just ventral to the infundibulum proliferate less (Rizzoti et al., 2004). As a consequence, the infundibulum appears stretched; it does not evaginate fully toward RP and *Fgf8* and *Bmp4* domains are expanded anteriorly. This ventral extension of infundibular signaling is proposed to induce more oral ectoderm toward an RP fate, as observed by ectopic anterior expression of the RP marker *Hesx1* in the oral ectoderm. Later, bifurcations appear in the dorsal region of RP (Rizzoti et al., 2004). Conditional homozygous deletion of *Sox2* in the ventral diencephalon has not yet been reported, but bifurcations are also observed in *Sox2*$^{+/-}$ embryos (Kelberman et al., 2006), which suggests that SOX2 performs the same role as SOX3. The bifurcations represent slight morphogenetic abnormalities; they persist as extra clefts in the adult gland but are not thought to affect pituitary secretions significantly (Rizzoti et al., 2004; Kelberman et al., 2006). An elegant study supervised by Larysa Pevny examined the consequences of further reducing the dose of SOX2 by using hypomorphic alleles of the gene. Patterning of the developing hypothalamus was perturbed by broadening of the floor and ectopic growth posterior to the infundibulum (Langer et al., 2012).

Several direct targets of SOX2 and SOX3 have been identified in the ventral diencephalon. These shed light on the complex role of both transcription factors and explain how the ventral diencephalon is patterned.

Expression of the secreted molecule Sonic Hedgehog (SHH) in the ventral diencephalon is dynamic. As RP develops, SHH is present anteriorly and laterally to the infundibulum, from which it is excluded (Manning et al., 2006; Trowe et al., 2013). Within the anterior domain, SHH is required directly for maintenance of the prospective anterior hypothalamus, and indirectly, because it is excluded from it, for correct formation of the infundibulum, which in turn is necessary for maintenance of RP (Zhao et al., 2012). In the absence of SHH, infundibular *Bmp4* and *Fgf10* patterns are expanded ventrally, resulting in an anterior shift and duplication of RP (Zhao et al., 2012). SOX2 and SOX3 are required to regulate *Shh* expression positively in this region by directly binding to a long-range 3′ enhancer of *Shh*, termed Brain Enhancer 2 (SBE2) (Zhao et al., 2012). Their activity on this enhancer is not redundant but rather dose-dependent, because alterations in the *Shh* expression pattern are observed in *Sox3*$^{Y/-}$ embryos and become more severe in compound *Sox3*$^{Y/-}$; *Sox2*$^{+/-}$ mutants (Zhao et al., 2012). In apparent contradiction, as we mentioned previously, SOX2 and SOX3 are also present in the infundibulum, but their direct target *Shh* is excluded from this structure, a necessary requirement for proper infundibular morphogenesis (Trowe et al., 2013).

What are the mechanisms underlying these complex expression patterns and interactions? The T-Box factors TBX2 and 3 are expressed in the ventral diencephalon midline, with TBX2 being restricted to the infundibulum (Pontecorvi et al., 2008). It was shown that these factors physically associate with SOX2 and SOX3, sequestering them away from the SBE2 enhancer, thereby preventing *Shh* expression in the infundibulum (Trowe et al., 2013). Indeed, when *Tbx3* is deleted, the *Shh* domain is expanded dorsally, the infundibulum still expresses *Fgf8* and *Bmp4*, but *Fgf10* is downregulated, there is more proliferation, and importantly, evagination fails to occur at 10.5 dpc. Consequently, RP fails to progress beyond this stage (Trowe et al., 2013). In chick, BMPs have been shown to induce expression of the transcription factor TBX2, which in this model also represses *Shh* expression, excluding it from the future infundibulum (Manning et al., 2006). The situation appears different in the presumptive posterior hypothalamus. Indeed, in *Sox2* hypomorphs there is expansion of the *Shh* expression pattern at the midline, posterior to the infundibulum, whereas the SHH antagonist *Gli3* is downregulated (Langer et al., 2012). To address these apparent differences it will be necessary to compare *Shh* patterns both anteriorly and posteriorly to the infundibulum in null, and probably compound *Sox2*;*Sox3* null, versus hypomorphic *Sox2* mutants. Alternatively, it is possible that different doses of SOX2 have contrasting effects, as demonstrated in the olfactory epithelium, where different amounts of the protein lead to differentiation into distinct neuronal types (Tucker et al., 2010).

Additional direct targets of SOXB1 proteins are connected with both cell proliferation and *Shh* regulation. SIX3 and SIX6, members of the *Drosophila Sine Oculis* transcription factors family, are associated with progenitor proliferation in the eye and pituitary (Li et al., 2002). SIX6 is excluded from the infundibulum whereas SIX3, whose expression is generally wider, appears to be present (Zhao et al., 2012; Oliver et al., 1995; Jeong et al., 2008). SOX2 and SOX3 bind directly to enhancers of *Six* and are required for *Six6* and *Six3* ventral diencephalon expression (Lee et al., 2012a, 2013; Zhao et al., 2012). Moreover, both SIX3 and SIX6 have been shown to regulate the *Shh* SBE2 enhancer directly (Jeong et al., 2008). Therefore, both directly and indirectly through SIX proteins, SOX2 and SOX3 are important activators of *Shh* expression, whose signaling activity is crucial for correct patterning of the anterior hypothalamus (Zhao et al., 2012; Jeong et al., 2008).

In addition, SOX2 has been shown to regulate *Six3.2* and *Nkx2.1* in Medaka fish (Beccari et al., 2012). *Nkx2.1* encodes for a transcription factor required for midline CNS development in the mouse forebrain (Kimura et al., 1996). When *Sox2* is deleted from the telencephalon, downregulation of both *Nkx2.1* and *Shh* is observed, along with severe defects (Ferri et al., 2013). It is not known whether SOX2 also activates *Nkx2.1* expression in the diencephalon, but it would be interesting to examine this issue because loss of *Nkx2.1* leads in particular to absence of the infundibulum (Kimura et al., 1996).

In Medaka and during *Xenopus* eye development, SOX2 activates expression of the transcription RAX (Beccari et al., 2012; Danno et al., 2008). It may be a relevant target in the developing murine hypothalamus because it is required for tanycyte differentiation, a specialized glial cell type endowed with stem cell properties that also display SOX2 and SOX3 expression (Miranda-Angulo et al., 2014).

In summary, the complex regulation of SOX2 and SOX3 in the ventral diencephalon underlies, at least partially, the antagonism between BMP and SHH in ventral diencephalon that is required for infundibular morphogenesis and hence RP positioning and development (Zhao et al., 2012; Trowe et al., 2013) (Figure 3). Later roles for SOX2 during hypothalamic neurogenesis have yet to be examined.

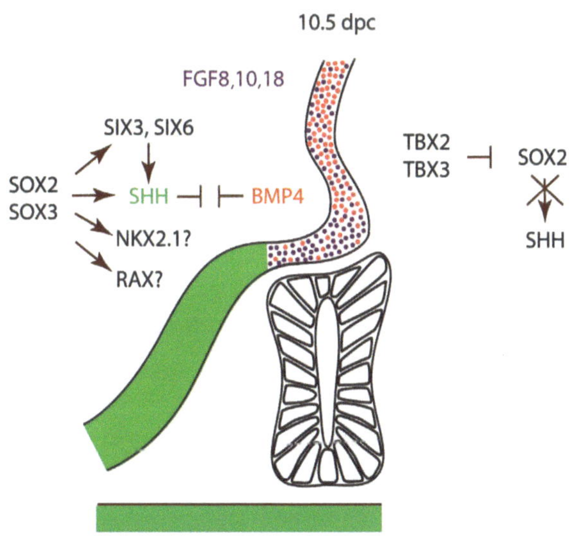

Figure 3 ***Role of SOX2 for ventral diencephalon patterning.*** SOX2 and SOX3 positively regulate the expression of *Shh* both directly and indirectly, because they also upregulate the expression of *Six3* and *Six6*, which in turn act on *Shh* enhancer SBE2 (Zhao et al., 2012). SHH must be excluded from the infundibulum for normal morphogenesis (Trowe et al., 2013); to prevent activation of its expression by SOX2 and SOX3, TBX2 and 3 physically interact with the two SOXB1 proteins, sequestering them away from the SBE2 enhancer (Trowe et al., 2013). In consequence, opposing activities of BMP4 and SHH within the ventral diencephalon pattern this region and participate in correct morphogenesis of the infundibulum and, accordingly, positioning of RP. Within the infundibulum, BMP4 and FGF8, 10, and 18 are required for development of RP.

SOX2 in the Oral Ectoderm

SOX2 and SHH are also expressed in the oral ectoderm, in which a regulatory connection similar to that in the ventral midline may be present. In contrast to mammals in which RP separates completely from the oral ectoderm, lower vertebrates maintain a connection, termed the buccohypophyseal canal. A similar connection, as well as abnormal fenestration of the underlying sphenoid bone, is seen in some mice (and humans) when the SHH pathway is deregulated or in certain *Sox2* mutant mice (Jayakody et al., 2012). In these mutants, the pituitary is ventrally displaced just above the oral cavity and sometimes connected to it. Therefore, SOX2 could be necessary to activate *Shh* expression in the oral ectoderm to promote normal basal skull and pituitary morphogenesis.

SOX2 Is Required for Proliferation of RP Progenitors

Early studies examining *Sox2* heterozygous embryos, which presented a slight pituitary hypoplasia, suggested that SOX2 may be associated with progenitor cell maintenance (Kelberman et al., 2006). More recently, conditional deletion of *Sox2* during pituitary development was performed using a *Hesx1*^{Cre} allele (Jayakody et al., 2012). *Hesx1* is present in the anterior neural plate, ridge, and primitive endoderm (Dattani et al., 1998); however, significant deletion of *Sox2* in RP is only observed later, at 12.5 dpc (Jayakody et al., 2012). Deletion is associated with a reduction in proliferation, and as a consequence fewer endocrine cells differentiate. Expression of the transcription factors PROP1, expressed in SOX2-positive progenitors (Yoshida et al., 2009), and PIT1 (also known as POU1F1), expressed in SOX2-negative committed cells (Fauquier et al., 2008), both of which are required for maintenance of somatotrophs, lactotrophs, and thyrotrophs, are downregulated in *Sox2* mutant embryos (Jayakody et al., 2012). The molecular mechanisms underlying the phenotype are not known and several pathways could be involved downstream of SOX2.

First, activation of a constitutively active form of β-catenin results in overproliferation (Gaston-Massuet et al., 2011; Andoniadou et al., 2013); therefore, the Wnt pathway could act downstream of, and/or along with SOX2 to regulate progenitor proliferation. Indeed, SOX proteins may regulate β-catenin/TCF transcriptional activities (Kormish et al., 2010). However, there appears to be little nuclear β-catenin present at this stage in RP (Brinkmeier et al., 2007), and at least initially, deletion of the gene in RP causes no defects (Olson et al., 2006). Its contribution to normal proliferative events is therefore unlikely.

Secondly, the NOTCH-1 receptor is a target of SOX2 in the retina (Taranova et al., 2006) and activation of the pathway is associated with maintenance of an uncommitted proliferative state in different developmental contexts, including the pituitary. More precisely, deletion of the transcription factor *Rbpj*, which acts downstream of the NOTCH signalling pathway, results in reduced progenitor proliferation (Zhu et al., 2006). Moreover, conditional deletion of either *Rbpj* or *Notch2* in RP also causes a down-regulation of

PROP1 (Zhu et al., 2006; Nantie et al., 2014), as does *Sox2* deletion (Jayakody et al., 2012). Finally, global impairment of endocrine differentiation results in upregulation of both NOTCH pathway components and SOX2 (Welcker et al., 2013). Therefore, NOTCH signaling is active early in RP and could be involved downstream of SOX2 to maintain progenitors in a proliferative undifferentiated state.

Finally, as mentioned earlier, SIX proteins are direct targets of SOX2 in the ventral diencephalon (Lee et al., 2012a, 2013; Zhao et al., 2012) and are also expressed in RP. They are associated with cell proliferation in the pouch because $Six6^{-/-}$ mutants display a hypoplastic pituitary (Li et al., 2002), which is similar to that in *Sox2* mutants (Jayakody et al., 2012). SIX6 has been suggested to promote proliferation by repressing expression of the cell cycle-negative regulator p27KIP1; such an interaction is demonstrated to underlie the $Six6^{-/-}$ retinal phenotype (Li et al., 2002). Interestingly, in *Sox2* mutant RP, a slight upregulation of P27 is observed (Jayakody et al., 2012). Therefore, loss of *Sox2* could result in loss of *Six6*, allowing upregulation of *p27*, which in turn would induce a reduction in cell proliferation.

In conclusion, SOX2 regulates early pituitary progenitor proliferation, allowing a sufficient pool of cells to be generated to give rise to all endocrine populations that differentiate before birth.

SOX2 IN THE POSTNATAL AXIS

SOX2 in Adult Pituitary Stem Cells

Proliferative potential decreases in late SOX2-positive embryonic progenitors, coincident with the upregulation of SOX9, whereas stem cells in the adult are more or less quiescent, only rarely undergoing either proliferation or differentiation under normal conditions (Fauquier et al., 2008; Rizzoti et al., 2013; Andoniadou et al., 2013). Importantly, fully differentiated endocrine cells also retain the capacity to divide, but similarly do so rarely. Because pituitary cell turnover is low, these latter self-replicating events are probably sufficient to cope with endocrine cell turnover, as was shown for corticotrophs (Langlais et al., 2013). However, in situations in which demand for new cells is stronger, such as a diphtheria toxin-based endocrine cell ablation model, SOX2-positive stem cells appear to be mobilized (Fu et al., 2012; Fu and Vankelecom, 2012). This is also seen in response to physiological demands. We decided to explore pituitary stem cell regenerative potential stimulated by pituitary target organ ablation, notably gonadectomy and adrenalectomy. These were known to induce a transient mitotic wave in the pituitary followed by generation of new hormonal cells, specifically those normally regulating the now ablated organ (gonadotrophs and corticotrophs, respectively). However, cell division is mainly observed in nonendocrine cells, which suggests that a population of progenitors is activated by organ ablation (Nolan and Levy, 2006). By combining this ablation model with lineage tracing, we have shown that adult pituitary stem cells proliferate and differentiate in the appropriate cell type in response to physiological demand (Rizzoti et al., 2013). This represented a step toward using or manipulating pituitary stem cells in treatment, in addition to providing a model to explore how physiological changes alter stem cell behavior.

Some cancers contain a subpopulation of cells with greater potential for cancer initiation and maintenance. These cells, called cancer stem cells, share some critical characteristics with normal stem cells, including the capacities for self-renewal, multilineage differentiation, and unlimited proliferation. At least two theories may explain their emergence: a stem cell that undergoes somatic alterations and becomes a tumor cell or a differentiated cell that acquires stem cell properties with malignant behavior in the tumorigenic process. Tumor-forming potential of the stem cell compartment in the pituitary has been examined. Sustained activation of the WNT pathway in pituitary embryonic progenitors, achieved by inducing expression of a degradation resistant form of β-catenin, induces the formation of tumorigenic clusters, resembling human adamantinomatous craniopharyngioma (Gaston-Massuet et al., 2011). Interestingly, some of the clusters retained SOX2 expression. Using a *Sox2*^{CreERT2} allele, it was shown that postnatal induction of the degradation-resistant form of β-catenin also induced tumor formation. Surprisingly, tumors were not composed of mutant cells. Instead, the mutant β-catenin–positive cells initially form clusters that subsequently induce in a paracrine fashion neighboring cells to form the tumors (Andoniadou et al., 2013). The physiological role of WNT signaling in pituitary stem cells is unknown, but it is associated with self-renewal in different stem cell populations. In another model of pituitary tumorigenesis, heterozygosity for the retinoblastoma *Rb* tumor suppressor gene, intermediate lobe tumors develop. Nestin and SOX2-positive cells are also located around tumorigenic nodules (Gleiberman et al., 2008). In a more recent report, a stem cell antigen-1 population was selected from dissected $Rb^{+/-}$ tumors. These cells could form SOX2-positive spheres in vitro, suggesting a progenitor identity, and therefore potentially the presence of cancer stem cells (Donangelo et al., 2014). However, it is unclear how these relate to the SOX2-positive cells surrounding the nodules in vivo.

The physiological or pathological role of SOX2 in these contexts is not known. Expression of SOX2 (and SOX9) correlates with pituitary stem cell identity; it is essentially the SOX2 (and SOX9) positive fraction, after sorting, that displays sphere or colony-forming ability in vitro, an assay reflecting progenitor potential (Fauquier et al., 2008; Andoniadou et al., 2012; Rizzoti et al., 2013). However, we do not know which aspect of stem cell identity SOX2 is regulating. The retina is an interesting model in which the role of SOX2 has been dissected. In the developing retina, SOX2 is required for both maintenance of proliferating progenitors and correct differentiation. NOTCH-1 has a key role in these events and is directly downstream of SOX2 (Taranova et al., 2006). In the postnatal retina, there is a population of dormant progenitors, the Müller glial cells. These express SOX2, and in sharp contrast to its earlier role, the protein is required to maintain quiescence. In its absence, there are abnormal cell divisions leading to retinal degeneration. The mechanisms underlying SOX2 activity in this context are not known but appear to be NOTCH pathway-independent (Surzenko et al., 2013). SOX2 is therefore versatile, probably through interactions with different partners and because the chromatin landscape evolves, allowing it to promote opposing actions (proliferation versus quiescence) in otherwise equivalent cells (retinal progenitors/stem cells), in different temporal contexts (embryo versus postnatal), although it is still mostly associated with maintenance of an undifferentiated, multipotent state.

A role for the NOTCH pathway in proliferation of adult pituitary progenitors has been proposed from experiments in vitro, using pituitary aggregates (Chen et al., 2006) and cell culture (Tando et al., 2013). More recently, conditional deletion of *Notch2* in embryonic progenitors demonstrated such a role in vivo. Although proliferation in the early embryo is not affected, it is reduced just after birth and SOX2; SOX9-positive progenitor numbers are decreased. In addition, the localization of SOX2- and SOX9-positive cells appears abnormal in the mutants (Nantie et al., 2014). Therefore, NOTCH-2-mediated signaling may be important for SOX2-positive stem cells early postnatally, during the growth phase observed during the first weeks of life when progenitors are highly proliferative (Gremeaux et al., 2012). It is not known whether NOTCH signaling is also involved later, in the adult stem cells, to regulate the balance between quiescence and mobilization induced by physiological challenges (Rizzoti et al., 2013), or in pathological situations (Andoniadou et al., 2013).

The balance between SOX2 and $P27^{kip}$ is likely to be important in the control of proliferation of pituitary stem cells, as suggested in the embryo (Jayakody et al., 2012), and also during tumorigenesis. Indeed, $P27^{kip}$ is not only a negative cell cycle regulator; it is also able to recruit co-repressors to inhibit transcription of target genes. It was shown to downregulate *Sox2* expression in embryonic stem cells (ESCs), and this interaction is relevant during ESC differentiation (Li et al., 2012a). In vivo, $p27^{kip}$ null mice display pituitary intermediate lobe tumors and a thicker stem cell layer lines the cleft. Deletion of one allele of *Sox2* on this null background rescues these phenotypes, highlighting the physiological relevance of the repressive action of P27 on *Sox2* expression, and presumably on proliferation of SOX2-positive stem cells in the pituitary cleft (Li et al., 2012a). It is interesting to examine the potential role of SOX2 in *p27* null intermediate lobe tumorigenesis and why the tumors are limited to this lobe. In the embryo, P27 is not initially expressed in RP but becomes ubiquitous later on (Bilodeau et al., 2009). Therefore, in addition to its role as a direct negative regulator of the cell cycle, the repressive action of P27 on *Sox2*, and potentially on other targets that promote cell proliferation, could be involved in suppressing the proliferative potential of the stem cells. Interestingly, in postmitotic cells in the developing ear, SOX2 activates the transcription of $p27^{kip}$ to maintain quiescence; therefore, the relationship between the two proteins is context dependent (Liu et al., 2012; see chapter 15).

In conclusion, since the initial characterization of its expression in adult pituitary stem cells/progenitors (Fauquier et al., 2008), SOX2 has been used as a marker for these cells in both physiological and pathological contexts. It clearly defines a heterogeneous population of cells, and thus it will be important to characterize the different subpopulations of SOX2-positive cells to better approach its role. In the postnatal retina, SOX2 is mostly restricted to dormant Müller glial cells, where it is required to maintain quiescence (Surzenko et al., 2013). In the pituitary, SOX2-positive cells include the population of S-100-positive glial-like cells, the folliculostellate cells, and it would be interesting to investigate whether there are parallels with the retina. It is also important to study its potential role during tumorigenesis because it is associated with this process in different genetic mouse models.

SOX2 in Adult Hypothalamic Stem Cells

A population of neural stem cells has been characterized in the hypothalamus (Lee et al., 2012b; Li et al., 2012b; Robins et al., 2013a,b; Haan et al., 2013). They are both neurogenic and gliogenic, located at the base of the third ventricle where they colocalize with α-tanycytes, a specialized glial cell type (Robins et al., 2013a). In addition, it has been suggested that some may reside in the parenchyma (Robins et al., 2013b). They are responsive to diet but there seems to be some variability between reports, perhaps linked to age of the animal, because a high-fat diet can induce neurogenesis in immature animals (Lee et al., 2012b) or impair it and be associated with inflammation later in life (Li et al., 2012b). It is important to examine the contribution of hypothalamic stem cells in other life-changing, physiological contexts, such as puberty and pregnancy, when the organism needs to adapt to and trigger a new physiological status, as in the case of puberty. In addition, it has been reported that neurogenesis, in particular in the hypothalamus, is reduced during aging, but the consequences of this are not known (Zhang et al., 2013).

The role of SOX2 in these cells has not been investigated. Maintenance of hippocampal stem cells relies on SOX2, in particular through *Shh* activation (Favaro et al., 2009). It is not known whether a similar interaction exists in hypothalamic stem cells, but their maintenance relies at least on SOX3 (Galichet et al., in preparation).

SOX2 AND HUMAN DISORDERS AFFECTING THE HYPOTHALAMO–PITUITARY AXIS

The infundibulum and RP develop at the midline and are sensitive to defects in ventral diencephalon and oral ectoderm patterning. In humans, several syndromes affecting the formation of midline structures have been described, with the earliest developmental perturbations leading to more extensive defects. Septo-optic dysplasia (1/10,000 live births) is such a syndrome. It can be caused by specific gene mutations, but it also has environmental origins (McCabe and Dattani, 2014). It is a heterogeneous condition and a diagnosis is made when at least two of the following features are present: optic nerve hypoplasia, midline brain defects, or hypothalamo-pituitary endocrine deficits. Mutations in both SOX2 and SOX3 are associated with variants of this syndrome (Kelberman et al., 2006, 2008) because they are linked with additional features. Indeed, heterozygous SOX2 mutations were initially associated with severe eye defects and mental retardation (Fantes et al., 2003) and the pituitary defects went unnoticed until it was realized that heterozygous mutant mice had a mild pan-hypopituitarism that prompted a more thorough analysis of patients (Kelberman et al., 2006). As we described above, perturbations in *Shh* regulation may at least partly explain the ventral diencephalic midline defects observed in these patients (Zhao et al., 2012). In agreement with this hypothesis, SOX2 is present in the ventral diencephalon in humans but intriguingly not in the infundibulum and posterior pituitary (Kelberman et al., 2008). Because of the importance of SOX2 and SOX3 during murine infundibular development, it would be interesting to characterize SOX3 expression in these territories to understand whether there is conservation or divergence in regulatory mechanisms. There are also some differences between the murine and human SOX2 heterozygous postnatal phenotypes (Kelberman et al., 2006). A different sensitivity to SOX2 dosage appears to exist, as already observed for the eye phenotype (Taranova et al., 2006). Both display pituitary hypoplasia, but in humans the deficits mainly affect the hypothalamo–pituitary–gonadal axis and present as delayed puberty, whereas in mice all pituitary hormones are affected (Kelberman et al., 2006). In humans, the delay in puberty is of hypothalamic origin, presumably reflecting the early requirement for SOX2 in the olfactory placode for GnRH neuron generation. Some patients have an isolated growth hormone deficiency, but it is not known whether this is of pituitary or hypothalamic origin (Jayakody et al., 2012).

Hypothalamic hamartoma, benign tumors developing around the base of the third ventricle, are also observed in patients carrying SOX2 mutations (Kelberman et al., 2006). These may result from disrupted SHH signaling, as suggested by the analysis of mouse *Sox2* hypomorphic mutants (Langer et al., 2012). In addition, two unrelated mutations in the human gene have been associated with slow-growing pituitary tumors (Alatzoglou et al., 2011). As we described earlier, ectopic activation of WNT signaling in mice results in the formation of tumors resembling human adamantinomatous craniopharyngioma. In agreement with this hypothesis, human tumors are characterized by β-catenin accumulation (Gaston-Massuet et al., 2011). In vitro, tests of the truncated mutant protein associated with a slow-growing pituitary tumor show an inability to antagonize β-catenin; therefore, tumors are proposed to result potentially from deregulated WNT signaling (Alatzoglou et al., 2011).

Pituitary tumors in humans are mostly nonmalignant but they can be secretory (mainly prolactinomas). An excess of pituitary hormones may lead to morbidity, as seen in Cushing disease (corticotroph tumors) where deregulated adrenocorticotropic hormone secretion leads to increased cortisol. To potentially characterize cancer stem cell populations in pituitary tumors, SOX2 has been used as a marker. A side population was selected from resected human pituitary adenomas; SOX2 was present and the cells could form spheres in vitro, which suggests the presence of a progenitor population in tumors (Mertens et al., 2015).

In conclusion, SOX2 mutations are associated with a wide range of phenotypes, from pituitary hypoplasia to tumorigenesis. This clear contrast may reflect the nature of the different mutations affecting the gene and the sensitivity of progenitors to the dose of SOX2. More work, such as modeling of the mutations in mice, is now required to better understand the involvement of normal and mutant forms of SOX2.

CONCLUSION

The role of SOX2 during development of both the hypothalamus and the pituitary has been the focus of recent investigations. These have demonstrated its crucial role during morphogenesis of the axis for patterning of the ventral diencephalon (Zhao et al., 2012; Trowe et al., 2013) and maintenance of RP progenitors (Jayakody et al., 2012). In contract, we know little about its function in hypothalamic and pituitary adult stem cells where its expression is maintained (Fauquier et al., 2008). It is important to understand its role in these newly characterized stem cells to approach the function of these populations in the adult axis, both in physiological and pathological situations but also for regenerative medicine. Current treatments for endocrine deficiencies, either replacement or substitution therapies, do not mimic endogenous secretions and are associated with side effects; therefore, regeneration of missing endocrine cells would represent a significant improvement. Reports using murine (Suga et al., 2011) and human stem cells (Dincer et al., 2013) have shown that endocrine cells can be differentiated from ESCs and transplanted, which demonstrate their therapeutic potential.

REFERENCES

Alatzoglou, K.S., Andoniadou, C.L., Kelberman, D., Buchanan, C.R., Crolla, J., Arriazu, M.C., Roubicek, M., Moncet, D., Martinez-Barbera, J.P., Dattani, M.T., 2011. SOX2 haploinsufficiency is associated with slow progressing hypothalamo-pituitary tumours. Hum. Mutat. 32, 1376–1380.

Alatzoglou, K.S., Webb, E.A., Le Tissier, P., Dattani, M.T., 2014. Isolated growth hormone deficiency (GHD) in childhood and adolescence: recent advances. Endocr. Rev. 35, 376–432.

Andoniadou, C.L., Gaston-Massuet, C., Reddy, R., Schneider, R.P., Blasco, M.A., Le Tissier, P., Jacques, T.S., Pevny, L.H., Dattani, M.T., Martinez-Barbera, J.P., 2012. Identification of novel pathways involved in the pathogenesis of human adamantinomatous craniopharyngioma. Acta Neuropathol. 124, 259–271.

Andoniadou, C.L., Matsushima, D., Mousavy Gharavy, S.N., Signore, M., Mackintosh, A.I., Schaeffer, M., Gaston-Massuet, C., Mollard, P., Jacques, T.S., Le Tissier, P., Dattani, M.T., Pevny, L.H., Martinez-Barbera, J.P., 2013. Sox2(+) stem/progenitor cells in the adult mouse pituitary support organ homeostasis and have tumor-inducing potential. Cell Stem Cell 13, 433–445.

Beccari, L., Conte, I., Cisneros, E., Bovolenta, P., 2012. Sox2-mediated differential activation of Six3.2 contributes to forebrain patterning. Development 139, 151–164.

Bedont, J.L., Newman, E.A., Blackshaw, S., 2015. Patterning, specification, and differentiation in the developing hypothalamus. Wiley Interdiscip. Rev. Dev. Biol. http://dx.doi.org/10.1002/wdev.187.

Bilodeau, S., Roussel-Gervais, A., Drouin, J., 2009. Distinct developmental roles of cell cycle inhibitors p57Kip2 and p27Kip1 distinguish pituitary progenitor cell cycle exit from cell cycle reentry of differentiated cells. Mol. Cell. Biol. 29, 1895–1908.

Brinkmeier, M.L., Potok, M.A., Davis, S.W., Camper, S.A., 2007. TCF4 deficiency expands ventral diencephalon signaling and increases induction of pituitary progenitors. Dev. Biol. 311, 396–407.

Chen, J., Crabbe, A., Van Duppen, V., Vankelecom, H., 2006. The notch signaling system is present in the postnatal pituitary: marked expression and regulatory activity in the newly discovered side population. Mol. Endocrinol. 20, 3293–3307.

Danno, H., Michiue, T., Hitachi, K., Yukita, A., Ishiura, S., Asashima, M., 2008. Molecular links among the causative genes for ocular malformation: Otx2 and Sox2 coregulate Rax expression. Proc. Natl. Acad. Sci. U.S.A. 105, 5408–5413.

Dattani, M.T., Martinez-Barbera, J.P., Thomas, P.Q., Brickman, J.M., Gupta, R., Martensson, I.L., Toresson, H., Fox, M., Wales, J.K., Hindmarsh, P.C., Krauss, S., Beddington, R.S., Robinson, I.C., 1998. Mutations in the homeobox gene HESX1/Hesx1 associated with septo-optic dysplasia in human and mouse. Nat. Genet. 19, 125–133.

Davis, S.W., Camper, S.A., 2007. Noggin regulates Bmp4 activity during pituitary induction. Dev. Biol. 305, 145–160.

Davis, S.W., Mortensen, A.H., Camper, S.A., 2011. Birthdating studies reshape models for pituitary gland cell specification. Dev. Biol. 352, 215–227.

Denef, C., 2008. Paracrinicity: the story of 30 years of cellular pituitary crosstalk. J. Neuroendocrinol. 20, 1–70.

Dincer, Z., Piao, J., Niu, L., Ganat, Y., Kriks, S., Zimmer, B., Shi, S.H., Tabar, V., Studer, L., 2013. Specification of functional cranial placode derivatives from human pluripotent stem cells. Cell Rep. 5, 1387–1402.

Donangelo, I., Ren, S.G., Eigler, T., Svendsen, C., Melmed, S., 2014. Sca1(+) murine pituitary adenoma cells show tumor-growth advantage. Endocr. Relat. Cancer 21, 203–216.

Donner, A.L., Episkopou, V., Maas, R.L., 2007. Sox2 and Pou2f1 interact to control lens and olfactory placode development. Dev. Biol. 303, 784–799.

Ekonomou, A., Kazanis, I., Malas, S., Wood, H., Alifragis, P., Denaxa, M., Karagogeos, D., Constanti, A., Lovell-Badge, R., Episkopou, V., 2005. Neuronal migration and ventral subtype identity in the telencephalon depend on SOX1. PLoS Biol. 3, e186.

Ericson, J., Norlin, S., Jessell, T.M., Edlund, T., 1998. Integrated FGF and BMP signaling controls the progression of progenitor cell differentiation and the emergence of pattern in the embryonic anterior pituitary. Development 125, 1005–1015.

Fantes, J., Ragge, N.K., Lynch, S.A., Mcgill, N.I., Collin, J.R., Howard-Peebles, P.N., Hayward, C., Vivian, A.J., Williamson, K., Van Heyningen, V., Fitzpatrick, D.R., 2003. Mutations in SOX2 cause anophthalmia. Nat. Genet. 33, 461–463.

Fauquier, T., Rizzoti, K., Dattani, M., Lovell-Badge, R., Robinson, I.C., 2008. SOX2-expressing progenitor cells generate all of the major cell types in the adult mouse pituitary gland. Proc. Natl. Acad. Sci. U.S.A. 105, 2907–2912.

Favaro, R., Valotta, M., Ferri, A.L., Latorre, E., Mariani, J., Giachino, C., Lancini, C., Tosetti, V., Ottolenghi, S., Taylor, V., Nicolis, S.K., 2009. Hippocampal development and neural stem cell maintenance require Sox2-dependent regulation of Shh. Nat. Neurosci. 12, 1248–1256.

Ferri, A., Favaro, R., Beccari, L., Bertolini, J., Mercurio, S., Nieto-Lopez, F., Verzeroli, C., La Regina, F., De Pietri Tonelli, D., Ottolenghi, S., Bovolenta, P., Nicolis, S.K., 2013. Sox2 is required for embryonic development of the ventral telencephalon through the activation of the ventral determinants Nkx2.1 and Shh. Development 140, 1250–1261.

Fu, Q., Vankelecom, H., 2012. Regenerative capacity of the adult pituitary: multiple mechanisms of lactotrope restoration after transgenic ablation. Stem Cells Dev. 21, 3245–3257.

Fu, Q., Gremeaux, L., Luque, R.M., Liekens, D., Chen, J., Buch, T., Waisman, A., Kineman, R., Vankelecom, H., 2012. The adult pituitary shows stem/progenitor cell activation in response to injury and is capable of regeneration. Endocrinology 153, 3224–3235.

Gaston-Massuet, C., Andoniadou, C.L., Signore, M., Jayakody, S.A., Charolidi, N., Kyeyune, R., Vernay, B., Jacques, T.S., Taketo, M.M., Le Tissier, P., Dattani, M.T., Martinez-Barbera, J.P., 2011. Increased wingless (Wnt) signaling in pituitary progenitor/stem cells gives rise to pituitary tumors in mice and humans. Proc. Natl. Acad. Sci. U.S.A. 108, 11482–11487.

Gleiberman, A.S., Michurina, T., Encinas, J.M., Roig, J.L., Krasnov, P., Balordi, F., Fishell, G., Rosenfeld, M.G., Enikolopov, G., 2008. Genetic approaches identify adult pituitary stem cells. Proc. Natl. Acad. Sci. U.S.A. 105, 6332–6337.

Gremeaux, L., Fu, Q., Chen, J., Vankelecom, H., 2012. Activated phenotype of the pituitary stem/progenitor cell compartment during the early-postnatal maturation phase of the gland. Stem Cells Dev. 21, 801–813.

Haan, N., Goodman, T., Najdi-Samiei, A., Stratford, C.M., Rice, R., El Agha, E., Bellusci, S., Hajihosseini, M.K., 2013. Fgf10-expressing tanycytes add new neurons to the appetite/energy-balance regulating centers of the postnatal and adult hypothalamus. J. Neurosci. 33, 6170–6180.

Hoefflin, S., Carter, D.A., 2014. Neuronal expression of SOX2 is enriched in specific hypothalamic cell groups. J. Chem. Neuroanat. 61-62, 153–160.

Hutton, S.R., Pevny, L.H., 2011. SOX2 expression levels distinguish between neural progenitor populations of the developing dorsal telencephalon. Dev. Biol. 352, 40–47.

Japon, M.A., Rubinstein, M., Low, M.J., 1994. In situ hybridization analysis of anterior pituitary hormone gene expression during fetal mouse development. J. Histochem. Cytochem. 42, 1117–1125.

Jayakody, S.A., Andoniadou, C.L., Gaston-Massuet, C., Signore, M., Cariboni, A., Bouloux, P.M., Le Tissier, P., Pevny, L.H., Dattani, M.T., Martinez-Barbera, J.P., 2012. SOX2 regulates the hypothalamic-pituitary axis at multiple levels. J. Clin. Invest. 122, 3635–3646.

Jeong, Y., Leskow, F.C., El-Jaick, K., Roessler, E., Muenke, M., Yocum, A., Dubourg, C., Li, X., Geng, X., Oliver, G., Epstein, D.J., 2008. Regulation of a remote Shh forebrain enhancer by the Six3 homeoprotein. Nat. Genet. 40, 1348–1353.

Kelberman, D., De Castro, S.C., Huang, S., Crolla, J.A., Palmer, R., Gregory, J.W., Taylor, D., Cavallo, L., Faienza, M.F., Fischetto, R., Achermann, J.C., Martinez-Barbera, J.P., Rizzoti, K., Lovell-Badge, R., Robinson, I.C., Gerrelli, D., Dattani, M.T., 2008. SOX2 plays a critical role in the pituitary, forebrain, and eye during human embryonic development. J. Clin. Endocrinol. Metab. 93, 1865–1873.

Kelberman, D., Rizzoti, K., Avilion, A., Bitner-Glindzicz, M., Cianfarani, S., Collins, J., Chong, W.K., Kirk, J.M., Achermann, J.C., Ross, R., Carmignac, D., Lovell-Badge, R., Robinson, I.C., Dattani, M.T., 2006. Mutations within Sox2/SOX2 are associated with abnormalities in the hypothalamo-pituitary-gonadal axis in mice and humans. J. Clin. Invest. 116, 2442–2455.

Kelberman, D., Rizzoti, K., Lovell-Badge, R., Robinson, I.C., Dattani, M.T., 2009. Genetic regulation of pituitary gland development in human and mouse. Endocr. Rev. 30, 790–829.

Khonsari, R.H., Seppala, M., Pradel, A., Dutel, H., Clement, G., Lebedev, O., Ghafoor, S., Rothova, M., Tucker, A., Maisey, J.G., Fan, C.M., Kawasaki, M., Ohazama, A., Tafforeau, P., Franco, B., Helms, J., Haycraft, C.J., David, A., Janvier, P., Cobourne, M.T., Sharpe, P.T., 2013. The buccohypophyseal canal is an ancestral vertebrate trait maintained by modulation in sonic hedgehog signaling. BMC Biol. 11, 27.

Kimura, S., Hara, Y., Pineau, T., Fernandez-Salguero, P., Fox, C.H., Ward, J.M., Gonzalez, F.J., 1996. The T/ebp null mouse: thyroid-specific enhancer-binding protein is essential for the organogenesis of the thyroid, lung, ventral forebrain, and pituitary. Genes Dev. 10, 60–69.

Kondoh, H., Uchikawa, M., Kamachi, Y., 2004. Interplay of Pax6 and SOX2 in lens development as a paradigm of genetic switch mechanisms for cell differentiation. Int. J. Dev. Biol. 48, 819–827.

Kormish, J.D., Sinner, D., Zorn, A.M., 2010. Interactions between SOX factors and Wnt/beta-catenin signaling in development and disease. Dev. Dyn. 239, 56–68.

Langer, L., Taranova, O., Sulik, K., Pevny, L., 2012. SOX2 hypomorphism disrupts development of the prechordal floor and optic cup. Mech. Dev. 129, 1–12.

Langlais, D., Couture, C., Kmita, M., Drouin, J., 2013. Adult pituitary cell maintenance: lineage-specific contribution of self-duplication. Mol. Endocrinol. 27, 1103–1112.

Lee, B., Rizzoti, K., Kwon, D.S., Kim, S.Y., Oh, S., Epstein, D.J., Son, Y., Yoon, J., Baek, K., Jeong, Y., 2012a. Direct transcriptional regulation of Six6 is controlled by SoxB1 binding to a remote forebrain enhancer. Dev. Biol. 366, 393–403.

Lee, D.A., Bedont, J.L., Pak, T., Wang, H., Song, J., Miranda-Angulo, A., Takiar, V., Charubhumi, V., Balordi, F., Takebayashi, H., Aja, S., Ford, E., Fishell, G., Blackshaw, S., 2012b. Tanycytes of the hypothalamic median eminence form a diet-responsive neurogenic niche. Nat. Neurosci. 15, 700–702.

Lee, B., Song, H., Rizzoti, K., Son, Y., Yoon, J., Baek, K., Jeong, Y., 2013. Genomic code for Sox2 binding uncovers its regulatory role in Six3 activation in the forebrain. Dev. Biol. 381, 491–501.

Li, H., Collado, M., Villasante, A., Matheu, A., Lynch, C.J., Canamero, M., Rizzoti, K., Carneiro, C., Martinez, G., Vidal, A., Lovell-Badge, R., Serrano, M., 2012a. p27(Kip1) directly represses Sox2 during embryonic stem cell differentiation. Cell Stem Cell 11, 845–852.

Li, J., Tang, Y., Cai, D., 2012b. IKKbeta/NF-kappaB disrupts adult hypothalamic neural stem cells to mediate a neurodegenerative mechanism of dietary obesity and pre-diabetes. Nat. Cell Biol. 14, 999–1012.

Li, X., Perissi, V., Liu, F., Rose, D.W., Rosenfeld, M.G., 2002. Tissue-specific regulation of retinal and pituitary precursor cell proliferation. Science 297, 1180–1183.

Liu, Z., Walters, B.J., Owen, T., Brimble, M.A., Steigelman, K.A., Zhang, L., Mellado Lagarde, M.M., Valentine, M.B., Yu, Y., Cox, B.C., Zuo, J., 2012. Regulation of p27Kip1 by Sox2 maintains quiescence of inner pillar cells in the murine auditory sensory epithelium. J. Neurosci. 32, 10530–10540.

Manning, L., Ohyama, K., Saeger, B., Hatano, O., Wilson, S.A., Logan, M., Placzek, M., 2006. Regional morphogenesis in the hypothalamus: a BMP-Tbx2 pathway coordinates fate and proliferation through Shh downregulation. Dev. Cell 11, 873–885.

McCabe, M.J., Dattani, M.T., 2014. Genetic aspects of hypothalamic and pituitary gland development. Handb. Clin. Neurol. 124, 3–15.

Mertens, F.M., Gremeaux, L., Chen, J., Fu, Q., Willems, C., Roose, H., Govaere, O., Roskams, T., Cristina, C., Becu-Villalobos, D., Jorissen, M., Vander Poorten, V., Bex, M., Van Loon, J., Vankelecom, H., 2015. Pituitary tumors contain a side population with tumor stem cell-associated characteristics. Endocr. Relat. Cancer, in press.

Miranda-Angulo, A.L., Byerly, M.S., Mesa, J., Wang, H., Blackshaw, S., 2014. Rax regulates hypothalamic tanycyte differentiation and barrier function in mice. J. Comp. Neurol. 522, 876–899.

Nantie, L.B., Himes, A.D., Getz, D.R., Raetzman, L.T., 2014. Notch signaling in postnatal pituitary expansion: proliferation, progenitors, and cell specification. Mol. Endocrinol. 28, 731–744.

Nolan, L.A., Levy, A., 2006. A population of non-luteinising hormone/non-adrenocorticotrophic hormone-positive cells in the male rat anterior pituitary responds mitotically to both gonadectomy and adrenalectomy. J. Neuroendocrinol. 18, 655–661.

Oliver, G., Mailhos, A., Wehr, R., Copeland, N.G., Jenkins, N.A., Gruss, P., 1995. Six3, a murine homologue of the sine oculis gene, demarcates the most anterior border of the developing neural plate and is expressed during eye development. Development 121, 4045–4055.

Olson, L.E., Tollkuhn, J., Scafoglio, C., Krones, A., Zhang, J., Ohgi, K.A., Wu, W., Taketo, M.M., Kemler, R., Grosschedl, R., Rose, D., Li, X., Rosenfeld, M.G., 2006. Homeodomain-mediated beta-catenin-dependent switching events dictate cell-lineage determination. Cell 125, 593–605.

Pearson, C.A., Placzek, M., 2013. Development of the medial hypothalamus: forming a functional hypothalamic-neurohypophyseal interface. Curr. Top. Dev. Biol. 106, 49–88.

Pearson, C.A., Ohyama, K., Manning, L., Aghamohammadzadeh, S., Sang, H., Placzek, M., 2011. FGF-dependent midline-derived progenitor cells in hypothalamic infundibular development. Development 138, 2613–2624.

Pontecorvi, M., Goding, C.R., Richardson, W.D., Kessaris, N., 2008. Expression of Tbx2 and Tbx3 in the developing hypothalamic-pituitary axis. Gene Expr. Patterns 8, 411–417.

Que, J., Okubo, T., Goldenring, J.R., Nam, K.T., Kurotani, R., Morrisey, E.E., Taranova, O., Pevny, L.H., Hogan, B.L., 2007. Multiple dose-dependent roles for Sox2 in the patterning and differentiation of anterior foregut endoderm. Development 134, 2521–2531.

Rawson, N.E., Lischka, F.W., Yee, K.K., Peters, A.Z., Tucker, E.S., Meechan, D.W., Zirlinger, M., Maynard, T.M., Burd, G.B., Dulac, C., Pevny, L., Lamantia, A.S., 2010. Specific mesenchymal/epithelial induction of olfactory receptor, vomeronasal, and gonadotropin-releasing hormone (GnRH) neurons. Dev. Dyn. 239, 1723–1738.

Rizzoti, K., Lovell-Badge, R., 2007. SOX3 activity during pharyngeal segmentation is required for craniofacial morphogenesis. Development 134, 3437–3448.

Rizzoti, K., Lovell-Badge, R., 2011. Regenerative Medicine: Organ recital in a dish. Nature 480(7375), 44–46.

Rizzoti, K., Brunelli, S., Carmignac, D., Thomas, P.Q., Robinson, I.C., Lovell-Badge, R., 2004. SOX3 is required during the formation of the hypothalamo-pituitary axis. Nat. Genet. 36, 247–255.

Rizzoti, K., Akiyama, H., Lovell-Badge, R., 2013. Mobilized adult pituitary stem cells contribute to endocrine regeneration in response to physiological demand. Cell Stem Cell 13, 419–432.

Robins, S.C., Stewart, I., Mcnay, D.E., Taylor, V., Giachino, C., Goetz, M., Ninkovic, J., Briancon, N., Maratos-Flier, E., Flier, J.S., Kokoeva, M.V., Placzek, M., 2013a. alpha-Tanycytes of the adult hypothalamic third ventricle include distinct populations of Fgf-responsive neural progenitors. Nat. Commun. 4, 2049.

Robins, S.C., Trudel, E., Rotondi, O., Liu, X., Djogo, T., Kryzskaya, D., Bourque, C.W., Kokoeva, M.V., 2013b. Evidence for NG2-glia derived, adult-born functional neurons in the hypothalamus. PLoS One 8, e78236.

Schlosser, G., Patthey, C., Shimeld, S.M., 2014. The evolutionary history of vertebrate cranial placodes II. Evolution of ectodermal patterning. Dev. Biol. 389, 98–119.

Soukup, V., Horacek, I., Cerny, R., 2013. Development and evolution of the vertebrate primary mouth. J. Anat. 222, 79–99.

Stevenson, E.L., Corella, K.M., Chung, W.C., 2013. Ontogenesis of gonadotropin-releasing hormone neurons: a model for hypothalamic neuroendocrine cell development. Front. Endocrinol. (Lausanne) 4, 89.

Suga, H., Kadoshima, T., Minaguchi, M., Ohgushi, M., Soen, M., Nakano, T., Takata, N., Wataya, T., Muguruma, K., Miyoshi, H., Yonemura, S., Oiso, Y., Sasai, Y., 2011. Self-formation of functional adenohypophysis in three-dimensional culture. Nature 480, 57–62.

Surzenko, N., Crowl, T., Bachleda, A., Langer, L., Pevny, L., 2013. SOX2 maintains the quiescent progenitor cell state of postnatal retinal Muller glia. Development 140, 1445–1456.

Takuma, N., Sheng, H.Z., Furuta, Y., Ward, J.M., Sharma, K., Hogan, B.L., Pfaff, S.L., Westphal, H., Kimura, S., Mahon, K.A., 1998. Formation of Rathke's pouch requires dual induction from the diencephalon. Development 125, 4835–4840.

Tando, Y., Fujiwara, K., Yashiro, T., Kikuchi, M., 2013. Localization of notch signaling molecules and their effect on cellular proliferation in adult rat pituitary. Cell Tissue Res. 351, 511–519.

Taranova, O.V., Magness, S.T., Fagan, B.M., Wu, Y., Surzenko, N., Hutton, S.R., Pevny, L.H., 2006. SOX2 is a dose-dependent regulator of retinal neural progenitor competence. Genes Dev. 20, 1187–1202.

Treier, M., Gleiberman, A.S., O'connell, S.M., Szeto, D.P., Mcmahon, J.A., Mcmahon, A.P., Rosenfeld, M.G., 1998. Multistep signaling requirements for pituitary organogenesis in vivo. Genes Dev. 12, 1691–1704.

Treier, M., O'connell, S., Gleiberman, A., Price, J., Szeto, D.P., Burgess, R., Chuang, P.T., Mcmahon, A.P., Rosenfeld, M.G., 2001. Hedgehog signaling is required for pituitary gland development. Development 128, 377–386.

Trowe, M.O., Zhao, L., Weiss, A.C., Christoffels, V., Epstein, D.J., Kispert, A., 2013. Inhibition of Sox2-dependent activation of Shh in the ventral diencephalon by Tbx3 is required for formation of the neurohypophysis. Development 140, 2299–2309.

Tucker, E.S., Lehtinen, M.K., Maynard, T., Zirlinger, M., Dulac, C., Rawson, N., Pevny, L., Lamantia, A.S., 2010. Proliferative and transcriptional identity of distinct classes of neural precursors in the mammalian olfactory epithelium. Development 137, 2471–2481.

Welcker, J.E., Hernandez-Miranda, L.R., Paul, F.E., Jia, S., Ivanov, A., Selbach, M., Birchmeier, C., 2013. Insm1 controls development of pituitary endocrine cells and requires a SNAG domain for function and for recruitment of histone-modifying factors. Development 140, 4947–4958.

Wood, H.B., Episkopou, V., 1999. Comparative expression of the mouse Sox1, Sox2 and Sox3 genes from pre-gastrulation to early somite stages. Mech. Dev. 86, 197–201.

Yoshida, S., Kato, T., Susa, T., Cai, L.Y., Nakayama, M., Kato, Y., 2009. PROP1 coexists with SOX2 and induces PIT1-commitment cells. Biochem. Biophys. Res. Commun. 385, 11–15.

Zhang, G., Li, J., Purkayastha, S., Tang, Y., Zhang, H., Yin, Y., Li, B., Liu, G., Cai, D., 2013. Hypothalamic programming of systemic ageing involving IKK-beta, NF-kappaB and GnRH. Nature 497, 211–216.

Zhao, L., Zevallos, S.E., Rizzoti, K., Jeong, Y., Lovell-Badge, R., Epstein, D.J., 2012. Disruption of SoxB1-dependent Sonic hedgehog expression in the hypothalamus causes septo-optic dysplasia. Dev. Cell 22, 585–596.

Zhu, X., Zhang, J., Tollkuhn, J., Ohsawa, R., Bresnick, E.H., Guillemot, F., Kageyama, R., Rosenfeld, M.G., 2006. Sustained notch signaling in progenitors is required for sequential emergence of distinct cell lineages during organogenesis. Genes Dev. 20, 2739–2753.

CHAPTER 15

SOX2 in Neurosensory Fate Determination and Differentiation in the Inner Ear

Kathryn S.E. Cheah[1], **Pin-Xian Xu**[2,3]
[1]Department of Biochemistry, Li Ka Shing Faculty of Medicine, The University of Hong Kong, Hong Kong, China; [2]Genetics and Genomic Sciences, Icahn School of Medicine at Mount Sinai, New York, NY, USA; [3]Developmental and Regenerative Biology, Icahn School of Medicine at Mount Sinai, New York, NY, USA

INNER EAR DEVELOPMENT

The mammalian inner ear is responsible for the sense of balance and hearing. Five sensory organs in the vestibular system detect changes in angular acceleration or gravity, whereas the organ of Corti in the coiled cochlea performs the auditory function. These sensory organs contain two major types of epithelial cells: hair cells and underlying supporting cells (Figure 1). Although the hair cells of vestibular and auditory systems of the ear are not identical, they act as mechanosensors to detect sound, gravity, and acceleration transmitted to the central nervous system through sensory neurons. These six sensory organs and their associated sensory neurons develop from common progenitors in the otic epithelium at the otocyst stage at around embryonic day 9.5 (E9.5) (Whitfield, 2015; Appler and Goodrich, 2011). The otocyst is derived from the otic placode-thickened ectoderm that forms next to the hindbrain at about E8.5 and gives rise to all inner ear structures including neuronal, sensory, and nonsensory structures. Precursor cells for these distinct inner ear structures are specified early at the otocyst stage and their correct location within the otocyst can be identified by landmark expression patterns of key transcriptional regulators that specify and direct their differentiation fate. Absence or partial loss of these cells or disorganization of their spatial organization can lead to impaired inner ear function (Groves et al., 2013).

The neuroblasts that form the vestibulocochlear ganglion (CVG) are the first lineage to be specified within the anteroventral region of the otocyst as marked by the expression of the basic helix-loop-helix (bHLH) proneural gene *Neurogenin 1* (*Neurog1*). The neuroblasts then delaminate and express the neuronal differentiation gene *Neurod1* to differentiate eventually into neurons that connect to the sensory hair cells in the inner ear and to the brain (Fritzsch et al., 2006; Sanchez-Calderon et al., 2007). The dorsal otocyst contributes the vestibular chamber whereas the ventral portion outgrows to form the cochlear duct. Fate-mapping studies and regionalized expression patterns of genes (e.g., *Bmp4*, *Lfng*, *Jag1*, and *Islet1*) reveal domains within the otocyst that mark compartments

http://dx.doi.org/10.1016/B978-0-12-800352-7.00015-3

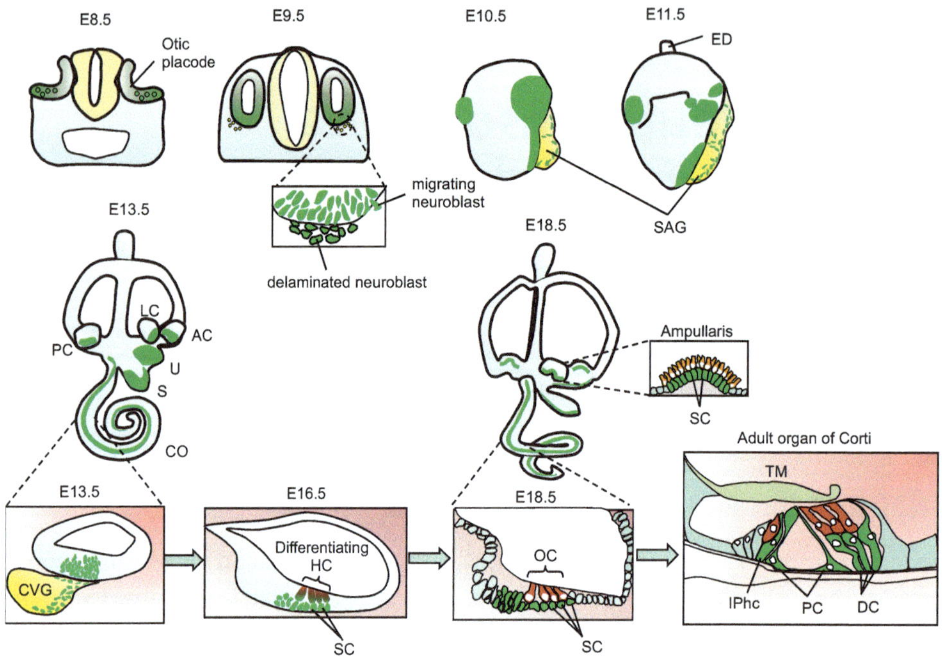

Figure 1 Sox2/*SOX2 expression during mouse inner ear development.* Diagrammatic representation of *Sox2*/SOX2 (green) expression (see main text for details). Note downregulation of *Sox2*/SOX2 in differentiating hair cells from E16.5 but sustained expression in supporting cells even in the adult inner ear. Abbreviations: AC: anterior crista; CO: cochlea; CVG: cochleovestibular ganglion; DC: Deiter's cell; ED: endolymphatic duct; HC: hair cells; iPhc: inner phalangeal cell; LC: lateral crista; OC: organ of Corti; PC: posterior crista; PC: pillar cell; SAG: statoacoustic ganglion; S: Saccule; TM: tectorial membrane; U: utricle.

within which local patterning and cell fate specification occur before emergence of the different cell types. Cell lineage tracing in lower vertebrates indicates that hair and supporting cells are clonally related, arising from a common progenitor (Fekete et al., 1998) and that otic neuroblasts and prosensory progenitors also share a common progenitor (Lang and Fekete, 2001; Appler and Goodrich, 2011). Correct regional gene expression is critical for generating distinct precursor cells in the inner ear. However, we have only rudimentary knowledge about the cellular hierarchies and regulatory networks that drive differential gene expression pattern in inner ear development and disease.

SOX2 EXPRESSION IN THE DEVELOPING INNER EAR

Sox2 is one of the earliest genes that mark the neurosensory progenitors within the otic epithelium (summarized in Figure 1) and its expression pattern in these cell types implicates its role early in the development of sensory neurons and prosensory organs and later in the specification and differentiation of hair cells and supporting cells. *Sox2* expression is first detected in the lateral edge of the otic placode at E8.5 in mouse

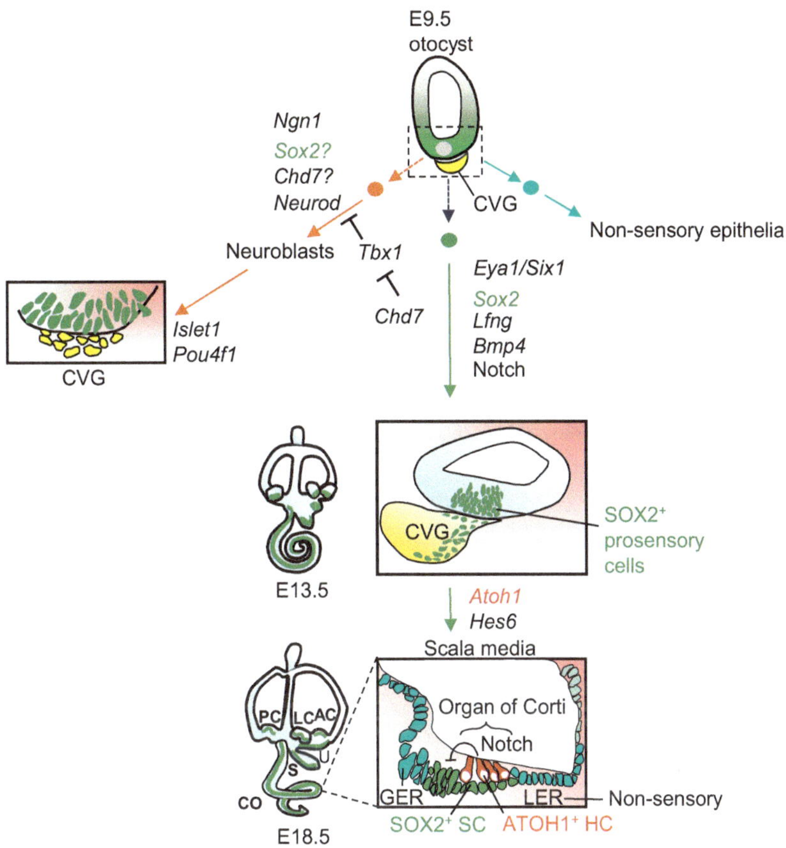

Figure 2 ***Lineage specification of otic progenitors.*** Cell lineages derived from sensory and prosensory progenitors arising from a common progenitor in the otic epithelium (OE): sensory neurons in cochleo-vestibular ganglion (CVG), hair cells (HC), supporting cells (SC) in sensory organs (utricle (u), saccule (S), cristae (PC and AC), Organ of Corti in cochlear (CO)). The other progenitors in the OE contribute to nonsensory supporting structures of the inner ear.

development (Wood and Episkopou, 1999; Uchikawa et al., 1999; Mak et al., 2009). At E9.5, when the otic vesicle closes and pinches off from the ectoderm, SOX2 is expressed in the ventral and lateral regions of the otocyst and migrating neuroblasts (Kiernan et al., 2005; Mak et al., 2009). At E10.5, SOX2 expression domain separates to form an anterior strip and a posterior spot. One day later, the anterior domain further segregates, with the dorsal domain marking presumptive anterior crista, lateral crista and the utricle, whereas the ventral domain gives rise to the future organ of Corti and the saccule. At E13.5, SOX2 expression is found in all sensory progenitor cells and some cells located in CVG (Puligilla et al., 2010; Mak et al., 2009). Once differentiation of progenitors into either hair or supporting cells is initiated, SOX2 expression is downregulated in the differentiating hair cells but is strongly maintained in supporting cells until the adult stage in all sensory organs (Hume et al., 2007; Mak et al., 2009) (Figures 2 and 3).

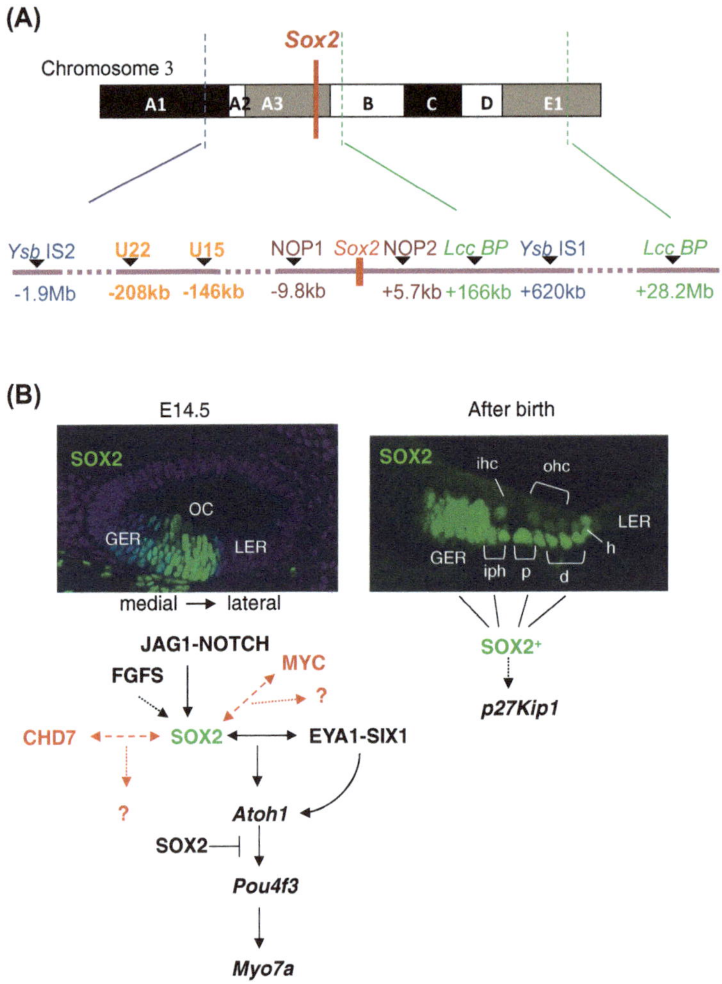

Figure 3 ***SOX2 gene regulatory network in development of the Organ of Corti.*** (A) Schematic diagram of mouse chromosome 3 showing the relative position of transgene insertion points (IS) in *Ysb*, the breakpoints (BP) in *Lcc*, two well-known *Sox2* regulatory elements, NOP1 and NOP2, and two otic enhancers active in chick (U22 and U15) that are conserved between chick and mouse (Uchikawa et al., 2003; Okamoto et al., 2015; Dong et al., 2002). Note mouse–chick conserved putative otic enhancers U22 and U15 have not been tested functionally in the mouse. (B) SOX2 (green) is expressed in all sensory progenitors in the primordial organ of Corti at E12.5–14.5 before their differentiation into either hair or supporting cells. Notch signaling is required to maintain *Sox2* expression domain within the sensory epithelium (Pan et al., 2010, 2013; Neves et al., 2011; Lanford et al., 1999), whereas Fgfs may have a role in maintaining *Sox2* expression levels (Ono et al., 2014). SOX2 interacts with transcription factors SIX1 and EYA1 to induce *Atoh*1 expression and hair cell differentiation (Ahmed et al., 2012a). During hair and supporting cell differentiation, SOX2 levels are largely reduced in the hair cells but strongly maintained in the supporting cells. Several studies have shown that SOX2 has an inhibitory effect on hair cell differentiation by antagonizing ATOH1 activity. In the supporting cells, SOX2 is required to maintain *p27Kip1* expression. SOX2 is also expressed in GER cells. Potential partners (in red) working with SOX2 in controlling prosensory development are SOX2 and CHD7 (based on embryonic stem cells) (Engelen et al., 2011) and SOX2 and MYC (based on *Myc* immortalized otic prosensory cells) (Kwan et al., 2014). For these partnerships validation of roles and downstream targets in sensorigenesis in the organ of Corti in vivo is still required (dashed/dotted red arrows).

This expression pattern is conserved between human and mouse and to a large extent in the chick (Neves et al., 2007). In humans, SOX2 expression has been found in the prosensory domain of developing cochlea at week 10 and 4 days (W10.4) and in hair cells and supporting cells at W14. By W19, SOX2 is downregulated in hair cells but remains in supporting cells (Locher et al., 2013).

SOX2 IS ESSENTIAL FOR PROSENSORY SPECIFICATION IN EARLY OTIC DEVELOPMENT

Hearing loss associated with human *SOX2* mutations implicates a role in inner ear development (Hagstrom et al., 2005). During inner ear morphogenesis, the process of prosensory specification defines the specific regions of the otic epithelium that will give rise to the six separate inner ear sensory organs. The essential role of *Sox2* in prosensory specification was discovered through studying two allelic mouse mutants with recessive deafness and balance impairment, *Light coat and circling* (*Lcc*/*Lcc*, completely deaf) induced by X-irradiation and Yellow submarine (*Ysb*/*Ysb*, severely hearing impaired) arising from transgene insertional mutagenesis (Dong et al., 2002; Kiernan et al., 2005). In *Lcc*/*Lcc* inner ears, all six sensory regions are absent, neither hair cells nor supporting cells differentiate, whereas the sensory epithelium is severely disrupted in *Ysb*/*Ysb*. Because the *Sox2* gene is intact in these mutants, these phenotypes are caused by the severely reduced (*Lcc*/*Lcc*) or reduced (*Ysb*/*Ysb*) expression of *Sox2*, specifically within the developing inner ear (Kiernan et al., 2005) because of X-ray or transgene insertion-induced inactivation of regulatory elements controlling the expression of *Sox2* in the inner ear. These mutants implicate a temporal and dose-dependent role for *Sox2* in determining prosensory development, similar to other developmental systems such as the eye, foregut, and pituitary (Dabdoub et al., 2008; Que et al., 2007; Taranova et al., 2006; Kelberman et al., 2006). The impact on both hair cell and supporting cell specification is consistent with an essential function for *Sox2* in the common progenitors for prosensory specification. Since this discovery, considerable effort has been made to understand the molecular controls, both upstream and downstream, as well as partner proteins in distinct cell types in the inner ear, as discussed below.

WIDELY SEPARATED DEOXYRIBONUCLEIC ACID ENHANCERS DIRECT OTIC *SOX2* EXPRESSION

Evidence points to distinct multiple enhancers that control *Sox2* expression in the inner ear, some of which act from long range (Figure 3(A)). The *Sox2* allelic mutants *Lcc* and *Ysb* arose because of disruption of regulatory elements located far downstream of the gene (Dong et al., 2002) (Cheah, unpublished data). Five phylogenetically conserved enhancers upstream and downstream of *Sox2* spanning over 50 kb of the chicken genome can direct *Sox2* expression in the neural plate of chicken embryos and neural tube in transgenic mice (Kamachi et al., 2009; Uchikawa et al., 2003; Zappone et al., 2000). Other enhancers exhibited distinct spatiotemporal specificities and activity corresponding to differing levels of *Sox2* expression at three stages of lens development. Therefore, separate elements are likely to be responsible for enhancing, maintaining and fine-tuning *Sox2* expression (Takemoto et al., 2006; Uchikawa et al., 2003). Two elements, NOP-1 (approximately 8 kb 5′) and NOP-2 (approximately 5 kb 3′), were individually active in early nasal and otic placodes. A study searching for enhancers in a 200-kb span of chicken genome encompassing the *Sox2* gene indicated the presence of at least four additional enhancers that are active in early inner ear development in the otic placode or vesicle (Okamoto et al., 2015) (refer to Chapter 7 for the variety of *Sox2*-associated enhancers). The NOP1 and NOP-2 enhancers are both intact in *Ysb* and *Lcc*, which suggests an underlying impact on other regulatory elements in these mutants. The differing degree of impact of the chromosomal changes in *Ysb* and *Lcc* on *Sox2* expression suggests that the level, temporal, and cell-specific otic transcription of *Sox2* is controlled by multiple long-range *cis*-regulatory elements. Whether the action of all of the chicken otic enhancers is functionally conserved in mammals is still unclear. It is also not known whether these elements continue to be active throughout all stages of *Sox2* expression or whether additional enhancer combinations come into play.

SOX2 AND INNER EAR NEUROGENESIS

The molecular controls regulating specification, delamination, and proliferation of otic neuroblasts and their differentiation and maturation are only partially understood (Maier et al., 2014; Appler and Goodrich, 2011). Various transcription factors are required for otic neurogenesis (Figure 2). As mentioned earlier, NEUROG1 is one of the earliest factors to be expressed in the neurogenic field in the otocyst and is essential for neuronal differentiation (Ma et al., 1998, 2000; Fritzsch et al., 2006). Other transcription factors important for otic neurogenesis are NEUROD1, TBX1, ISLET1, ASCL1 (Nishimura et al., 2014; Coate and Kelley, 2013), EYA1 (Zou et al., 2004), and SIX1 (Zheng et al., 2003). In the hypomorphic $Sox2^{Lcc/Lcc}$ mutants, sensory neurons form initially but are reduced and rapidly disappear by E15.5 (Puligilla et al., 2010) (Cheah and Fritzsch, unpublished work), which suggests that *Sox2* is not essential for the specification of neuronal precursors but is required for their maintenance. Neves and colleagues used the chick system to show that forced *Sox2* expression in the chick otic cup induces $TUJ1^{+}$ and $ISLET1^{+}$ neurogenic patches within neurosensory region of the otic epithelium (Neves et al., 2011). Moreover, forced expression of *Sox2* in the otic placode induced the expression of *Neurog1* and *Neurod* (Neves et al., 2012). To definitively address the role of *Sox2* in the specification of otic neuronal precursors, it will be necessary to ablate *Sox2* at the otic field/placode stage. Unfortunately, because of limitations in Cre reagents, it has not been possible to address the impact of complete loss of expression on the induction of neurogenesis.

Overexpression of *Sox2* in mouse cochlear epithelium can induce ectopic neuronal development (Puligilla et al., 2010; Evsen et al., 2013). Forced expression of *Sox2* in mouse cochlear nonsensory greater epithelial ridge (GER) cells adjacent to the sensory epithelium, which can be induced to differentiate into either hair cells or spiral neurons, induces immature neurons but is insufficient to induce *Neurog1* or *Neurod1* expression and neuronal differentiation to generate a mature neuron (Puligilla et al., 2010). In the mouse, SOX2 cooperates with the transcriptional complex EYA1–SIX1 as well as the SWI/SNF chromatin regulators to induce *Neurog1* and *Neurod1* activation to convert cochlear nonneuronal GER cells into mature spiral neurofilament-positive neurons (Ahmed et al., 2012b). How the chromatin-remodeling complex is recruited to chromatin and the direct targets of SOX2 and its partner proteins in neuronal differentiation are interesting areas for future studies.

SOX2 AND SENSORY CELL FATE SPECIFICATION AND DIFFERENTIATION IN THE COCHLEAR EPITHELIUM

Although the *Lcc* and *Ysb* mutants demonstrate the essential role of *Sox2* in sensory development, insights into the mechanistic basis for the lack of sensory specification in the absence of *Sox2* are limited. Below we will specifically discuss the possible role of *Sox2* and its interaction with other regulators during the formation and differentiation of the organ of Corti.

Nonsensory cells develop within the GERs and lesser epithelial ridges flanking the organ of Corti (Figure 2). SOX2 is expressed in the nascent organ of Corti and GER cells (Figure 3(B)). The progenitors within the nascent organ of Corti are induced to either adopt a hair cell fate or supporting cell fate, a process mediated by NOTCH1 signaling (Lanford et al., 1999) and the bHLH transcription factor ATOH1 (Bermingham et al., 1999). ATOH1 is essential for hair cell differentiation and its expression is activated in the base between E13.5 and E14.5 and reaches the apex at approximately E17.5 to induce hair cell differentiation from the base toward the apex (Chen et al., 2002). In the differentiated organ of Corti, there are three rows of outer hair cells and one row of inner along with distinct subtypes of underlying supporting cells such as inner pillar cells, outer pillar cells and Deiter's cells (Figure 2). SOX2 expression disappears from differentiating hair cells but is strongly maintained in supporting cells and GER cells at postnatal stages (Figure 3(B)).

Studies have shown that in the mouse, high levels of SOX2 have an inhibitory effect on hair cell differentiation by antagonizing ATOH1 activity (Dabdoub et al., 2008). In the supporting cells, SOX2 has been shown to be necessary for the maintenance of *p27Kip1* expression in pillar cells (a subtype of supporting cells) because deletion of *Sox2* in these cells at postnatal stages resulted in loss of *p27kip1* expression (Liu et al., 2012). In addition, studies have suggested that SOX2 may regulate *Prox1* in the supporting cells (Dabdoub et al., 2008). However, it remains unclear why SOX2 levels are highly maintained in supporting cells even at the adult stage, which do not reenter the cell cycle to replace lost hair cells.

Sensory hair cell fate specification is associated with the expression of *Atoh1*, which was found to be activated in a subpopulation of the prosensory progenitors that express multiple cellular factors, including SOX2, in a basal-to-apical and medial-to-lateral (inner to outer hair cells) gradient (Figures 1 and 2). Therefore, it is clear that SOX2 alone is insufficient to control the gradient pattern of *Atoh1* activation and hair cell differentiation in the cochlea and that interaction(s) with unique cofactor(s) is required for directing the progenitors to adopt a hair cell fate. Although *Atoh1* expression is extremely low or absent in the early otic vesicle stage, a mouse *Atoh1* enhancer reporter has been found to be active in the $SOX2^+$ domain when electroporated into the chick otic vesicle (Neves et al., 2012). SOX2 was shown to bind directly to the mouse *Atoh1* enhancer and *Sox2* overexpression in the chick otic vesicle transiently activates its expression. SOX2 chromatin immunoprecipitation (ChIP)-polymerase chain reaction analyses on otic vesicles showed binding not only to *Atoh1*, but also to several genes known to antagonize *Atoh1* expression, such as *Neurog1* and *Neurod* (Neves et al., 2012, 2013). These studies implicate a role of SOX2 in differentially regulating neurogenesis versus sensorigenesis during development in the chick. Several recent studies have revealed the importance of SOX2 in inducing hair cell fate, as discussed below.

THE *SOX2* GENE REGULATORY NETWORK: COOPERATION WITH SIGNALING PATHWAYS AND PARTNERS

The downregulation of *Sox2* as hair cells differentiate suggests that the dose of SOX2 and its temporal regulation is critical for hair cell differentiation. In carrying out its role, the SOX2 regulatory network interacts with several signaling pathways: Notch, FGF, and Hedgehog (Whitfield, 2015). In the ear, loss of its expression in *Jag1* conditional mutants suggests that SOX2 acts after the early specification of prosensory patches (Kiernan et al., 2006; Pan et al., 2010, 2013). Fibroblast growth factor (Fgf) signaling is essential for the development of hair cells: Mutants null for components of the pathway such as *Fgf receptor 1* (*Fgfr1*), *Fgf20*, and *Fgfr1–2/3* display deranged loss of hair cells (Huh et al., 2013; Ono et al., 2014; Pirvola et al., 2002). In loss of function *Fgfr1–2/3* mutants, *Sox2* expression is not maintained, which suggests that these factors may act upstream of *Sox2* to maintain its expression (Ono et al., 2014) (Figure 3).

Several studies have revealed its partner proteins during inner ear development. As discussed in Chapter 8, a distinct characteristic of SOX proteins is that they form a complex with partner proteins to function as either a transcriptional repressor or activator, depending on binding of their partner transcription factors to the deoxyribonucleic acid (DNA) adjacent to a functional SOX-binding site.

Coloboma, heart defect, atresia choanae (CHARGE) syndrome is associated with retarded growth and development and a variety of phenotypes affecting many organs including the eye, heart, gut, genitourinary axis, and ear. The chromatin remodeling adenosine triphosphatase CHD7 associated with CHARGE syndrome has been identified as a SOX2 transcriptional partner (Engelen et al., 2011). SOX2 and CHD7 interact directly and ChIP assays revealed that they both overlap in their binding to the genome. Loss of *Chd7* in mice also results in reduced *Sox2* expression (Hurd et al., 2010, 2012). Hedgehog signaling has a role in controlling the timing of hair cell differentiation and cell cycle exit (Tateya et al., 2013; Bok et al., 2013). The discovery that SOX2 and CHD7 regulate a common set of genes including *Jag1, Gli3*, and *Mycn* (Engelen et al., 2011; Hurd et al., 2010, 2012) suggests that the SOX2–CHD7 partnership controls inner ear development in concert by regulating notch and hedgehog signaling and cell proliferation.

SOX2 IN HAIR CELL FATE INDUCTION IN CONCERT WITH PARTNERS

Cell cycle control is critical in the differentiation of sensory and nonsensory cells in the inner ear (Schimmang and Pirvola, 2013). During cochlear outgrowth from the ventral otocyst, the progenitors proliferate to expand; after reaching a certain number, they exit the cell cycle from the apex toward the base of the cochlear duct between E12.5 and E14.5 to form the prosensory domain, the primordium of the organ of Corti distinctively marked by the expression of cell cycle inhibitor p27KIP1 (Ruben, 1967; Chen and Segil, 1999; Lee et al., 2006). Several studies implicate a role for SOX2 partnerships in controlling hair cell differentiation via regulation of the cell cycle. SOX2 directly activates *p27Kip1* in postmitotic inner pillar cells whereas p27KIP1 directly represses *Sox2* during embryonic stem cell differentiation (Liu et al., 2012; Li et al., 2012). *Sox2* and *p27Kip1* may act in an autoregulatory loop to control the induction of hair cell differentiation. *Myc* (*c-myc*) and *Mycn* are expressed in the inner ear. Loss of *Mycn* function in the inner ear causes abnormalities in proliferation and differentiation of both sensory and nonsensory cells (Kopecky et al., 2011; Dominguez-Frutos et al., 2011). A study on multipotent otic progenitor cells derived by immortalization of SOX2 expressing neurosensory precursors that are competent for hair-cell and supporting-cell differentiation has implicated an SOX2–MYC partnership in regulating the cell cycle and proliferation in the induction of prosensory development and the initiation of hair cell differentiation (Kwan et al., 2014). Chromatin immunoprecipitation sequencing (ChIP-seq) and transcriptome (ribonucleic acid (RNA)-seq) analyses showed a large overlap (about 85%) between SOX2 and MYC in binding target promoters. It was particularly interesting that SOX2 and MYC co-occupied common promoters of genes (such as cyclin-dependent kinase genes) that control cell cycle progression and the initiation of hair cell differentiation. It would be important to validate the SOX2–MYC partnership in regulating cell cycle control in otic sensorigenesis in vivo.

During inner ear hair cell specification, EYA1 and SIX1 have been identified as partner transcription factors of SOX2, and these three factors form a complex to activate *Atoh1* transcription to induce hair cell fate. EYA1 is a member of the Eyes absent (EYA) protein family and is a transcriptional coactivator that affects transcription indirectly by interacting with DNA-binding proteins such as SIX1 and SOX2 (Ahmed et al., 2012a; Xu et al., 1997a,b; Pignoni et al., 1997). SIX1 belongs to the Sine oculis (SO/SIX) protein family that represents a divergent group of the homeodomain (HD) superfamily containing the DNA-binding HD and the SIX-specific domain that participates in protein–protein interactions (Pignoni et al., 1997). Depending on the context, SO/SIX proteins can act as either transcriptional activators (Ohto et al., 1999; Silver et al., 2003) or repressors (Li et al., 2002). However, EYA–SIX interactions lead to formation of a bipartite transcription factor, leading to potent transcriptional activation, even under conditions in which SIX alone acts as a repressor (Ohto et al., 1999; Silver et al., 2003; Li et al., 2003; Ahmed et al., 2012a). EYA–SIX interactions also lead to nuclear translocation of EYA in cultured cells (Ohto et al., 1999; Buller et al., 2001) and developmental organs (Xu et al., 2014).

Eya1 and *Six1* are coexpressed with *Sox2* in the neurosensory progenitors from the placodal stage and in the cochlear epithelium until E12.5, and from E12.5 these three factors show differential pattern of expression. During hair cell specification, *Six1* was found to be activated in a subpopulation of prosensory progenitors that expresses both *Eya1* and *Sox2* from a basal-to-apical and medial-to-lateral direction, a pattern similar to that of *Atoh1* expression (Ahmed et al., 2012a). This unique pattern of *Six1* expression and its onset of expression earlier than that of *Atoh1* suggest that *Atoh1* activation might be achieved by combinatorial regulation of *Sox2, Eya1*, and *Six1* but SIX1 might serve as a critical positive regulator for controlling the developmental pattern of *Atoh1* expression in the cochlear duct. Evidence from gain-of-function studies indicates that SOX2, SIX1, and EYA1 proteins directly interact to control *Atoh1* expression and thereby specify hair cell fate (Ahmed et al., 2012a).

Several groups carried out electroporation experiments in different systems to examine the requirement of SOX2 in hair cell induction. When *Sox2* is electroporated alone into mouse cochlear sensory epithelium, it inhibits hair cell formation because of its antagonistic effect against ATOH1 (Dabdoub et al., 2008). Forced expression of *Sox2* in cochlear GER cells fails to induce hair cell fate (Puligilla et al., 2010). In zebrafish, *sox2* is not required for initial development of the sensory epithelium, but co-misexpression of *sox2* with *atoh1* enhances the competence of the otic epithelium to respond to ATOH1 (Sweet et al., 2011). Myosin VIIa (Myo7a) is a marker for differentiating hair cells. Forced *Sox2* expression in the chick otic cup induces not only $TUJ1^+$ and $ISLET1^+$ neurogenic patches but also $MYO7A^+$ ectopic hair cells within neurosensory or nonsensory region of the otic epithelium (Neves et al., 2011). Although high levels of SOX2 are known to inhibit hair cell differentiation by antagonizing ATOH1, this study suggests sufficiency of SOX2 in inducing not only neuronal but also sensory hair cell differentiation in the chick. However, these studies do not address whether SOX2 induces sensory cell development via directly activating *Atoh1* expression and whether the ectopic $MYO7A^+$ cells are also $ATOH1^+$. Whether the regulatory mechanisms are conserved between chick and mouse is also unclear.

The sufficiency of SOX2 with its partner proteins EYA1 and SIX1 in hair cell induction was addressed using a mouse cochlear explant system (Ahmed et al., 2012a). When *Sox2* was electroporated alone or in combination with *Eya1, Six1*, or both together into nonsensory epithelial cells within the GER of embryonic cochlea, no ectopic $MYO7A^+$ hair cells were induced. This was consistent with the observation that a high dose of SOX2 inhibits hair cell formation. Forced expression of both *Eya1/Six1* without SOX2 in the GER cells induced ectopic $MYO7A^+$ hair cells (Ahmed et al., 2012a). However, this ability of *Eya1–Six1* in hair cell induction requires low endogenous SOX2 activity, which is present in the GER cells, to activate *Atoh1*.

The direct molecular link between EYA1, SIX1, SOX2, and *Atoh*1 has been uncovered (Figure 4). Deoxyribonucleic acid-binding sites for SOX2 and SIX1 were found in the 1.4-kb *Atoh1* enhancer, which is located approximately 3.4 kb 3′ of the *Atoh1* coding sequence and contains two conserved elements (element A and B) and can direct expression of reporter transgenes to the inner ear hair cells (Helms et al., 2000; Chow et al., 2006). A functional SOX2-binding element was found to be located in the conserved element A (Figure 4(A)), which is accompanied by a binding consensus site for the *Drosophila* So protein. Electroporation of *Sox2* alone into the cochlear GER cells with a reporter that contains the SOX2- and So-binding sites was able to weakly transactivate the reporter and this function was mediated through the SOX2-binding site. As discussed in Chapter 8, the binding of SOX protein alone to DNA does not exert their gene regulatory functions, which can only be achieved by forming transcriptional complexes with partner transcription factors (Kamachi and Kondoh, 2013); the weak activation of the reporter induced by forced *Sox2* expression might depend on interaction with endogenous EYA1 and SIX1 activity in the GER cells. Co-misexpression of *Sox2* with *Eya1* and *Six1* led to a large increase in reporter activity and these proteins direct interact in vitro and in vivo, which further indicates that SOX2 forms a transcriptional complex with EYA1 and SIX1 to activate *Atoh1*. In contrast to the element A, the conserved enhancer B was robustly activated by EYA1/SIX1 via the SIX1-binding site and addition of SOX2 did not appear to have an effect on this element (Figure 4(B)). Based on distinct responsiveness of the *Atoh1* enhancers to different combinations of SOX2, EYA1, and SIX1, it has been proposed that enhancer A may be critical for mediating

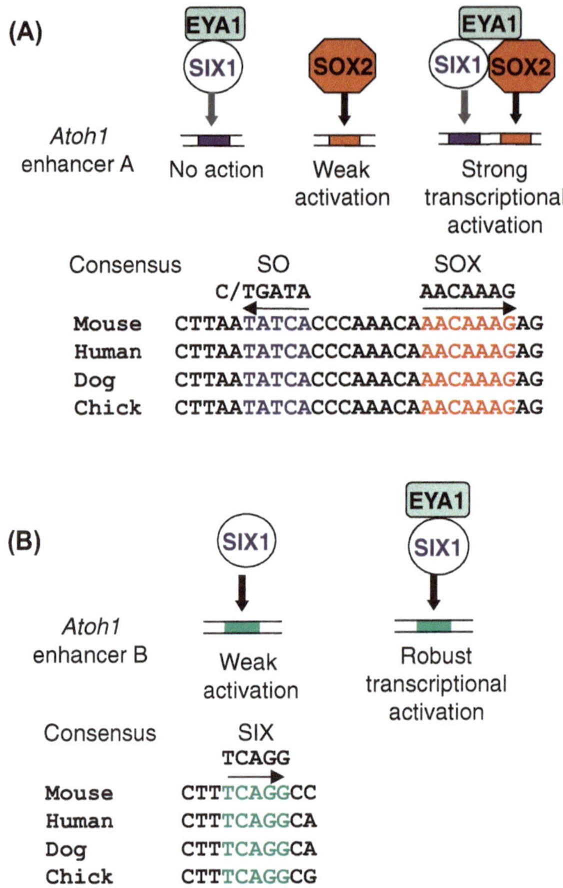

Figure 4 ***Co-regulation of* Atoh1 *transcription by SOX2, EYA1, and SIX1.*** (A) Functional binding sequences of SOX2–SIX1–EYA1 complexes were identified in the *Atoh1* enhancer A region. The SOX2-binding sequences match the consensus, whereas the SIX1-binding sequences differ from the mammalian Six-binding consensus, but conform to the consensus for the *Drosophila* So protein. Gel-shift assay confirmed binding of SIX1 to this consensus sequences (Ahmed et al., 2012a). (B) Functional binding sequences of SIX1–EYA1 complex were identified in the *Atoh1* enhancer B region. SOX2 has no effect on this enhancer region (Ahmed et al., 2012a).

initial *Atoh1* activation, which could be stimulated by SIX1 in cooperation with SOX2 and EYA1 within the organ of Corti. In contrast, enhancer B on its own is likely to be essential for the maintenance or upregulation of *Atoh1* expression to promote hair cell differentiation. This explains why EYA1 and SIX1 but not SOX2 are abundantly expressed in the hair cells even after birth. In the absence of SIX1, EYA1/SOX2 alone may not lead to the formation of an active complex capable of transcriptional activation, which explains why in the primordial organ of Corti, *Atoh1* activation is not initiated in all prosensory precursors that express both EYA1 and SOX2.

PERSPECTIVE AND REGENERATIVE MEDICINE FOR THE INNER EAR

The mammalian cochlear hair cells are known to lose proliferative capacities after they enter the differentiation stage. Unlike those in nonmammalian vertebrates such as birds and fish, lost hair cells cannot be replaced through proliferation and transdifferentiation of surrounding supporting cells. $SOX2^{+}OCT4^{+}$ stem cells isolated from human fetal cochleae and human embryonic stem cell–derived otic progenitors can be induced to differentiate into sensory hair cells and neurons (Boddy et al., 2012; Chen et al., 2007, 2009, 2012). Despite progress in the quest for stem cell-based therapies for the human inner ear, there are considerable challenges (Okano and Kelley, 2012). Because SOX2 is a key factor controlling sensory cell development in the inner ear, better knowledge of hair cell specification and differentiation in relation to the expression of SOX2 with its partner proteins would be extremely important and beneficial for understanding of normal development of the cochlea and how to restore hearing in damaged cochlea. Identification of the critical role of SOX2 and its partner proteins has begun to elucidate the molecular events controlling neuronal or hair cell specification. However, significant questions remain about the downstream targets of SOX2 and the full range of SOX2 partners: how it interacts with its partner proteins to exert positional cues to regulate specification, proliferation, differentiation, and maintenance, and what intrinsic and extrinsic mechanisms control the levels of these transcription factors. SOX2 expression is maintained in supporting cells in the adult inner ear, which, unlike those of nonmammalian vertebrates, remain quiescent even after sensory hair cell damage or loss. Although studies reported that SOX2 might regulate *p27Kip1* expression in the supporting cells to maintain a quiescent state (Liu et al., 2012), the nature and molecular controls of supporting cell differentiation during development and the significance of persistent SOX2 expression in those cells are not well understood. To date, we do not fully understand why $SOX2^{+}$ supporting cells do not respond to damage to induce re-expression of developmental regulatory genes important in embryonic development and growth. The greatest challenge to achieving transdifferentiation and regeneration is how to reactivate the molecular pathways that control hair cell specification, differentiation, maintenance, and function during development. Activation of *Atoh1* can induce hair cell fate but ATOH1 alone is apparently insufficient to promote hair cell differentiation, maintenance, and function. The mechanism governing transdifferentiation in lower vertebrates is still not well understood, although studies in chicks suggest that ATOH1 is involved in hair cell regeneration (Cafaro et al., 2007). As described above, SOX2 alone is insufficient to activate robust *Atoh1* transcription and it requires cofactors EYA1 and SIX1, both of which are not expressed in supporting cells in the ear. After *Atoh1* activation, *Sox2* also needs to be downregulated for hair cell differentiation to occur. Distinct regulatory mechanisms for controlling *Sox2* levels at different stages and cell types, the combinatorial code of target specificity defined by SOX2 and its distinct cofactors, as well as chromatin regulators and its transcriptional activation and repression activities in the inner ear have left much to be revealed. Future studies with stage- and cell type–specific ChIP-seq and RNA-seq analysis and proteomics are essential for a molecular understanding of SOX2-regulatory networks in the different phases of sensory development in the inner ear and how these are disrupted in disease. A comprehensive understanding of growth control in the neurosensory progenitors, the rules that govern cellular behavior of $SOX2^+$ supporting cells in adult ear, and the nature of molecular factors necessary for inducing competency of cells to respond to damage or an instructive signal for inner ear sensory cell formation would be important in inner ear research with implications in disease such as sensorineural hearing loss.

ACKNOWLEDGMENTS

We thank Michael Kwong for making the figures. KSEC is supported by the Research Grants Council of Hong Kong (Projects #761910 and #17102914) and PXX is supported by grants from NIH and NYSTEM (C029566).

REFERENCES

Ahmed, M., Wong, E.Y., Sun, J., Xu, J., Wang, F., Xu, P.X., 2012a. Eya1-Six1 interaction is sufficient to induce hair cell fate in the cochlea by activating Atoh1 expression in cooperation with Sox2. Dev. Cell 22, 377–390.

Ahmed, M., Xu, J., Xu, P.X., 2012b. EYA1 and SIX1 drive the neuronal developmental program in cooperation with the SWI/SNF chromatin-remodeling complex and SOX2 in the mammalian inner ear. Development 139, 1965–1977.

Appler, J.M., Goodrich, L.V., 2011. Connecting the ear to the brain: molecular mechanisms of auditory circuit assembly. Prog. Neurobiol. 93, 488–508.

Bermingham, N.A., Hassan, B.A., Price, S.D., Vollrath, M.A., Ben-Arie, N., Eatock, R.A., Bellen, H.J., Lysakowski, A., Zoghbi, H.Y., 1999. Math1: an essential gene for the generation of inner ear hair cells. Science 284, 1837–1841.

Boddy, S.L., Chen, W., Romero-Guevara, R., Kottam, L., Bellantuono, I., Rivolta, M.N., 2012. Inner ear progenitor cells can be generated in vitro from human bone marrow mesenchymal stem cells. Regen. Med. 7, 757–767.

Bok, J., Zenczak, C., Hwang, C.H., Wu, D.K., 2013. Auditory ganglion source of Sonic hedgehog regulates timing of cell cycle exit and differentiation of mammalian cochlear hair cells. Proc. Natl. Acad. Sci. U.S.A. 110, 13869–13874.

Buller, C., Xu, X., Marquis, V., Schwanke, R., Xu, P.X., 2001. Molecular effects of Eya1 domain mutations causing organ defects in BOR syndrome. Hum. Mol. Genet. 10, 2775–2781.

Cafaro, J., Lee, G.S., Stone, J.S., 2007. Atoh1 expression defines activated progenitors and differentiating hair cells during avian hair cell regeneration. Dev. Dyn. 236, 156–170.

Chen, P., Johnson, J.E., Zoghbi, H.Y., Segil, N., 2002. The role of Math1 in inner ear development: uncoupling the establishment of the sensory primordium from hair cell fate determination. Development 129, 2495–2505.

Chen, P., Segil, N., 1999. p27(Kip1) links cell proliferation to morphogenesis in the developing organ of Corti. Development 126, 1581–1590.

Chen, W., Cacciabue-Rivolta, D.I., Moore, H.D., Rivolta, M.N., 2007. The human fetal cochlea can be a source for auditory progenitors/stem cells isolation. Hear. Res. 233, 23–29.

Chen, W., Johnson, S.L., Marcotti, W., Andrews, P.W., Moore, H.D., Rivolta, M.N., 2009. Human fetal auditory stem cells can be expanded in vitro and differentiate into functional auditory neurons and hair cell-like cells. Stem Cells 27, 1196–1204.

Chen, W., Jongkamonwiwat, N., Abbas, L., Eshtan, S.J., Johnson, S.L., Kuhn, S., Milo, M., Thurlow, J.K., Andrews, P.W., Marcotti, W., Moore, H.D., Rivolta, M.N., 2012. Restoration of auditory evoked responses by human ES-cell-derived otic progenitors. Nature 490, 278–282.

Chow, L.M., Tian, Y., Weber, T., Corbett, M., Zuo, J., Baker, S.J., 2006. Inducible Cre recombinase activity in mouse cerebellar granule cell precursors and inner ear hair cells. Dev. Dyn. 235, 2991–2998.

Coate, T.M., Kelley, M.W., 2013. Making connections in the inner ear: recent insights into the development of spiral ganglion neurons and their connectivity with sensory hair cells. Semin. Cell Dev. Biol. 24, 460–469.

Dabdoub, A., Puligilla, C., Jones, J.M., Fritzsch, B., Cheah, K.S., Pevny, L.H., Kelley, M.W., 2008. Sox2 signaling in prosensory domain specification and subsequent hair cell differentiation in the developing cochlea. Proc. Natl. Acad. Sci. U.S.A. 105, 18396–18401.

Dominguez-Frutos, E., Lopez-Hernandez, I., Vendrell, V., Neves, J., Gallozzi, M., Gutsche, K., Quintana, L., Sharpe, J., Knoepfler, P.S., Eisenman, R.N., Trumpp, A., Giraldez, F., Schimmang, T., 2011. N-myc controls proliferation, morphogenesis, and patterning of the inner ear. J. Neurosci. 31, 7178–7189.

Dong, S., Leung, K.K., Pelling, A.L., Lee, P.Y., Tang, A.S., Heng, H.H., Tsui, L.C., Tease, C., Fisher, G., Steel, K.P., Cheah, K.S., 2002. Circling, deafness, and yellow coat displayed by yellow submarine (ysb) and light coat and circling (lcc) mice with mutations on chromosome 3. Genomics 79, 777–784.

Engelen, E., Akinci, U., Bryne, J.C., Hou, J., Gontan, C., Moen, M., Szumska, D., Kockx, C., Van Ijcken, W., Dekkers, D.H., Demmers, J., Rijkers, E.J., Bhattacharya, S., Philipsen, S., Pevny, L.H., Grosveld, F.G., Rottier, R.J., Lenhard, B., Poot, R.A., 2011. Sox2 cooperates with Chd7 to regulate genes that are mutated in human syndromes. Nat. Genet. 43, 607–611.

Evsen, L., Sugahara, S., Uchikawa, M., Kondoh, H., Wu, D.K., 2013. Progression of neurogenesis in the inner ear requires inhibition of Sox2 transcription by neurogenin1 and neurod1. J. Neurosci. 33, 3879–3890.

Fekete, D.M., Muthukumar, S., Karagogeos, D., 1998. Hair cells and supporting cells share a common progenitor in the avian inner ear. J. Neurosci. 18, 7811–7821.

Fritzsch, B., Beisel, K.W., Hansen, L.A., 2006. The molecular basis of neurosensory cell formation in ear development: a blueprint for hair cell and sensory neuron regeneration? BioEssays 28, 1181–1193.

Groves, A.K., Zhang, K.D., Fekete, D.M., 2013. The genetics of hair cell development and regeneration. Annu. Rev. Neurosci. 36, 361–381.

Hagstrom, S.A., Pauer, G.J., Reid, J., Simpson, E., Crowe, S., Maumenee, I.H., Traboulsi, E.I., 2005. SOX2 mutation causes anophthalmia, hearing loss, and brain anomalies. Am. J. Med. Genet. A 138A, 95–98.

Helms, A.W., Abney, A.L., Ben-Arie, N., Zoghbi, H.Y., Johnson, J.E., 2000. Autoregulation and multiple enhancers control Math1 expression in the developing nervous system. Development 127, 1185–1196.

Huh, S.H., Narhi, K., Lindfors, P.H., Haara, O., Yang, L., Ornitz, D.M., Mikkola, M.L., 2013. Fgf20 governs formation of primary and secondary dermal condensations in developing hair follicles. Genes Dev. 27, 450–458.

Hume, C.R., Bratt, D.L., Oesterle, E.C., 2007. Expression of LHX3 and SOX2 during mouse inner ear development. Gene Expression Patterns 7, 798–807.

Hurd, E.A., Micucci, J.A., Reamer, E.N., Martin, D.M., 2012. Delayed fusion and altered gene expression contribute to semicircular canal defects in Chd7 deficient mice. Mech. Dev. 129, 308–323.

Hurd, E.A., Poucher, H.K., Cheng, K., Raphael, Y., Martin, D.M., 2010. The ATP-dependent chromatin remodeling enzyme CHD7 regulates pro-neural gene expression and neurogenesis in the inner ear. Development 137, 3139–3150.

Kamachi, Y., Iwafuchi, M., Okuda, Y., Takemoto, T., Uchikawa, M., Kondoh, H., 2009. Evolution of non-coding regulatory sequences involved in the developmental process: reflection of differential employment of paralogous genes as highlighted by Sox2 and group B1 Sox genes. Proc. Jpn. Acad. Ser. B Phys. Biol. Sci. 85, 55–68.

Kamachi, Y., Kondoh, H., 2013. Sox proteins: regulators of cell fate specification and differentiation. Development 140, 4129–4144.

Kelberman, D., Rizzoti, K., Avilion, A., Bitner-Glindzicz, M., Cianfarani, S., Collins, J., Chong, W.K., Kirk, J.M., Achermann, J.C., Ross, R., Carmignac, D., Lovell-Badge, R., Robinson, I.C., Dattani, M.T., 2006. Mutations within Sox2/SOX2 are associated with abnormalities in the hypothalamo-pituitary-gonadal axis in mice and humans. J. Clin. Invest. 116, 2442–2455.

Kiernan, A.E., Pelling, A.L., Leung, K.K., Tang, A.S., Bell, D.M., Tease, C., Lovell-Badge, R., Steel, K.P., Cheah, K.S., 2005. Sox2 is required for sensory organ development in the mammalian inner ear. Nature 434, 1031–1035.

Kiernan, A.E., Xu, J., Gridley, T., 2006. The Notch ligand JAG1 is required for sensory progenitor development in the mammalian inner ear. PLoS Genet. 2, e4.

Kopecky, B., Santi, P., Johnson, S., Schmitz, H., Fritzsch, B., 2011. Conditional deletion of N-Myc disrupts neurosensory and non-sensory development of the ear. Dev. Dyn. 240, 1373–1390.

Kwan, K.Y., Shen, J., Corey, D.P., 2014. C-MYC transcriptionally amplifies SOX2 target genes to regulate self-renewal in multipotent otic progenitor cells. Stem Cell Rep. 4 (1), 47–60.

Lanford, P.J., Lan, Y., Jiang, R., Lindsell, C., Weinmaster, G., Gridley, T., Kelley, M.W., 1999. Notch signalling pathway mediates hair cell development in mammalian cochlea. Nat. Genet. 21, 289–292.

Lang, H., Fekete, D.M., 2001. Lineage analysis in the chicken inner ear shows differences in clonal dispersion for epithelial, neuronal, and mesenchymal cells. Dev. Biol. 234, 120–137.

Lee, Y.S., Liu, F., Segil, N., 2006. A morphogenetic wave of p27Kip1 transcription directs cell cycle exit during organ of Corti development. Development 133, 2817–2826.

Li, H., Collado, M., Villasante, A., Matheu, A., Lynch, C.J., Canamero, M., Rizzoti, K., Carneiro, C., Martinez, G., Vidal, A., Lovell-Badge, R., Serrano, M., 2012. p27(Kip1) directly represses Sox2 during embryonic stem cell differentiation. Cell Stem Cell 11, 845–852.

Li, X., Oghi, K.A., Zhang, J., Krones, A., Bush, K.T., Glass, C.K., Nigam, S.K., Aggarwal, A.K., Maas, R., Rose, D.W., Rosenfeld, M.G., 2003. Eya protein phosphatase activity regulates Six1-Dach-Eya transcriptional effects in mammalian organogenesis. Nature 426, 247–254.

Li, X., Perissi, V., Liu, F., Rose, D.W., Rosenfeld, M.G., 2002. Tissue-specific regulation of retinal and pituitary precursor cell proliferation. Science 297, 1180–1183.

Liu, Z., Walters, B.J., Owen, T., Brimble, M.A., Steigelman, K.A., Zhang, L., Mellado Lagarde, M.M., Valentine, M.B., Yu, Y., Cox, B.C., Zuo, J., 2012. Regulation of p27Kip1 by Sox2 maintains quiescence of inner pillar cells in the murine auditory sensory epithelium. J. Neurosci. 32, 10530–10540.

Locher, H., Frijns, J.H., van Iperen, L., de Groot, J.C., Huisman, M.A., Chuva de Sousa Lopes, S.M., 2013. Neurosensory development and cell fate determination in the human cochlea. Neural Dev. 8, 20.

Ma, Q., Anderson, D.J., Fritzsch, B., 2000. Neurogenin 1 null mutant ears develop fewer, morphologically normal hair cells in smaller sensory epithelia devoid of innervation. J. Assoc. Res. Otolaryngol. 1, 129–143.

Ma, Q., Chen, Z., del Barco Barrantes, I., de la Pompa, J.L., Anderson, D.J., 1998. neurogenin1 is essential for the determination of neuronal precursors for proximal cranial sensory ganglia. Neuron 20, 469–482.

Maier, E.C., Saxena, A., Alsina, B., Bronner, M.E., Whitfield, T.T., 2014. Sensational placodes: neurogenesis in the otic and olfactory systems. Dev. Biol. 389, 50–67.

Mak, A.C., Szeto, I.Y., Fritzsch, B., Cheah, K.S., 2009. Differential and overlapping expression pattern of SOX2 and SOX9 in inner ear development. Gene Expression Patterns 9, 444–453.

Neves, J., Kamaid, A., Alsina, B., Giraldez, F., 2007. Differential expression of Sox2 and Sox3 in neuronal and sensory progenitors of the developing inner ear of the chick. J. Comp. Neurol. 503, 487–500.

Neves, J., Parada, C., Chamizo, M., Giraldez, F., 2011. Jagged 1 regulates the restriction of Sox2 expression in the developing chicken inner ear: a mechanism for sensory organ specification. Development 138, 735–744.

Neves, J., Uchikawa, M., Bigas, A., Giraldez, F., 2012. The prosensory function of Sox2 in the chicken inner ear relies on the direct regulation of Atoh1. PLoS One 7, e30871.

Neves, J., Vachkov, I., Giraldez, F., 2013. Sox2 regulation of hair cell development: incoherence makes sense. Hear. Res. 297, 20–29.

Nishimura, K., Weichert, R.M., Liu, W., Davis, R.L., Dabdoub, A., 2014. Generation of induced neurons by direct reprogramming in the mammalian cochlea. Neuroscience 275, 125–135.

Ohto, H., Kamada, S., Tago, K., Tominaga, S.I., Ozaki, H., Sato, S., Kawakami, K., 1999. Cooperation of six and eya in activation of their target genes through nuclear translocation of Eya. Mol. Cell. Biol. 19, 6815–6824.

Okamoto, R., Uchikawa, M., Kondoh, H., 2015. Sixteen additional enhancers associated with the chicken Sox2 locus outside the central 50-kb region. Dev. Growth Differ. 57, 24–39.

Okano, T., Kelley, M.W., 2012. Stem cell therapy for the inner ear: recent advances and future directions. Trends Amplif. 16, 4–18.

Ono, K., Kita, T., Sato, S., O'Neill, P., Mak, S.S., Paschaki, M., Ito, M., Gotoh, N., Kawakami, K., Sasai, Y., Ladher, R.K., 2014. FGFR1-Frs2/3 signalling maintains sensory progenitors during inner ear hair cell formation. PLoS Genet. 10, e1004118.

Pan, W., Jin, Y., Chen, J., Rottier, R.J., Steel, K.P., Kiernan, A.E., 2013. Ectopic expression of activated notch or SOX2 reveals similar and unique roles in the development of the sensory cell progenitors in the mammalian inner ear. J. Neurosci. 33, 16146–16157.

Pan, W., Jin, Y., Stanger, B., Kiernan, A.E., 2010. Notch signaling is required for the generation of hair cells and supporting cells in the mammalian inner ear. Proc. Natl. Acad. Sci. U.S.A. 107, 15798–15803.

Pignoni, F., Hu, B., Zavitz, K.H., Xiao, J., Garrity, P.A., Zipursky, S.L., 1997. The eye-specification proteins So and Eya form a complex and regulate multiple steps in Drosophila eye development. Cell 91, 881–891.

Pirvola, U., Ylikoski, J., Trokovic, R., Hebert, J.M., Mcconnell, S.K., Partanen, J., 2002. FGFR1 is required for the development of the auditory sensory epithelium. Neuron 35, 671–680.

Puligilla, C., Dabdoub, A., Brenowitz, S.D., Kelley, M.W., 2010. Sox2 induces neuronal formation in the developing mammalian cochlea. J. Neurosci. 30, 714–722.

Que, J., Okubo, T., Goldenring, J.R., Nam, K.T., Kurotani, R., Morrisey, E.E., Taranova, O., Pevny, L.H., Hogan, B.L., 2007. Multiple dose-dependent roles for Sox2 in the patterning and differentiation of anterior foregut endoderm. Development 134, 2521–2531.

Ruben, R.J., 1967. Development of the inner ear of the mouse: a radioautographic study of terminal mitoses. Acta Otolaryngol. (Suppl. 220), 1–44.

Sanchez-Calderon, H., Milo, M., Leon, Y., Varela-Nieto, I., 2007. A network of growth and transcription factors controls neuronal differentation and survival in the developing ear. Int. J. Dev. Biol. 51, 557–570.

Schimmang, T., Pirvola, U., 2013. Coupling the cell cycle to development and regeneration of the inner ear. Semin. Cell Dev. Biol. 24, 507–513.

Silver, S.J., Davies, E.L., Doyon, L., Rebay, I., 2003. Functional dissection of eyes absent reveals new modes of regulation within the retinal determination gene network. Mol. Cell. Biol. 23, 5989–5999.

Sweet, E.M., Vemaraju, S., Riley, B.B., 2011. Sox2 and Fgf interact with Atoh1 to promote sensory competence throughout the zebrafish inner ear. Dev. Biol. 358, 113–121.

Takemoto, T., Uchikawa, M., Kamachi, Y., Kondoh, H., 2006. Convergence of Wnt and FGF signals in the genesis of posterior neural plate through activation of the Sox2 enhancer N-1. Development 133, 297–306.

Taranova, O.V., Magness, S.T., Fagan, B.M., Wu, Y., Surzenko, N., Hutton, S.R., Pevny, L.H., 2006. SOX2 is a dose-dependent regulator of retinal neural progenitor competence. Genes Dev. 20, 1187–1202.

Tateya, T., Imayoshi, I., Tateya, I., Hamaguchi, K., Torii, H., Ito, J., Kageyama, R., 2013. Hedgehog signaling regulates prosensory cell properties during the basal-to-apical wave of hair cell differentiation in the mammalian cochlea. Development 140, 3848–3857.

Uchikawa, M., Ishida, Y., Takemoto, T., Kamachi, Y., Kondoh, H., 2003. Functional analysis of chicken Sox2 enhancers highlights an array of diverse regulatory elements that are conserved in mammals. Dev. Cell 4, 509–519.

Uchikawa, M., Kamachi, Y., Kondoh, H., 1999. Two distinct subgroups of Group B Sox genes for transcriptional activators and repressors: their expression during embryonic organogenesis of the chicken. Mech. Dev. 84, 103–120.

Whitfield, T.T., 2015. Development of the inner ear. Curr. Opin. Genet. Dev. 32, 112–118.

Wood, H.B., Episkopou, V., 1999. Comparative expression of the mouse Sox1, Sox2 and Sox3 genes from pre-gastrulation to early somite stages. Mech. Dev. 86, 197–201.

Xu, J., Wong, E.Y., Cheng, C., Li, J., Sharkar, M.T., Xu, C.Y., Chen, B., Sun, J., Jing, D., Xu, P.X., 2014. Eya1 interacts with Six2 and Myc to regulate expansion of the nephron progenitor pool during nephrogenesis. Dev. Cell 31, 434–447.

Xu, P.X., Cheng, J., Epstein, J.A., Maas, R.L., 1997a. Mouse Eya genes are expressed during limb tendon development and encode a transcriptional activation function. Proc. Natl. Acad. Sci. U.S.A. 94, 11974–11979.

Xu, P.X., Woo, I., Her, H., Beier, D.R., Maas, R.L., 1997b. Mouse Eya homologues of the Drosophila eyes absent gene require Pax6 for expression in lens and nasal placode. Development 124, 219–231.

Zappone, M.V., Galli, R., Catena, R., Meani, N., De Biasi, S., Mattei, E., Tiveron, C., Vescovi, A.L., Lovell-Badge, R., Ottolenghi, S., Nicolis, S.K., 2000. Sox2 regulatory sequences direct expression of a (beta)-geo transgene to telencephalic neural stem cells and precursors of the mouse embryo, revealing regionalization of gene expression in CNS stem cells. Development 127, 2367–2382.

Zheng, W., Huang, L., Wei, Z.B., Silvius, D., Tang, B., Xu, P.X., 2003. The role of Six1 in mammalian auditory system development. Development 130, 3989–4000.

Zou, D., Silvius, D., Fritzsch, B., Xu, P.X., 2004. Eya1 and Six1 are essential for early steps of sensory neurogenesis in mammalian cranial placodes. Development 131, 5561–5572.

CHAPTER 16

SOX2 in the Skin

Natacha A. Agabalyan, Andrew Hagner, Waleed Rahmani, Jeff Biernaskie
Department of Comparative Biology and Experimental Medicine, Faculty of Veterinary Medicine and Alberta Children's Hospital Research Institute, Cumming School of Medicine, University of Calgary, Calgary, Albert, Canada

The skin functions as the body's primary barrier to the external environment, organized to repel pathogens and other environmental insults. Depending on the anatomical location, the skin can also be accompanied by a variety of ectodermal growths and glands such as hair and nails; mammary, sebaceous, and sweat glands; as well as a variety of immune and nerve cells. Altogether, the skin serves as more than just a physical impediment: it provides an immune barrier and has an important role in sensation, thermoregulation, nutrient absorption, fluid retention, and vitamin D synthesis. Maintenance of this dynamic organ and its diverse functions requires a continuous and coordinated turnover of somatic cells.

The skin can be divided into two developmentally distinct layers: the thin, dense outermost layer called the epidermis and the thick, supportive, elastic inner layer called the dermis. Because of the rate of cellular turnover in mammalian skin, it is perhaps not surprising that several stem cell populations have been identified, including those of the basal epidermis and in the bulge of the hair follicle. It might then follow that SOX2, a transcription factor often associated with stem cell self-renewal in other cell types such as embryonic stem cells, may also have a similar role in the stem cells of the skin. This has never been definitively shown in vivo, however.

Several stem cell populations have been identified and have been shown to be responsible for maintaining cellular turnover in the skin, specifically in the basal epidermis and in the hair follicle (Arnold et al., 2011; Fuchs and Raghavan, 2002; Halprin, 1972; Okubo et al., 2006; Tsai et al., 2010). Interestingly, SOX2, a crucial cell fate determinant with a role in developmental contexts and in adult stem cell maintenance in several different organs, is expressed in both the epidermal and the dermal layers of the skin (Arnold et al., 2011; Atit et al., 2006; Blanpain and Fuchs, 2009; Biernaskie et al., 2009; Clavel et al., 2012; Driskell et al., 2009; Fuchs and Horsley, 2008; Haeberle et al., 2004; Sennett and Rendl, 2012). In the epidermis, SOX2 is expressed in a subset of basal epidermal keratinocytes, mechanosensory cells called Merkel cells, and in specialized glial cells surrounding cutaneous nerve terminals. In the dermis, is exclusively localized to specialized mesenchymal cells comprising the dermal papilla (DP) and dermal sheath (DS) of hair follicles. Interestingly, SOX2 expression within this niche is dynamic but is sustained from throughout hair follicle morphogenesis and into adulthood. Although the role of SOX2 in the developing

Sox2
http://dx.doi.org/10.1016/B978-0-12-800352-7.00016-5

and adult skin has not been conclusively established, it is thought to be involved in several skin diseases including fibrogenesis and several forms of skin cancer.

Sox2 EXPRESSION IN THE EPIDERMIS

The stratified, cornified epithelial layers that make up the adult mammalian epidermis arise from a single-cell-thick layer of neuroectoderm, beginning from mouse embryonic day 8 (E8) onward. After exposure to Wnt signaling, these multipotent early epithelial cells express bone morphogenic proteins (BMPs), which are critical for the spatial patterning of hair follicles particularly through BMP4 (Botchkarev and Sharov, 2004; Botchkarev et al., 2001). Developing epidermis is enclosed by a temporary squamous layer called the periderm (mouse E9 and later), which is shed during the final stages of epidermal differentiation. The surface ectoderm develops into the basal epidermal keratinocyte layer, which lies adjacent to the basement membrane, maintains a population of keratinocyte stem cells, and gives rise to the stratified squamous layers of mature epidermis.

Basal Epidermal Cells

In human skin, rare SOX2-expressing cells have been reported in the basal epidermis at 11–12 weeks of gestation, in association with the newly budding hair germ (Laga et al., 2010). As the hair follicle matures to the hair peg stages (13–24 weeks), SOX2-expressing cells remain in this region. However, in adult skin, epidermal SOX2 expression is restricted to the proximal tips of rete ridges, the epidermal thickenings located between hair follicles (Figure 1) (Laga et al., 2010). In mouse skin, SOX2 expression is not observed in either embryonic or adult epidermal cells (Driskell et al., 2009).

Sensory nerve endings terminating around the bulge region of the hair follicle (Figure 1) also contain a subset of unique SOX2-expressing cells in both human and mouse skin (Biernaskie et al., 2009; Laga et al., 2010; Johnston et al., 2013). Using the SOX2-enhanced green fluorescent protein (EGFP) knockin mouse combined with *Wnt1*Cre lineage tracing, these cells are thought to be of neural crest origin in mice (Johnston et al., 2013). This work has also highlighted the importance of SOX2-expressing cells in wound healing, which will be discussed in more detail in the final section of this chapter.

Mechanosensory Merkel Cells

SOX2 is also expressed in mechanosensory cells of the human and mouse epidermis, known as Merkel cells (Figure 1) (Biernaskie et al., 2009; Driskell et al., 2009; Laga et al., 2010; Haeberle and Lumpkin, 2008). These neuroendocrine cells are located next to the large sensory guard hairs of the mouse and express voltage-gated ion channels as part of their role in mechanotransduction (Lumpkin and Caterina, 2007; Lumpkin et al., 2010;

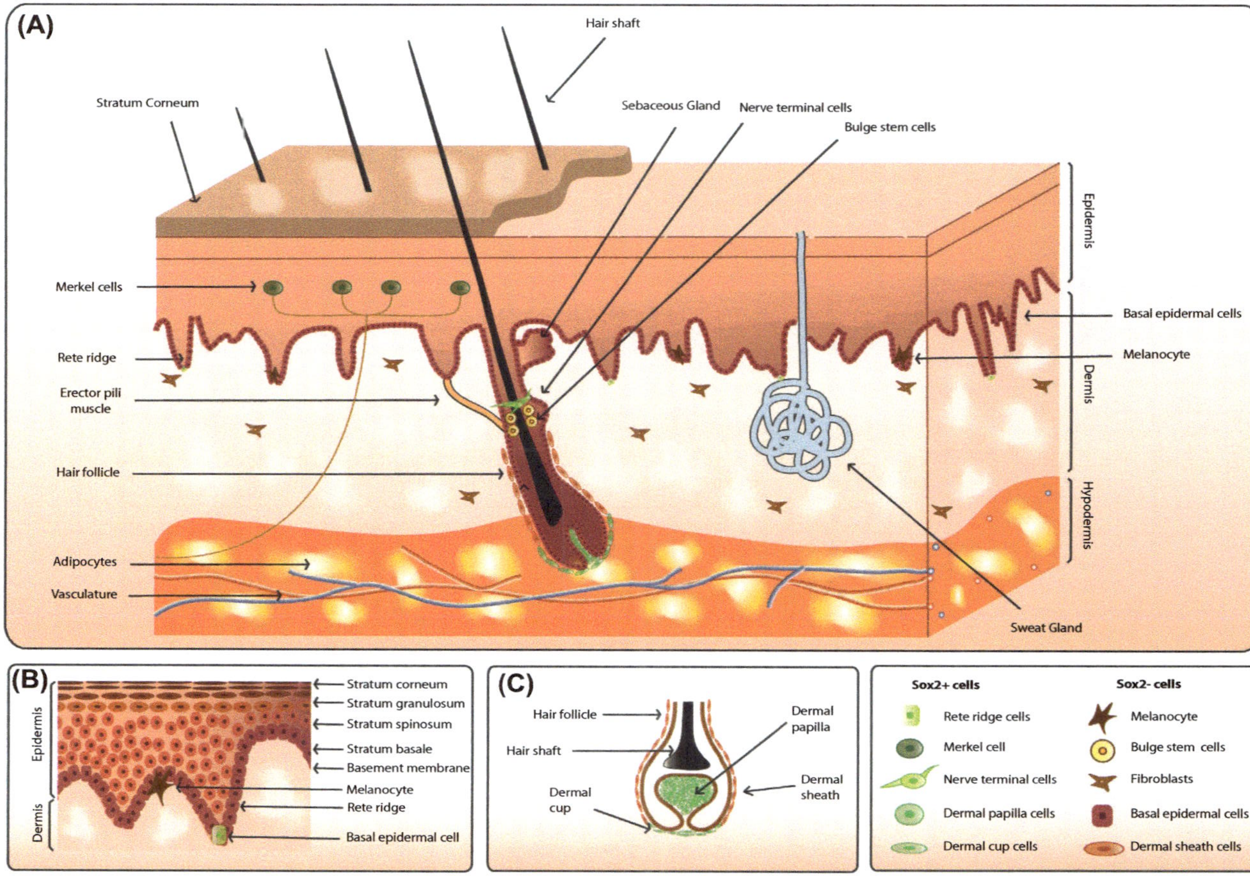

Figure 1 ***SOX2 expression in the adult human skin.*** (A) SOX2-expressing cells are found in both the epidermal and dermal compartments of adult human skin, including nerve terminal cells around the bulge region and cells of the hair follicle. (B) In the epidermis, $SOX2^+$ cells are found in basal epidermal cells of the rete ridges. Merkel cells, mechanosensory cells found in the epidermis are also $SOX2^+$. (C) In the dermis, $SOX2^+$ cells are mainly confined to the dermal papilla and dermal cup. *Adapted from Oswald (2009).*

Piskorowski et al., 2008; Haeberle et al., 2004). Merkel cells have been shown to be part of the epidermal lineage (Morrison et al., 2009), originating from keratin 14^+ (K14) cells adjacent to the epidermal basement membrane, and subsequently losing K14 expression upon differentiation (Van Keymeulen et al., 2009; Woo et al., 2010; Morrison et al., 2009). In mouse skin, $SOX2^+$ Merkel cells, characterized by K8 expression, have been found as early as E14.5, with most $SOX2^+/K8^+$ cells appearing at E15.5 (Lesko et al., 2013).

To assess the function of SOX2 within the Merkel cell population in the skin, SOX2 was conditionally deleted in epithelial cell progenitors using a *K14*Cre-*Sox2*$^{fl/fl}$ mouse model. Loss of SOX2 resulted in overall reduction in the number of Merkel cells; however, no effect on cell proliferation, apoptosis, hair follicle type, or synaptic activity was observed (Bardot et al., 2013; Lesko et al., 2013). Comparison of wild-type and mutant skin showed that loss of SOX2 and subsequent reduction in Merkel cell number did not affect skin innervation, because there was no appreciable change in nerve density (Lesko et al., 2013). Therefore, within the epidermal lineage, SOX2 appears to function as a critical regulator of Merkel cell differentiation.

It is known that *Atho1* is critical in specifying the Merkel cell lineage (Maricich et al., 2009). Recent research has linked this transcription factor with the POLYCOMB complex, in particular its subunits EZH1 and EZH2, which regulate both *Atho1* expression and SOX2 expression. Ablation of EZH1 and EZH2 in mouse skin showed an increase in Merkel cell number, likely owing to increased differentiation of progenitor cells toward the Merkel cell lineage. Ablation of *Sox2* in EZH1/2 KO skin rescued this effect, confirming that Polycomb-repressed *Sox2* maintains the undifferentiated state of epidermal stem/progenitor cells. In the proposed mechanism, *Polycomb* acts as a repressor of both *Sox2* and *Athol*. When this repression is lost, epidermal stem cells differentiate along the Merkel cell lineage. It has also been shown that *Sox2* activity further promotes ATHO1 expression, inducing continued differentiation (Bardot et al., 2013).

Some $SOX2^+$ cells in the tips of rete ridges of the human epidermis also express microphthalmia-associated transcription factor (MITF), a critical regulator of melanocyte development. This may indicate a role for SOX2 within the melanocyte lineage, or that $SOX2^+$ Merkel cells give rise to melanocytes. This possibility is interesting because most melanomas seem to originate in the epidermis, although rigorous study of this $SOX2^+/MITF^+$ population has yet to be performed (Laga et al., 2010).

Sox2 EXPRESSION IN THE HAIR FOLLICLE

Mammalian dorsal dermis is derived from the dermomyotome of the developing embryo, whereas ventral and limb dermis develops from the lateral plate and craniofacial dermis comes from the neural crest (Couly et al., 1992; Osumi-Yamashita et al., 1994). Early transplantation work using epithelial and mesenchymal tissue in both mice and chickens has established the role of the mesenchyme as a driver of integumental development through epithelial–mesenchymal interactions, both in patterning of hair placodes (nascent hair follicles) and determining their lineage commitment (Hardy, 1992; Olivera-Martinez et al., 2004; Hamburger and Hamilton, 1992; Mauger, 1972). In the dermis, SOX2 is not expressed by interfollicular dermal fibroblasts, but is exclusively found in the specialized inductive cells within the DP and lower DS of the hair follicle.

Sox2 Expression in Dermal and Hair Follicle Morphogenesis

In early stages of human skin development (6–12 weeks of gestation), SOX2-expressing cells are scattered throughout the dermis. Interaction between naive fibroblasts and overlying epithelial cells results in the formation of hair placodes, focal thickenings of the basal epidermal layer in which cells become oriented vertically (approximately E13.5 in mice and 11–12 weeks of gestation in humans). Consequently, these naive dermal fibroblasts respond to cues from the overlying placodes and aggregate to form the $SOX2^+$ dermal condensate in both rodents and human skin (Sennett and Rendl, 2012; Biernaskie et al., 2009; Driskell et al., 2009; Laga et al., 2010). The dermal condensate is a primitive form of the DP responsible for stimulating the overlying epithelial cells to proliferate rapidly, yielding matured invaginations called the hair germ and subsequently forming the hair peg. The hair germ and pegs continue to develop into more mature hair follicles under the careful orchestration of the $SOX2^+$ dermal condensate and subsequent DP. Postnatally and during most of adulthood, SOX2-expressing cells are restricted to the DP and the immediately adjacent DS (Figure 2(A) and (B)) (Atit et al., 2006; Biernaskie et al., 2009; Laga et al., 2010).

Sox2 Expression During Postnatal Hair Follicle Cycling

The mammalian hair follicle has the ability to undergo regular cycles of regeneration, dictated by signaling between the DP and epithelial stem/progenitor cells in the bulge region and secondary germ (Schmidt-Ullrich and Paus, 2005; Fuchs, 2007; Jahoda and Reynolds, 1996; Paus and Cotsarelis, 1999). The hair cycle can be separated into three major stages: growth phase (anagen), degeneration (catagen), and quiescence (telogen) (Stenn and Paus, 2001; Paus and Foitzik, 2004). During early anagen, DP cells signal to bulge stem cells to initiate rapid epithelial cell expansion and subsequent DS proliferation (Jahoda and Oliver, 1984; Greco et al., 2009). The expansion of matrix cells surrounds the DP and the hair follicle extends deeper into the dermis, such that the DP of fully grown follicles are situated nearer or adjacent to the lipid-rich hypodermis, depending on the follicle type (Cotsarelis et al., 1990; Stenn and Paus, 2001). Studies using a *Sox2*–EGFP knockin mouse showed that in adult mouse skin, *Sox2* expression was detectable only during anagen growth phase (Biernaskie et al., 2009). During hair follicle regression and/or telogen resting state, DP and DS cells appear to downregulate expression of *Sox2*. Widespread cell death in epithelial layers of the hair causes follicle collapse to form an epithelial strand, which draws the DP back into proximity of the bulge

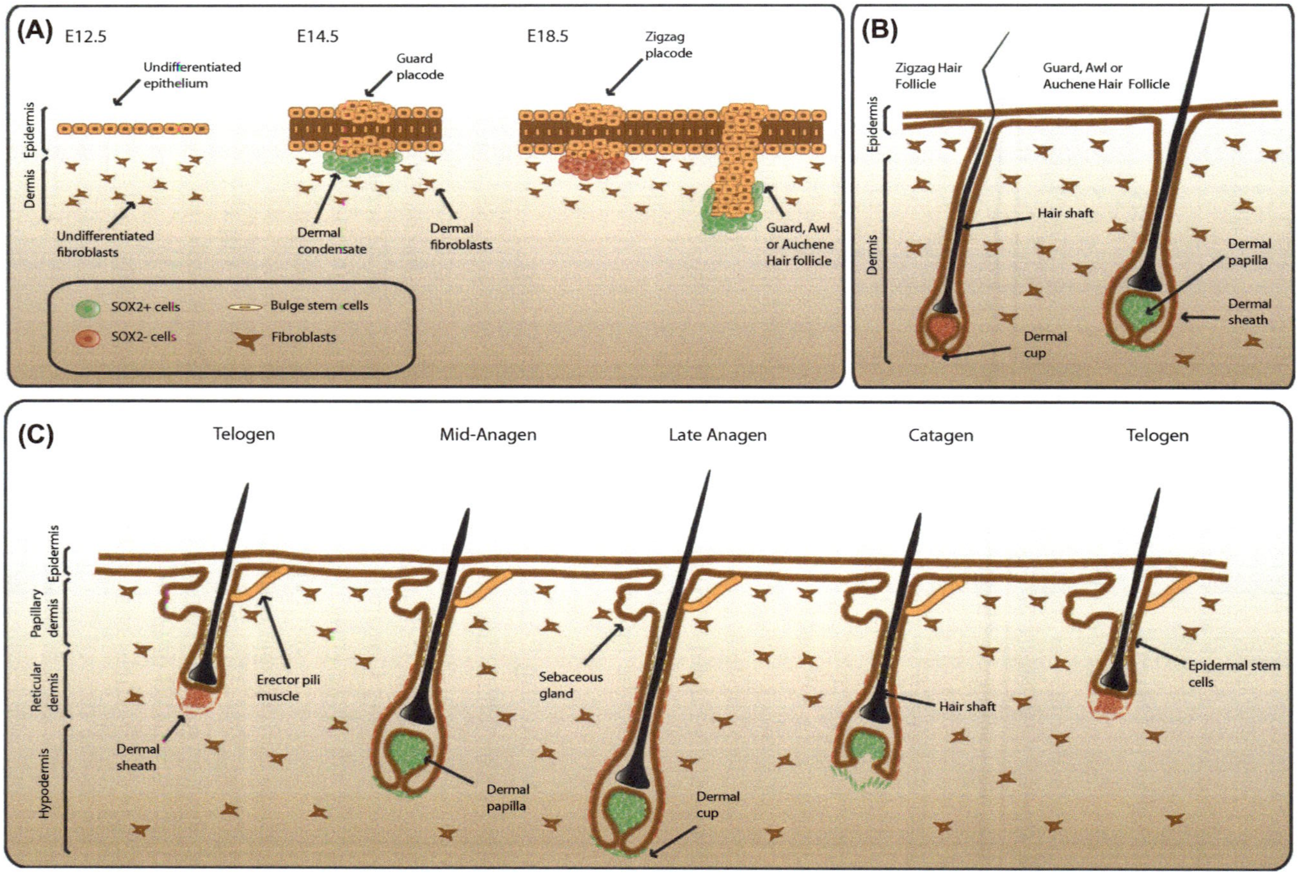

Figure 2 ***Sox2 expression in hair follicle morphogenesis and cycling in mouse skin.*** (A) $SOX2^+$ cells are found in the dermal condensate, aggregating below guard, awl, and auchene hair placodes during hair follicle morphogenesis. The condensate later develops into a hair peg, which continues to express $SOX2^+$ cells at its base. Zigzag hair placodes and pegs do not contain $SOX2^+$ cells. (B) In the adult hair follicle, $SOX2^+$ cells are found in the dermal papilla and dermal cup of guard, awl, and auchene hair follicles. These regions of the zigzag hair follicle are $SOX2^-$. (C) During hair follicle cycling, the $SOX2^+$ dermal papilla and dermal cup cells undergo changes. At telogen (resting phase), both dermal papilla and dermal cup cells are $SOX2^-$. During anagen (growth phase), SOX2 is expressed in these cells but not in the dermal sheath. This expression is maintained during catagen (degeneration phase) before being downregulated again for the next telogen phase. *Adapted from Driskell et al (2011).*

(Botchkarev and Kishimoto, 2003). The resultant telogen follicle maintains downregulated levels of *Sox2* and remains quiescent for a time (Figure 2(C)).

Early attempts to determine the role of SOX2 in the hair follicle showed that its expression in the DP is strongly associated with specific hair follicle types in mice (Driskell et al., 2009). There are six major types of mouse hair follicles: whisker, guard, awl, auchene, zigzag, and tail (Müller-Röver et al., 2001; Paus and Cotsarelis, 1999). In the back skin, four of these six hair follicle types (guard, awl, auchene, and zigzag) are present and can be distinguished on the basis of length, number of medulla cells, and presence of bends or kinks in the shaft. Guard hairs, the longest, develop during the first wave of hair follicle morphogenesis at E13–14.5. Awl hairs are shorter than guard hairs and form in the second wave of hair morphogenesis at E15–16.5. Auchene hairs are similar to awl hairs except that they have one kink in the hair fiber. Finally, zigzag hairs have at least two bends in the hair shaft, are the most abundant in the back skin, and develop in the final wave of hair morphogenesis at E18.5 (Müller-Röver et al., 2001; Paus and Cotsarelis, 1999). Using a *Sox2*–EGFP knockin reporter strain, Driskell et al. reported that guard hair follicles have entirely $SOX2^+$ DP and awl/auchene DP are heterogeneous for SOX2 and CD133 expression, whereas zigzag hair follicle lack SOX2 expression (Figure 2(A) and (B)). When DP cells were sorted using fluorescence-activated cell sorting for the various subpopulations of SOX2- and CD133-expressing cells, each showed distinct gene expression profiles. More interestingly, each subpopulation exhibited different capacities for hair follicle induction. That is, ex vivo hair follicle reconstitution assays revealed that only $SOX2^+$ DP cells were able to induce awl/auchene follicles, which suggested that SOX2 has a dose-dependent role in hair type specification (Driskell et al., 2009).

Attempts to uncover a direct role for SOX2 in hair follicle specification instead yielded an interesting modulatory role for $SOX2^+$ DP cells in promoting differentiation of hair shaft progenitors (Clavel et al., 2012). Using the *Tbx18*Cre line, *Sox2* was deleted specifically in early embryonic DP precursor cells and early postnatal hair follicles. Surprisingly, ablation of *Sox2* in embryonic DP cells had no effect on the normal development of hair follicles or the progression through the adult hair cycle. Moreover, the total number and distribution of all the hair follicles types remained unchanged. However, the authors noted a delay in postnatal hair shaft outgrowth in guard, awl, and auchene (but not zigzag) hair types. This result was compelling because SOX2 is exclusively expressed by the hair follicle types affected and not in zigzag follicles. Further experiments showed that this effect was not the result of a developmental delay. Instead, ablation of *Sox2* seemed to alter the migration speed of differentiating hair shaft progenitors in the epithelial compartment directly surrounding the DP niche. Transcriptional profiling revealed that *Sox2* ablation resulted in increased BMP6 and decreased expression of Sostdc1, a BMP inhibitor and direct transcriptional target of *Sox2*. The increased BMP signaling was demonstrated to inhibit matrix cell migration and thereby reduce the rate of hair growth. Overall, the authors propose a model in which SOX2 functions as a key regulator of hair fiber growth by controlling hair shaft progenitor migration through modulation of BMP-mediated mesenchymal–epithelial cross-talk. This model is corroborated in part by the fact that SOX2-negative zigzag hair follicles have a naturally higher BMP signaling activity (indicated by pSMAD1,5 immunohistochemistry) and therefore exhibit a shorter phenotype.

Together with findings that hair follicles generating zigzag hairs can be modified to generate large hairs (or the reverse) later, depending on the number of cells within the DP (Chi et al., 2013; Rahmani et al., 2014), it seems that the expression of SOX2 in three of the four hair types is coincidental rather than causal.

Questions have also been raised as to whether SOX2 expression is an intrinsic property of DP cells or environmentally regulated (Driskell et al., 2011). In vivo, $SOX2^+$ DP cells were shown to be least proliferative, and yet upon transplantation, these cells make the greatest contribution to the dermis, which suggests an inverse relationship between proliferation and trichogenicity. Furthermore, $SOX2^+/CD133^+$ DP cells exhibit remarkable proliferative capacity in vitro, which suggests that this highly proliferative phenotype is restricted by the local environment in the DP.

Because these experiments provide strong evidence for a functional role for SOX2 in the hair follicle, a closer look at the signaling pathways involved in hair regulation is warranted. In the murine hair follicle, inhibition of Wnt signaling in the hair follicle epithelium, as exemplified by K5-driven cKO of *β-catenin* or through ectopic expression of the secreted WNT-inhibitor DKK1, interferes with growth-phase initiation and matrix cell proliferation and maturation (Choi et al., 2013). The close association of these phenotypes with those of the above BMP-mediated interference with matrix cell differentiation are some of many examples of the importance of these mutually opposed and interdependent canonical pathways in regulating hair growth in the skin.

Whereas the direct connection between Wnt/β-catenin signaling and SOX2 has remained elusive in the skin, Wnt signaling has been shown to lie upstream of SOX2 expression in other related systems. For example, Wnt/β-catenin is involved in fate decisions of sensory hair cells of the inner ear, in the developing retina, and in taste bud development. *Lcc/Lcc* and *Ysb/Ysb* mutant mice were found to lack or contain dysregulated inner ear epithelium and sensory hair cells, which was linked to an absence or reduced expression of SOX2, respectively (Kiernan et al., 2005). In the tongue, where it was found that SOX2 is also abundantly expressed in developing taste buds, hypomorphic *Sox2* mice failed to develop mature taste buds. K5-driven overexpression of *Sox2* in the epithelium lacked keratinized filiform papillae, which pointed to a role for SOX2 in specifying papillary fate (Okubo et al., 2006). Inhibiting canonical Wnt signaling within the developing *Xenopus* retina using morpholinos against the *Xfzd5* receptor also prevents SOX2 expression, and diverted differentiation of neural progenitors fourfold toward a glial fate (Van Raay et al., 2005). This finding was reinforced by Taranova and colleagues, who also identified a role for Notch signaling downstream of SOX2 in repressing fate specification of retinal ganglion cells by comparing several *Sox2* cKO mouse genotypes that simulate *Sox2* ablation in a concentration-dependent manner (Taranova et al., 2006). This is interesting in the context of hair follicle DP maintenance, because there is substantial overlap between gene expression by the DP and those traditionally associated with neuronal lineages (Rendl et al., 2005). However, the sum of these experiments, although providing critical new insights, has yet to establish a comprehensive understanding of the regulation of Sox2 and its molecular control within the skin.

Recent work has pointed out that DP cell numbers, which correlate with SOX2 expression, have an important role in determining hair type. Because SOX2 is not expressed in the DP of zigzag hair follicles or induced ectopic follicles and it does not seem to be required to induce guard or zigzag hairs upon transplantation, it has been suggested that the expression of Sox2 in DP cells may not make a remarkable functional contribution to the DP (Collins et al., 2012). When the number of DP cells was carefully ablated, a change in hair type was observed at particular threshold. It may be that the number of cells in the DP determines the level of SOX2 expression in a hair follicle, which in turn affects hair type. Recent evidence for the role of SOX2 in regulating hair outgrowth may shed some light on the link between these observed effects.

Sox2 Expression in Hair Induction

The DP is the focus of intense interest primarily because of the large body of evidence supporting its role as the inductive center for hair follicle development and growth. Removal of DP cells by either genetic ablation or targeted laser ablation results in failure of the hair follicle to reenter anagen (Chi et al., 2013; Rompolas et al., 2012). Moreover, transplantation of the DP or lower DS into nonhairy skin is sufficient to induce hair growth (McElwee et al., 2003). However, questions remain as to how the DP is maintained, because the number of cells within this compartment fluctuates with the hair cycle and DP cells rarely express markers of proliferation or apoptosis. Indeed, there is strong circumstantial evidence for cell movement between the DP and the DS. When the DP is removed via microdissection, the remaining follicle regenerates only when the lower DS remains intact, which indicates that the mitotically inactive DP cells may instead be maintained by a population of neighboring DS cells (McElwee et al., 2003). More recent in vivo genetic fate mapping studies show that a population of $SOX2^{+}/\alpha SMA^{+}$ cells in the DS region immediately adjacent to the DP contains a dermal stem cell population. These hair follicle dermal stem cells (hfDSCs) are capable of contributing cells to the DP, generating the DS, and are maintained in this niche over many hair follicle growth cycles (Biernaskie et al., 2009; Rahmani et al., 2014). Indeed, hfDSC progeny that differentiate and eventually reconstitute the regenerating DS exhibit a loss of SOX2 expression as they exit the dermal cup region. Contrastingly, SOX2 expression is maintained in progeny that are recruited into the DP, which is consistent with previous work showing that SOX2 is an important contributor to normal DP function (Clavel et al., 2012). Interestingly, it appears that hfDSC progeny that are recruited to the DP can also reenter the stem cell pool after hair follicle regression (Rahmani et al., 2014). Based on its role in preventing differentiation in ES cells and neural stem cells, one could speculate that SOX2 may similarly act to prevent differentiation to a DS fate, thereby permitting adoption of the stem cell function once they have emigrated out of the DP microenvironment.

Sox2 Expression in Skin-Derived Precursors

Investigations into an accessible source of autologous, multipotent precursor cells led to the discovery of skin-derived precursors (SKPs) (Toma et al., 2001), a population of dermal cells that can be isolated from embryonic and adult mammalian skin. These precursor cells exhibit long-term self-renewal in vitro and are capable of generating cells that express markers associated with immature neurons, dermal fibroblasts, peripheral glial cells, smooth muscle cells, and adipocytes in vitro (Toma et al., 2001; Fernandes et al., 2004). Subsequent experiments demonstrated that SKPs, grown in conditions that support gliogenesis, could give rise to proliferative Schwann cells capable of remyelinating axons after transplantation to the injured peripheral and central nervous system (McKenzie, 2006; Biernaskie et al., 2007).

The endogenous origin of SKPs can be traced to $SOX2^+$ mesenchymal cells found in the DP and DS within the developing and adult hair follicle (Biernaskie et al., 2009). Interestingly, DP cells express genes associated with neural crest stem cells, including SOX9, nestin, p75, snail, and slug, suggesting either a neural crest origin or recapitulation of a neural crest-like expression profile within this specific follicular niche. By prospectively isolating $SOX2GFP^+$ cells, Biernaskie and colleagues showed that SKPs were enriched in this mesenchymal population (Biernaskie et al., 2009). Using serial hair follicle reconstitution assays, SKPs were shown to function as dermal stem cells, clonally generating dermal cells and inducing hair morphogenesis. Green fluorescent protein–tagged $SOX2^+$ SKP cells were shown to home to and repopulate the hair follicle dermal niche (DP and DS) while maintaining their ability to self-renew after subsequent isolation and expansion. Finally, it was found that these $SOX2^+$ dermal cells share a similar transcriptional profile as cultured SKPs, based on comparative microarray analysis. The authors were thus able to define $SOX2^+$ dermal precursors, located in the DP and lower DS in neonatal mice, as the cells of origin for cultured SKPs.

Although these studies hinted at a possible role for SOX2 in maintaining an undifferentiated dermal stem cell state, studies provided more direct evidence for such a role. Tamoxifen inducible deletion of *Sox2* in SKPs isolated from *Sox2*$^{flox/flox}$ mouse skin revealed a marked decline in colony formation as well as a reduction in individual SKPs colony size (Suflita et al., 2013). Unfortunately, these experiments examined only the effects of *Sox2* deletion in primary SKPs, and so it remains unclear as to whether SOX2 is involved in promoting proliferation or continued self-renewal, which would require serial, clonal expansion of SKP in the absence of SOX2. Interestingly, deletion of FOXD3 in SKPs, another neural crest–associated gene, was found to block SKP formation almost entirely. This study suggests that SOX2 may regulate proliferation/self-renewal of SKPs in vitro. Future studies will need to determine whether SOX2 is an active participant in regulating dermal stem cell behavior in vivo and whether SOX2 interacts with FOXD3.

SOX2-expressing cells in the DP are mitotically inactive, but both microdissected DP cells and SKPs are capable of dramatic expansion when removed from their hair follicle niche and grown in vitro (in the presence of fibroblast growth factor). This indicates that the hair follicle microenvironment has an essential role in modulating SOX2 function within the hair follicle mesenchyme.

SOX2 EXPRESSION IN SKIN CANCERS AND WOUNDS

As mentioned, SOX2 expression is found across multiple cellular lineages in both mouse and human skin. Considering its expression pattern and role within the skin, there is growing interest in determining the function of SOX2 not only during skin homeostasis, but more important, during tumorigenesis and after cutaneous injury. The remainder of this chapter will explore recent investigations into understanding the activity of SOX2 in melanoma and nonmelanoma cancers, healing skin wounds, and the progression of skin fibrosis.

Sox2 in Skin Cancers

Skin cancers are the most common type of cancer among fair-skinned populations around the world and can be divided into two major types: melanoma and nonmelanoma (Narayanan et al., 2010). Melanoma skin cancers are primarily caused by excessive ultraviolet exposure causing dysfunction in pigment-producing melanocyte cell populations in the skin. They are extremely aggressive, highly heterogeneous, and resistant to conventional therapies at advanced stages (Santini et al., 2014). Moreover, they contribute to nearly 75% of all skin cancer deaths within the United States (Narayanan et al., 2010). Nonmelanoma skin cancers, on the other hand, encompass a variety of skin neoplasms that include cutaneous lymphomas, adnexal tumors, Merkel cell carcinoma (MCC), basal cell carcinoma (BCC), and squamous cell carcinomas (SCC) (Madan et al., 2010). Although BCC and SCC have the highest incidence rates in the world and can lead to great disfigurement when located on the head and neck, they are usually associated with a low mortality rate (Ahmed et al., 2008). Generally, SCC is more aggressive and less frequent than BCC.

A common theme in the discussion regarding SOX2 and skin cancers, both melanoma and nonmelanoma, is that SOX2 is not an artifact of cancer, but rather a central transcription factor responsible for supporting key features of the carcinoma such as growth, proliferation, and invasiveness. During embryonic and neural development, SOX2 is interlocked within a genomic network that is essential to the proliferative and migratory nature of the cells at play. However, when reactivated in adulthood, this genomic network is responsible for many of the carcinomas across the body, including the skin (Schoenhals et al., 2009).

Sox2 in Melanoma

Until recently, the mechanisms behind the initiation and invasiveness of melanoma were poorly understood. Investigation into a role for SOX2 in melanoma began with a detailed analysis of its expression pattern in primary human malignant melanomas. Immunohistochemical analysis revealed that ~35% of human malignant melanoma samples were positive for SOX2 expression, compared to ~16% of nevi (Laga et al., 2010). Moreover, the A2058 human melanoma cell line stained robustly for SOX2, whereas reverse transcriptase–polymerase chain reaction (RT-PCR) confirmed a ~250-fold higher expression of Sox2 than in SK-MEL-5 cells (Laga et al., 2010). Within the same study, the authors asked whether SOX2 expression correlated with tumor thickness and patient survival time. Using a univariate logistic regression analysis, they found that $SOX2^{+}$ melanomas were more likely to be 1 mm thicker than $SOX2^{-}$ melanomas but did not significantly affect survival time.

A deeper analysis of SOX2 in melanomas revealed three primary roles: tumor initiation and growth, self-renewal of melanoma-initiating cells (MICs), and tumor invasiveness. When *SOX2* was knocked out in SSM2c, M26c, M5, A375, and A2058 human melanoma cells using short hairpin RNA (shRNA), analysis of the growth kinetics in all lines revealed diminished cell proliferation and increased apoptosis. Transplantation of SOX2-deficient cell lines into nude mice formed tumors that were significantly smaller than control lines (Laga et al., 2010; Santini et al., 2014). Melanoma-initiating cells have an unlimited self-renewal capacity, the potential to initiate and maintain tumor growth, and can grow as spheres in vitro (Schatton et al., 2008); however, when *SOX2* was silenced using the same shRNA, smaller MIC spheres were formed. When *SOX2* was overexpressed, the number and size of MIC spheres increased. Collectively, these results suggest that SOX2 has a role in both MIC self-renewal and proliferation (Santini et al., 2014). Moreover, the fact that Sox2 depleted MICs can initiate tumor growth suggests that SOX2 may be more important in the transit amplifying/progenitor pool rather than the tumor-initiating cells (TICs). SOX2 was also shown to contribute to melanoma cell invasion in a conventional in vitro Matrigel assay. Silencing *SOX2* significantly enhanced human melanoma cell invasion (by 380%) (Girouard et al., 2012).

Molecular descriptions of the contribution of SOX2 to melanoma growth, self-renewal, and invasiveness have been further developed. It was shown that Hedgehog-GLI (HH-GLI) signaling drives self-renewal and tumorigenicity of human MICs (Santini et al., 2012).

The same authors proceeded to ask whether the HH-GLI pathway was an upstream regulator of *SOX2*, because both GLI1 and GLI2 were found to bind to the *SOX2* promoter at about 300 base pairs upstream to the transcriptional start site. When the HH-GLI pathway was inhibited by silencing *Smoothened* or *Gli1*, Sox2 protein and mRNA expression were nearly abolished in SSM2c melanoma cells (Santini et al., 2014). Conversely, activation of the HH-GLI pathway by silencing Patched showed increased expression of *SOX2* and, interestingly, increased the number and proliferation of MIC spheres. Silencing *SOX2* and *Patched* abolished the effect of the HH-GLI pathway in melanoma cells (Santini et al., 2014). These results suggest that growth and self-renewal of MICs are regulated by the HH-GLI pathway via activation of *SOX2*. In a separate study, it was shown that silencing *SOX2* using shRNA significantly reduced transcription of matrix metalloproteinase-3 (*MMP3*) mRNA. When *MMP3* was silenced, the in vitro invasiveness of melanoma cells in Matrigel was reduced by ~55%. These results were reinforced by histological observations showing the proximity of MMP3 and SOX2 expression in human melanomas and xenografts (Girouard et al., 2012). Therefore, MMP3 appears downstream of *SOX2* and important for melanoma invasion.

Sox2 in Squamous Cell Carcinoma

Squamous cell carcinomas are abnormal growths arising from the outermost layer of the epidermis. In normal adult skin, the epidermal layer lacks SOX2 expression. With the exception of Merkel cells, most preneoplastic lesions and invasive SCCs in humans have a heterogeneous expression of SOX2 (Boumahdi et al., 2014; Siegle et al., 2014). Hyperplastic lesions, induced with 7,12-dimethylbenz-alpha-anthracene/12-*O*-tetradecanoyl-phorbol-13-acetate (DMBA/TPA), focally express SOX2 in suprabasal cells of the epidermis (Boumahdi et al., 2014). Using transgenic models and human melanoma cells lines, SOX2 was shown to activate SSCs and support their growth, survival, and maintenance (Boumahdi et al., 2014; Siegle et al., 2014). For example, *Sox2*–EGFP mice were used to show that the Sox2^{+} population of an invasive SCC had the greatest capacity to induce a secondary tumor after transplantation (Boumahdi et al., 2014). Also, silencing *Sox2* with shRNA significantly reduced SCC growth within a xenograft transplant, increased levels of apoptosis, and decreased proliferation (Siegle et al., 2014). More convincingly, the time for tumor induction after the application of DMBA/TPA increased by several months after *Sox2* was specifically deleted in K14^{+} cells using *K14Cre-Sox2*$^{fl/fl}$ mice (Boumahdi et al., 2014). It was also convincingly shown that SOX2 was a necessary component of the SCC tumor using several other genetic tools. When Sox2^{+} cells were ablated with diphtheria toxin (*Sox2*CreER–RosaDTA) or *Sox2* itself was deleted (*K14Cre-Sox2*$^{fl/f}$) after formation of the tumor, there was an almost complete regression of both benign papillomas and invasive SCCs (Boumahdi et al., 2014). Another approach employed the use of transplanted lentivirus-transfected tumor initiating cells with doxycycline-inducible shRNA against SOX2 (shSOX2) in nude mice. After the tumor was established, induction of shSOX2 with doxycycline significantly reduced the growth of the tumor grafts (Siegle et al., 2014).

To understand why SOX2 is expressed in SCCs and not in the healthy epidermis, Siegle and colleagues compared the methylation status of the *Sox2* promoter between SOX2$^+$ TICs and wild-type hair follicle stem cell (HFSC)-derived keratinocytes. The analysis revealed that the repressive histone H3 lysine 27 trimethylation (H3K27me3) mark was significantly reduced in SOX2$^+$ TICs compared with HFSC-derived keratinocytes (Siegle et al., 2014). This finding suggests a unique level of epigenetic regulation responsible for de novo SOX2 expression in SCC. It was shown using RT-PCR and chromatin immunoprecipitation–quantitative polymerase chain reaction that *SOX2* directly regulates the downstream expression of key genes responsible for cancer cell proliferation, stemness, chromatic regulation, and metabolism (Siegle et al., 2014; Boumahdi et al., 2014). Some of these genes included *CD133* and the proangiogenic factors *Vegfa* and *Nrp1*. These bioinformatic analyses were supported by experiments showing that SCC growth can be blocked if *Nrp1* is silenced with shRNAs (Siegle et al., 2014).

Merkel Cell Carcinomas

Merkel cell carcinoma is a rare and highly aggressive form of nonmelanoma skin cancer. Approximately 80% of MCCs are associated with the newly identified Merkel cell polyomavirus, but the cell of origin for this rare carcinoma remains controversial (Tilling and Moll, 2012). When nine human samples were stained for SOX2, all showed diffuse expression in more than 50% of the cells (Laga et al., 2010). Whether Sox2 has a role in the formation, maintenance, or aggressiveness of MCCs remains unknown. However, in light of the studies that showed a significant reduction in the number of Merkel cells after the conditional deletion of *Sox2* in K14Cre cells (Bardot et al., 2013; Lesko et al., 2013), one could speculate that MCCs might arise from a dysregulation of SOX2 in K14$^+$ cells.

Wound Healing and Fibrosis

Skin wound healing relies heavily on the collaborative efforts of many different tissues and cell lineages occurring over multiple overlapping phases (Martin, 1997). To regenerate damaged structures within the skin, multiple cell populations within the site of injury flood the wound bed. For example, the newly formed dermis is repopulated by migratory fibroblasts from the intact dermis at the wound margins, circulating bone marrow derived fibrocytes, vasculature-associated pericytes, and possibly even multipotent stem cells residing within the dermis (Abe et al., 2001; Hinz et al., 2007). Although much work has delineated the origin and fate of these migrant cells, it remains important to understand them further. As illustrated with skin cancers, studying the expression of SOX2 within cutaneous wounds has strengthened the proposed theme of this chapter: In the skin, SOX2 expression is an active participant in the context of proliferation, differentiation, and migration.

The first evidence for the role of $Sox2^+$ cells in repairing damaged skin came from a study that closely examined the hair follicle DS and papilla in *Sox2*–EGFP mice (Biernaskie et al., 2009). Normally, Sox2 expression in healthy skin is restricted to the DP and lower DS of some hair follicles, with no expression in the interfollicular dermis. However, 72 h after wounding the back skin of *Sox2*–EGFP mice, GFP^+ cells were found in the wound site. The authors went on to show that hair follicle–associated $Sox2^+$ cells generate multipotent SKPs in vitro and induce hair growth when transplanted. Wounding near a small patch of induced hair caused GFP^+ cells to migrate into the newly generated dermis, which suggests a role for hair follicle–associated $Sox2^+$ cells in wounding. Although novel, a more stringent lineage trace was needed to pinpoint the exact origin and fate of $Sox2^+$ cells after wounding. To address this, Johnston and colleagues crossed the *Sox2*–EGFP mouse with a *Wnt1*Cre–R26TdTomato or *Dermo1*Cre–R26TdTomato mouse to mark neural crest and dermal progeny, respectively. Nine days after injury, almost all Sox2–$EGFP^+$ cells in the wound expressed $p75NTR^+$ and $S100\beta^+$, whereas cells traced by *Wnt1*Cre only expressed $TdTomato^+$ (Johnston et al., 2013). From this finding, the authors proposed that $Sox2^+$ cells scattered within a cutaneous wound do not originate from DP or Merkel cells, because neither are of neural crest origin. Rather, they likely derive from nerve terminal cells and damaged cutaneous nerve cells, because they constitute the only Sox2-expressing neural crest–derived cells upon injury (cutaneous nerve cells express Sox2 only after injury). To discern which of these two populations is responsible for contributing cells into the wound, the *Sox2*CreERT2–R26TdTomato lineage tracing method was used. In this system, reporter expression is induced only after tamoxifen application. This method also took advantage of the delayed Sox2 expression in cutaneous nerve cells. Few $TdTomato^+$ cells were seen within the wound when tamoxifen was applied 3 weeks before wounding. However, when tamoxifen was applied after injury, many $TdTomato^+$ cells were seen scattered throughout the regenerating dermis just as in the *Sox2*–EGFP mouse. Altogether, the authors concluded that $Sox2^+$ cells within the healing wound were derived mostly from cutaneous nerve cells of injured nerves and partially from nerve terminal cells. Although this study clarified the developmental origin of these $Sox2^+$ wound cells, the vague description of cutaneous nerve cells and lack of experimental consideration for blood-derived cells has left the field wanting for a more accurate anatomical origin.

To address the functional role of Sox2 within the wound, Johnston and colleagues ubiquitously knocked out *Sox2* in 9-month-old $Sox2^{fl/fl}$:$R26^{CreERT2/+}$ mice and performed punch wounds 5 weeks later. Key features of the wound healing process, such as the epithelial gap, wound width, dermal tissue regeneration, and number of $Ki67^+$ proliferating cells, were detrimentally altered compared with control (Johnston et al., 2013). Similar deficits were seen in *Sox2* heterozygous mice, and the authors concluded that neural crest–derived $Sox2^+$ cells are necessary for skin repair. Although Sox2 seems to be a requirement for proper skin repair, it is premature to conclude that the deficits in skin repair resulted from local Sox2 deficits when *Sox2* was knocked out ubiquitously.

Acute wounds normally heal in an orderly manner; however, wounds can often lead to hyperproliferation or fibrotic scar, in which the tissue is replaced by dysfunctional extracellular matrix. Clinical examples include keloids, hypertrophic scars, and scleroderma (Diegelmann and Evans, 2004). In wild-type mice, application of bleomycin reliably induced fibrotic skin lesions consisting of Sox2$^+$ cells (Liu et al., 2014). Conditional knockout in *Sox2*Cre-expressing cells of CCN2, a matricellular protein connective tissue growth factor responsible for directly promoting fibroblast adhesion, rendered mice resistant to bleomycin-induced skin fibrosis. The clinical implication here, according to the authors, is that therapeutically blocking the recruitment of Sox2$^+$ cells through CCN2 may be an effective approach in combating fibrotic skin disease. However, a robust analysis of the role of Sox2 and CCN2 in the progression of fibrosis remains lacking. Moreover, confusion remains as to origin and fate of these Sox2$^+$ cells. Liu and colleagues claimed that the Sox2$^+$ cells in bleomycin-induced fibrotic lesions originate from the hair follicle mesenchyme based on SOX2 expression in the DP before bleomycin treatment. In light of work by Johnston and colleagues showing a neural crest origin for these Sox2$^+$ cells, it may be that there are multiple origins for these cells, and additional work will be required to gain a complete understanding of Sox2 in skin injury.

CONCLUSION

Although the transcription factor *Sox2* is generally thought of as a regulator of stem cell maintenance and regulation of differentiation, its role in orchestrating the many cells of the skin may be more complex. Sox2 is expressed in a variety of cell types in both human and mouse skin. In healthy mouse epidermis, epidermal cells do not express Sox2. However, in human epidermis, rare Sox2$^+$ cells are first seen in the basal layer at 11–12 weeks of gestation, closely apposed to newly budding hair germs. In the adult human, SOX2 expression is restricted to tips of rete ridges. Hair follicle–associated nerve terminal cells and intraepidermal Merkel cells in both human and mouse skin also show robust SOX2 expression. In the dermis, SOX2 expression is exclusively found in the DP and lower DS of guard, awl, and auchene anagen hair follicles. Finally, self-renewing SKPs can be derived from Sox2$^+$ cells of the dermis.

Closer examination of Sox2 in these various cell populations revealed roles similar to those found within embryonic stem cells and neural stem cells. For example, the conditional knockout of *Sox2* in K14Cre cells showed that Sox2 is important in the differentiation of K14$^+$ cells into K8$^+$ Merkel cells. Deleting *Sox2* in Tbx18Cre DP cells uncovered a much sought-after role in hair regeneration rather than in hair type specification, by promoting differentiation of hair shaft progenitors through the modulation of BMP-mediated mesenchymal-epithelial cross-talk. Finally, deletion of *Sox2* in SKPs reduced the number and size of spheres in vitro, providing evidence for a role for Sox2 in proliferation and self-renewal, which is otherwise repressed in the tightly regulated DP microenvironment.

In diseased and wounded skin, strong SOX2 expression was detected in melanomas, SCCs, MCCs, cutaneous wounds, and fibrotic lesions. In melanomas and SCCs, Sox2 was determined to have an important role in tumor initiation, growth, maintenance, and invasiveness. Both cutaneous wounds and fibrotic lesions lead to an influx of Sox2^{+} cells within the regenerating dermis. Although these cells seem to originate from a neural crest origin, finding the anatomical source and understanding the heterogeneity of SOX2-expressing cells contributing to wounds will require further study. Indeed, the function of Sox2 in the wound remains inconclusive, particularly because injury may induce SOX2 expression in otherwise negative interfollicular dermal cells. However, early attempts have made inroads into understanding the importance of Sox2 in the progression of wound healing and fibrosis.

While the fields of pluripotent stem cell and cancer research continue to drive much of our understanding of the functions of important factors such as Sox2, the varied collection of cell types within the mammalian skin that express Sox2 serves as an excellent model system to understand its diverse functional repertoire. Certainly, Sox2 is an important player in multiple cutaneous lineages and in hair follicle regeneration, wound healing, and oncogenesis. Future studies examining Sox2 function will undoubtedly improve skin health and development of therapies to maintain or restore normal skin function.

REFERENCES

Abe, R., Donnelly, S.C., Peng, T., Bucala, R., Metz, C.N., 2001. Peripheral blood fibrocytes: differentiation pathway and migration to wound sites. J. Immunol. 166 (12), 7556–7562.

Ahmed, A.H., Soyer, H.P., Saunders, N., Boukamp, P., Roberts, M.S., 2008. Non-melanoma skin cancers. Drug Discov. Today Dis. Mech. 5 (1), e55–e62.

Arnold, K., Sarkar, A., Yram, M.A., Polo, J.M., Bronson, R., Sengupta, S., Seandel, M., Geijsen, N., Hochedlinger, K., 2011. Sox2(+) adult stem and progenitor cells are important for tissue regeneration and survival of mice. Cell Stem Cell 9 (4), 317–329.

Atit, R., Sgaier, S.K., Mohamed, O.A., Taketo, M.M., Dufort, D., Joyner, A.L., Niswander, L., Conlon, R.A., 2006. Beta-catenin activation is necessary and sufficient to specify the dorsal dermal fate in the mouse. Dev. Biol. 296 (1), 164–176.

Bardot, E.S., Valdes, V.J., Zhang, J., Perdigoto, C.N., Nicolis, S., Hearn, S.A., Silva, J.M., Ezhkova, E., 2013. Polycomb subunits Ezh1 and Ezh2 regulate the Merkel cell differentiation program in skin stem cells. EMBO J. 32 (14), 1990–2000.

Biernaskie, J., Paris, M., Morozova, O., Fagan, B.M., Marra, M., Pevny, L., Miller, F.D., 2009. SKPs derive from hair follicle precursors and exhibit properties of adult dermal stem cells. Cell Stem Cell 5 (6), 610–623.

Biernaskie, J., Sparling, J.S., Liu, J., Shannon, C.P., Plemel, J.R., Xie, Y., Miller, F.D., Tetzlaff, W., 2007. Skin-derived precursors generate myelinating Schwann cells that promote remyelination and functional recovery after contusion spinal cord injury. J. Neurosci. 27 (36), 9545–9559.

Blanpain, C., Fuchs, E., 2009. Epidermal homeostasis: a balancing act of stem cells in the skin. Nat. Rev. Mol. Cell Biol. 10 (3), 207–217.

Botchkarev, V.A., Kishimoto, J., 2003. Molecular control of epithelial-mesenchymal interactions during hair follicle cycling. J. Invest. Dermatol. Symp. Proc. Soc. Invest. Dermatol. Inc. Eur. Soc. Dermatol. Res. 8 (1), 46–55.

Botchkarev, V.A., Sharov, A.A., 2004. BMP signaling in the control of skin development and hair follicle growth. Differentiation 72 (9–10), 512–526.

Botchkarev, V.A., Botchkareva, N.V., Nakamura, M., Huber, O., Funa, K., Lauster, R., Paus, R., Gilchrest, B.A., 2001. Noggin is required for induction of the hair follicle growth phase in postnatal skin. FASEB J. Off. Publ. Fed. Am. Soc. Exp. Biol. 15 (12), 2205–2214.

Boumahdi, S., Driessens, G., Lapouge, G., Rorive, S., Nassar, D., Le Mercier, M., Delatte, B., Caauwe, A., Lenglez, S., Nkusi, E., Brohee, S., Salmon, I., Dubois, C., del Marmol, V., Fuks, F., Beck, B., Blanpain, C., 2014. SOX2 controls tumour initiation and cancer stem-cell functions in squamous-cell carcinoma. Nature 511 (7508), 246–250.

Chi, W., Wu, E., Morgan, B.A., 2013. Dermal papilla cell number specifies hair size, shape and cycling and its reduction causes follicular decline. Development 140 (8), 1676–1683.

Choi, Y.S., Zhang, Y., Xu, M., Yang, Y., Ito, M., Peng, T., Cui, Z., Nagy, A., Hadjantonakis, A.-K., Lang, R.A., Cotsarelis, G., Andl, T., Morrisey, E.E., Millar, S.E., 2013. Distinct functions for Wnt/β-catenin in hair follicle stem cell proliferation and survival and interfollicular epidermal homeostasis. Cell Stem Cell 13 (6), 720–733.

Clavel, C., Grisanti, L., Zemla, R., Rezza, A., Barros, R., Sennett, R., Mazloom, A.R., Chung, C.-Y., Cai, X., Cai, C.-L., Pevny, L., Nicolis, S., Ma'ayan, A., Rendl, M., 2012. Sox2 in the dermal papilla niche controls hair growth by fine-tuning BMP signaling in differentiating hair shaft progenitors. Dev. Cell 23 (5), 981–994.

Collins, C.A., Jensen, K.B., MacRae, E.J., Mansfield, W., Watt, F.M., 2012. Polyclonal origin and hair induction ability of dermal papillae in neonatal and adult mouse back skin. Dev. Biol. 366 (2), 290–297.

Cotsarelis, G., Sun, T.T., Lavker, R.M., 1990. Label-retaining cells reside in the bulge area of pilosebaceous unit: implications for follicular stem cells, hair cycle, and skin carcinogenesis. Cell 61 (7), 1329–1337.

Couly, G.F., Coltey, P.M., Le Douarin, N.M., 1992. The developmental fate of the cephalic mesoderm in quail-chick chimeras. Development 114 (1), 1–15.

Diegelmann, R.F., Evans, M.C., 2004. Wound healing: an overview of acutre, fibrotic and delayed healing. Front. Biosci. 9, 283–289.

Driskell, R.R., Clavel, C., Rendl, M., Watt, F.M., 2011. Hair follicle dermal papilla cells at a glance. J. Cell Sci. 124 (Pt 8), 1179–1182.

Driskell, R.R., Giangreco, A., Jensen, K.B., Mulder, K.W., Watt, F.M., 2009. Sox2-positive dermal papilla cells specify hair follicle type in mammalian epidermis. Development 136 (16), 2815–2823.

Fernandes, K.J.L., McKenzie, I.A., Mill, P., Smith, K.M., Akhavan, M., Barnabé-Heider, F., Biernaskie, J., Junek, A., Kobayashi, N.R., Toma, J.G., Kaplan, D.R., Labosky, P.A., Rafuse, V., Hui, C.-C., Miller, F.D., 2004. A dermal niche for multipotent adult skin-derived precursor cells. Nat. Cell Biol. 6 (11), 1082–1093.

Fuchs, E., 2007. Scratching the surface of skin development. Nature 445 (7130), 834–842.

Fuchs, E., Horsley, V., 2008. More than one way to skin. Genes Dev. 22 (8), 976–985.

Fuchs, E., Raghavan, S., 2002. Getting under the skin of epidermal morphogenesis. Nat. Rev. Genet. 3 (3), 199–209.

Girouard, S.D., Laga, A.C., Mihm, M.C., Scolyer, R.A., Thompson, J.F., Zhan, Q., Widlund, H.R., Lee, C.-W., Murphy, G.F., 2012. SOX2 contributes to melanoma cell invasion. Lab. Invest. 92 (3), 362–370.

Greco, V., Chen, T., Rendl, M., Schober, M., Pasolli, H.A., Stokes, N., Cruz-Racelis, dela, J., Fuchs, E., 2009. A two-step mechanism for stem cell activation during hair regeneration. Cell Stem Cell 4 (2), 155–169.

Haeberle, H., Lumpkin, E.A., 2008. Merkel cells in somatosensation. Chemosens. Percept. 1 (2), 110–118.

Haeberle, H., Fujiwara, M., Chuang, J., Medina, M.M., Panditrao, M.V., Bechstedt, S., Howard, J., Lumpkin, E.A., 2004. Molecular profiling reveals synaptic release machinery in Merkel cells. Proc. Natl. Acad. Sci. U.S.A. 101 (40), 14503–14508.

Halprin, K.M., 1972. Epidermal 'turnover time'–a re-examination. Br. J. Dermatol. 86 (1), 14–19.

Hamburger, V., Hamilton, H.L., 1992. A series of normal stages in the development of the chick embryo. Dev. Dyn. 195 (4), 231–272.

Hardy, M.H., 1992. The secret life of the hair follicle. Trends Genet. 8 (2), 55–61.

Hinz, B., Phan, S.H., Thannickal, V.J., Galli, A., Bochaton-Piallat, M.L., Gabbiani, G., 2007. The myofibroblast: one function, multiple origins. Am. J. Pathol. 170 (6), 1807–1816.

Jahoda, C.A., Oliver, R.F., 1984. Changes in hair growth characteristics following the wounding of vibrissa follicles in the hooded rat. J. Embryol. Exp. Morphol. 83, 81–93.

Jahoda, C.A., Reynolds, A.J., 1996. Dermal-epidermal interactions. Adult follicle-derived cell populations and hair growth. Dermatol. Clin. 14 (4), 573–583.

Johnston, A.P.W., Naska, S., Jones, K., Jinno, H., Kaplan, D.R., Miller, F.D., 2013. Sox2-mediated regulation of adult neural crest precursors and skin repair. Stem Cell Rep. 1 (1), 38–45.
Kiernan, A.E., Pelling, A.L., Leung, K.K.H., Tang, A.S.P., Bell, D.M., Tease, C., Lovell-Badge, R., Steel, K.P., Cheah, K.S.E., 2005. Sox2 is required for sensory organ development in the mammalian inner ear. Nature 434 (7036), 1031–1035.
Laga, A.C., Lai, C.-Y., Zhan, Q., Huang, S.J., Velazquez, E.F., Yang, Q., Hsu, M.-Y., Murphy, G.F., 2010. Expression of the embryonic stem cell transcription factor SOX2 in human skin: relevance to melanocyte and merkel cell biology. Am. J. Pathol. 176 (2), 903–913.
Lesko, M.H., Driskell, R.R., Kretzschmar, K., Goldie, S.J., Watt, F.M., 2013. Sox2 modulates the function of two distinct cell lineages in mouse skin. Dev. Biol. 382 (1), 15–26.
Liu, S., Herault, Y., Pavlovic, G., Leask, A., 2014. Skin progenitor cells contribute to bleomycin-induced skin fibrosis. Arthritis Rheumatol. 66 (3), 707–713.
Lumpkin, E.A., Caterina, M.J., 2007. Mechanisms of sensory transduction in the skin. Nature 445 (7130), 858–865.
Lumpkin, E.A., Marshall, K.L., Nelson, A.M., 2010. Review series: the cell biology of touch. J. Cell Biol. 191 (2), 237–248.
Madan, V., Lear, J.T., Szeimies, R.-M., 2010. Non-melanoma skin cancer. Lancet 375 (9715), 673–685.
Maricich, S.M., Wellnitz, S.A., Nelson, A.M., Lesniak, D.R., Gerling, G.J., Lumpkin, E.A., Zoghbi, H.Y., 2009. Merkel cells are essential for light-touch responses. Science 324 (5934), 1580–1582.
Martin, P., 1997. Wound healing – aiming for perfect skin regeneration. Science 276 (5309), 75–81.
Mauger, A., 1972. The role of somitic mesoderm in the development of dorsal plumage in chick embryos. J. Embryol. Exp. Morphol. 28 (2), 313–341.
McElwee, K.J., Kissling, S., Wenzel, E., Huth, A., Hoffmann, R., 2003. Cultured peribulbar dermal sheath cells can induce hair follicle development and contribute to the dermal sheath and dermal papilla. J. Invest. Dermatol. 121 (6), 1267–1275.
McKenzie, I.A., 2006. Skin-derived precursors generate myelinating Schwann cells for the injured and dysmyelinated nervous system. J. Neurosci. 26 (24), 6651–6660.
Morrison, K.M., Miesegaes, G.R., Lumpkin, E.A., Maricich, S.M., 2009. Mammalian Merkel cells are descended from the epidermal lineage. Dev. Biol. 336 (1), 76–83.
Müller-Röver, S., Handjiski, B., van der Veen, C., Eichmüller, S., Foitzik, K., McKay, I.A., Stenn, K.S., Paus, R., 2001. A comprehensive guide for the accurate classification of murine hair follicles in distinct hair cycle stages. J. Invest. Dermatol. 117 (1), 3–15.
Narayanan, D.L., Saladi, R.N., Fox, J.L., 2010. Ultraviolet radiation and skin cancer. Int. J. Dermatol. 49 (9), 978–986.
Okubo, T., Pevny, L.H., Hogan, B.L.M., 2006. Sox2 is required for development of taste bud sensory cells. Genes Dev. 20 (19), 2654–2659.
Olivera-Martinez, I., Thélu, J., Dhouailly, D., 2004. Molecular mechanisms controlling dorsal dermis generation from the somitic dermomyotome. Int. J. Dev. Biol. 48 (2–3), 93–101.
Osumi-Yamashita, N., Ninomiya, Y., Doi, H., Eto, K., 1994. The contribution of both forebrain and mid-brain crest cells to the mesenchyme in the frontonasal mass of mouse embryos. Dev. Biol. 164 (2), 409–419.
Oswald, R., 2009. How your skin works. Available from: http://health.howstuffworks.com/skin-care/information/anatomy/skin.htm [15June 2015].
Paus, R., Cotsarelis, G., 1999. The biology of hair follicles. N. Engl. J. Med. 341 (7), 491–497.
Paus, R., Foitzik, K., 2004. In search of the 'hair cycle clock': a guided tour. Differentiation 72 (9–10), 489–511.
Piskorowski, R., Haeberle, H., Panditrao, M.V., Lumpkin, E.A., 2008. Voltage-activated ion channels and Ca(2+)-induced Ca (2+) release shape Ca (2+) signaling in Merkel cells. Pflügers Arch. Eur. J. Physiol. 457 (1), 197–209.
Rahmani, W., Abbasi, S., Hagner, A., Raharjo, E., Kumar, R., Hotta, A., Magness, S., Metzger, D., Biernaskie, J., 2014. Hair follicle dermal stem cells regenerate the dermal sheath, repopulate the dermal papilla and modulate hair type. Dev. Cell 31 (5), 543–558.
Rendl, M., Lewis, L., Fuchs, E., 2005. Molecular dissection of mesenchymal-epithelial interactions in the hair follicle. PLoS Biol. 3 (11), e331.

Rompolas, P., Deschene, E.R., Zito, G., Gonzalez, D.G., Saotome, I., Haberman, A.M., Greco, V., 2012. Live imaging of stem cell and progeny behaviour in physiological hair-follicle regeneration. Nature 487 (7408), 496–499.

Santini, R., Pietrobono, S., Pandolfi, S., Montagnani, V., D'Amico, M., Penachioni, J.Y., Vinci, M.C., Borgognoni, L., Stecca, B., 2014. SOX2 regulates self-renewal and tumorigenicity of human melanoma-initiating cells. Oncogene 33 (38), 4697–4708.

Santini, R., Vinci, M.C., Pandolfi, S., Penachioni, J.Y., Montagnani, V., Olivito, B., Gattai, R., Pimpinelli, N., Gerlini, G., Borgognoni, L., Stecca, B., 2012. Hedgehog-GLI signaling drives self-renewal and tumorigenicity of human melanoma-initiating cells. Stem Cells 30 (9), 1808–1818.

Schatton, T., Murphy, G.F., Frank, N.Y., Yamaura, K., Waaga-Gasser, A.M., Gasser, M., Zhan, Q., Jordan, S., Duncan, L.M., Weishaupt, C., Fuhlbrigge, R.C., Kupper, T.S., Sayegh, M.H., Frank, M.H., 2008. Identification of cells initiating human melanomas. Nature 451 (7176), 345–349.

Schmidt-Ullrich, R., Paus, R., 2005. Molecular principles of hair follicle induction and morphogenesis. BioEssays News Rev. Mol. Cell. Dev. Biol. 27 (3), 247–261.

Schoenhals, M., Kassambara, A., De Vos, J., Hose, D., Moreaux, J., Klein, B., 2009. Embryonic stem cell markers expression in cancers. Biochem. Biophys. Res. Commun. 383 (2), 157–162.

Sennett, R., Rendl, M., 2012. Mesenchymal-epithelial interactions during hair follicle morphogenesis and cycling. Semin. Cell Dev. Biol. 23 (8), 917–927.

Siegle, J.M., Basin, A., Sastre-Perona, A., Yonekubo, Y., Brown, J., Sennett, R., Rendl, M., Tsirigos, A., Carucci, J.A., Schober, M., 2014. SOX2 is a cancer-specific regulator of tumour initiating potential in cutaneous squamous cell carcinoma. Nat. Commun. 5, 4511.

Stenn, K.S., Paus, R., 2001. Controls of hair follicle cycling. Physiol. Rev. 81 (1), 449–494.

Suflita, M.T., Pfaltzgraff, E.R., Mundell, N.A., Pevny, L.H., Labosky, P.A., 2013. Ground-state transcriptional requirements for skin-derived precursors. Stem Cells Dev. 22 (12), 1779–1788.

Taranova, O.V., Magness, S.T., Fagan, B.M., Wu, Y., Surzenko, N., Hutton, S.R., Pevny, L.H., 2006. SOX2 is a dose-dependent regulator of retinal neural progenitor competence. Genes Dev. 20 (9), 1187–1202.

Tilling, T., Moll, I., 2012. Which are the cells of origin in merkel cell carcinoma. J. Skin Cancer 2012, 6.

Toma, J.G., Akhavan, M., Fernandes, K.J., Barnabé-Heider, F., Sadikot, A., Kaplan, D.R., Miller, F.D., 2001. Isolation of multipotent adult stem cells from the dermis of mammalian skin. Nat. Cell Biol. 3 (9), 778–784.

Tsai, S.-Y., Clavel, C., Kim, S., Ang, Y.-S., Grisanti, L., Lee, D.-F., Kelley, K., Rendl, M., 2010. Oct4 and klf4 reprogram dermal papilla cells into induced pluripotent stem cells. Stem Cells 28 (2), 221–228.

Van Keymeulen, A., Mascre, G., Youseff, K.K., Harel, I., Michaux, C., De Geest, N., Szpalski, C., Achouri, Y., Bloch, W., Hassan, B.A., Blanpain, C., 2009. Epidermal progenitors give rise to Merkel cells during embryonic development and adult homeostasis. J. Cell Biol. 187 (1), 91–100.

Van Raay, T.J., Moore, K.B., Iordanova, I., Steele, M., Jamrich, M., Harris, W.A., Vetter, M.L., 2005. Frizzled 5 signaling governs the neural potential of progenitors in the developing *Xenopus* retina. Neuron 46 (1), 23–36.

Woo, S.-H., Stumpfova, M., Jensen, U.B., Lumpkin, E.A., Owens, D.M., 2010. Identification of epidermal progenitors for the Merkel cell lineage. Development 137 (23), 3965–3971.

CHAPTER 17

SOX2 in the Development and Maintenance of the Trachea, Lung, and Esophagus

Neng Chun Wong[1,3], Parth Armin[1], Ming Jiang[3], Wei-Yao Ku[2], Ian Jacobs[2], Jianwen Que[2,3,*]

[1]Department of Biology, University of Rochester, Rochester, NY, USA; [2]Department of Biomedical Genetics, University of Rochester Medical Center, Rochester, NY, USA; [3]Division of Digestive and Liver Diseases and Columbia Center for Human Development, Department of Medicine, Columbia University, New York, NY, USA

*Corresponding author: BB 8-810, 650 West 168th Street, New York, NY 10032. Email: jq2240@cumc.columbia.edu

INTRODUCTION

The trachea and esophagus are two individual tubes that connect to the lungs and stomach, respectively. All of these organs are generated during development from the anterior portion of a single lumen tube termed the foregut. Intimate interaction between the endoderm and mesoderm is necessary for specification of the anterior foregut into the different domains that eventually give rise to these organs. Intriguingly, studies of mouse genetic models suggest that the foregut tube exhibits dorsal-ventral patterning of signaling molecules and transcription factors during the specification (Jacobs et al., 2012). Disruption of this patterning can lead to abnormal formation of the trachea, lung, and esophagus.

Extensive proliferation and differentiation are required for epithelial morphogenesis in the organs derived from the anterior foregut. When the trachea, lung, and esophagus are established, the epithelium lining these tissues is simple columnar. However, they take on different differentiation programs that generate cells with distinct characteristics. For example, cells lining the trachea and lung remain as simple columnar. By contrast, the cells lining the esophagus become stratified squamous epithelia. The transcription factor SOX2 has previously been shown to regulate the self-renewal of embryonic stem cells and early embryogenesis (Avilion et al., 2003; Masui et al., 2007; Que et al., 2009; Arnold et al., 2011; Sarkar and Hochedlinger, 2013). Genetic studies revealed that it also has critical roles in foregut morphogenesis at different stages. Before separation, SOX2 is enriched at the dorsal epithelium and is required for separation of the trachea and esophagus and remains essential for subsequent differentiation of epithelial cells in the trachea, lung, and esophagus. Lineage tracing studies in adult animals demonstrate that SOX2 marks stem/progenitor cell populations in multiple tissues including the esophagus and SOX2 continues to have important roles in the maintenance of epithelial homeostasis in these tissues (Arnold et al., 2011). Abnormal SOX2 protein

Sox2
http://dx.doi.org/10.1016/B978-0-12-800352-7.00017-7

levels are associated with the onset of multiple diseases including malignancies. These findings and their relevance to human health will be discussed in this chapter.

SOX2 IN SEPARATION OF THE TRACHEA AND ESOPHAGUS

Generation of the trachea and esophagus from the early foregut is initiated at 4 weeks of gestation in humans and occurs between embryonic day (E)9.5 and E11.5 in mouse embryos. This morphogenetic event requires dorsal-ventral patterning of signaling molecules and transcription factors (Jacobs et al., 2012). Disruption of relevant signaling pathways (e.g., Wnt and bone morphogenetic protein [Bmp]) or transcription factor activities leads to the formation of esophageal atresia with/without fistula (EA/TEF) in which the esophagus is ended with a blind end sac and a communication develops between the trachea and stomach. Esophageal atresia with/without fistula is a relatively common birth defect with an incidence rate of 1/3500. Notably, EA/TEF is often accompanied by developmentally abnormal lungs.

Before separation, SOX2 is expressed in the endoderm of the anterior foregut from the pharyngeal arches down to the region that will be the future stomach. SOX2 is preferentially expressed in the dorsal epithelial cells of the unseparated foregut tube at E9.5, in direct contrast to the transcription factor NKX2.1, which is expressed in the ventral epithelium (Figure 1) (Que et al., 2007; Harris-Johnson et al., 2009). Dorsal-ventral patterning of these two transcription factors is required for foregut separation. Down-regulation of SOX2 protein to near 5% of the wild-type level leads to the formation of EA/TEF in *Sox2*$^{GFP/COND}$ hypomorphic mutants, which harbor one *Sox2* null allele (SOX2-GFP) and one hypomorphic allele (*Sox2-COND* or *Sox2-loxp*^{Pevny}) (Taranova et al., 2006; Que et al., 2007). Conversely, loss of *Nkx2.1* also leads to EA/TEF

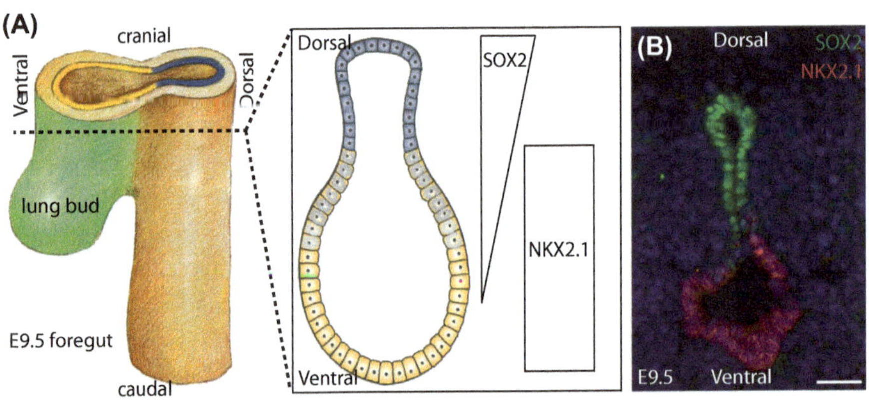

Figure 1 Dorsal-ventral patterning of SOX2 and NKX2.1 is critical for generating the esophagus and trachea from the anterior foregut. (A) Schematic diagram indicates the dorsal-ventral expression pattern of SOX2 and NKX2.1. (B) SOX2 and NKX2.1 are highly expressed in the epithelium of the dorsal and ventral foregut at E9.5, respectively. Nuclei are counterstained with DAPI. Scale bar: 50 μm.

formation (Minoo et al., 1999). These results suggest that SOX2 and NKX2.1 are essential to specify the esophageal and tracheal domains, respectively. Downregulation or deletion of either molecule leads to expansion of domains marked by the other. Consequently, epithelial cells in the TEF of *Sox2*$^{GFP/COND}$ hypomorphic mutants express high levels of NKX2.1, which suggests that in the absence of a sufficiently high level of SOX2, NKX2.1 expression expands dorsally and reprograms the dorsal epithelium to a respiratory fate (Que et al., 2007). In addition, cartilage nodules are ectopically present in the TEF. By contrast, SOX2 is highly expressed in all of the epithelial cells of the unseparated foregut tube of *Nkx2.1* null mutants and smooth muscle cells surround the single lumen tube, a feature characteristic of the esophagus (Que et al., 2007).

We have previously shown that fibroblast growth factor 10 (FGF10) inhibits SOX2 expression in the mouse foregut in organ culture (Que et al., 2007). FGF10 is enriched in the mesenchyme of the ventral foregut before foregut separation initiates. However, foregut separation proceeds normally in *Fgf10* null mutants, despite a lack of lung morphogenesis (Ramasamy et al., 2007; El Agha et al., 2012). These findings suggest that other signaling molecules are involved in the regulation of SOX2/NKX2.1 patterning in the early foregut (Min et al., 1998; Que et al., 2007). Domyan et al. showed that the Bmp signaling pathway suppresses SOX2 expression at the ventral side of the foregut before separation (Figure 2). Simultaneous deletion of BMP receptors, such as *Bmpr1a* and *1b*, blocks generation of the esophagus and trachea in *Shh-Cre;Bmpr1a*$^{loxp/-}$*;Bmp r1b*$^{-/-}$ mutants. Consistently, inactivation of *Sox2* rescues the formation of the trachea and NKX2.1 expression in *Shh-Cre;Bmpr1a*$^{loxp/-}$*;Bmpr1b*$^{-/-}$*;Sox2*$^{loxp/loxp}$ mutants (Domyan et al., 2011). Furthermore, the Wnt signaling pathway is preferentially active in the ventral foregut and deletion of *Wnt2/2b* or the downstream mediator *ß-catenin* results in an unseparated foregut (Harris-Johnson et al., 2009; Goss et al., 2009). These mutants lack development of the lung because the respiratory marker NKX2.1 cannot

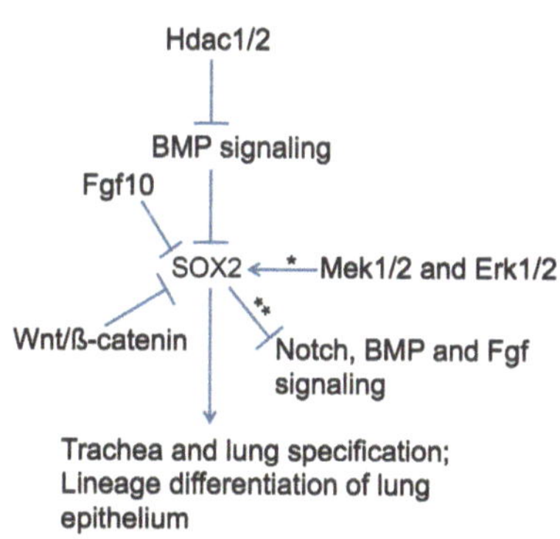

Figure 2 Multiple signaling pathways have been implicated in the regulation of SOX2 during specification of the respiratory system. *Mek1/2 and Erk1/2 are required for the maintenance but not initiation of SOX2 expression. **SOX2 overexpression suppresses the activities of Notch, Bmp, and Fgf signaling pathways. See text for details.

be detected in the ventral foregut at any time point. By contrast, SOX2 expression domain expands into this region (Harris-Johnson et al., 2009).

SOX2 IN LUNG DEVELOPMENT AND MAINTENANCE

Overview of Development of the Respiratory System

The trachea connects to the lung through two primary (main) bronchi. The primary bronchi branch into the lung in a stereotypical manner to form a treelike structure composed of smaller bronchi, bronchioles, terminal bronchioles, and respiratory bronchioles. The respiratory bronchioles transition into alveolar ducts that open into atria that communicate with alveolar sacs, with the latter terminating into alveoli. From primary bronchi to alveolar ducts, there is a dramatic increase in the number of airways and a progressive decrease in the diameter of each airway (Figure 4). For convenience, the primary bronchi, smaller bronchi, and bronchioles are collectively named proximal airways, whereas the terminal and respiratory bronchioles and alveolar ducts are named distal airways. Along these airways runs a network of blood vessels that communicates with capillaries surrounding the alveoli to accomplish the activity of gas exchange.

Much of our initial understanding of lung development is derived through anatomical and histological characterization of mouse lungs. The development of the lung is initiated at E9.0–E9.5 in mice and 3–4 weeks in humans and can be divided into five stages. At the initial stage, termed the embryonic stage (E9.0–E11.5 in mice and 3–6 weeks in humans), primitive bronchial buds begin to form at the ventral site of foregut and enlarge to form two primary lung buds. At the pseudoglandular stage (E11.5–E16.5 in mice and 6–16 weeks in humans), the two distinct lung buds continue to branch out and form a treelike structure. The epithelial cells lining the proximal airways exhibit signs of differentiation. The first specified lineage is the neuroendocrine cell (Cutz et al., 1985; McDowell et al., 1985; Borges et al., 1997). Epithelial cells largely remain as uncommitted progenitor cells at this stage although most epithelial cells express p63, a marker used to detect basal cells in the adult trachea (Rock et al., 2009). Differentiated epithelium other than neuroendocrine cells can be readily detected with corresponding markers at E14.5; they include ciliated (FOXJ1 and acetylated tubulin positive), Clara (SCGB1A1 positive; also called Cc10), minor populations of neuroendocrine cells (CGRP positive), and goblet cells (Agr2 positive) (Morrisey and Hogan, 2010). At the canalicular stage (E16.5–E17.5 in mice and 16–28 weeks in humans), the size of airways increases and respiratory units begin to form. These units are composed of a respiratory bronchiole leading to alveolar ducts and terminal sacs (primitive alveoli) with associated capillaries. At the saccular stage (E17.5 to postnatal [P]5 in mice and 28 weeks to term in humans), the terminal sacs are surrounded by capillaries as they continue to develop and fine-tune. The type I and II (AEC1 [T1 alpha-positive] and AEC2 [surfactant protein C–positive]) epithelial cells lining the terminal air sacs start to produce large amounts of surfactant proteins, preparing the lung for its first breath after birth. At the alveolar stage (P5–P30 in mice and in the first few years after birth in humans), the final stage, the terminal air sacs (alveoli) are subdivided by the ingrowth of ridges or crests known as the secondary septae. Throughout this period, the alveoli continue to multiply through increasing secondary septae while the AEC1 and AEC2 cells expand and differentiate.

Multiple signaling pathways including Bmp, Shh, and Wnt have been shown to regulate lung morphogenesis at different developmental stages (Cardoso and Lu, 2006; Morrisey and Hogan, 2010). The Hippo pathway has also been implicated in the regulation of lung morphogenesis. Interestingly, several of these pathways converge in their functions through SOX2, controlling multiple aspects of lung morphogenesis at different stages: for example, setting up the proximal-distal boundary and modulating epithelial differentiation. In adults, SOX2 remains to be expressed in the airway epithelium especially in progenitor cells and is tightly controlled by relevant pathways (Que et al., 2009; Tompkins et al., 2009; Arnold et al., 2011). Deregulated SOX2 protein levels are associated with pathological changes such as squamous cell metaplasia and tumors.

SOX2 and SOX9 in the Specification of Proximal–Distal Airways

SOX2 is expressed at low levels when the lung buds initiate from the ventral foregut endoderm. Shortly afterward, the levels of SOX2 increase in the proximal airways (from trachea to the bronchioles), leading to a gradient that is high in the proximal and low in the distal airways. In contrast, the distal airway epithelium expresses high levels of SOX9 (Figure 3). This expression pattern is maintained at the canalicular stage. Whereas SOX2 continues to be expressed in the proximal airways, the expression of SOX9 in the

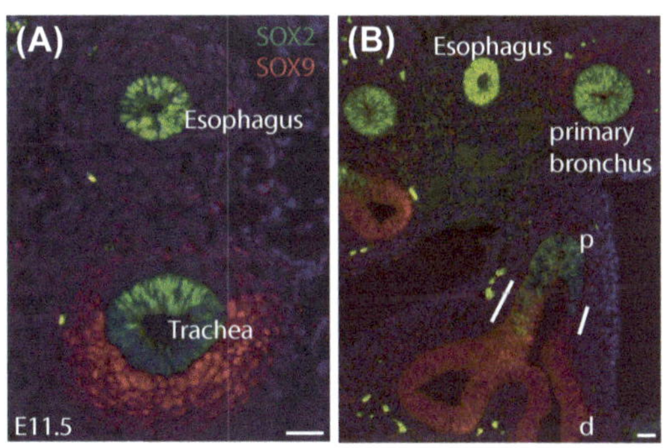

Figure 3 SOX2 and SOX9 expression in the esophagus, trachea, and lung at E11.5. (A) SOX2 remains highly expressed in the esophagus, as opposed to lower expression in the trachea. Sox9 is expressed in the ventral mesenchyme of the trachea where the C-ring cartilage develops. (B) Gradient expression pattern of SOX2 and SOX9 in the airways. SOX2 is highly expressed in the main bronchus, in contrast to its low expression in the more distal part of the developing airways. SOX9 is highly expressed in the distal airways. Note some of the epithelial cells express both SOX2 and SOX9 as indicated by the white line. Abbreviation: p, proximal; d, distal. Scale bar: 50 μm.

peripheral lung is gradually reduced at the saccular and alveolar stages (Chang et al., 2013). The boundary between SOX2 and SOX9 in the early lung seems to correspond to the future bronchoalveolar duct junction (BADJ) in the adult. A study revealed that setting up such a boundary requires the Hippo pathway (Mahoney et al., 2014).

Activation of the Hippo pathway leads to the inhibition of cell growth, differentiation, and cell death. This pathway was first defined in *Drosophila* when screening for tumor suppressors. It was shown that inactivation of genes including *Warts*, *Hippo*, *Salvador*, or *Mats* results in robust tissue overgrowth in flies and these genes were subsequently identified as the major components of the Hippo pathway. The pathway consists of a core kinase cascade. Phosphorylation of Warts by Hippo promotes the nuclear translocation of Warts, which in turn phosphorylates and inactivates the transcriptional co-activator Yorkie. Yorkie is unable to bind deoxyribonucleic acid (DNA) by itself but can bind to the transcription factor Scalloped and regulates genes (e.g., cyclin E) that are involved in cell growth and inhibition of apoptosis (Yu and Guan, 2013). Interestingly, Yes-associated protein (YAP), an ortholog of Yorkie, has a unique expression pattern in the proximal and distal airways: nuclear localization co-labeled with SOX9 at distal lung epithelial cells and cytoplasmic localization co-labeled with SOX2 at proximal lung epithelial cells (Mahoney et al., 2014). YAP and PDZ-binding motif (TAZ), another transcriptional co-activator downstream of the Hippo pathway, does not bind to DNA, but instead binds to the promoters of target genes by interacting with transcription factors including TEA domain family members (TEAD) 1–4 (Vassilev et al., 2001; Goulev et al., 2008; Wu et al., 2008; Zhang et al., 2008; Zhao et al., 2008). YAP/TAZ also interacts with the transcription factors downstream of the transforming growth factor-β (TGF-β) signaling pathway. For example, YAP/TAZ can bind to SMAD1 (Alarcón et al., 2009) and SMAD2/3 (Varelas et al., 2008) to regulate the transcription of genes involved in proliferation, differentiation, and development.

In the absence of YAP, the conducting airways are greatly reduced and its differentiated epithelial cells are substituted with less differentiated cells and mucus-producing cells. By contrast, distal airways are expanded at the expense of the proximal progenitor cells that normally contribute to large airways. Consistently, only a few SOX2-positive cells are present in the most proximal epithelial tubules, whereas strong SOX9 staining is observed throughout the epithelium of the dilated structures of mutant lungs (Mahoney et al., 2014). Therefore, it appears that Hippo/Yap regulates SOX2 expression and programs the SOX2-positive progenitor cells into proximal airway epithelium. In addition, the Yap binding partners TEAD transcription factors are enriched at the *Sox2* promoter in the E12.5 lungs. This raises the possibility that TEADs and Yap cooperate to regulate *Sox2* and its downstream target genes. Enrichment of TEADs in the *Sox2* promoter is also found in the adult airway epithelium (Mahoney et al., 2014), which suggests that Yap continues to have roles in maintaining the epithelium in the adult airways.

SOX2 Regulates Epithelial Differentiation in the Developing Lung

Sox2 loss-of-function studies also revealed important roles for this transcription factor in lung development, especially for epithelial morphogenesis during the later stages (Table 1). Conditional deletion of *Sox2* with *Nkx2.5-Cre* exhibits Cre activities in both the epithelium and mesenchyme of the E9.5 ventral foregut and results in mutants that succumb to death at birth owing to respiratory distress. The trachea is truncated and the cartilage rings are abnormally formed. SOX2 is not expressed in the mesenchyme, which suggests that SOX2 indirectly regulates mesenchymal differentiation in addition to its

Table 1 Genetic manipulation affecting SOX2 expression and lung/tracheal development

Mouse line	Lung/tracheal phenotype	SOX2 levels	Changes of lung/tracheal epithelium	References
Nkx2.5-Cre;Sox2$^{COND/COND}$	Short trachea, long bronchi, abnormal cartilage	Deleted	Decrease number of basal, ciliated and Clara cells. Increase in mucus-producing cells	Que et al. (2009)
Shh-Cre;Bmpr1a$^{loxp/-}$;Bmpr1b$^{-/-}$	Tracheal agenesis	Increased	Increase in the number of p63+ cells	Domyan et al. (2011)
Rosa26-rtTA;Tet-Fgf10	Abnormal proximal and distal patterning	Decreased	Decrease in the number of SOX2+ proximal airway epithelium	Volckaert et al. (2013)
Shh-Cre;ß-catenin$^{loxp/loxp}$	Tracheal and lung agenesis; No lung buds form	ND	Expansion of SOX2+ epithelium into the ventral foregut at E9.5	Harris-Johnson et al. (2009)
Shh-Cre;Mek1$^{loxp/loxp}$;Mek2$^{-/-}$	Shortened trachea; reduced cartilage rings, lung agenesis	Decreased	Reduced proliferation Devoid of basal, goblet and ciliated cells, with only a few Clara cells	Boucherat et al. (2014)
Spc-Cre;ß-catenin^{exon3} (activated form)	Lung hypoplasia	Decreased	Decrease in the number of ciliated and Clara cells; many cells do not express markers labeling differentiated lung epithelial lineages at E18.5	Hashimoto et al. (2012)
Shh-Cre;Hdac1$^{loxp/loxp}$;Hdac2$^{loxp/loxp}$.	Inhibition of lung branching morphogenesis	Decreased	Decreased number of Clara cells, ciliated cells, and neuroendocrine cells	Wang et al. (2013)
Spc-rtTA;myc-Sox2	Lung hypoplasia	Increased	Increased number of p63+ cells; expansion of Clara cells into enlarged air sacs	Gontan et al. (2008)

Continued

Table 1 Genetic manipulation affecting SOX2 expression and lung/tracheal development—cont'd

Mouse line	Lung/tracheal phenotype	SOX2 levels	Changes of lung/ tracheal epithelium	References
Spc-Cre; Rosa26R-Sox2-IRES-GFP	Regions of epithelial hyperplasia in bronchi, bronchioles and alveoli	Increased	Ectopic p63+ cells in alveolar region	Lu et al. (2010)
Scgb1a1-CreER;Rosa26R-Sox2-IRES-GFP	Hyperplasia of bronchial epithelium	Increased	Hyperplastic cells are positive for ciliated marker FOXJ1 in alveoli	Lu et al. (2010)
Scgb1a1-Cre;Sox2$^{loxp/loxp}$	Not reported	Decreased	Reduced number of goblet, Clara, and ciliated cells; reduced cell proliferation	Tompkins et al. (2009)
Spc-rtTA;myc-Hif3a	Decreased number of alveolar spaces and reduced branching	Increased	Reduced numbers of Clara cells, alveolar epithelial type I and type II cells; aberrant basal cells were observed in distal lung	Huang et al. (2013)
Scgb1a1(Cc10)-rtTa$^{tg/-}$/Teto-Cre$^{tg/-}$/ Foxm1$^{loxp/loxp}$	Not reported	Decreased	Reduced number of Clara and ciliated cells; presence of alveolar type II cells in conducting airways	Balli et al. (2013)
Shh-Cre;Yap$^{loxp/loxp}$	Lung hypoplasia with dilated cyst-like structures	Decreased	Expansion of SOX9+ distal lung epithelium	Mahoney et al. (2014)

role in regulating the epithelium. The numbers of basal, ciliated, and Clara cells are decreased in the mutant trachea and lung, whereas the number of mucus-producing cells is increased. Interestingly, a considerably large population of cells expresses no differentiation markers that label different lung cell lineages (Que et al., 2009). In a separate study, removal of *Sox2* with *Scgb1a1-Cre* to target Clara cells and their derivatives (e.g., ciliated cells) also results in abnormal differentiation of the epithelium (Tompkins et al., 2009; Rawlins and Perl, 2012). Cre recombinase activity in the *Scgb1a1-Cre* mouse line can be detected in the conducting airway epithelium from E16 to E17 and after birth (Reynolds et al., 2008; Zemke et al., 2009). *Sox2* deletion in the *Scgb1a1-Cre;Sox2$^{loxp/loxp}$* mutants, which contains the *Sox2-loxp^{Lang}* conditional allele, does not lead to lethality, but the epithelium lining the conducting airways presents cuboidal-to-squamous metaplasia

(Tompkins et al., 2009). These epithelial cells are stained negatively for any differentiation markers that label different lung cell lineages. Nevertheless, Nkx2.1 remains to be expressed in these cells, which suggests that loss of SOX2 does not alter their lung epithelium identity. The authors further showed that SOX2 activates the transcription of *Scgb1a1*, *FoxJ1*, and *Agr2*, differentiation markers characteristic of Clara, ciliated, and goblet cells, respectively. In addition, SOX2 forms a protein complex with SMAD3 and suppresses the SMAD3-mediated TGF-β pathway in human bronchial epithelial cells. The authors therefore suggest that deletion of *Sox2* leads to unchecked TGF-β signaling, resulting in reduction in cell proliferation (Alejandre-Alcázar et al., 2008; Tompkins et al., 2009).

Changes in the activities of SOX2 upstream regulators also lead to abnormal differentiation of the lung epithelium (Figure 2; Table 1). For example, ectopic Wnt signaling activities through the use of the conditional $\beta\text{-}catenin^{Exon3}$ allele suppress SOX2 expression. Similar to $NKX2.5\text{-}Cre;Sox2^{COND/COND}$ mutants, the epithelial cells lining the bronchioles do not express markers for any known differentiated cell types (Hashimoto et al., 2012). In addition, loss of the transcription factor FOXM1, which has been shown to regulate the transcription of SOX2 and SCGB1A1 in Clara cells, also results in abnormal differentiation of the epithelial cells and an increased number of goblet cells (Ustiyan et al., 2012). These findings show that SOX2 is required for the differentiation of epithelial progenitor cells during lung development. A study examining the impact of epigenetic regulation in lung development suggests that Bmp signaling inhibits SOX2 expression in the growing distal tips (Wang et al., 2013). The epigenetic regulation enzymes histone deacetylase 1 and 2 (Hdac1 and Hdac2) modify the expression of Bmp4. Lungs deficient in *Hdac1* and *Hdac2* have increased levels of BMP4 and decreased SOX2 expression in the proximal airways, subsequently resulting in branching defects and truncation of the trachea (Wang et al., 2013). Moreover, loss of *Hdac1/2* in Clara cells affects regeneration after injury and extinguishes the capability of Clara cells to repopulate ciliated cells. This phenotype is recapitulated upon progressive deletion of *Sox2* in Clara cells after birth (Tompkins et al., 2009).

Regulation of SOX2 expression by the ERK/(mitogen-activated protein (MAP) kinase is also critical for epithelial differentiation in the developing lung (Boucherat et al., 2014) (Figure 2). Simultaneous deletion of the two ERK/MAP kinase genes, *Mek1* and *Mek2*, which encode dual-specificity kinases responsible for ERK/MAP kinase activation in the epithelium, leads to lung agenesis with the loss of SOX2 expression at E18.5. *Shh-Cre* is used for the conditional deletion and is active in the epithelium when lung buds initiate (Rodriguez et al., 2010; Harris-Johnson et al., 2009). Intriguingly, SOX2 is expressed in all of the tracheal epithelium in $Shh\text{-}Cre;Mek1^{loxp/loxp};Mek2^{-/-}$ mutants at E14.5, although the expression level is reduced. Therefore, it appears that the ERK/MAPK pathway is required for the maintenance but not for the initiation of SOX2 expression. The epithelium becomes simple squamous cells devoid of basal, goblet, and ciliated cells, with only a few Clara cells after losing *Mek1* and *Mek2* (Boucherat et al., 2014). Of note is that no lung phenotype was observed in *Shh-Cre;Mek1*$^{loxp/+}$;*Mek2*$^{-/-}$ and *Shh-Cre;Mek1*$^{loxp/loxp}$; *Mek2*$^{+/-}$ embryos or mice, which suggests that the presence of one functional *Mek* allele in lung epithelium is sufficient for normal lung development. Consistent with the important role of the ERK/MAPK pathway in lung development, deletion of the downstream effectors *Erk1* and *Erk2* in the epithelium (*Shh-Cre*; *Erk1*$^{-/-}$; *Erk2*$^{loxp/loxp}$) recapitulates the phenotypes exhibited in the E18.5 *Shh-Cre;Mek1*$^{loxp/loxp}$;*Mek2*$^{-/-}$ embryos: lung agenesis, abnormal tracheal epithelial differentiation (Boucherat et al., 2014).

Genetic gain-of-function studies also shed important insight into the role for SOX2 in lung development (Table 1). Multiple studies have shown that SOX2 overexpression leads to defects in branching morphogenesis and epithelial cell differentiation. For example, SOX2 overexpression in the *Spc-rtTA;oTet-myc-Sox2* mutants blocks branching morphogenesis and results in a marked decrease in the number of airways (Gontan et al., 2008). The transgenic *Spc-rtTA* allele starts to be active in the lung epithelium at around E11.5–E12.5. When doxycycline water was fed starting at E6.5, the resulting lung is composed of enlarged airspaces lined with dysplastic epithelium and the animals die at birth. Further analysis revealed that SOX2 overexpression reduces Fgf and Bmp signaling activities as the levels of their downstream mediators, phosphorylated ERK1/2 and pSMAD1/5/8, are decreased. In addition, SOX2 overexpression promotes a premature commitment of epithelial cells toward differentiation, leading to an increase in the number of neuroendocrine cells, p63-positive basal cells, and cells expressing gastric/intestinal proteins (Gontan et al., 2008). In a separate study, SOX2 overexpression in *Spc-Cre;Rosa26R-Sox2-IRES-GFP* mice resulted in respiratory distress a few weeks after birth. The transgenic *Spc-Cre* is active in the lung epithelium at around E10.5 (Okubo and Hogan, 2004). SOX2 overexpression leads to regions of epithelial hyperplasia in airways in the presence of ectopic p63-positive cells. Moreover, in some mutants adenocarcinoma develops in the peripheral region of the lung (Lu et al., 2010). These results further emphasize the important roles for SOX2 in regulating epithelial proliferation and differentiation during development.

SOX2 Is Involved in Epithelial Maintenance and Tumorigenesis in the Adult Respiratory System

SOX2 continues to have important roles in the adult respiratory system. The gradient expression of SOX2 is maintained in all of the epithelial cells lining the trachea and airways up to the BADJ (Gontan et al., 2008; Que et al., 2009) (Figure 4). Loss of *Sox2* in the adult trachea reduces the proliferation of basal progenitor cells and impairs regeneration of the epithelium after SOX2-induced injury (Que et al., 2009). Conversely, conditional overexpression of SOX2 using *Scgb1a1-CreER* in the adult Clara cells and their derivatives promotes the proliferation of epithelial cells, especially in the distal

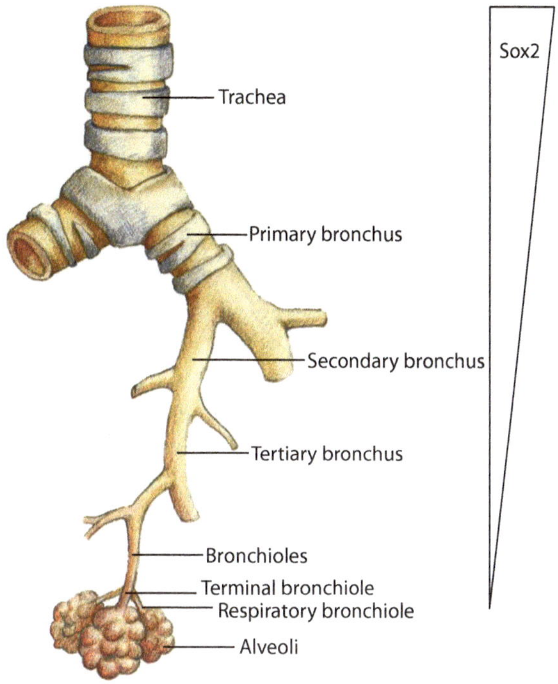

Figure 4 The gradient expression pattern of Sox2 is maintained in the adult respiratory system.

bronchioles and adjacent alveoli with increased levels of the cell cycle regulator cyclin D1 (Lu et al., 2010). Whereas extensive epithelial hyperplasia occurs in the mutant lungs with heterozygotic overexpression of SOX2, homozygotic overexpression results in more severe hyperplasia both at the BADJ and in the alveoli. In addition, SOX2 overexpression induces the formation of pseudostratified epithelium at the BADJ where ectopic p63-positive cells underlie columnar cells. Notably, these p63-positive cells do not express the basal cell markers keratin 14 and T1 alpha (also known as nerve growth factor receptor). As time goes on, adenocarcinoma develops in the alveoli of around half of SOX2 homozygotic overexpressing mice. Interestingly, tumor cells do not have Clara or alveolar cells, but instead contain a significant proportion of ciliated cells, and about 50% of cells express p63 and other squamous cell genes (Lu et al., 2010).

SOX2 overexpression seems to suppress the activities of the Notch signaling pathway. The transcript levels of *Hes1*, *Notch1*, and *Notch2* were significantly downregulated upon SOX2 overexpression. Chromatin immunoprecipitation (ChIP) further demonstrated that SOX2 binds the regulatory sequences of *HES1*, *NOTCH1*, and *NOTCH2* in SOX2-overexpressing H520 human lung cancer cells (Xu et al., 2014). Interestingly, missense and nonsense mutations in NOTCH and receptors are associated with the development of squamous cell tumors in the lung (Wang et al., 2011). In animal models, inhibition of Notch signaling leads to alveolar hyperplasia with ectopic expression of squamous markers (Xu et al., 2014). These findings suggest that SOX2 overexpression-induced epithelial hyperplasia and tumors exhibit squamous cell characteristics although the tumor is adenocarcinoma. They also suggest that tumorigenesis in these mutants is partially mediated through inhibition of the Notch signaling pathway. Intriguingly, the authors also demonstrated that SOX2 overexpression is able to suppress K-RasG12D–induced tumor formation in the *Scgb1a1-CreER;LSL-K-RasG12D;Rosa26R-Sox2-IRES-GFP* compound mouse line. Tumors formed in the alveolar regions are markedly smaller than age-matched *Scgb1a1-CreER;LSL-K-RasG12D* mice (Xu et al., 2014). In addition, in the alveolar hyperplastic regions of compound mice but not *Scgb1a1-CreER;LSL-K-RasG12D* mice there are clusters of cells positive for the squamous cell markers Keratin 5 (KRT5) and p63 (Xu et al., 2014).

Squamous cell carcinoma (SCC) of the lung is the second most common subtype of lung cancers. Gene amplification or overexpression of *Sox2* has been frequently identified in human SCC including lung SCC (Bass et al., 2009; Hussenet et al., 2010; Brcic et al., 2012; Cancer Genome Atlas Research, 2012). However, as mentioned above, SOX2 overexpression leads to the formation of adenocarcinoma with the presence of basal-like p63-positive cells in the peripheral lung (Lu et al., 2010), which suggests that input from other genes is required for SOX2-mediated SCC formation. Mukhopadhyay et al. (Mukhopadhyay et al., 2014) performed a pilot screen and identified that the loss of *LKB1* (also known as serine/threonine kinase 11) cooperates with lentivirus-mediated SOX2 overexpression to drive SCC formation in the mouse lungs. LKB1 is a tumor suppressor that has been implicated in the regulation of metabolism, cell polarity, and growth (Ji et al., 2007; Shackelford and Shaw, 2009; Perez-Moreno et al., 2012). In agreement with this, mTOR and Stat3, two major pathways relevant to metabolism and cell growth control, are activated in lung SCCs developed in the *Lenti-Sox2-Cre;Lkb1$^{loxp/loxp}$* mutants (Mukhopadhyay et al., 2014). Loss-of-function mutation of LKB1 has been found in 5–19% of human SCCs (Ji et al., 2007; Shackelford and Shaw, 2009; Perez-Moreno et al., 2012). In particular, 55% of the human SCCs with strong SOX2 expression also activated the mTOR pathway (Mukhopadhyay et al., 2014), raising the possibility of using mTOR inhibitors to treat SOX2-overexpressing lung SCC. One unanswered question in this study is the cell of origin for SCCs. Use of cell type-specific SOX2 overexpression combined with the loss of *Lkb1* will be useful for addressing the issue.

SOX2 IN MORPHOGENESIS AND MAINTENANCE OF ESOPHAGUS

Overview of Esophageal Development

Once the esophagus is established from the anterior foregut, it undergoes extensive morphogenesis in both the epithelium and mesenchyme, eventually giving rise to an epithelial tube surrounded by layers of muscle. The epithelium in the adult esophagus is stratified squamous composed of undifferentiated basal cells and differentiated suprabasal cells located atop of the basal cells. Nevertheless, the epithelium lining the nascent esophagus is composed of a simple columnar epithelium that is gradually replaced by a stratified squamous epithelium (Yu et al., 2005). Meanwhile, the mesenchymal cells proliferate and differentiate into multiple layers of muscle cells (Jacobs et al., 2012).

The epithelium of the mouse esophagus exhibits dynamic changes throughout development. At the initial stage of the transition (E11.5–E14.5), the epithelium stratifies without exhibiting apparent differentiation. Some epithelial cells facing the lumen are ciliated and are stained positively with acetylated tubulin and FOXJ1 antibodies (Que, 2015). Proliferative cells are found throughout the epithelium at this stage. From E14.5 to E18.5, the top layers of epithelium start to differentiate and express the squamous cell markers Loricrin and Involucrin (Rodriguez et al., 2010). The proliferative cells can be found only in the bottom layers of the epithelium at this stage. The numbers of ciliated cells increase and appear to be evenly distributed along the whole esophagus. The epithelial cells also produce high-molecular-weight glycoproteins, which are positive for periodic acid–Schiff (PAS) staining, especially in the abdominal segments of the esophagus. After birth, the numbers of ciliated cells decrease rapidly but are still present in patches 4 days after birth (Raymond et al., 1991). The top layer of the epithelium starts to be keratinized at around 9 days after birth. Of note, human esophageal epithelium is not keratinized at any stage of development. Also, the esophageal side of the stomach termed forestomach is lined with the stratified squamous epithelium like the esophagus in mice (Jacobs et al., 2012).

SOX2 Regulates Epithelial Morphogenesis in Developing Esophagus

SOX2 expression is dynamic during morphogenesis of the epithelium in the developing esophagus. When the esophagus is established from the foregut, SOX2 is expressed highly in the simple columnar epithelium, accompanied by high levels of p63 (Rodriguez et al., 2010). Beginning at E14.5, the top layers of epithelium start to lose expression of SOX2 and p63, and at birth expression of these transcription factors is maintained exclusively by basal cells, as are high levels of the intermediate filament protein KRT5 (Rodriguez et al., 2010). Deletion of the *p63* gene blocks the conversion of simple columnar to stratified squamous epithelia and many of the columnar cells are ciliated at birth (Daniely et al., 2004). By contrast, decreased SOX2 protein levels do not affect the conversion. When the levels of SOX2 protein are reduced to about 30% of wild type, the epithelium is able to stratify in both the esophagus and forestomach. However, the epithelium is disorganized and produces large amounts of glycoproteins that are positive for PAS staining. Some of the epithelial cells also express the mucin protein MUC5AC that is normally found in the glandular stomach (hindstomach) (Que et al., 2007), which suggests that gene and protein is essential for the differentiation of epithelial progenitor cells during morphogenesis of the esophagus.

SOX2 in Maintenance and Tumorigenesis of Adult Esophagus

SOX2 is exclusively expressed in basal progenitor cells of the adult esophagus and forestomach in mice (Arnold et al., 2011; Liu et al., 2013). In the human esophagus, the SOX2 expression domain is relatively broader, although basal cells still have the highest expression levels (Liu et al., 2013). Thus far, no functional study has been reported regarding the role of SOX2 in maintenance of the adult esophagus, although decreased or loss of SOX2 expression has been found in esophageal diseases such as Barrett's esophagus (Chen et al., 2008). On the other hand, *Sox2* gene amplification and overexpression have been closely linked to the formation of esophageal SCC (ESCC) (Bass et al., 2009). We have also established a mouse model to test whether SOX2 overexpression is a driver or passenger for ESCC formation. Our studies revealed that SOX2 cooperates with inflammatory signaling to drive malignant transformation of basal progenitor cells.

When SOX2 is conditionally overexpressed in the basal cells of the esophagus and forestomach of the *Krt5-CreER;Rosa26R–Sox2-IRES-GFP* compound mutants, the epithelium becomes hyperplastic with increased proliferation and decreased differentiation. Intriguingly, SCCs develop in the forestomach but not the esophagus in the mutants. Subsequent analysis revealed that cooperation between SOX2 overexpression and inflammation is required for the initiation of malignancy (Liu et al., 2013; Bass and Wang, 2013). The evidence suggests that the loss of differentiation and the protective keratin layer in the forestomach epithelium upon SOX2 overexpression leads to inflammation, presumably after acid-induced injury. Consistently, high levels of interleukin (IL)-1β and IL-6 were detected in the hyperplastic epithelium in the forestomach but not the esophagus a few weeks after SOX2 induction. Interleukin-6 is a potent activator of the Stat3 signaling pathway, which cooperates with SOX2 overexpression to drive the transformation of basal progenitor cells, resulting in the formation of SCC. In keeping with this line, co-overexpression of SOX2 and constitutively active STAT3 is able to transform human telomerase reverse-transcriptase immortalized esophageal epithelial cells, which have basal cell-like characteristics (Harada et al., 2003; Liu et al., 2013). Of relevance to clinical observations, patients with high levels of SOX2 and phosphorylated STAT3 have a dismal 5-year survival rate compared with patients with SOX2 overexpression only (Liu et al., 2013).

CONCLUSION REMARKS AND FUTURE DIRECTION

SOX2 is required for multiple steps in formation of the trachea, lung, and esophagus. During these processes, SOX2 is complementarily expressed with other transcription factors, for example, NKX2.1 and SOX9. Disruption of this expression pattern leads to changes in epithelial identities. Intriguingly, SOX2 and the transcription factor CDX2 are also expressed complementarily in the development of gastrointestinal (GI) organs; SOX2 expression is high in the upper GI tract and CDX2 expression is high in the

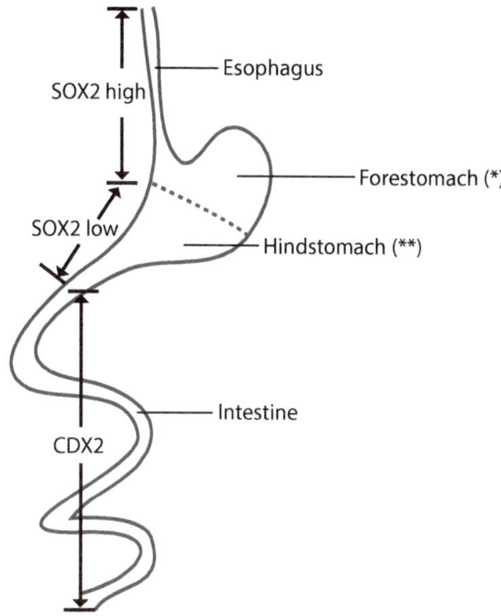

Figure 5 The schematic diagram indicates that SOX2 and CDX2 are expressed complementarily in the gastrointestinal tract during development. Whereas SOX2 is highly expressed in the developing esophagus and mouse forestomach where the epithelium is also stratified and squamous, it is also expressed at low levels in the hindstomach (glandular stomach), where the epithelium is simple columnar. The boundary between SOX2 and CDX2 is located at the pylorus. *Forestomach does not exist in humans. **The epithelium lining the hindstomach is similar to the counterpart in the human stomach.

lower GI tract. Deletion of *Cdx2* leads to ectopic expression of SOX2 and genes marking stratified squamous cells (Gao et al., 2009) (Figure 5). Therefore, these findings suggest that SOX2 serves as a master regulator of epithelial cells where it is highly expressed. Nevertheless, we do not know exactly how SOX2 mediates this function during the development of the trachea, lung, and esophagus. Similarly, we now know that *SOX2* is an oncogene involved in the formation of cancers in the lung and esophagus, but exactly how SOX2 exerts its oncogenic function remains largely unknown. Therefore, future endeavors will include the combination of genetic tools with biochemistry approaches (e.g., SOX2 ChIP-sequencing) to identify SOX2 downstream targets that control organ specification, epithelial differentiation, and maintenance.

ACKNOWLEDGMENTS

We thank Dr Wellington V. Cardoso at the Columbia Center for Human Development, Columbia University, for critical reading and discussion of the chapter. We also thank our colleagues in the Que laboratory for helpful discussion. Our work described in this chapter is supported by R01DK100342, March of Dimes 1-FY14-283, and NYSTEM C029555.

REFERENCES

Alarcón, C., Zaromytidou, A.-I., Xi, Q., Gao, S., Yu, J., Fujisawa, S., Barlas, A., Miller, A.N., Manova-Todorova, K., Macias, M.J., Sapkota, G., Pan, D., Massagué, j., 2009. Nuclear CDKs drive smad transcriptional activation and turnover in BMP and TGF-β pathways. Cell 139, 757–769.

Alejandre-Alcázar, M.A., Michiels-Corsten, M., Vicencio, A.G., Reiss, I., Ryu, J., De Krijger, R.R., Haddad, G.G., Tibboel, D., Seeger, W., Eickelberg, O., Morty, R.E., 2008. TGF-β signaling is dynamically regulated during the alveolarization of rodent and human lungs. Dev. Dyn. 237, 259–269.

Arnold, K., Sarkar, A., Yram, M.A., Polo, J.M., Bronson, R., Sengupta, S., Seandel, M., Geijsen, N., Hochedlinger, K., 2011. Sox2(+) adult stem and progenitor cells are important for tissue regeneration and survival of mice. Cell Stem Cell 9, 317–329.

Avilion, A.A., Nicolis, S.K., Pevny, L.H., Perez, L., Vivian, N., Lovell-Badge, R., 2003. Multipotent cell lineages in early mouse development depend on SOX2 function. Genes Dev. 17, 126–140.

Bass, A.J., Wang, T.C., 2013. An inflammatory situation: SOX2 and STAT3 cooperate in squamous cell carcinoma initiation. Cell Stem Cell 12, 266–268.

Bass, A.J., Watanabe, H., Mermel, C.H., Yu, S., Perner, S., Verhaak, R.G., Kim, S.Y., Wardwell, L., Tamayo, P., Gat-Viks, I., Ramos, A.H., Woo, M.S., Weir, B.A., Getz, G., Beroukhim, R., O'kelly, M., Dutt, A., Rozenblatt-Rosen, O., Dziunycz, P., Komisarof, J., Chirieac, L.R., Lafargue, C.J., Scheble, V., Wilbertz, T., Ma, C., Rao, S., Nakagawa, H., Stairs, D.B., Lin, L., Giordano, T.J., Wagner, P., Minna, J.D., Gazdar, A.F., Zhu, C.Q., Brose, M.S., Cecconello Jr. U.R, I., Marie, S.K., Dahl, O., Shivdasani, R.A., Tsao, M.-S., Rubin, M.A., Wong, K.K., Regev, A., Hahn, W.C., Beer, D.G., Rustgi, A.K., Meyerson, M., 2009. SOX2 is an amplified lineage-survival oncogene in lung and esophageal squamous cell carcinomas. Nat. Genet. 41, 1238–1242.

Borges, M., Linnoila, R.I., Van De Velde, H.J., Chen, H., Nelkin, B.D., Mabry, M., Baylin, S.B., Ball, D.W., 1997. An achaete-scute homologue essential for neuroendocrine differentiation in the lung. Nature 386, 852–855.

Boucherat, O., Nadeau, V., Berube-Simard, F.A., Charron, J., Jeannotte, L., 2014. Crucial requirement of ERK/MAPK signaling in respiratory tract development. Development 141, 3197–3211.

Brcic, L., Sherer, C.K., Shuai, Y., Hornick, J.L., Chirieac, L.R., Dacic, S., 2012. Morphologic and clinicopathologic features of lung squamous cell carcinomas expressing Sox2. Am. J. Clin. Pathol. 138, 712–718.

Balli, D., Ustiyan, V., Zhang, Y., Wang, I.C., Masino, A.J., Ren, X., Whitsett, J.A., Kalinichenko, V.V., Kalin, T.V., 2013. Foxm1 Transcription Factor Is Required for Lung Fibrosis and Epithelial–to–mesenchymal Transition.

Cancer Genome Atlas Research Network, 2012. Comprehensive genomic characterization of squamous cell lung cancers. Nature 489, 519–525.

Cardoso, W.V., Lu, J., 2006. Regulation of early lung morphogenesis: questions, facts and controversies. Development 133, 1611–1624.

Chang, D.R., Martinez Alanis, D., Miller, R.K., Ji, H., Akiyama, H., Mccrea, P.D., Chen, J., 2013. Lung epithelial branching program antagonizes alveolar differentiation. Proc. Natl. Acad. Sci. 110, 18042–18051.

Chen, X., Qin, R., Liu, B., Ma, Y., Su, Y., Yang, C.S., Glickman, J.N., Odze, R.D., Shaheen, N.J., 2008. Multilayered epithelium in a rat model and human Barrett's esophagus: similar expression patterns of transcription factors and differentiation markers. BMC Gastroenterol. 8, 1.

Cutz, E., Gillan, J.E., Bryan, A.C., 1985. Neuroendocrine cells in the developing human lung: morphologic and functional considerations. Pediatr. Pulmonol. 1, S21–S29.

Daniely, Y., Liao, G., Dixon, D., Linnoila, R.I., Lori, A., Randell, S.H., Oren, M., Jetten, A.M., 2004. Critical role of p63 in the development of a normal esophageal and tracheobronchial epithelium. Am. J. Physiol. Cell Physiol. 287, C171–C181.

Domyan, E.T., Ferretti, E., Throckmorton, K., Mishina, Y., Nicolis, S.K., Sun, X., 2011. Signaling through BMP receptors promotes respiratory identity in the foregut via repression of Sox2. Development 138, 971–981.

El Agha, E., Al Alam, D., Carraro, G., Mackenzie, B., Goth, K., De Langhe, S.P., Voswinckel, R., Hajihosseini, M.K., Rehan, V.K., Bellusci, S., 2012. Characterization of a novel fibroblast growth factor 10 (Fgf10) knock-in mouse line to target mesenchymal progenitors during embryonic development. PLoS One 7, e38452.

Gao, N., White, P., Kaestner, K.H., 2009. Establishment of intestinal identity and epithelial-mesenchymal signaling by Cdx2. Dev. Cell 16, 588–599.

Gontan, C., De Munck, A., Vermeij, M., Grosveld, F., Tibboel, D., Rottier, R., 2008. Sox2 is important for two crucial processes in lung development: branching morphogenesis and epithelial cell differentiation. Dev. Biol. 317, 296–309.

Goss, A.M., Tian, Y., Tsukiyama, T., Cohen, E.D., Zhou, D., Lu, M.M., Yamaguchi, T.P., Morrisey, E.E., 2009. Wnt2/2b and beta-catenin signaling are necessary and sufficient to specify lung progenitors in the foregut. Dev. Cell 17, 290–298.

Goulev, Y., Fauny, J.D., Gonzalez-Marti, B., Flagiello, D., Silber, J., Zider, A., 2008. SCALLOPED interacts with YORKIE, the nuclear effector of the hippo tumor-suppressor pathway in Drosophila. Curr. Biol. 18, 435–441.

Harada, H., Nakagawa, H., Oyama, K., Takaoka, M., Andl, C.D., Jacobmeier, B., Von Werder, A., Enders, G.H., Opitz, O.G., Rustgi, A.K., 2003. Telomerase induces immortalization of human esophageal keratinocytes without p16INK4a inactivation. Mol. Cancer Res. 1, 729–738.

Harris-Johnson, K.S., Domyan, E.T., Vezina, C.M., Sun, X., 2009. beta-Catenin promotes respiratory progenitor identity in mouse foregut. Proc. Natl. Acad. Sci. U.S.A. 106, 16287–16292.

Hashimoto, S., Chen, H., Que, J., Brockway, B.L., Drake, J.A., Snyder, J.C., Randell, S.H., Stripp, B.R., 2012. β-Catenin–SOX2 signaling regulates the fate of developing airway epithelium. J. Cell Sci. 125, 932–942.

Hussenet, T., Dali, S., Exinger, J., Monga, B., Jost, B., Dembelé, D., Martinet, N., Thibault, C., Huelsken, J., Brambilla, E., Du Manoir, S., 2010. *SOX2* is an oncogene activated by recurrent 3q26.3 amplifications in human lung squamous cell carcinomas. PLoS One 5, e8960.

Huang, Y., Kapere Ochieng, J., Kempen, M.B.-V., Munck, A.B.-D., Swagemakers, S., Van Ijcken, W., Grosveld, F., Tibboel, D., Rottier, R.J., 2013. Hypoxia inducible factor 3α plays a critical role in alveolarization and distal epithelial cell differentiation during mouse lung development. PLoS One 8, e57695.

Jacobs, I.J., Ku, W.Y., Que, J., 2012. Genetic and cellular mechanisms regulating anterior foregut and esophageal development. Dev. Biol. 369, 54–64.

Ji, H., Ramsey, M.R., Hayes, D.N., Fan, C., Mcnamara, K., Kozlowski, P., Torrice, C., Wu, M.C., Shimamura, T., Perera, S.A., Liang, M.C., Cai, D., Naumov, G.N., Bao, L., Contreras, C.M., Li, D., Chen, L., Krishnamurthy, J., Koivunen, J., Chirieac, L.R., Padera, R.F., Bronson, R.T., Lindeman, N.I., Christiani, D.C., Lin, X., Shapiro, G.I., Janne, P.A., Johnson, B.E., Meyerson, M., Kwiatkowski, D.J., Castrillon, D.H., Bardeesy, N., Sharpless, N.E., Wong, K.K., 2007. LKB1 modulates lung cancer differentiation and metastasis. Nature 448, 807–810.

Liu, K., Jiang, M., Lu, Y., Chen, H., Sun, J., Wu, S., Ku, W.Y., Nakagawa, H., Kita, Y., Natsugoe, S., Peters, J.H., Rustgi, A., Onaitis, M.W., Kiernan, A., Chen, X., Que, J., 2013. Sox2 cooperates with inflammation-mediated Stat3 activation in the malignant transformation of foregut basal progenitor cells. Cell Stem Cell 12, 304–315.

Lu, Y., Futtner, C., Rock, J.R., Xu, X., Whitworth, W., Hogan, B.L., Onaitis, M.W., 2010. Evidence that SOX2 overexpression is oncogenic in the lung. PLoS One 5, e11022.

Mahoney, John E., Mori, M., Szymaniak, Aleksander D., Varelas, X., Cardoso, Wellington V., 2014. The hippo pathway effector yap controls patterning and differentiation of airway epithelial progenitors. Dev. Cell 30, 137–150.

Masui, S., Nakatake, Y., Toyooka, Y., Shimosato, D., Yagi, R., Takahashi, K., Okochi, H., Okuda, A., Matoba, R., Sharov, A.A., Ko, M.S., Niwa, H., 2007. Pluripotency governed by Sox2 via regulation of Oct3/4 expression in mouse embryonic stem cells. Nat. Cell Biol. 9, 625–635.

McDowell, E.M., Newkirk, C., Coleman, B., 1985. Development of hamster tracheal epithelium: I. A quantitative morphologic study in the fetus. Anat. Rec. 213, 429–447.

Min, H., Danilenko, D.M., Scully, S.A., Bolon, B., Ring, B.D., Tarpley, J.E., Derose, M., Simonet, W.S., 1998. Fgf-10 is required for both limb and lung development and exhibits striking functional similarity to Drosophila branchless. Genes Dev. 12, 3156–3161.

Minoo, P., Su, G., Drum, H., Bringas, P., Kimura, S., 1999. Defects in tracheoesophageal and lung morphogenesis in Nkx2.1(−/−) mouse embryos. Dev. Biol. 209, 60–71.

Morrisey, E.E., Hogan, B.L., 2010. Preparing for the first breath: genetic and cellular mechanisms in lung development. Dev. Cell 18, 8–23.

Mukhopadhyay, A., Berrett, K.C., Kc, U., Clair, P.M., Pop, S.M., Carr, S.R., Witt, B.L., Oliver, T.G., 2014. Sox2 cooperates with Lkb1 loss in a mouse model of squamous cell lung cancer. Cell Rep. 8, 40–49.

Okubo, T., Hogan, B.L., 2004. Hyperactive Wnt signaling changes the developmental potential of embryonic lung endoderm. J. Biol. 3, 11.

Perez-Moreno, P., Brambilla, E., Thomas, R., Soria, J.C., 2012. Squamous cell carcinoma of the lung: molecular subtypes and therapeutic opportunities. Clin. Cancer Res. 18, 2443–2451.

Que, J., 2015. The initial establishment and epithelial morphogenesis of the esophagus: a new model of tracheal-esophageal separation and transition of simple columnar into stratified squamous epithelium in the developing esophagus. Wiley Interdiscip. Rev. Dev. Biol. 4, 419–430

Que, J., Luo, X., Schwartz, R.J., Hogan, B.L., 2009. Multiple roles for Sox2 in the developing and adult mouse trachea. Development 136, 1899–1907.

Que, J., Okubo, T., Goldenring, J.R., Nam, K.T., Kurotani, R., Morrisey, E.E., Taranova, O., Pevny, L.H., Hogan, B.L., 2007. Multiple dose-dependent roles for Sox2 in the patterning and differentiation of anterior foregut endoderm. Development 134, 2521–2531.

Ramasamy, S.K., Mailleux, A.A., Gupte, V.V., Mata, F., Sala, F.G., Veltmaat, J.M., Del Moral, P.M., De Langhe, S., Parsa, S., Kelly, L.K., Kelly, R., Shia, W., Keshet, E., Minoo, P., Warburton, D., Bellusci, S., 2007. Fgf10 dosage is critical for the amplification of epithelial cell progenitors and for the formation of multiple mesenchymal lineages during lung development. Dev. Biol. 307, 237–247.

Rawlins, E.L., Perl, A.K., 2012. The a"MAZE"ing world of lung-specific transgenic mice. Am. J. Respir. Cell Mol. Biol. 46, 269–282.

Raymond, C., Anne, V., Millane, G., 1991. Development of esophageal epithelium in the fetal and neonatal mouse. Anat. Rec. 230, 225–234.

Reynolds, S.D., Zemke, A.C., Giangreco, A., Brockway, B.L., Teisanu, R.M., Drake, J.A., Mariani, T., Di, P.Y.P., Taketo, M.M., Stripp, B.R., 2008. Conditional stabilization of β-catenin expands the pool of lung stem cells. Stem Cells 26, 1337–1346.

Rock, J.R., Onaitis, M.W., Rawlins, E.L., Lu, Y., Clark, C.P., Xue, Y., Randell, S.H., Hogan, B.L., 2009. Basal cells as stem cells of the mouse trachea and human airway epithelium. Proc. Natl. Acad. Sci. U.S.A. 106, 12771–12775.

Rodriguez, P., Da Silva, S., Oxburgh, L., Wang, F., Hogan, B.L., Que, J., 2010. BMP signaling in the development of the mouse esophagus and forestomach. Development 137, 4171–4176.

Sarkar, A., Hochedlinger, K., 2013. The sox family of transcription factors: versatile regulators of stem and progenitor cell fate. Cell Stem Cell 12, 15–30.

Shackelford, D.B., Shaw, R.J., 2009. The LKB1-AMPK pathway: metabolism and growth control in tumour suppression. Nat. Rev. Cancer 9, 563–575.

Taranova, O.V., Magness, S.T., Fagan, B.M., Wu, Y., Surzenko, N., Hutton, S.R., Pevny, L.H., 2006. SOX2 is a dose-dependent regulator of retinal neural progenitor competence. Genes Dev. 20, 1187–1202.

Tompkins, D.H., Besnard, V., Lange, A.W., Wert, S.E., Keiser, A.R., Smith, A.N., Lang, R., Whitsett, J.A., 2009. Sox2 is required for maintenance and differentiation of bronchiolar clara, ciliated, and goblet cells. PLoS One 4, e8248.

Ustiyan, V., Wert, S.E., Ikegami, M., Wang, I.C., Kalin, T.V., Whitsett, J.A., Kalinichenko, V.V., 2012. Foxm1 transcription factor is critical for proliferation and differentiation of Clara cells during development of conducting airways. Dev. Biol. 370, 198–212.

Varelas, X., Sakuma, R., Samavarchi-Tehrani, P., Peerani, R., Rao, B.M., Dembowy, J., Yaffe, M.B., Zandstra, P.W., Wrana, J.L., 2008. TAZ controls Smad nucleocytoplasmic shuttling and regulates human embryonic stem-cell self-renewal. Nat. Cell Biol. 10, 837–848.

Vassilev, A., Kaneko, K.J., Shu, H., Zhao, Y., Depamphilis, M.L., 2001. TEAD/TEF transcription factors utilize the activation domain ofYAP65, a Src/Yes-associated protein localized in the cytoplasm. Genes Dev. 15, 1229–1241.

Volckaert, T., Campbell, A., De Langhe, S., 2013. c-Myc regulates proliferation and *Fgf10* expression in airway smooth muscle after airway epithelial injury in mouse. PLoS One 8, e71426.

Wang, N.J., Sanborn, Z., Arnett, K.L., Bayston, L.J., Liao, W., Proby, C.M., Leigh, I.M., Collisson, E.A., Gordon, P.B., Jakkula, L., Pennypacker, S., Zou, Y., Sharma, M., North, J.P., Vemula, S.S., Mauro, T.M., Neuhaus, I.M., Leboit, P.E., Hur, J.S., Park, K., Huh, N., Kwok, P.-Y., Arron, S.T., Massion, P.P., Bale, A.E., Haussler, D., Cleaver, J.E., Gray, J.W., Spellman, P.T., South, A.P., Aster, J.C., Blacklow, S.C., Cho, R.J., 2011. Loss-of-function mutations in Notch receptors in cutaneous and lung squamous cell carcinoma. Proc. Natl. Acad. Sci. 108, 17761–17766.

Wang, Y., Tian, Y., Morley, M.P., Lu, M.M., Demayo, F.J., Olson, E.N., Morrisey, E.E., 2013. Development and regeneration of Sox2+ endoderm progenitors are regulated by a Hdac1/2-Bmp4/Rb1 regulatory pathway. Dev. Cell 24, 345–358.

Wu, S., Liu, Y., Zheng, Y., Dong, J., Pan, D., 2008. The TEAD/TEF family protein scalloped mediates transcriptional output of the hippo growth-regulatory pathway. Dev. Cell 14, 388–398.

Xu, X., Huang, L., Futtner, C., Schwab, B., Rampersad, R.R., Lu, Y., Sporn, T.A., Hogan, B.L.M., Onaitis, M.W., 2014. The cell of origin and subtype of K-Ras-induced lung tumors are modified by Notch and Sox2. Genes Dev. 28, 1929–1939.

Yu, F.-X., Guan, K.-L., 2013. The Hippo pathway: regulators and regulations. Genes Dev. 27, 355–371.

Yu, W.Y., Slack, J.M., Tosh, D., 2005. Conversion of columnar to stratified squamous epithelium in the developing mouse oesophagus. Dev. Biol. 284, 157–170.

Zemke, A.C., Teisanu, R.M., Giangreco, A., Drake, J.A., Brockway, B.L., Reynolds, S.D., Stripp, B.R., 2009. β-Catenin is not necessary for maintenance or repair of the bronchiolar epithelium. Am. J. Respir. Cell Mol. Biol. 41, 535–543.

Zhang, L., Ren, F., Zhang, Q., Chen, Y., Wang, B., Jiang, J., 2008. The TEAD/TEF family of transcription factor scalloped mediates hippo signaling in organ size control. Dev. Cell 14, 377–387.

Zhao, B., Ye, X., Yu, J., Li, L., Li, W., Li, S., Yu, J., Lin, J.D., Wang, C.-Y., Chinnaiyan, A.M., Lai, Z.-C., Guan, K.-L., 2008. TEAD mediates YAP-dependent gene induction and growth control. Genes Dev. 22, 1962–1971.

INDEX

Note: Page numbers followed by "f" indicate figures and "t" indicate tables.

E

F

G

H

T

V

W

X

Y

Z

FSC
www.fsc.org
MIX
Papier | Fördert gute Waldnutzung
FSC® C083411

Kolibri 360 GmbH
Ferdinand-Jühlke-Straße 7
99095 Erfurt, Deutschland
produktsicherheit@kolibri360.de